Thermodynamics for Chemical Engineering Undergraduates

Geoffrey L. Price

Professor Emeritus
Cain Department of Chemical Engineering
Louisiana State University

Professor Emeritus
Russell School of Chemical Engineering
University of Tulsa

formerly:

Robert Hughes Harvey Endowed Professor
Cain Department of Chemical Engineering
Louisiana State University

Wayne Banes Rumley Chair and Professor
Russell School of Chemical Engineering
University of Tulsa

March, 2021

Edition 1, second printing, copyright 2021

All rights reserved. This publication may not be copied in whole or part either digitally or in any type of hardcopy without prior written consent of the author of this book.

ISBN: 978-0-578-87883-6

Thermodynamics for Chemical Engineering Undergraduates

Preface.. vii

Forward to Chemical Engineering Students.. x

List of Symbols... xii

1. Introductory Material... 1
   a. Overview of Thermodynamics... 1
   b. Units and Conversions... 1
   c. Common Thermodynamic Units.. 4
   d. Chapter 1 Example Problems.. 6
   e. References.. 7

2. The First Law of Thermodynamics.. 8
   a. Basic Considerations.. 8
   b. Internal Energy and other Forms of the Energy of a System.... 9
   c. Energy Balance and the First Law... 10
   d. Enthalpy and the Energy Balance for Flow Systems................ 10
   e. The Concepts of Reversibility and Irreversibility..................... 12
   f. State Functions... 14
   g. Constant P and Constant V Processes and the Definition of Heat Capacity.... 15
   h. Other Forms of the First Law.. 19
   i. Chapter 2 Example Problems.. 19

3. Volumetric Properties of Pure Fluids.. 25
   a. Real Substances, Liquids and Phases....................................... 25
   b. The Phase Rule.. 28
   c. The Definition of an Ideal Gas and The Ideal Gas Temperature Scale........ 29
   d. The Relationship of P-V-T to Energy for non-Ideal Gases....... 32
   e. Equations of State for Real Gases... 32
      i. The Compressibility Factor.. 32
      ii. The Virial Equation of State... 33
      iii. Van der Waals Equation... 34
      iv. Other Cubic Equations of State..................................... 36
      v. Higher Order Equations of State.................................... 37
      vi. Corresponding States and Generalized Correlations.... 38
   f. Correlations for Liquid Volumes... 40
   g. The Steam Tables.. 40
   h. Simple Processes with Ideal Gases... 42
   i. Chapter 3 Example Problems.. 45
   j. References.. 57

4. Thermochemistry 59
   a. Sensible Heat Effects 59
   b. Heats of Formation, Reaction and Combustion 61
      i. Generalized Stoichiometry 61
      ii. Enthalpies of Formation 62
      iii. Enthalpies of Reaction 63
      iv. Heats of Combustion 65
   c. Adiabatic Flame Temperature 66
   d. Chapter 4 Example Problems 68

5. The Second Law of Thermodynamics 76
   a. The Nature of Entropy and the Second Law 76
   b. Heat Engines and Statement of the Second Law of Thermodynamics 80
   c. Carnot Cycle 84
   d. Definition of Entropy 87
   e. Third Law of Thermodynamics 88
   f. Entropy Change for an Ideal Gas 88
   g. Lost Work 90
   h. Chapter 5 Example Problems 92

6. Relationships Among Thermodynamic Properties for Pure Fluids 97
   a. Fundamental Relationships 97
   b. Maxwell Relations 98
   c. T, P, and V Explicit Forms of the Energy Functions 99
   d. Residual Properties 101
      i. Residual Properties from Equations of State 102
      ii. Residual Properties from Corresponding States 105
   e. Two Phase Systems and the Clapeyron Equation 106
   f. Chapter 6 Example Problems 108
   g. References 119

7. Macroscopic Balances for Process Equipment 120
   a. Conservation of Mass 120
   b. First Law Balance 121
   c. Second Law Analysis of Processes 123
   d. Irreversibility, Exergy, and Availability 124
   e. Steady-State Flow Devices 127
      i. Bernoulli's Equation 127
      ii. Maximum Velocity in a Pipe 128
      iii. Valves 129
      iv. Heat Exchangers 131
      v. Pumps 132
      vi. Compressors 133
      vii. Multi-Stage Compression Processes 135

		viii.	Turbines 137
		ix.	Thermal Efficiencies of Pumps, Compressors, and Turbines......... 137
	f.	Chapter 7 Example Problems 138

8. Processes for the Production of Power 147
    a. Steam Rankine Cycle 147
    b. Otto Engine 149
    c. Diesel Engine 153
    d. Brayton Cycle 154
    e. Jet Engine 156
    f. Chapter 8 Example Problems 157

9. Refrigeration and Liquification Cycles 163
    a. Vapor Compression Refrigeration 163
    b. Absorption Refrigeration 166
    c. Joule-Thompson Liquification 167
    d. Chapter 9 Example Problems 169

10. Fugacity of Pure Fluids and Ideal Mixtures 175
    a. Characterization of Processes Moving Toward Equilibrium 175
    b. Relationships Among Thermodynamic Properties
           for Multi-Component Systems 176
    c. Definition of Fugacity and Activity and the Phi-Gamma Formulation of VLE . 179
    d. Raoult's Law 183
    e. Fugacities of Pure Components 189
        i. Fugacities of Pure Gases 189
        ii. Fugacities of Pure Liquids and Solids (Condensed Phases) 190
    f. Properties of Mixtures of Ideal Gases 192
    g. Chapter 10 Example Problems 193

11. Fugacity of Mixtures of Real Gases 203
    a. Partial Molar Gibbs Free Energy of Mixtures 203
    b. Fugacity Coefficients for Simple Cases 205
    c. Evaluation of Fugacity Coefficients in Mixtures Using EOS 206
    d. Fugacity Coefficients from Generalized Correlations 210
    e. Fugacity Coefficients in the Phi-Gamma Formulation of VLE 212
    f. Chapter 11 Example Problems 213
    g. References 222

12. Activity and Fugacity of Liquid Mixtures 223
    a. More on Partial Molar Properties 223
    b. Excess Properties 226
    c. Partial Molar Excess Properties 228
    d. Regular Solutions 229
    e. Modified Raoult's Law Solutions 231

	f.		Higher Order Excess Gibbs Free Energy Functions...................233
	g.		Excess Gibbs Free Energy Functions Derived from Molecular Arguments ... 234
	h.		Determination of Parameters for Activity Coefficient Correlations.........237
	i.		Azeotropes.................241
	j.		Thermodynamic Consistency...................242
		i.	The Differential Test.........................242
		ii.	The Integral Test...........243
	k.		More on the Gibbs-Duhem Equation 246
	l.		Chapter 12 Example Problems 247
	m.		References....................254
13.			Reaction Equilibrium.......................255
	a.		The Reaction Coordinate and Extent of Reaction 255
	b.		Gibbs Free Energy of Reaction...................257
	c.		The Reaction Equilibrium Constant...................258
	d.		Gaseous Reaction Mixtures 260
	e.		The Van't Hoff Equation 261
	f.		Liquid and Heterogeneous Reactions...................262
	g.		Multiple Reactions...................263
	h.		Chapter 13 Example Problems 265
	i.		References....................275
14.			Other Topics in Phase Equilibrium...................278
	a.		Liquid-Liquid Equilibrium...................278
	b.		Osmotic Pressure 285
	c.		Vapor Pressure from an EOS 287
	d.		Chapter 14 Example Problems 289
	e.		References....................292
15.			Contents of the Appendix...................293
	a.		Appendix A: Conversion Factors and Values of R........................295
	b.		Appendix B: Heat Capacities of Pure Substances 296
	c.		Appendix C: Critical Constants 298
	d.		Appendix D: Lee-Kesler Tables...................301
		i.	Table D.1 - z^0...................301
		ii.	Table D.2 - z^1...................301
		iii.	Table D.3 - h^{R0}/RT_c...................302
		iv.	Table D.4 - h^{R1}/RT_c...................302
		v.	Table D.5 - s^{R0}/R...................303
		vi.	Table D.6 - s^{R1}/R...................303
		vii.	Table D.7 - φ^0...................304
		viii.	Table D.8 - φ^1...................304
	e.		Appendix E: Antoine constants for vapor pressure of selected components ... 305
	f.		Appendix F: Enthalpies and Free Energies of Formation...................307

Properties of Selected Pure Components:
- g. Appendix G: Steam... 309
  - i. G.1 - Saturated Steam Temperature Increment................... 309
  - ii. G.2 - Saturated Steam Pressure Increment...................... 311
  - iii. G.3 - Superheated Steam....................................... 313
  - iv. G.4 - Compressed Liquid Water.................................. 324
  - v. P-H Diagram for Steam 326
  - vi. Mollier Diagram for Steam..................................... 327
  - vii. T-S Diagram for Steam.. 328
- h. Appendix H: Properties of R-134a................................. 329
  - i. H.1 - Saturated R-134a... 329
  - ii. H.2 - Superheated R-134a....................................... 331
  - iii. H.3 - P-H Diagram for R-134a.................................. 334
- i. Appendix I: Properties of R-23................................... 335
  - i. I.1 - Saturated R-23... 335
  - ii. I.2 - Superheated R-23 337
  - iii. I.3 - P-H Diagram for R-23.................................... 344
- j. Appendix J: Properties of Ammonia................................ 345
  - i. J.1 - Saturated Ammonia.. 345
  - ii. J.2 - Superheated Ammonia 347
  - iii. J.3 - P-H Diagram for Ammonia................................. 359
- k. Appendix K: Properties of Nitrogen............................... 360
  - i. K.1 - Saturated Nitrogen....................................... 360
  - ii. K.2 - Superheated Nitrogen 362
- l. Appendix L: Properties of Oxygen................................. 376
  - i. L.1 - Saturated Oxygen... 376
  - ii. L.2 - Superheated Oxygen 377
- m. Appendix M: Properties of Carbon Dioxide......................... 380
  - i. M.1 - Saturated Carbon Dioxide................................. 380
  - ii. M.2 - Superheated Carbon Dioxide 381
- n. Appendix N: Properties of Methane................................ 387
  - i. N.1 - Saturated Methane.. 387
  - ii. N.2 - Superheated Methane 388
- o. Appendix O: Properties of Hydrogen............................... 391
  - i. O.1 - Saturated Hydrogen....................................... 391
  - ii. O.2 - Superheated Hydrogen 392
- p. Appendix P: Properties of R-11................................... 395
  - i. P.1 - Saturated R-11 395
  - ii. P.2 - Superheated R-11... 397

16. Subject Index 400

Preface

Thermodynamics is a science. Much like calculus or chemistry for instruction of students, the fundamental subject material for undergraduate instruction has not changed much over the decades. However, computational tools have changed drastically since the later half of the 20th century, and authors of thermodynamics books have taken it upon themselves to introduce their own specific computational methods into their thermodynamics books. This is the first reason I have felt compelled to write a new volume which discusses only the science. Sure, modern computational techniques make it possible to see advanced thermodynamic calculations come to life for students, but this does not change the science. And as the computational methods advance, the computational methods included in their books by the authors become antiquated while the underlying science does not. This just confuses students, and in many instances it just adds a layer of old style computational work which is now irrelevant that students must learn. You will not find specific software or any specific computational methods mentioned in this textbook. That does not mean that I don't use computations in conjunction with teaching because I do. But I keep up with the methods that are available, choose computational methods that fit in with the current instruction the students I am teaching have received (which are usually many more than one), and bring the science from the book to the computations that are most relevant in the modern instructional climate. During my tenure as a University Professor while teaching Thermodynamics, I have used Fortran, Basic, Lotus 123, Visual Basic, C, C^{++}, Matlab, LabView, to current choices Excel, Aspen HYSYS. Students in my classes get computationally oriented problems to do which, as you can see from the list of computation packages, has evolved and instruction in the most convenient methods that are currently available for doing the computations. The science of Thermodynamics has not changed during that time period.

For the popular textbooks these days, authors participate in a circular firing squad, trying to add more and heap more into their books, which is irrelevant in terms of the science, trying to get noticed and get their textbooks adopted. The large publishing houses contribute their economic interests (which in all fairness is their business) instead of supporting the science. New editions with enhanced problem sets with solutions and new discussion about current events clouds the science. What I would say to the authors of these books is that you should teach the science in the science and engineering courses, then add courses that have focus in current events such as environmental and ethics courses to your degree program. Such courses can change with the time while the science doesn't change. You don't need to generate a new edition (which delights the publishers) to hit the most relevant current topics because the science hasn't changed. I have also noted that these thermodynamics books that add current events discussion suffer from their sophomoric nature. Their opinion, not scientific perspective, are sophomoric at best and totally wrong at the worse. Some books (which I have been using for instruction) however, go the opposite direction and load their books with theory way beyond that which is necessary for an undergraduate engineering education. In doing so, they cloud the foundational material undergraduate students need and also leave out a logical sequence for presenting material to a student who is just beginning to learn. Don't get me wrong, these are very good books indeed, but they lack proper logistics for undergraduate teaching.

In my 40+ years of teaching thermodynamics, I have taught five different flavors of thermodynamics courses. I have taught an introductory thermodynamics course to engineering students which included chemical, electrical, petroleum, and engineering physics students. I have also taught Thermodynamics I and II to chemical engineering students only, and an Equilibrium Thermodynamics course, very similar to the Thermodynamics II, course to chemical engineering students. Finally, I have taught graduate thermodynamics to beginning chemical engineering graduate students at two different universities. But I have no special training in thermodynamics, as my graduate studies were outside of thermodynamics. So I have learned along the way while teaching, and this perspective has helped me organize thermodynamics instruction into a logical sequence. I find the thermodynamics course taught to all engineering students as both having some superfluous material which some engineering instructors feel is needed because students don't understand the underlying chemistry while also leaving out important material. Case in point for the superfluous material is the use of ideal gas constants that are specific to individual gases because they are written per unit mass rather than per mole of gas. Students (apparently) don't understand how to get from mass to moles and vice versa. If engineering students don't know this much chemistry, they shouldn't be allowed to have an engineering degree. The topic of mixtures is also generally left out of courses such as this, and I find this acceptable because this subject does need a modest understanding of chemistry. I think its okay not to expect a high level of chemistry skill from engineering students other than chemical engineering students. The reason that I bring all of this up is because I have tried to separate the subject of mixtures so that the book is completely suitable for teaching a course too all engineering students by judicious choice of chapters of the book to study. For a first course in thermodynamics for students in all engineering disciplines, study chapters 1 - 3 and 5 - 9. These chapters do not require any of the other material (particularly that on mixtures) in this book as background material. This copy of the book you are reading is entitled "Thermodynamics for Chemical Engineering Undergraduates", but I plan to offer another flavor which drops chapters 4 and 10 - 14 and is entitled "Thermodynamics for Engineering Undergraduates" which will hopefully be available soon. Perhaps the title won't scare off engineering educators other than chemical engineers. For an introductory course for chemical engineering students, study the same chapters as the course for all engineering students but include chapter 4 in sequence right after chapter 3. My opinion is that chemical engineering students have the background to learn some extra material in the introductory thermodynamics course because their chemistry instruction will include some of the same topics and students will already have some foundation.

A second semester thermodynamics class for chemical engineering students should include the remaining chapters in the book. Review material at the beginning of such of class could touch briefly on important review material from chapters 2 - 6, then the basic course includes chapters 10 - 14 taken in sequence. Again, these chapters have been ordered and material logically sequenced so students can follow right along.

Graduate students should have a separate textbook. Advanced material for graduate students does not exist in this book, but this book can serve as a foundational book for more advanced texts for teaching.

The point of this book is to teach enough of the science of thermodynamics for undergraduate engineering students to have the understanding they need to continue into the upper level engineering courses - nothing more. I have tried to judiciously choose material which accomplishes this goal. As such, you may find a lot of material that other textbooks have that this one doesn't. For example, this book doesn't have anything on Statistical Thermodynamics other than a mention here and there regarding how some things have become known. My opinion is that undergraduate engineering students don't need to understand Statistical Thermodynamics to continue to upper level classes, and they certainly shouldn't spend a lot of time on that material if they don't have a mastery of phase equilibrium. Chemical engendering students get enough instruction in this area in their chemistry classes and other engineering students don't even study mixtures or are not even required to have thermodynamics in their undergraduate studies. What do they need Statistical Thermodynamics for in their higher level courses?

This book has enough material to present a full introductory course in thermodynamics to all engineering students, and enough material for chemical engineers to complete a full second semester of thermodynamics. If I added more material, I could not include it in a two-semester sequence of undergraduate thermodynamics instruction while doing what is here justice. I have taught these courses for 40+ years. I know what can be fit in to these courses with proper instruction and instructional support, and what I have here is more than enough. If you are instructing students and you feel like some added material is necessary, you should have the ability to include it yourself (even if you have to drop some material in this book). You can also give students assignments in other topics I haven't covered here. For example, you could ask students to look up rocket engines or the Claude Process for liquification. Students will have the underlying thermodynamic ability to understand other material such as these and they don't have to buy a new book to find the information. But I am not suggesting "my way or the highway". Please let me know if you think things should be added to this text and I will definitely consider it in future editions.

For thermodynamic properties of pure substances, I have relied upon NIST programs and data. Like computer calculations and changing software, this is another item that is fluid with respect to time. As measurement and computational accuracies improve, the values for thermodynamic properties change, though only slightly over time. New definitions for standard states, such as the change from 1 atm to 1 bar for the standard state pressure, change pure component values slightly. Instructors need to be flexible with changes like this and instill in their students a sense of checking out the numbers and pursuing the most accurate data when it matters. Usually, for engineering calculations, changes such as these make little difference because the 4th digit of accuracy is not sought, and engineering students at the stage of taking thermodynamics need to start to get a grip on this reality. The advent of the internet provides copious amounts of data I never had as a student, and one might ask why a compendium of numbers exist in the appendix of this book in the first place when the numbers (for the most part) are available via search engines. The answer is that having a set of numbers as a primary standard allows students to check their answers, especially to homework and example problems more easily. Also, my own personal way of administering exams is to tell students to bring a hardcopy of their book so that I can count on them having a consistent set of properties from which to work and I have made it a practice of always giving open book exams so they have the data available.

I couldn't write a book without acknowledging the wonderful and lasting friendship of the other faculty members I have served with at both Louisiana State University and The University of Tulsa. I am not going to mention any special names because those of you who are special know who you are and I would hate to slight more that a couple of people I have worked with. My sons, Jerry, Joey and David are an inspiration to me (they are all engineers to their father's delight), and my wife, Judy deserves special acknowledgment for sharing her home office with me while I wrote most of this book. Most of all, Judy allows me to love her deeply and it means the world to me.

Finally, I am going to plagiarize a line from my PhD Dissertation from 1979:

TO THE GLORY OF GOD AND JESUS CHRIST MY PERSONAL SAVOR

Geoffrey L. Price
Spring 2021

Forward for Chemical Engineering Students

Congratulations on your wise choice of getting into the chemical engineering profession! I have never thought even for a second that I chose the wrong profession. I have been challenged and inspired from the first day of my undergraduate studies to the final course I taught and even into my retirement as writing this book has also been a challenge while also being very satisfying in a sense of writing it all down in the way I developed it. I have honed this instructional material with students and their success in the chemical engineering curriculum in mind.

I offer this encouragement because chemical engineering is a difficult discipline and is certainly not for the "slackers". However, the professional and financial rewards are worth it if you have the aptitude, and not everyone does. You have a lot of work to do before you "walk across the stage" and thermodynamics certainly is part of that work to be done.

In this book, I have tried to focus on things you will need to know as you advance through your chemical engineering curriculum. Your instructor might add more material because I certainly can't cover it all, but I believe I have covered everything in thermodynamics you will need for your upper level chemical engineering classes. If you are using this book as a supplement to a more mainstream textbook, I hope you will find the material I present here as a logical sequence of topics. I have been very careful to treat each subject before using that subject material in more advanced material in later chapters. Nonetheless, I have assumed you know appropriate topics in chemistry and mathematics which you should have covered in your freshman year. Some students may even have an extensive enough background in chemistry and mathematics from their high school studies to take their first course in thermodynamics in their freshman year.

The way to study in this book is as follows. Read the chapter, work through the example problems. Use your computing resources such as a calculator, spreadsheet, or whatever you want to use to verify some of the calculations and confirm numbers that have been read out of property tables as you work through the problems. I recommend that you save your calculations because a lot of the work to be done builds on calculations previously done. If you don't understand why particular equations are being used, go back to the relevant section and re-read. If you are still having trouble, contact your instructor. Then, do the student problems. Once you understand the example problems, it should be relatively clear how to do the student problems. Many of the student problems can be solved by following a similar example problem. I have found over the years that students say that they understand the example problems when they actually don't understand. Seeing what equation someone else looked up in an example problem is not understanding. Knowing the theory and the governing equations is the understanding level which is needed. Typically, students pick up simpler concepts such as the 1st Law of Thermodynamics easily but have more difficulty with concepts such as fugacity and activity. Re-reading the theory sections after seeing the scope of the calculations is often very useful in such situations.

I hope you will not forget the subjects covered early in the book as you are proceeding to new areas. This book builds and I have been especially careful not to jump around. I also hope that you recognize the organization of the chapters. The book builds from the 1st Law to the 2nd Law and from there into the application of these laws in things from industrial equipment to mixtures. In all cases, we treat simpler items first before moving to more complex items. One of the reasons I have written this book is because I find the mainstream textbooks on thermodynamics poorly organized with irrelevant items such as current events and theory more complex than needed for undergraduates hiding the important foundational theory. I teach the material covered in this book in two 3-hour courses (semester system) and there is plenty in this book to fill the two courses when support items such as examinations, computational lectures, etc. are included.

A word about significant digits used in this book is warranted. You will find that most calculations I have done are done to 4 or perhaps 5 significant digits. This is not because I feel that the accuracy is to that level, but because I want you to be able to verify where my calculations come from, and I don't want to lose a lot of significant digits to rounding error. Indeed, I doubt there is a single calculation in this book that I would place confidence in to 4 digits of accuracy. Most of the calculations in this book are ± 5% at best. However, I think that if I pull a number from the steam tables which is entered there as 1.5816, it seems to me I should use that number in my calculations rather than 1.58 so it's clear to you where it came from. Always think about the accuracy of the final result when it you are applying it to something that matters. In the same vein, it is absurd to report 10 or 15 digits when doing engineering calculations in almost all cases.

Please know that there are copious quantities of more information available to you. This book should serve as a guide for you to help you learn, but as with all chemical engineering subject areas, there is much more to learn and is available to you. Once you graduate, unless you go to graduate school, you probably won't do any more of the complex calculations that are laid out in this book, just as you will likely not solve complex mathematics problems anymore. The point is that you need to understand foundational material so you know where to look when the time comes and you do have a complex problem to solve and something happens - such as your computer-aided design program is stuck in a infinite loop because something is not quite specified correctly. Also, I would advise you that data that is available in the open literature can change with better measurements and if there are items where lives or fortunes depend upon your calculations, please seek multiple sources before you put your stamp of approval on it.

The other thing I wanted to accomplish with this book is to offer a book of reasonable cost. Right now, I am only trying to recover the costs of publishing, copyright, ISBN registration, and to pay royalties for other's copyrighted material. I hope it is valuable to you and that you will recommend it to others. My effort will be wasted unless students use it.

Geoffrey L. Price
Spring 2021

List of Symbols

A	Area
A	Helmholtz Free Energy
A	parameter in equations of state
A	chemical species
A	parameter in heat capacity equation
a	molar Helmholtz Free Energy
a	acceleration
a	parameter in equations of states
a	acceleration
a	activity
B	parameter in equations of state
B	parameter in heat capacity equation
B	second virial coefficient, volume expansion
B'	second virial coefficient, pressure expansion
b	parameter in equations of states
C	parameter in heat capacity equation
C	third virial coefficient, volume expansion
C'	third virial coefficient, pressure expansion
C_p	constant pressure heat capacity
C_v	constant volume heat capacity
D	parameter in heat capacity equation
D	fourth virial coefficient, volume expansion
D'	fourth virial coefficient, pressure expansion
E	energy content
E	parameter in heat capacity equation
F	Force
F	Friction factor
F	an arbitrary function
F	degrees of freedom
f	fugacity
f	function of
G	Gibbs free energy
g	gaseous
g	molar Gibbs free energy
g	acceleration of gravity
g_c	dimensional constant = 32.174 lbm-ft/lbf-s^2
H	enthalpy
h	molar enthalpy
h	specific energy
h	height
i	irreversibility
K	equilibrium constant
k	Henry's Law constant

KE	kinetic energy
ke	molar kinetic energy
ke	specific kinetic energy
m	mass
m	moles of aqueous ions per mole of solute
N	number of components
n	moles
PE	potential energy
pe	molar potential energy
pe	specific potential energy
P	pressure
Q	heat
q	heat per unit mole or per unit mass
q	effective volume
R	universal gas constant
R	reservoir (thermal)
r	radius
r	compression ratio
S	entropy
s	molar entropy
s	solid
T	temperature, in °C = centigrade, K = kelvin, °F = Fahrenheit, °R = Rankine
U	internal energy
u	molar internal energy
u	specific internal energy
u	velocity (when used with kinetic energy calculations)
v	molar volume
v	specific volume
v	velocity (when used with kinetic energy calculations)
V	volume
W	work
w	work per unit mole or unit mass
x	liquid mole fraction
x	distance
x	quality (of steam or other component)
y	vapor mole fraction
z	solid mole fraction
z	volume fraction
z	compressibility factor
z	height

Greek Letters

α	parameter in the Peng-Robinson equation
α	arbitrary phase (usually used as a superscript)
β	arbitrary phase (usually used as a superscript)
γ	activity coefficient

γ	ratio of C_p to C_v
Δ	change in a property or energy function
ζ	extent of reaction
η	cycle efficiency
η	thermal efficiency of devices producing work
η	thermal efficiency of devices requiring work
θ	arbitrary temperature scale
μ	chemical potential
μ	Joule-Thompson coefficient
ν	generalized stoichiometric coefficient
π	number of phases
π	osmotic pressure
ρ	density
φ	fugacity coefficient
ψ	exergy
ω	acentric factor
ω	coefficient of performance

Subscripts

γ	in terms of activity coefficients - used in equilibrium constants
φ	in terms of fugacity coefficients - used in equilibrium constants
°	initial
a	in terms of activity - used in equilibrium constants
abs	absorbed
c	critical property
C	cold
f	formation
f	in terms of fugacity - used in equilibrium constants
G	gas
gen	generated
H	hot
i	component i
in	inlet or beginning conditions
L	liquid
max	maximum
mix	mixing
net	net amount
out	outlet or exiting conditions
r	reaction
r	reduced property
rev	reversible
s	as in W_s, shaft work
sys	system
T	temperature
tot	total
V	vapor

x	in terms of liquid mole fraction - used in equilibrium constants
y	in terms of vapor mole fraction - used in equilibrium constants

Superscripts

o	standard state
∞	infinite dilution
APP	apparent
az	azeotrope
c	condensed phase
E	excess
el	electronic energy contribution
IG	ideal gas
ke	kinetic energy contribution
rot	rotational energy contribution
R	residual
s	saturated
sat	saturated
tot	total
vap	vaporization
vib	vibrational energy contribution

Overscript

.	Rate per unit time
_	partial molar

Chapter 1 - Introductory Material

1.a Overview of Thermodynamics

The term "thermodynamics" derives from Greek words for "heat + motion". As such, it might imply a study of how fast heat flows, but to engineers, a study of how fast heat flows is generally termed "heat transfer". Instead, the term "thermodynamics" should be considered more like "heat + distribution" or even better "energy + distribution". In this book, we focus on equilibrium or quasi-equilibrium states and how these states differ in energy content. Since in the theoretical sense, an infinite amount of time is required to reach equilibrium, the study of transitions from one equilibrium state to another theoretically takes an infinite amount of time, but this is what is meant by "thermodynamics" in the classical sense. The study of how fast one state transitions to another by the transfer of heat is a separate engineering science.

Thermodynamics is a fundamental science and it has broad applications to virtually everything in the universe. It can be applied at both the microscopic (molecular) and macroscopic (large collections of molecules) levels. However, as an introductory book, we will focus more on macroscopic levels though in several instances we will consider qualitatively how properties of molecules collectively lead to macroscopic states.

In this book, we focus on the first and second laws of thermodynamics and the applications of these two laws to chemical and mechanical processes. The first law of thermodynamics could be called the law of conservation of energy, and the reader may have encountered this law elsewhere. The second law of thermodynamics gives rise to the concept of entropy which is a property that can be determined for a state of system. A description of entropy is best left to more formal treatments which begin later in this book, but as a rough overview, we can think of entropy as a measure of the random nature of a system. The ramifications of this overview are left to later treatments in this book.

1.b Units and Conversions

Some fundamental units are length, time, and mass. SI (Système International or International System of Units) units are respectively meters (m), seconds (s), and kilograms (kg). Other units can be derived from these. For example, units of force can be derived from Newton's law:

$$F = ma \qquad 1.1$$

where F is force, m is mass, and a is acceleration. In SI, force is generally given in newtons (N) and if that is the case, m is generally in kg and acceleration is in m/s^2. In English units (also often called field units), force is usually in lbf (pounds of force), and if mass is in lbm (pounds of mass), the acceleration is 32.174 feet per second squared (ft/s^2). Note that a pound mass (lbm) is not the same as a pound force (lbf). A pound force is a force that accelerates a one pound mass at 32.174 feet per second squared. Thus, in standard gravity, one lbm (pound mass) weighs one lbf (pound of force). However, one kilogram does not weigh one newton in standard gravity. Because the concepts are fundamentally different in SI and English units, we define a conversion factor. In English units:

$$g_c = 32.174 \frac{\text{lbm} \cdot \text{ft}}{\text{lbf} \cdot \text{s}^2} \qquad 1.2$$

and in SI units:

$$g_c = 1 \frac{\text{kg} \cdot \text{m}}{\text{N} \cdot \text{s}^2} \qquad 1.3$$

Students tend to get confused because in SI units g_c has a value of 1, but it still exists. However, just as you would never cancel N and kg when they appear as a ratio in the dimensions of an equation, never cancel lbf and lbm. In either case, use g_c. Note that you are using g_c whenever you substitute one N for one kg·m/s². Similarly, you can substitute one lbf for 32.174 lbm·ft/s².

Temperature is a common property about which engineering students probably already have a general knowledge. Common temperature units derive from centuries ago and they began with the relative scales Fahrenheit (°F) and Celsius (°C). Fahrenheit derived his scale from the human body temperature which he measured as 100°F and the lowest temperature he could reach in the lab which was a salt-ice mixture that he measured as 0°F. Though we know these two anchor points to be inaccurate today, the Fahrenheit scale is still popular in field measurements and widely used on a daily basis in cooking and weather reports. Today, the scale can be derived from the Kelvin scale (see below).

The Celsius scale was derived from the freezing points and boiling points of water in air at atmospheric pressure, and these points were originally defined as 0°C and 100°C, respectively. In years past, the scale was known as the centigrade scale, but the term "Celsius" has become the scientific consensus. Still, students may find references to a centigrade scale in older books and articles. The Kelvin scale was originally derived from the Celsius scale, but the Celsius scale is now derived from the Kelvin scale.

The absolute scales, Rankine (°R) and Kelvin (K) are absolute temperature scales based upon "absolute zero" which is known as the lowest possible temperature. Note that formally, the degrees symbol " ° " is used with all the scales except Kelvin. This is because one kelvin (not one degree Kelvin) is formally defined by SI as a unit of temperature, just like one meter is defined as a unit of length. The SI system does not formally recognize the other temperature scales as fundamental units.

Temperature scales and their definitions have gone through numerous changes and refinements over the years. These refinements are important for extreme accuracy and reproducibility and are left to the reader to research if desired. The stories are historical and extremely interesting but beyond our needs from an engineering standpoint as we can define all the scales from the Kelvin scale. Currently, the SI definition of the kelvin is derived from the Boltzmann constant $k = 1.380649 \times 10^{-23}$ kg·m²/K·s² but an understanding of how this relates to our common perception of temperature and temperature scales requires an understanding of statistical mechanics. A slightly older SI definition of the Kelvin scale is more understandable and perfectly useful for engineering calculations and concepts. In this slightly older definition, the Kelvin scale is defined such that 0 K is absolute zero and 273.16 K is the triple point of water is the point at which liquid, vapor and solid states of pure water co-exist. We define one kelvin as 1/273.16 units of this scale. Temperatures higher than 273.16 K go in kelvins above that point.

To an astute reader, the question of what we mean by "1/273.16 units of this scale" is a very good question. Do we do this by dividing a column of mercury evenly in length in a thermometer? If so, why? What do we do outside the range of temperatures where mercury, or other thermometer fluids are liquids? We will again take up this definition at a later point in this book in the context of an ideal gas and further define the Kelvin scale.

From the Kelvin scale, we define:

$$T(°C) = T(K) - 273.15 \qquad 1.4$$

This places the ice-point (the point at which solid and liquid water co-exist in atmospheric air) at 0°C or 273.15 K. This scale also places the boiling point of water at atmospheric pressure at 100°C or 373.15 K. Note that the triple-point and the ice-points of water differ by 0.01°C, but this difference is very often small enough to be ignored for engineering calculations.

The Rankine and Fahrenheit scales are defined, respectively, as:

$$T(°R) = 1.8 \times T(K) \quad \text{and} \quad T(°F) = 1.8 \times [T(K) - 273.15] + 32$$

From these definitions, we readily find the ice point is 32°F or 491.67°R. The boiling point of water at atmospheric pressure is 212°F or 671.67°R.

We can also find interconversion equations from scales other than K.

$$T(°R) = T(°F) + 459.67 \qquad \text{and} \qquad T(°F) = 1.8 \times T(°C) + 32$$

These equations are often already familiar to engineering and science students.

Pressure derives from force per unit area, and like temperature, there are also absolute and relative pressure units. Absolute pressure can also be thought of as a pressure relative to a perfect vacuum. Common units are pascals (Pa) = 1 N/m², bar = 10^5 Pa, atm (atmospheres, which is the pressure of the atmosphere under standard atmospheric conditions), and psia = lbf/in² (pounds force per square inch). The "a" at the end of "psia" refers to "absolute" meaning the pressure is in absolute units or is relative to a perfect vacuum. Two of the relative scales are psig which is the same as "psi gauge", and barg which is the same as "bar gauge". Both of these measurements refer to pressure above atmospheric pressure as is typically measured by some "gauge" operating in the atmosphere. Think of a tire and the air pressure inside. The pressure we normally refer to as tire pressure is pressure above the atmosphere which surrounds it. If we use a tire pressure gauge to measure the pressure in the tire, it might be say 35 psig, meaning "35 pounds per square inch above the pressure of the atmosphere". That is what the tire pressure gauge measures, not the absolute pressure. In converting from gauge pressure to absolute pressure, we really can only do the conversion approximately (but close enough for almost all engineering calculations on Earth) unless we know the exact atmospheric pressure where the gauge is being used. However, all we need to do is add atmospheric pressure to gauge pressure to get absolute pressure. If atmospheric pressure is 14.7 psia or 1.013 bar (which is standard atmospheric pressure):

$$P\ (psia) = P\ (psig) + \text{atmospheric pressure} \approx P\ (psig) + 14.7$$

$$P\ (bar) = P\ (barg) + \text{atmospheric pressure} \approx P\ (barg) + 1.013$$

SI units such as pascals and bar automatically assume absolute pressure so they do not include the "a" appendage as is generally done in the English system.

1.c   Common Thermodynamic Units

The common thermodynamic units we use in this book derive from energy, and there are two basic kinds we will use, thermal and mechanical energy. Thermal energy is also loosely called heat and mechanical energy is loosely called work.

Originally in the English system, the BTU (British Thermal Unit) was defined as the amount of energy required to heat one lbm of water one °F, but several variants exist depending on the initial temperature of the water, though usually 60°F is used and the differences in the initial temperature yield very small differences in the absolute size of 1 BTU. As with other units, BTU's are now better defined relative to the more formal SI units. Similar to the BTU in concept, the calorie (cal) was originally defined as the amount of energy required to heat one gram of water 1 °C. 1 BTU is about 252.2 cal. Note that in the dietary world, the term "calorie" which is used for the energy content of foods is actually a kilocalorie (kcal).

Mechanical energy or work derives from force exerted through a distance. Since force could vary during the time it is exerted, we generally write:

$$W = \int_{x_1}^{x_2} F\,dx$$

In thermodynamics, we are often interested in work done by expansion or compression of a gas. Consider what happens if we use a piston/cylinder arrangement to expand/compress the gas. Since pressure is force per unit area:

$$W = \int_{x_1}^{x_2} PA\,dx$$

If the piston moves the distance dx, then we have changed the volume of gas the piston/cylinder contains, and assuming A is a constant, Adx = dV. Therefore, we find:

$$W = \int_{V_1}^{V_2} P dV \qquad 1.5$$

This expression is used widely in thermodynamics.

Joule's experiments in the 19th century established the equivalence of mechanical energy to thermal energy, and he published a value of 778.24 ft-lbf as being equivalent to 1 BTU [1]. Today, the equivalent has been refined to 1 BTU = 778.169 ft-lbf. The fundamental SI unit of energy is named after Joule, and 1 J = 1 N·m. Joules are used for both mechanical and thermal forms of energy in the SI system, and as mechanical and thermal energy are equivalent, in theory we can use any units for either form. However, in practice in the English system, we normally use ft-lbf for mechanical energy and BTU for thermal energy. The calorie was used for thermal energy in the old metric systems, but it has been replaced by the use of J in the SI system.

Power is defined as rate of production of energy and has units of energy per unit time. The fundamental unit in SI is the watt (W) which is 1 J/s. BTU/s, BTU/min and BTU/hr are common units for thermal power in English units, but the common unit for mechanical work is horsepower (hp). 1 hp = 550 ft-lbf/s = 745.7 W. As the name implies, the horsepower unit was originally derived from how much work a horse was able to exert over a period of time.

If we lift a mass in a gravitational field, we are doing a kind of work as this height displacement is a force through a distance. However, we generally refer to this as the mass gaining *potential energy* (PE). If we lift a mass to height z:

$$PE = \frac{g}{g_c} mz \qquad 1.6$$

Also consider *kinetic energy* (KE) as:

$$KE = \frac{mu^2}{2g_c} \qquad 1.7$$

The symbol "u" in equation 1.7 is velocity, not to be confused with internal energy which will be defined later in this book. PE and KE are types of energy that are important to many thermodynamic considerations.

Suppose you have 1 kg of water at 25°C and 1 bar pressure in a container and then another 1 kg in a separate container at 25°C and 1 bar. If the two are put together into a single container, V doubles. Since V is dependent on the total mass of the system, it is termed an *extensive* variable. But the 2 kg mixture does not have twice the pressure or temperature of the original separate 1 kg systems. T and P are termed *intensive* variables which are not dependent on the size of the system.

Extensive variables such as V (we will later define many other extensive variables) can be written per unit mole, termed *molar variables* or per unit of mass termed *specific variables*. Thus,

$V/n = v$ (molar volume) or $V/m = v$ (specific volume, the same symbol as molar volume). The units in the equation tell us which units are being used. Note that the molar and specific equivalents of extensive variables are intensive variables. For example, the molar volume of water is not dependent on the size of the system. The reciprocal of specific volume is *density*, $\rho = 1/v$.

1.d   Chapter 1 Example Problems

1.1   Calculate PE for a 2 kg mass (4.409 lbm) lifted 5 m (16.40 ft). Use standard gravity as $9.806 \text{ m/s}^2 = 32.174 \text{ ft/s}^2$. Do the calculation in both SI units (N, kg, m, and s) and English units (lbm, lbm, ft and s) and compare the results.

Solution:

In SI units:

$$PE = \frac{9.806 \text{ m}/\text{s}^2}{1 \text{ kg} \cdot \text{m}/\text{N} \cdot \text{s}^2} 2 \text{ kg} \cdot 5 \text{ m} = \mathbf{98.06 \text{ N} \cdot \text{m}}$$

In English units:

$$PE = \frac{32.174 \text{ ft}/\text{s}^2}{32.174 \text{ lbm} \cdot \text{ft}/\text{lbf} \cdot \text{s}^2} 4.409 \text{ lbm} \cdot 16.40 \text{ ft} = \mathbf{72.31 \text{ ft} \cdot \text{lbf}}$$

If we convert 72.31 ft·lbf to N·m, we find these to be the same within rounding error.

1.2   What is the kinetic energy of a 2 kg mass (4.409 lbm) traveling at 5 m/s (16.40 ft/s)? Do the calculation in both SI and English units and compare the results.

Solution:

In SI units:

$$KE = \frac{2 \text{ kg} \cdot 5^2 \text{ m}^2/\text{s}^2}{2 \cdot 1.0 \frac{\text{kg} \cdot \text{m}}{\text{N} \cdot \text{s}^2}} = \mathbf{25.0 \text{ N} \cdot \text{m}}$$

In English units:

$$KE = \frac{4.409 \text{ lbm} \cdot 16.40^2 \text{ ft}^2/\text{s}^2}{2 \cdot 32.174 \frac{\text{lbm} \cdot \text{ft}}{\text{lbf} \cdot \text{s}^2}} = \mathbf{18.43 \text{ ft} \cdot \text{lbf}}$$

If you convert 18.43 ft·lbf to N·m we find these to be the same within rounding error.

1.3 Compute the work done when 100 cm³ of a gas is heated at constant pressure of 1 bar until it reaches 125 cm³.

Solution:

Equation 1.5 is appropriate. In this case, P is constant so it can come outside the integral yielding:

$$W = P\int_{V_1}^{V_2} dV = P(V_2 - V_1) = 1\,\text{bar}(125 - 100)\,\text{cm}^3$$

$$W = 25\,\text{cm}^3 \cdot \text{bar} = \mathbf{2.5\ J}$$

1.4 Convert the following:
a. 50°C = what in K?
b. 100°F = what in R?
c. 166 R = what in K?
d. 344°C = what in °F?
e. 344K = what in °F?

Solution:

a. 50 + 273 = **312 K**
b. 100 + 460 = **560 R**
c. 166 /1.8 = **92.2 K**
d. 344x1.8 + 32 = **651.2°F**
e. 1.8x344 - 460 = **159.2°F**

Student Problems:

1. What is the potential energy (PE) of 100 lbm sack of concrete lifted 6 feet up by a forklift? Provide answers in both ft-lbf and J.

2. Does the potential energy of a 100 lbm sack of concrete at 6 feet above the ground differ from a 100 lbm barbell at the same height?

3. A 100 kg sprinter is running at 8 m/s. What is the sprinter's kinetic energy in J and in ft-lbm?

1.e   References

1. Mayer, J.L. and Joule, J. P.; "Equivalent mécanique de la chaleur. Chaleur spécifique des gaz sous volume constant", Journal des débats politiques et littéraires, 8 June (1854).

Chapter 2 - The First Law of Thermodynamics

2.a   Basic Considerations

In order to properly define and apply the First Law of Thermodynamics, some definitions are necessary. Very important to the discussion is the definition of a *system*. A system is defined as any collection of mass or volume in space on which we choose to focus our attention. The system may be enclosed in an impermeable boundary or the boundary may be imaginary. A system is therefore a subset of the universe, and how we choose to define the system is very important to applying the First Law of Thermodynamics. We define *surroundings* as everything in the universe that surrounds the system. With these definitions, it should be clear that the universe is the sum of the system and surroundings.

A *process* is a series of chemical or mechanical operations applied to a system. The *state* of a system refers to the physical properties that characterize the system. The physical properties defining the state of system especially include temperature and pressure but are not at all limited to those quantities.

Simply stated, the First Law of Thermodynamics says that energy is neither created or destroyed, but it can change forms. We should note through Einstein's theory of relativity, mass and energy can be interconverted, and this can have an impact on how we treat a system with respect to the First Law of Thermodynamics. We should include mass as energy if we are dealing with nuclear reactions and we are converting mass to energy and vice versa, but basically, in this book, we are not dealing with nuclear reactions and we don't have to consider this possible impact.

Mathematically, for any process, the First Law of Thermodynamics is:

$$\Delta E_{universe} = 0 \qquad \qquad 2.1$$

We use the change in energy in equation 2.1 as we are considering changes that take place when a process occurs. As we have defined a system and surroundings making up the universe, then:

$$\pm \Delta E_{system} \pm \Delta E_{surroundings} = 0 \qquad \qquad 2.2$$

Here, we have included the $\pm$ signs to indicate that we have not yet defined the direction of adding to or subtracting from the system/surroundings as positive or negative energy. The choices will be discussed below.

Now, it is easiest to consider the surroundings as not being a system with a corresponding state. If we consider the surroundings like a system, we will need to compute the energy content of a very large collection of molecules indeed! Instead, we consider the change in energy content of the surroundings to be the energy that leaves or is added to the system. There are two ways that the system can accept or transmit energy to or from the surroundings and these are through heat and/or work. Note that heat (Q) and work (W) are transitionary in this sense. Neither can be stored. From equation 2.2 we can therefore write:

$$\pm \Delta E_{system} = \mp Q \mp W \qquad 2.3$$

Another way to look at equation 2.3 is that whatever energy leaves the system arrives at the surroundings or anything that leaves the surroundings arrives at the system. Also, we do not write ΔQ or ΔW in the equation because they are transitionary. We do not consider that the surroundings have less heat or work when transferred, and we consider each in the context of an overall amount that has been transferred.

2.b Internal Energy and other Forms of the Energy of a System

Let's now consider the energy of the system. It is possible that the system is translating in space or is in a gravitational field. Think of a baseball hit by a batter as the system. The velocity of the baseball represents *kinetic energy* (KE), and as it rises or falls, its *potential energy* (PE) in the gravity of the earth rises and falls. In this process, KE is converted to PE as the ball rises, and vice versa as the ball falls. We will separate these two forms of energy that the system possess as separate parts of the overall energy of the system, but the energy that is internal to the baseball is another form of energy. A warm ball on a hot day has more energy than a cold ball on a cold day.

Next, we define the *internal energy*, U, as all the rest of the energy the system possesses. Let's dissect this a bit. Suppose the system is a gas. Then the molecules of the gas are randomly bouncing around the container and each molecule individually has a KE and the sum of KE over all the molecules of the system are part of U. Note that it is different from the kinetic energy of the system which has been defined as translation of the system as a whole. Also, individual molecules that make up the gas may have electronic, rotational, chemical bond or vibrational energy, and there may be mutual interactions of one molecule with another. All of these forms of energy combine to make up U. The internal energy of a gas compared to a liquid would include the equivalent of the heat required to make the liquid into a gas (i.e., the heat of vaporization), and *etc*. All energy the system possesses other than KE as a translation of the system as a whole, and PE of the system as a whole in a gravitational field is included in U. Thus:

$$\Delta E_{system} = \Delta U + \Delta KE + \Delta PE \qquad 2.4$$

This suggests that if energy is added to a system, it can be distributed into only three "containers". Energy added to the system must increase U, KE, PE, or some of each. As we have suggested earlier, Q and W are not part of the energy of the system. To better understand this, consider the following. If heat is added to a pot of water on a stove, under this definition of U, it would be inaccurate to say the water possesses heat. Instead, the hot water has more internal energy than cold water or U of hot water is larger than U of cold water. To make hot water from cold water, we transfer heat to the cold water, but once transferred, the hot water possess a higher value of U, not Q.

Another very important aspect of this discussion is that U is determined solely by the state of the system. If we know the temperature and pressure of a lbm of water (about a pint by volume) we are heating on a stove, we have fixed U for the water as the system. Every lbm of water in a pans all over the universe at the same T and P have the same U as does a lbm of water in a lake, or in a pipe on the way to a boiler, or wherever it might be as long as it is at the same T and P. Like Joule's experiments showing the equivalence of heat and work as forms of energy, we could have rapidly stirred the water and gotten to the same T and P by adding work instead of heat and the water would have the same U as every other lbm of water at the same T and P.

2.c Energy Balance and the First Law

Now we are ready to put these concepts together, and decide upon the sign convention regarding what is positive heat and work. Combining equations 2.4 and 2.3, we get:

$$\Delta U + \Delta KE + \Delta PE = \mp Q \mp W$$

We define $+Q$ as heat added to a system and $+W$ as work done by (leaving) the system. This yields:

$$\boxed{\Delta U + \Delta KE + \Delta PE = Q - W} \qquad 2.5$$

The choice of signs coupled with the definitions of energy being added or removed for Q and W is arbitrary, and textbooks commonly use both this and other sign conventions. Engineers developed the form given in 2.5 in the context of conversion of heat into work especially in steam engines and is therefore especially appropriate for this book.

As we will see below, equation 2.5 is most easily applied to static systems, which are systems that are not flowing. An example is a piston and cylinder device that contains a gas. However, we should emphasize that this equation applies universally to all processes.

2.d Enthalpy and the Energy Balance for Flow Systems

Consider now the process depicted in Figure 2.d.1. Imagine some process being performed on 1 kg of fluid as a system that is processed by some equipment. We follow the 1 kg of fluid entering at point 1 and leaving at point 2. We have depicted a pump for the process equipment, but it could include any items such as a compressor, turbine, boiler, or etc. All the heat that is added (which could come from multiple sources) is shown as Q coming into the processing device. Similarly, the sum of all work leaving from the process

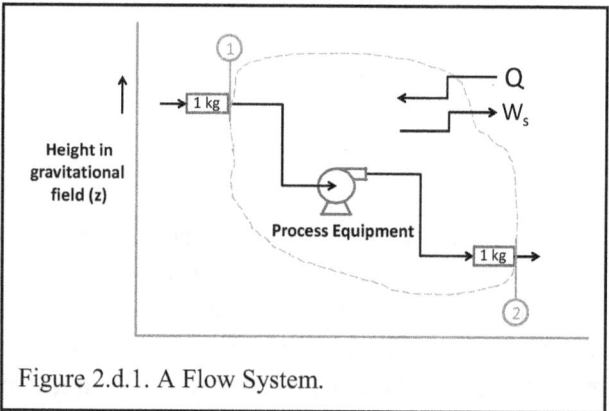

Figure 2.d.1. A Flow System.

equipment is shown as W_s. This work has the subscript "s" because it is termed *shaft work* and it is most often applied to rotating shafts where work is done by pumps, compressors, turbines, etc. Note that if we have a single pump like that depicted in Figure 2.d.1, work is being done on the system, so W_s for such a situation would be negative since we define positive work as being work done by a system. However, we could just as well have the system do work on the surroundings in which case the work would be positive.

Now, there is less obvious work being done by the system other than W_s and this work is termed *flow work*. As the 1 kg of fluid exits at point 2, the system pushes against the surroundings. Similarly, the surroundings do work at point 1 to push the fluid into process. Therefore the total work done by the system is that done by the equipment, W_s, plus the flow work, or:

$$W = W_s + W_{flow\ work} \qquad 2.6$$

Thus, the flow work is the work associated with getting the fluid in and out of the process. Let's consider how we can characterize the flow work in simple terms. From equation 1.5 we know that for P-V work:

$$W = \int_{V_{in}}^{V_{out}} P dV$$

In this case, P is constant at P_1 entering the system at point 1 and also at P_2 exiting the system at point 2. Therefore:

$$W_{entering} = P_1 \int_{V_{in}}^{V_{out}} dV = P_1(V_{out} - V_{in})$$

$$W_{leaving} = P_2 \int_{V_{out}}^{V_{in}} dV = P_2(V_{out} - V_{in})$$

As the fluid is pushed into the equipment, the volume of the fluid, V_1, is the volume displaced so $V_{out} - V_{in} = V_1$ and similarly, $V_2 = V_{out} - V_{in}$ is displaced in leaving the process.

$$W_{entering} = P_1 V_1 \qquad W_{leaving} = P_2 V_2$$

Now consider the signs. The fluid does work pushing out into the surroundings at point 2 so $W_{leaving}$ is positive. However, $W_{entering}$ is negative because the surroundings does work on the system pushing it in, or the system does negative work on the surroundings. Thus, using our formal sign convention:

$$W_{flow\ work} = P_2 V_2 - P_1 V_1 \qquad 2.7$$

Next, write a First Law for this process including the flow work:

$$\Delta U + \Delta KE + \Delta PE = Q - (W_s + P_2 V_2 - P_1 V_1)$$

and rearrange to:

$$(U_2 + P_2 V_2) - (U_1 + P_1 V_1) + \Delta KE + \Delta PE = Q - W_s$$

Note that the grouping U + PV appears twice, and it appears often in thermodynamics. This grouping is all the energy that a system contains (U) plus the energy (in the form of work) associated with its volume and pressure (PV). The system does not contain the PV energy, but it represents an energy term that is associated with a flowing fluid.

Thus, we define *enthalpy* (H) as:

$$\boxed{H = U + PV}$$ 2.8

and the First Law of Thermodynamics rearranged for a flowing system is:

$$\boxed{\Delta H + \Delta KE + \Delta PE = Q - W_s}$$ 2.9

The term ΔH includes the flow work, so we don't have to include that term elsewhere. In equation 2.9, the term W_s is all the work done by the system except the flow work. Note that in equation 2.5, W refers to all the work and must include the flow work if we apply it to a flowing system. So, equations 2.5 and 2.9 are both perfectly general equations but equation 2.5 applies more easily to static (non-flow) systems and equation 2.9 applies more easily to flowing systems.

2.e   The Concepts of Reversibility and Irreversibility

A very important concept in thermodynamics is the *reversible* process which is a limit that is extremely useful for thermodynamic calculations. In practice, only *irreversible* processes (all processes that are not reversible) exist, but development of many concepts depend upon the theoretical reversible process. Consider the processes depicted in Figure 2.e.1.

Look first at the process in row a). A weight compresses a spring initially, and the process is to slide the weight off to a shelf on the side and let the spring go unrestrained. The spring would leap wildly, vibrate and likely bang against the walls dissipating the energy

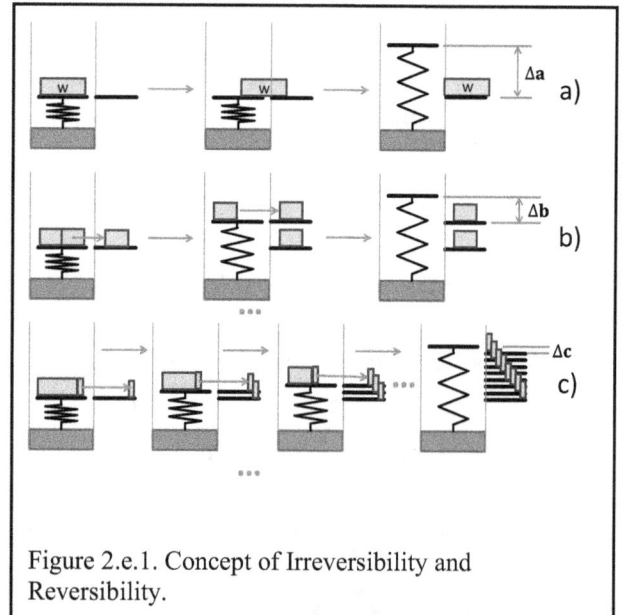

Figure 2.e.1. Concept of Irreversibility and Reversibility.

the spring had stored. Eventually, the spring comes back to an uncompressed equilibrium state. There is no potential energy increase of the weight.

Now refine the process to that in row b) and note that we only slide half the weight off first and let the spring increase the potential energy of the second half of the weight. As the spring uncompresses, it is restrained first by half the weight, then when the second half of the weight is removed, the spring is unrestrained but only has half the energy to dissipate compared to the process in row a). The potential energy of half the weight has been increased, so some of the initial energy stored in the compressed spring in its initial state has been converted to potential energy of half the weight.

Finally, consider dividing the weight further and further as in the process in row c). The forces restraining the spring are never much different than the spring is exerting. Note that the potential energy of the many pieces of the weight have been increased, and if we were to continue dividing the weight we would reach a limit where all the energy in the spring as initially compressed would be converted to potential energy of an infinite number of differentially small pieces of the weight.

Now, consider reversing each process. In the process in row a), in order to go in reverse, the first thing needed is to apply an external force to compress the spring such that the whole weight could be slid horizontally to restore the apparatus to it's initial position. To start the reverse process, the external force would have to compress the spring the distance Δa as shown on row a). Since we have to apply an external force and do work on the spring to make the process reverse, this process is irreversible. To reverse the process in row b), the spring would initially be compressed a much smaller distance, Δb, than in process a), then half the weight could be slid over so that less external force is required to complete the compression of the spring, slide the second half of the weight over, and restore the apparatus to the initial state. Still, the process requires external work and is irreversible. Finally, in process c), only a small external force is needed to start the process and eventually restore the apparatus to it's initial state. Imagine reversing the process in the limit when an infinite number of differentially small pieces of weight must be restored in sequence to recompress the spring and restore the full weight on the spring. In such a case, only a differential amount of external force (zero in the infinite limit) would be required to reverse the process, and we would term such a process a "reversible process" assuming there is no friction to divert some of the spring's initial energy to frictional heat.

Consider a piston and cylinder that contains a gas. Take just the gas as the system, and consider the amount of work which might be done in a process utilizing the same piston/cylinder arrangement to expand/compress the gas. To compute W, it is important to recognize that we need to use the *pressure of the gas* in the calculation of:

$$W = \int PdV \qquad 2.10$$

Furthermore, we need for the pressure of the gas to be uniform throughout the system, and we need an exact mapping of P to V. Otherwise, evaluation of the integral in the form of equation 2.10 is not analytically possible.

Now suppose the piston/cylinder apparatus exhibits friction between the piston and cylinder. In such a case, an external pressure applied to the piston to compress the gas would have to be larger than the pressure of the gas making up the system to overcome the friction. There would be work done in excess of that computed by equation 2.10 that would likely go to heat the piston and/or cylinder. Likewise for expansion of the gas, the pressure of the gas in the system would have to be larger than the external pressure applied to the piston, and less work would be done on the surroundings than the result given by 2.10 with the same result of heating the piston/cylinder. In either case, the overall process including the piston/cylinder arrangement would be irreversible.

Therefore, we generally imagine a piston/cylinder arrangement that is frictionless if we want to do thermodynamic calculations, and we take the process as being slow so that the external pressure applied to the piston is only differentially smaller or larger than the pressure of the gas at any instant, and the pressure of the gas is uniform throughout. All of these attributes are required if the process of adding (or subtracting) work is reversible.

Now, suppose we use a hotplate to heat the gas by placing a piston/cylinder apparatus on the hotplate. Though the reason is not apparent yet, we also require that heat added to the system be added with the temperature of the device providing heat being only differentially higher than the temperature of the system. Otherwise, the process is irreversible.

Note that even though frictionless apparatuses don't exist, and trying to add heat to an apparatus via a temperature that is only differentially greater than the temperature of the system is impossible, we can imagine these devices and processes for the purposes of calculations. Also note that the system doesn't know whether it has been operated upon by a reversible or irreversible apparatus. Therefore, we can imagine apparatuses that don't exist, apply them to imagine a change in a system, and compute whatever changes in the state variables we want. These changes in the system will be the same regardless of what apparatus was used in the process.

In practice, all processes are irreversible to some extent but the limit yielding a reversible process is apparent. Reversible processes are characterized by:

1. The absence of friction.
2. Differentially small differences in driving forces. This includes not only physical forces but also temperature differences driving the flow of heat.
3. Slow processes (because the driving forces must be differentially small).
4. Processes which can be reversed to restore the universe exactly back to it's initial condition without applying external work to reverse the process..

These characteristics and the definition of the reversible process will be applied often in this book.

2.f   State Functions

We have already mentioned the state of system as being the condition of system as characterized by the physical properties. We now carry this further to consider the thermodynamic quantities that characterize the state of the system. Consider some of the well known properties of a system especially temperature, pressure, internal energy and enthalpy. Variables such as these have values that are a result of the state of the system and are therefore termed *state functions*. We have already noted that U, for example, is determined by the state of the system, not by how the system got to be in its current state. T, P and H are the same in this regard as U.

The fact that these physical properties are state functions means that we can record the values of these functions in tables or relate them in figures. The steam table could not exist if T, P, U, and H are dependent on whether steam came from some process "A" or another process "B". If the properties were not state functions, we would have to define a path that we used to arrive at the state of the system.

In this book, only two things we deal with often are not state functions, and these are Q and W. The amount of heat and amount of work added is dependent on the process. An example relates to Joule's experiments where W could be substituted for Q to arrive at the same change in a system state. So, it's easiest just to remember that all the properties we deal with are state functions except Q and W.

2.g   Constant P and Constant V Processes and the Definition of Heat Capacity

Consider a process that takes place in a rigid container (a rigid container cannot change volume) at constant volume. Assume that ΔKE and ΔPE are zero for this process. Then:

$$\Delta U = Q - W$$

or, dividing through by n to put the equation on a molar basis:

$$\Delta u = q - w \qquad 2.11$$

We will often write the First Law in the form in equation 2.11 without comment, and the reader should know that we are assuming no significant KE and PE effects for that process.

We know that $dv = 0$ for a constant volume process, so:

$$w = \int P dv = 0$$

which leaves:

$$\Delta u = q \qquad \text{(Constant volume, reversible) 2.12}$$

Now suppose we have a frictionless piston/cylinder arrangement containing a fluid which we will treat as the system, and we reversibly heat the fluid holding the pressure constant. We can accomplish this by letting the piston slowly move up while restraining the piston with a constant pressure from the surroundings. Then:

$$\Delta u = q - \int P dv = q - P\Delta v = q - (P_2 v_2 - P_1 v_1)$$

We can arrange the equation this way because $P_2 = P_1$. Further rearrangement yields:

$$(u_2 + P_2 v_2) - (u_1 + P_1 v_1) = q$$

or

$$\Delta h = q \qquad \text{(Constant pressure, reversible) 2.13}$$

From this analysis, we define two heat capacities, C_v (*constant volume heat capacity*) and C_p (*constant pressure heat capacity*) roughly for now then we will get a rigorous definition below. These properties are defined such that:

$$q = \int C_v dT = \Delta u \qquad \text{(Constant volume, reversible) 2.14}$$

$$q = \int C_p dT = \Delta h \qquad \text{(Constant pressure, reversible) 2.15}$$

As we will make clearer later, C_v and C_p are functions of temperature and pressure, though they are often relatively weak functions of T. Under conditions where heat capacity is nearly constant or over a small temperature range, equations 2.14 and 2.15 can be integrated approximately to:

$$q = C_v (T_2 - T_1) \qquad \text{(approximate) 2.16}$$

$$q = C_p (T_2 - T_1) \qquad \text{(approximate) 2.17}$$

Note that q depends on whether we have constant volume or constant pressure process, so this further illustrates that it is not a state function. Though it is not a state function, we can still compute q without approximation under many conditions.

Consider now a function which is dependent on two independent variables:

$$F = f(x, y) \qquad 2.18$$

If F is a state function or exact function:

$$dF = \left(\frac{\partial F}{\partial x}\right)_y dx + \left(\frac{\partial F}{\partial y}\right)_x dy \qquad 2.19$$

In these equations, the subscripts on the partial derivatives indicate the variable which is being held constant while taking the derivative. Though these were mathematically determined when we wrote equation 2.18, thermodynamicists always add these subscripts because it is not always clear from which function the total derivative is derived. This is because in thermodynamics, there are many variables to choose from such as T, P, H, U, etc. Mathematicians usually leave the subscripts off, so equation 2.19 may seem new but is not to most engineering students.

Let's choose:

$$u = f(T, v) \quad \text{then} \quad du = \left(\frac{\partial u}{\partial T}\right)_v dT + \left(\frac{\partial u}{\partial v}\right)_T dv$$

Now apply to a constant volume process so $dv = 0$:

$$du = \left(\frac{\partial u}{\partial T}\right)_v dT \qquad \text{(constant v) 2.20}$$

Clearly, in light of equations 2.14 and 2.20:

$$C_v = \left(\frac{\partial u}{\partial T}\right)_v \qquad 2.21$$

A similar development can be applied to enthalpy:

$$h = f(T, P) \quad \text{and} \quad dh = \left(\frac{\partial h}{\partial T}\right)_P dT + \left(\frac{\partial h}{\partial P}\right)_T dP$$

Apply to a constant pressure process so $dP = 0$:

$$dh = \left(\frac{\partial h}{\partial T}\right)_P dT \qquad \text{(constant P) 2.22}$$

and

$$C_p = \left(\frac{\partial h}{\partial T}\right)_P \qquad 2.23$$

Equations 2.21 and 2.23 give the formal definitions of C_v and C_p, respectively, and from these equations we better understand that:

$$\Delta u = \int_{T_1}^{T_2} C_v dT \qquad \text{(constant v) 2.24}$$

and

$$\Delta h = \int_{T_1}^{T_2} C_p dT \qquad \text{(constant P) 2.25}$$

Comparing equations 2.24 and 2.25 to equations 2.14 and 2.15, respectively, note that the restriction of application to reversible processes has been removed. Equations 2.24 and 2.25 contain only state functions, so they are independent of the process and can be applied to any constant volume or pressure processes as appropriate. However, the equations (2.14 and 2.15):

$$q = \int_{T_1}^{T_2} C_v dT \text{ (constant v, reversible)} \qquad \text{and} \qquad q = \int_{T_1}^{T_2} C_p dT \text{ (constant P, reversible)}$$

are restricted to reversible processes only, and to constant volume or constant pressure as appropriate. Remember that q is not a state function and must have a process defined in order to compute it.

We have already integrated equations 2.14 and 2.15 approximately for the case where C_v and C_p are constant. This is simple and the results are given in equations 2.16 and 2.17. What do we do when we want a more accurate representation for the case where C_v and C_p are not constant with temperature? To do this, we need an equation to represent the temperature functionality of heat capacity. First, we should note that there is no good theoretical way to get the temperature functionality of heat capacities for all types of molecules that are accurate over a wide range of temperatures so there is no single equation to use and thermodynamicists have relied instead on heat capacity correlations with temperature. Generally, these are polynomial in nature because heat capacity changes smoothly with temperature and are not strong functions (usually) of temperature. Constants for the equation are fitted to heat capacity data, and the polynomials do a good job.

Many forms have been used and thermodynamic textbooks have given constants for many different equations. In this book, constants for the heat capacity equation:

$$C_p^{IG}/R = A + BT + CT^2 + DT^3 + E/T^2 \quad \text{where T is in K} \qquad 2.26$$

are given in Appendix B. We will later pick up the reasoning for the superscript "IG" on C_p when we cover ideal gases. Though T must be in K in equation 2.26, you can get C_p in whatever units you want by choosing an appropriate value of R.

Equation 2.26 can be integrated easily to yield:

$$\int_{T_1}^{T_2} \frac{C_p^{IG}}{R} dT = A(T_2 - T_1) + \frac{B}{2}(T_2^2 - T_1^2) + \frac{C}{3}(T_2^3 - T_1^3) + \frac{D}{4}(T_2^4 - T_1^4) - E\left(\frac{1}{T_2} - \frac{1}{T_1}\right) \quad 2.27$$

2.h Other Forms of the First Law

We have written two basic forms of the First Law:

$$\boxed{\Delta U + \Delta KE + \Delta PE = Q - W} \quad 2.5 \qquad \boxed{\Delta H + \Delta KE + \Delta PE = Q - W_s} \quad 2.9$$

Each term in both equation 2.5 and 2.9 carries the units of energy such as BTU or kJ. These equations may also be written per unit mass or per unit mole, simply by diving through by mass or moles:

$$\boxed{\Delta u + \Delta ke + \Delta pe = q - w} \quad 2.28 \qquad \boxed{\Delta h + \Delta ke + \Delta pe = q - w_s} \quad 2.29$$

Each term in equation 2.28 and 2.29 now carry units of energy per unit mass, also known as specific properties (e.g., specific internal energy), or molar properties (e.g., molar enthalpy) depending upon whether the equation has been divided through by mass or moles. We do not give different symbols for molar or specific properties, so students should be careful to make sure all terms in these equations have the same units. Note that when we consider q as being, for example, a molar q, that is heat added *per unit mole of working fluid*.

We can also write these equations as rate equations. Consider a flowing system in which we have a molar flowrate, $\dot{m}$, which would have units, for example, of moles/sec. If we multiply $\dot{m}$ by the molar enthalpy we get a rate of flow of energy, also known as power, moles/s x J/mole = J/s = W (watts). Let's apply this concept to equations 2.28 and 2.29:

$$\boxed{\dot{m}\Delta u + \dot{m}\Delta ke + \dot{m}\Delta pe = \dot{Q} - \dot{W}} \quad 2.30 \qquad \boxed{\dot{m}\Delta h + \dot{m}\Delta ke + \dot{m}\Delta pe = \dot{Q} - \dot{W}_s} \quad 2.31$$

Note that in equations 2.30 and 2.31, $\dot{Q}$, $\dot{W}$, and $\dot{W}_s$ have units of power such as BTU/s or kJ/s = kW.

2.i Chapter 2 Example Problems

2.1 One lbm of water goes over a 100 ft waterfall and lands in a the river at the base of the falls. No energy is exchanged between the water and the surroundings.

a. What is the potential energy of the water at the top of the falls relative to base of the falls?
b. What is the kinetic energy of the water just before it strikes the bottom, and what is it's potential energy at that point?

c. After the water enters the river at the bottom and has essentially zero kinetic and potential energy, what change has occurred in it's state?
d. Use equation 2.24 to estimate the temperature change of the water in the river at the bottom of the falls.

Solution:

a. Take the 1 lbm of water as the system

$$PE = mz\frac{g}{g_c} = (1 \text{ lbm})(100 \text{ ft})\left(\frac{32.17 \text{ ft/sec}^2}{32.17 \frac{\text{lbm} \cdot \text{ft}}{\text{lbf} \cdot \text{sec}^2}}\right) = 100 \text{ ft} \cdot \text{lbf}$$

b. Since $Q = W = 0$ because no energy is exchanged with the surroundings and $\Delta U = 0$ because the state of the water does not change (it is at the same T and P) while it is falling, the first law yields:

$\Delta U + \Delta KE + \Delta PE = Q - W$
$0 + \Delta KE + \Delta PE = 0 - 0$

So the potential energy is -100 ft-lbf because it is 100 ft *below* where it started, so:

$\Delta KE - 100 \text{ ft} \cdot \text{lbf} = 0$
$\Delta KE = 100 \text{ ft} \cdot \text{lbf}$

This means the water gained velocity (makes sense) as it reaches the bottom of the waterfall.

c. Going from just before the system enters the river at the bottom then into the river, the distance fallen is essentially zero so $\Delta PE = 0$, and there is no exchange of energy with the surroundings as it enters the river so $\Delta U = 0$ and $Q = W = 0$. $\Delta KE = -100$ ft - lbf because it goes from a high velocity to approximately zero velocity. Thus:

$\Delta U + (-100 \text{ ft-lbf}) = 0$
$\Delta U = \mathbf{100 \text{ ft-lbf}}$

d. $\Delta U = 100 \text{ ft} \cdot \text{lbf} = \int_{T_1}^{T_2} C_v dT \approx C_v(T_2 - T_1) = C_v \Delta T$

We need a value for C_v for liquid water at atmospheric conditions. We find the following entry for *liquid* water in Appendix B:

$$\frac{C_{P,298}}{R} = 9.072$$

With R = 1.986 BTU/lbmole-°R (Appendix A), $C_{P,298}$ = 18.02 BTU/lbmole-°R. For water, the molecular weight is 18.02 g/mole = 18.02 lbm/lbmole. So, in terms of per unit mass, $C_{P,298}$ = 1.00 BTU/lbm-°R, which of course we knew because the definition of a BTU is the amount of energy required to raise the temperature of one lbm of water 1°F (or °R). Converting 100 ft-lbf to BTU, we find ΔU = 0.129 BTU. This gives:

0.129 BTU = 1.00 BTU/lbm-°R ΔT
ΔT = **0.129°R = 0.129°F**.

Therefore, the river at the bottom of the 100 ft waterfall is 0.129°F degrees hotter than the water in the river leading to the falls. Of course, this would be difficult to measure, but the First Law requires that the potential energy of the water at the top of the waterfall is not lost, but is conserved. The water gains this energy through viscous dissipation upon slamming into the river at the bottom with a high velocity. What we should see from this exercise is that energy is transformed from one term to another in the First Law. Potential energy is converted to kinetic energy then to internal energy. If we caught the water in a bucket that was attached to rope and pulley, we could have done work instead of converting the energy to internal energy.

2.2   A 2000 lbm car is at the top of a 200 ft hill traveling at 60 mph. A stop sign is at the bottom of the hill. If the car stops at the stop sign, determine:
a.   the internal energy change of the car if the car is well insulated.
b.   the heat transferred to the surroundings if the car is isothermal.

Solution:

a.   Take the car as the system. $\Delta U + \Delta KE + \Delta PE = Q - W$

The car does no work, so W = 0. As stated in the problem, the car is well insulated which would keep heat from flowing in/out of the system, so Q = 0. 60 mph = 88 ft/sec and the car ends at 0 mph at the stop sign. Thus:

$$\Delta KE = \frac{2000 \text{ lbm} \left(0^2 - 88^2 \text{ ft}^2/\text{sec}^2\right)}{2 \cdot 32.17 \dfrac{\text{lbm} \cdot \text{ft}}{\text{lbf} \cdot \text{sec}^2}} = \mathbf{-2.41 \times 10^5 \text{ ft} \cdot \text{lbf}}$$

$$\Delta PE = \frac{2000 \text{ lbm} \left(0 - 200 \text{ft}\right) \cdot 32.17 \text{ ft}/\text{sec}^2}{32.17 \dfrac{\text{lbm} \cdot \text{ft}}{\text{lbf} \cdot \text{sec}^2}} = \mathbf{-4.00 \times 10^5 \text{ ft} \cdot \text{lbf}}$$

$\Delta U = 2.41 \times 10^5 + 4.00 \times 10^5 = 6.41 \times 10^5$ ft / lbf = **823.5 BTU**

This energy would be in the brakes of the car. As we know, automotive brakes get hot when they are used.

b. Since the car is isothermal and at atmospheric pressure, $\Delta U = 0$. W is still zero. However, heat is now transferred so the terms in the First Law are:

$$0 - 2.41 \times 10^5 - 4.00 \times 10^5 = Q - 0$$

$$Q = -6.41 \times 10^5 \text{ ft} \cdot \text{lbf} = \textbf{-823.5 BTU}$$

The negative sign indicates that the direction of the heat flow is out of the car (system). In this case, the brakes have cooled back to their initial temperature so the state of the car has not changed nor it's internal energy increased.

In actual practice, we would generally experience something between the two solutions given in this problem. As the car goes down the hill, the brakes are probably being used continuously, and as the car works it's way down the hill, the brakes would be getting hotter and hotter but continuously cooling at the same time. At the bottom of the hill, the brakes would likely still be hot, but not as hot as we would observe if the process was like the process described in part a).

2.3 Compute ΔU, ΔH, Q, W, and W_s when 0.004 moles of N_2 gas at 300 K and 100 cm³ is heated reversibly at constant pressure of 1 bar until it reaches 375 K and 125 cm³.

Solution:

Since this process is reversible and constant pressure, $\Delta H = Q = \int C_p dT$. We will take C_P constant so $\Delta H = Q = C_P (T_2 - T_1) = C_P \cdot 75$ K. Now we need a value for C_P. This can be found in Appendix B in the entry for nitrogen. We have choices, but we will use $C_{P,298}/R = 3.4882$ and we will learn to use the equation form for C_P (equation 2.26) later. What value of R do we use?

We will use R = 8.314 J/mole · K and let's watch our units. $C_{P,298}$ = 3.4882 · 8.314 J/mole · K = 29.0 J/mole · K. (Did you note that C_P and R have the same units?) We need to multiply by moles to get this in J, so 29.0 J/mole · 0.004 moles = 0.116 J/K. $\Delta \textbf{H} = \textbf{Q} = \textbf{0.116 J/K} \cdot \textbf{75 K} = \textbf{8.70 J}$.

By equation 2.9, $\textbf{W}_\textbf{s} = \textbf{0}$ since $\Delta H = Q$. This is by definition of W_s which is shaft work (virtually always a continuously rotating shaft), not P-V work.

We computed W previously for this case in the example problems in chapter 1 so $\textbf{W} = \textbf{2.5 J}$. Using $\Delta U = Q - W$, $\Delta \textbf{U} = \textbf{8.70 J} - \textbf{2.5 J} = \textbf{6.2 J}$.

2.4 A 1500 kg car is traveling on a flat surface at 100 km/h. Find the kinetic energy. Assuming no frictional losses so that kinetic energy is transformed to potential energy, if the car then coasts up a hill, how high will it go vertically before it comes to rest?

Solution:

Standard kinetic energy of the mass is

$KE = \frac{1}{2} m V^2 = \frac{1}{2} \times 1500 \text{ kg} \times [(100 \times 1000)/3600]^2 \text{ m}^2/\text{s}^2$
$KE = \frac{1}{2} \times 1500 \times 27.778^2 \text{ N·m} = 578712 \text{ J} = 578.7 \text{ kJ}$

For no frictional losses, KE is converted to PE, so

$KE = PE = m \cdot g \cdot h$
$578712 \text{ J} = 1500 \text{ kg} \times 9.807 \text{ m/s}^2 \times h$
h=39.3 m

2.5 At a park, atmospheric pressure is 1025 mbar. You climb a hill up to 350 m elevation and you later dive 12 m down in the ocean. What pressure do you feel at each place? Assume the density of water is 1000 kg/m³ and the density of air is 1.18 kg/m³.

Solution:

$\Delta P = \rho \cdot g \cdot h$

On the hill: $P = P_0 - \Delta P = 1025 \times 100 - 1.18 \times 9.81 \times 350 = 98448.47 \text{ Pa} = 98.45 \text{ kPa}$
In the ocean: $P = P_0 + \Delta P = 1025 \times 100 + 1000 \times 9.81 \times 12 = 220220 \text{ Pa} = 220.2 \text{ kPa}$

2.6 Find C_p in J/mole·K for the following at 300 K and 500 K using equation 2.26 and the equation constants given in Appendix B. What do you conclude about assuming C_p being constant over this temperature range?
a. methane
b. nitrogen
c. n-octane
d. o-xylene

Solution:

a. methane
$C_{p,300K} = 8.314 \text{ J/mole·K} (0.1591 + 0.0124190 \cdot 300 - 4.4710 \times 10^{-6} \cdot 300^2 + 5.7324 \times 10^{-10} \cdot 300^3 + 70944.1/300^2 =$ **35.63 J/mole·K**

$C_{p,500K} = 8.314 \text{ J/mole·K} (0.1591 + 0.0124190 \cdot 500 - 4.4710 \times 10^{-6} \cdot 500^2 + 5.7324 \times 10^{-10} \cdot 500^3 + 70944.1/500^2 =$ **46.61 J/mole·K**

similarly:

b. nitrogen $C_{p,300K} = 28.99$ J/mole·K $C_{p,500K} = 29.60$ J/mole·K

c. n-octane $C_{p,300K} = 189.3$ J/mole·K $C_{p,500K} = 286.1$ J/mole·K
d. o-xylene $C_{p,300K} = 133.8$ J/mole·K $C_{p,500K} = 203.9$ J/mole·K

Though there is too little data here for firm conclusions, we will learn that simple molecules like N_2 have C_p's that are much more constant that more complex molecules which fits with the observation above. We also note that sometimes, it would be quite inaccurate to assume C_p's are constant over this temperature range.

Student Problems

1. Water in a garden hose is flowing at 40 lbm/min and the hose has a diameter such that the velocity of the water is 12 ft/s. Water leaving the hose impinges on a wall at 7 ft above the ground so that it's velocity is effectively zero, then it runs down the wall to ground where it collects in a puddle. What is the kinetic energy of the water as it leaves the hose in ft-lbf/s and kW (= 1 kJ/s)? What is the potential energy of the water where it strikes the wall in ft-lbf/sec and kW? What is the change in U of the water from the time it leaves the hose to the time it collects in the puddle in ft-lbf/s and kW?

2. Given that $C_v/R = 2.4882$ for N_2 gas, compute ΔU, ΔH, Q, W, and W_s when 0.004 moles of N_2 gas at 300 K and 100 cm³ is heated reversibly at constant volume until it reaches 375 K.

Chapter 3 - Volumetric Properties of Pure Fluids

3.a Real Substances, Liquids and Phases

Consider the P-V-T behavior of a pure substance beginning with a P-T plane, see Figure 3.a.1. In this figure, note that the V axis would be perpendicular to the P-T plane and come out of the plane of the paper. On the P-T plane consider the important features. At high temperatures and low pressures, a pure substance exhibits vapor (gaseous) behavior. This is noted as "Vapor Region". Similarly, at moderate temperature and pressure, a liquid exists in the "Liquid Region", and at low temperature and moderate to high pressure, a solid exists in the "Solid Region". Delineating the vapor region from the liquid region is the "vapor-liquid equilibrium curve". At a particular temperature and pressure that exists

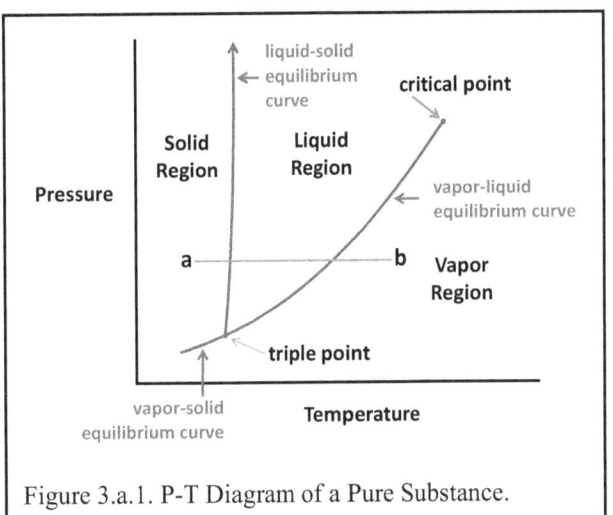

Figure 3.a.1. P-T Diagram of a Pure Substance.

on this curve, some vapor and some liquid would coexist and these two phases would be in equilibrium with each other. Similarly, solid and liquid phases coexist and are at equilibrium along the liquid-solid equilibrium curve and vapor and solid phases coexist along the vapor-solid equilibrium curve. Note also the triple point where all three phases solid, liquid and vapor coexist.

Consider a solid starting at point "a" on Figure 3.a.1. For a more specific example, consider water as a pure solid and in such a case, point "a" might be at say 1 atm and -20°C. We know pure (water) ice exists at that temperature and pressure. Now, consider heating the ice while holding the pressure constant at 1 atm. Pure solid ice would continue to exist but would increase in temperature without a phase change until the liquid-solid equilibrium curve is reached at very close to 0°C. Then, some liquid would begin to form but solid would also still exist. This is the point often referred to as the ice point, and would be quite similar to a glass of ice water. Initially, only a small amount of liquid would exist, but as we continued to add heat, we would slowly convert solid water into liquid water. It is very important to note that neither the temperature nor the pressure would change during this phase change process. We simply convert solid to liquid until the solid disappears and we have pure liquid water at 0°C. The amount of heat added from the point that the first infinitesimally small drop of water formed until the last infinitesimally small flake of solid disappeared would be the heat of liquification of solid water.

We would then continue to heat the pure liquid water at 0°C toward point "b". The temperature would rise from 0°C along segment a-b until we reached the vapor-liquid equilibrium curve very close to 100°C. This would be the *normal boiling point* of water since the pressure is 1 atm. Other pure substances would exhibit a different boiling temperature at 1 atm, but the temperature would be referred to the normal boiling point of that substance as long as the pressure was 1 atm.

Once the first infinitesimally small bubble of vapor appeared in the liquid (still assuming water) at 100°C and 1 atm, we would continue to add heat, but neither the temperature nor pressure would change while the liquid is converted to vapor. Note that the volume would increase greatly as we are converting a liquid into a vapor. This process of converting liquid into vapor would be similar to boiling water on a stove. Once the last infinitesimally small droplet of liquid is boiled off and we continue to add heat, the pure vapor would increase in temperature at constant pressure until we arrive at point "b" at say about 120°C using the example of water as the substance. The amount of heat added from the first bubble to the last droplet would be the heat of vaporization for the substance at the particular pressure with which we are dealing.

Now, consider a process by which we follow along the vapor-liquid equilibrium curve starting at a temperature and pressure below the point labeled "critical point". The vapor-liquid equilibrium curve is also called the *saturation curve* and properties which exist along this curve receive the supercript "s" for saturation. A common example is P^s which is the *saturation pressure* anywhere along this curve. Saturation pressure is also known as *vapor pressure* and the two terms are used interchangeably. Note that vapor pressure is a function of T. Now, the process of following the saturation curve could be accomplished by putting pure liquid into a piston and cylinder with all the air expelled, heating until the boiling point is reached and some vapor has formed, then locking the piston in place so the volume is constant and continuing to heat the vapor-liquid mixture toward the critical point. Liquid is being converted to vapor as it is heated, and the pressure naturally goes up because the volume of the vapor is greater than the volume of the liquid, but the volume of the mixture is constant so the pressure must go up in tandem with temperature. Both temperature and pressure are changing in a prescribed fashion as the substance follows the vapor-liquid equilibrium curve toward the critical point as long as both vapor and liquid coexist in the apparatus. As we are following this curve, the closer we get to the critical point, we would observe the density of the liquid and density of the vapor getting closer and closer to being the same. At the critical point, the density of the two phases become identical, and we find that we can longer identify a liquid existing, nor in exact language, does a vapor exist. Instead, the mixture becomes a supercritical fluid. When an infinitesimally small amount of heat is added to a fluid at its critical point it becomes a supercritical fluid. The supercritical fluid looks and acts like a vapor in many respects but has a density closer to that of a liquid and often can dissolve solids like a liquid. Any fluid at a temperature above its critical temperature cannot be converted to liquid no matter how much pressure might be applied. The density of the fluid may become as high or even higher than the typical liquid, but it does not become a liquid. The molecules making up a fluid above its critical temperature have enough kinetic energy to break any of the energetic attractions of molecules which lead to the formation of a liquid. All pure substances exhibit critical point behavior (unless they chemically break down or undergo other chemical conversions below the critical temperature).

Properties of a substance at its critical point are known as *critical properties* and these properties receive the "c" subscript as in T_c, P_c, and v_c. Critical temperatures and pressures have been measured for a very wide range of substances and are tabulated extensively. We also give a special designation to *reduced properties* which are defined as the ratio of a property to the critical property. Reduced properties receive the subscript "r" as in $P/P_c = P_r$ and $T/T_c = T_r$.

Consider the liquid-solid equilibrium curve in Figure 3.a.1. As shown in the figure, the curve is nearly a vertical line but tips slightly to the right. Thus, it is possible to be in the liquid region near the liquid-solid equilibrium curve and convert that liquid into a solid by applying a large pressure increase while holding the temperature constant. This concept is in line with Le Chatelier's principle that says that a system will move toward a state that relieves external forces applied to it. As shown in Figure 3.a.1, the solid would have a higher density than the liquid and external pressure would tend to favor the higher density material. Most pure substances fall into this category where the solid is more dense than the liquid. However, water is the one common substance for which this is not true. Liquid water exhibits its highest density at about 4°C and it is more dense than solid water (ice) around this temperature. This is why ice floats on liquid water. Also, this property tells us that the liquid-solid equilibrium curve in Figure 3.a.1 would be tilted slightly to the left if the diagram was for water. Le Chatelier's principle still holds and says that ice is converted to liquid water by applying enough pressure in the region of the P-T plane near the liquid-solid equilibrium curve.

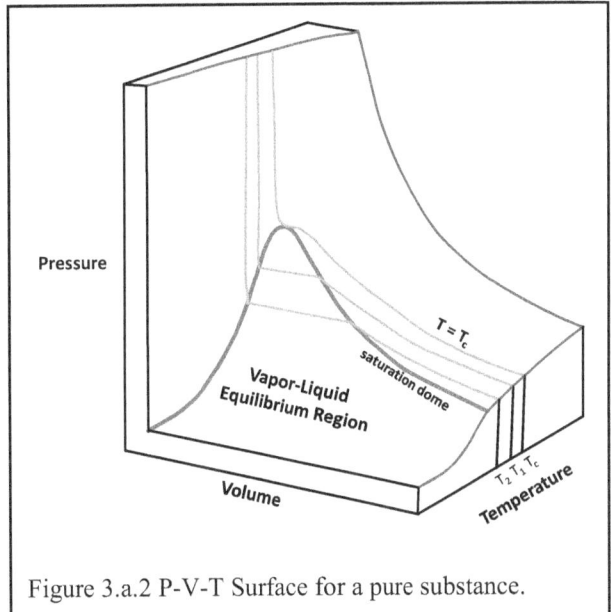

Figure 3.a.2 P-V-T Surface for a pure substance.

Remember that in Figure 3.a.1, the V axis is perpendicular to the P-T plane and comes out of the plane of the paper. Now, imagine taking the P-T plane in Figure 3.a.1 and rotating the entire plane about the P axis to begin to bring the V axis into focus as in Figure 3.a.2 where a schematic of the P-V-T surface is depicted. Ignore the solid region (it has been removed in Figure 3.a.2) and focus on how the vapor-liquid equilibrium curve would look as we continue to rotate about the P axis until we are looking directly at the P-V plane as in Figure 3.a.3. At any given point on the vapor-liquid equilibrium curve in Figure 3.a.1, both vapor and liquid can coexist, so depending on the relative amounts of liquid and vapor, the volume can be anything from the volume of pure liquid to the volume of pure vapor at that temperature and pressure. What we observe on the P-V plane is a *saturation dome*. The left side of the saturation dome in Figure 3.a.3 is the P-V relationship for the pure liquid. It is almost vertical at low temperatures because the volume of a liquid is not a strong function of temperature or pressure. The liquid side exists on the left side of the dome up to the critical point at the top of the dome. At the critical point, the volume of the liquid and volume of vapor become identical, so on the dome to the right of the critical point, vapor exists. The left side of the saturation dome represents *saturated liquid*, and the right side represents *saturated vapor*.

Note three *isotherms* (curves of constant temperature) in Figures 3.a.2 and 3.a.3. These represent the P-V relationship that would exist on a constant temperature slice of the P-T plane. The isotherm marked as $T = T_c$ is the slice that goes through the critical temperature and is termed the *critical isotherm*. Also depicted are isotherms at T_1 which is less than the critical temperature, and at T_2 which is lower than T_1. Note regions where the isotherms T_1 and T_2 are under the saturation dome, pressure is constant as the volume changes because the saturated liquid and vapor coexist in equilibrium with each other, but the relative amounts of liquid and vapor can change, thus changing the volume. More specifically, on Figure 3.a.3 consider what the point at "a" on the T_2 isotherm between points "b" and "c" represents. The volume at point "c" is the saturated vapor volume at T_2 while the volume at point "b" is the saturated liquid at T_2. The volume at point "a" is made up of part saturated liquid volume (V_L^s) and part saturated vapor volume (V_G^s) at T_2. In fact, V_a is a linear combination of V_L^s and V_G^s such that:

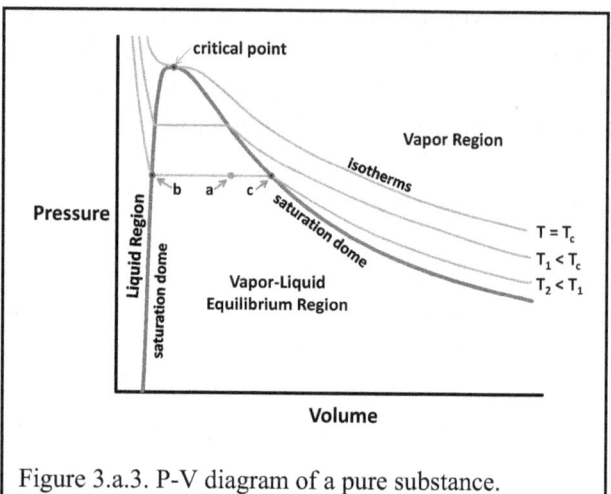

Figure 3.a.3. P-V diagram of a pure substance.

$$V_a = (1-x)V_L^s + xV_G^s \qquad 3.1$$

In equation 3.1, x is the fraction of vapor also known as *quality*. Furthermore, we can use the *lever arm rule* to compute x in this case as:

$$x = \frac{\overline{ab}}{\overline{bc}} \quad \text{and} \quad (1-x) = \frac{\overline{ac}}{\overline{bc}}$$

Equation 3.1 and the lever arm rule apply throughout the vapor-liquid equilibrium region under the saturation dome.

3.b   The Phase Rule

We have already encountered the suggestion that if T and P are known for a system consisting of a pure substance, the state of the system is determined. Consider this suggestion more fully, and in light of Figures 3.a.1, 3.a.2 and 3.a.3, suppose the pressure is specified as well as specifying that we have a saturated vapor. In such a case, note the temperature is known by these two specifications since the temperature is mapped exactly to the pressure by knowing that the state of the system lies on the vapor-liquid equilibrium curve or on the saturation dome. Apparently then, it is not required that we know both T and P, but knowing P and that the substance is saturated is sufficient. A more general rule known as the phase rule quantifies the number of degrees of freedom for a wide range of both pure substances and mixtures. The phase rule is:

$$F = 2 - \pi + N \qquad 3.2$$

In equation 3.2, π is the number of phases, N is the number of components, and F is the degrees of freedom which is also the number of properties which can be independently manipulated. Applying this to a single phase with one component, $\pi = 1$, N = 1 (pure component), then F = 2 so T and P can be independently varied. Thus, for example, we can have pure water at 1 atm and 25°C and also at 2 atm and 50°C. If we require a saturated vapor, note that this requires $\pi = 2$ as there is an infinitesimally small amount of saturated liquid in a saturated vapor. Thus, $\pi = 2$, $N = 1$, at that leaves $F = 1$. Specifying P alone (or T alone) in addition to requiring the substance be saturated specifies the state of the system.

Consider the triple point of a pure substance. Then $\pi = 3$, $N = 1$ so $F = 0$. There is one unique T, P, v, h, u, ... for a substance at its triple point. This is the main reason triple points, especially the triple point of water, are used as temperature reference points.

For systems with multi components, the composition of the system becomes one of the possible degrees of freedom. Thus, salt and water mixtures have different boiling points at atmospheric pressure depending on the amount of salt dissolved in the water. Taking boiling salt-water mixtures as an example, $\pi = 2$, $N = 2$ so $F = 2$. If for example $P = 1$ atm is specified, T can vary and will vary with composition.

One other important aspect of this analysis is that the properties we are dealing with as degrees of freedom must be intensive properties. Recall that T and P are naturally intensive properties, but V, U and H are not. However, $V/n = v$, $U/n = u$, and $H/n = h$ and these are intensive properties. To understand this a little better, note that we can make the size of a cell in which we reach the triple point of water large or small, so the total volume of the liquid phase of water, V, could be large or small. However, the molar volume, v, would be the same regardless of the size of the triple point cell. Also important is to recognize that T and P do not have to be specified as properties. We could just as well specify h and P, or u and T, or u and v, or any two intensive properties to specify the state of a system if $F = 2$.

3.c   The Definition of an Ideal Gas and The Ideal Gas Temperature Scale

Consider a system composed of a gas on the molecular level. Each molecule (or atom) of gas moves around in some container and each molecule has it's own kinetic energy characteristic of it's velocity. The molecules collide with other molecules or the walls of the container bouncing off and interacting energetically during the collisions. Furthermore, molecules composed of two or more atoms possess rotational energy, bond vibrational energy, and chemical bond energy. Summing all the types of energy over all the molecules gives the internal energy of the system.

The smaller the volume of the system and the higher the pressure means a greater likelihood of collisions resulting in higher overall molecular interaction energies for the system. Conversely, larger volumes and lower pressures result in a diminished number of collisions and therefore a diminished molecular interaction energy. Consider then a system consisting of a gas at a low pressure and high volume such that molecular interactions are energetically insignificant or the molecules (e.g., noble gases) may not have a propensity to interact energetically and molecular interactions are insignificant. In fact, consider a limit where:

1. Interactions of molecules are energetically zero.
2. The molecules are point sources so that the volume they occupy is zero.

Suppose we have a piston and cylinder arrangement containing a system consisting of one mole of a gas, then we place the entire apparatus into a bath at the triple point of water so that we hold the temperature constant at the triple point of water. Beginning at a high pressure, we measure P and v of the gas and record the product Pv then continue to record Pv as we increase the volume thus lowering the pressure. This process is depicted for two hypothetical gases in Figure 3.b.1. What we observe is that as $P \to 0$, the product Pv approaches a single value regardless of what gas we used for the system. This is true for real gases as well as the hypothetical gases we have used for illustrative purposes in Figure 3.b.1. As $P \to 0$, $v \to \infty$ so our requirements for ideal gas behavior are satisfied. Molecular interactions go to zero, and the molecular volumes of the individual molecules approach point sources compared to v. A general principle to remember is that *all gases approach ideal gas behavior, or become ideal gases, as pressure goes to zero.*

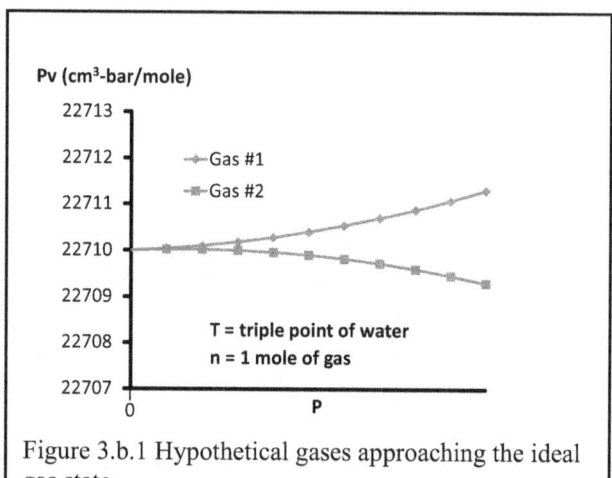

Figure 3.b.1 Hypothetical gases approaching the ideal gas state.

Now, let's finish the formal definition of the Kelvin temperature scale that we started in chapter 1, section 1.b. We define the Kelvin temperature scale as being proportional to Pv for an ideal gas. The proportionality factor we call "R", so from Figure 3.b.1 we have:

$$22710 \text{ cm}^3 \cdot \text{bar/mole} = R \cdot 273.16 \text{ K} \quad \text{or} \quad R = 83.14 \text{ cm}^3 \cdot \text{bar/mole} \cdot \text{K} \qquad 3.3$$

Note that the defined value of 273.16K has been used for the triple point of water. R is known as the gas constant, and it is a universal constant. The units in the numerator of R are energy units and $cm^3 \cdot bar$ can be converted to any other energy units. For example, $1 \text{ cm}^3 \cdot \text{bar} = 0.1 \text{ J}$ so R is also 8.314 J/mole-K.

Now consider how an unknown temperature can determined using an *ideal gas thermometer*. Suppose we want to measure the normal boiling point of some pure substance, so we take our piston and cylinder arrangement with 1 mole of gas, immerse the apparatus in the boiling substance, measure Pv as a function of P and extrapolate to zero pressure where we get some value at the intercept we'll call "a". We now get the measured T from:

$$T(K) = \frac{a(cm^3 \cdot bar/mole)}{83.14(cm^3 \cdot bar/mole \cdot K)}$$

This is how we properly and formally divide 273.16 kelvins at the triple point of water into even increments of 1 kelvin, and that is by using the ideal gas thermometer.

Many gases exhibit behavior very close to an ideal gas under a range of low to moderate pressures and moderate to high temperatures. We therefore define an ideal gas as a gas that obeys the P-V-T relationship:

$$PV = nRT \quad \text{or} \quad Pv = RT \qquad 3.4$$

A P-V-T relationship for a gas is also known as an *equation of state (EOS)* and the ideal gas law is the simplest EOS. We will further develop properties of an ideal gas in this book and refer especially to the ideal gas for establishing a reference point for properties of real gases.

Consider the relationship of C_p to C_v for and ideal gas. Starting with equation 2.8, divide the equation through by n to make it on a per mole basis:

$$h = u + Pv$$

But for an ideal gas, Pv=RT, so:

$$h = u + RT \qquad \text{(ideal gas) 3.5}$$

or, in differential form:

$$dh = du + RdT \qquad \text{(ideal gas) 3.6}$$

From equations 2.24 and 2.25, we see that:

$$dh = C_p\, dT \text{ (constant P)} \quad \text{and} \quad du = C_v\, dT \text{ (constant v)} \qquad 3.7 \text{ and } 3.8$$

As we will see later in this book, equations 3.7 and 3.8 apply to all process for ideal gases without restrictions of constant P or v. This is not obvious and will require a working knowledge of the Second Law of Thermodynamics to prove, so we leave its proof to later in this book. With equations 3.7 and 3.8, equation 3.6 yields:

$$C_p\, dT = C_v\, dT + R\, dT \qquad \text{(ideal gas) 3.9}$$

or:

$$C_p^{IG} = C_v^{IG} + R \qquad \text{(ideal gas) 3.10}$$

We have added the superscript "IG" to C_p and C_v in equation 3.10 and we will generally designate C_p and C_v with these superscripts for ideal gases in the future. Equation 3.10 tells us that we need only the value of C_p or C_v, not both, for ideal gases. The conventional way is to tabulate values of C_p and if C_v is needed, it is obtained from equation 3.10. Finally, observe that equation 3.10 does not suggest that C_p and C_v are independent of T, but that whatever temperature functionality C_p has, its value simply differs from C_v by R over the entire temperature function.

The enthalpy, h, and the internal energy, u, of ideal gases provide an extremely useful platform from which h, u and other properties of real gases can be built. Therefore, we generally are only interested in C_p (and C_v through equation 3.10) for real gases when they are extrapolated to their ideal gas state. The extrapolation could be done in a fashion similar to the extrapolation of Pv to zero pressure like given in Figure 3.b.1. In Chapter 6, we will build a theory to compute the difference between h and u for the ideal gas versus the real gas. Appendix B provides values of C_p for the *ideal gas state*.

3.d The Relationship of P-V-T to Energy for non-Ideal Gases

Imagine a real gas that is at a state where molecular interactions are significant energetically and imagine that the molecules are of a size that make the collective volume of the molecules significant compared to the overall volume of the gas. Place the gas, which we will treat as a system, in a frictionless piston and cylinder device and allow the piston to move freely but restrain the piston with an external pressure equal to the pressure of the gas so the gas remains at constant pressure. Now, imagine we have two switches. One switch turns off the molecular interactions and the other in the off state instantly causes all the molecules to have zero volume. These switches are obviously hypothetical, but the actions resulting from the use of these switches form an important mental exercise.

Assume both switches are off so the gas assumes the properties of an ideal gas. Now, switch on the molecular interactions. The molecules attract each other, and the overall result will be that the volume of the system goes down in response. This results in the piston moving to accommodate the smaller volume of the gas at constant pressure, and the surroundings do work on the system. This represents PV work that the surroundings do on the system, and a First Law consideration tells us that the internal energy of the system must have increased by the amount of PV work done by the surroundings. This increase in internal energy obviously equals the energy due to the molecular interactions. Therefore and most importantly for this mental exercise, the internal energy of the system relates to the P-V-T behavior of a gas through the molecular properties of the gas.

Now, turn the molecular interaction switch off and turn the molecular size switch on. This time, the volume of the gas increases and does work on the surroundings compared to the ideal gas state. The PV work done on the surroundings relates to the volume increase caused by a non-zero molecular size. We again reach the conclusion that the P-V-T behavior of a real gas relates to the internal energy of the gas through the molecular properties of the gas. We will quantify this relationship later in this book, but we now see the importance of being able to characterize the P-V-T behavior of real gases as well as ideal gases.

3.e Equations of State for Real Gases

3.e.i The Compressibility Factor

Equation 3.4 for an ideal gas can be corrected for any real gas by defining the *compressibility factor*, z, as:

$$z = \frac{Pv}{RT} \quad \text{or} \quad Pv = zRT \qquad 3.11$$

Clearly:

$$z = \frac{v^{actual}}{v^{ideal\ gas}}$$

Observed values for the compressibility factor are often near 1.0 when a gas is near ideal gas behavior, but the compressibility factor at the critical point (z_c, the *critical compressibility factor*) is generally between about 0.23 to 0.31. Liquids can also be assigned values of z as the molar volume of the liquid can be used in equation 3.11 as well as gaseous volumes. Liquid volumes are often $0.05 < z < 0.1$.

Equation 3.11 works for every possible situation, but the key to using the equation is to have an accurate method for determining values of z. Below, we discuss several possibilities.

3.e.ii   The Virial Equation of State

Theoretical justifications form the basis of an exact representation of P-V-T behavior for real gases in the form an infinite series known as the *virial equation*:

$$z = \frac{Pv}{RT} = 1 + B'P + C'P^2 + D'P^3 + \cdots \qquad 3.12$$

In equation 3.12, the terms in P represent interactions of molecules with B' representing the magnitude of interactions of two molecules, C' three molecules, D' four molecules, etc. If we divide equation 3.12 by v/RT, then back substitute P, we can collect the right hand side of the infinite series in the powers of 1/v instead of P:

$$z = 1 + B/v + C/v^2 + D/v^3 + \cdots \qquad 3.13$$

Algebraic relationships exist between the two sets of constants. Thus:

$$B' = \frac{B}{RT} \qquad C' = \frac{C - B^2}{(RT)^2} \qquad D' = \frac{D - 3BC + 2B^2}{(RT)^3} \qquad \text{etc...}$$

In the infinite limit, equations 3.12 and 3.13 are identical and are an exact representation of the P-V-T behavior of any gas. However, theoretical analysis shows that B, C, D,... are functions of T, but there is no exact theoretical representation of how the constants depend on T. Thus in application, equations 3.12 and 3.13 are not simple to use for representation of P-V-T behavior. However, they provide a theoretical basis from which other equations can be evaluated, and truncated forms of equations 3.12 and 3.13 can provide very useful engineering approximations to actual P-V-T behavior.

3.e.iii Van der Waals Equation

Van der Waals [1] proposed that the ideal gas EOS could better represent actual P-V-T behavior by modifying the ideal gas equation to incorporate interactions of molecules and incorporating the fact that molecules are not point sources. He wrote:

$$\left(P + \frac{a}{v^2}\right)(v-b) = RT \qquad 3.14$$

In equation 3.14, a and b are constants not dependent on P, v or T. The term a/v^2 modifies the pressure to include interactions of pairs of molecules and the term (v - b) represents the volume that molecules do not occupy. These two modifications were designed to correct the assumptions written into the ideal gas equation.

Equation 3.14 can be arranged so that it is explicit in pressure:

$$P = \frac{RT}{v-b} - \frac{a}{v^2} \qquad 3.15$$

or it can be arranged as a polynomial in volume:

$$v^3 - \left(b + \frac{RT}{P}\right)v^2 + \frac{a}{P}v - \frac{ab}{P} = 0 \qquad 3.16$$

Because the highest term is v^3 in equation 3.16, van der Waals equation is termed a *cubic equation of state*. The compressibility factor form is often written:

$$z = \frac{v}{v-b} - \frac{a}{RTv}$$

Consider how a and b in equation 3.15 might be determined. If we were to make two measurements of P, V and T for a real gas, then we could write:

$$P_1 = \frac{RT_1}{v_1 - b} - \frac{a}{v_1^2} \quad \text{and} \quad P_2 = \frac{RT_2}{v_2 - b} - \frac{a}{v_2^2}$$

These two equations could be solved simultaneously for a and b. Using these values of a and b, Van der Waals equation would then give back the measured values of P_1, v_1, and T_1 and P_2, v_2, and T_2 exactly and would be expected to be accurate in the region around points 1 and 2. But because of the limited accuracy of Van der Waals equation, it could not be expected to extrapolate accurately very far outside the region near points 1 and 2.

Rather than making experimental measurements, a method which is widely used to determine a and b is to get them from the critical point. At the critical point:

$$P_c = \frac{RT_c}{v_c - b} - \frac{a}{v_c^2} \qquad 3.17$$

Also, theoretical analysis, which is beyond the scope of this book, shows that at the critical point:

$$\left.\frac{\partial P}{\partial v}\right)_T = \left.\frac{\partial^2 P}{\partial v^2}\right)_T = 0 \qquad 3.18 \text{ and } 3.19$$

Three equations (3.17, 3.18, and 3.19) cannot all be satisfied at once by choice of only two adjustable parameters, a and b. The usual convention is to choose a and b to satisfy equations 3.18 and 3.19. The result of this calculation yields:

$$a = \frac{27R^2 T_c^2}{64 P_c} \qquad \text{and} \qquad b = \frac{RT_c}{8P_c} \qquad 3.20 \text{ and } 3.21$$

Using the values of a and b determined from the critical constants in equations 3.20 and 3.21, equation 3.17 then yields:

$$v_c = \frac{3RT_c}{8P_c} \qquad \text{or} \qquad z_c = 3/8 \qquad 3.22$$

Accordingly, using this method for determination of a and b, the value of z_c given by Van der Waals equation is too high as most substances have critical compressibility factors in the 0.23 to 0.31 range. However, other choices for a and b would give incorrect predictions of liquid formation near the critical point. Thus, Van der Waals equation often yields poor results. Nonetheless, Van der Waals equation is an equation that exhibits the general properties that are needed to give non-trivial solutions for real gases. For example, it readily predicts the formation of liquids from gases whereas the ideal gas law cannot. However, it does not result in high accuracy though it does improve on the ideal gas equation in regions near where constants a and b are known to give an accurate representation of P-V-T behavior. We will use Van der Waals equation solely as a relatively simple tool for instruction and we will never rely on its application for accurate P-V-T relationships over extended regions because much better equations exist. The same methods we use in the application of Van der Waals equation can be readily extended to more versatile equations that we will see below. However, considerably more mathematical effort is generally required for the more versatile equations, and we will leave the details of development of those equations for more accurate implementation to others.

3.e.iv Other Cubic Equations of State

Clearly, there are numerous refinements which could be made to improve Van der Waals equation, and numerous authors have studied and improved on Van der Waals equation. Walas [2] has a very good collection and many historical notes regarding development of these equations and the interested reader is referred there. We will mention only a few that are considered worthy of utility for engineering calculations, beginning with the Peng-Robinson equation:

$$P = \frac{RT}{v-b} - \frac{a}{v(v+b)+b(v-b)} \qquad 3.23$$

$$a = 0.45724 \frac{R^2 T_c^2 \alpha}{P_c} \qquad b = 0.07780 \frac{RT_c}{P_c}$$

$$\alpha = \left[1 + \left(0.37464 + 1.54226\omega - 0.26992\omega^2\right)\left(1 - T_r^{0.5}\right)\right]^2$$

The parameter ω is the *acentric factor* which is a parameter widely tabulated along with the critical constants T_c and P_c for individual substances. The acentric factor relates to the attractive forces of molecules via the vapor pressure of the substance in relation to it's critical point. ω is defined by:

$$\omega = -1 - \log_{10}\left(\frac{P^s\big|_{T_r=0.7}}{P_c}\right) \qquad 3.24$$

As with Van der Waals equation, the parameters a and b for the Peng-Robinson equation have been determined using equations 3.18 and 3.19. The Peng-Robinson equation then yields $z_c = 0.307$ which is much better than the value from Van der Waals equation. The polynomial form of the Peng-Robinson equation is:

$$z^3 - (1-B)z^2 + (A - 3B^2 - 2B)z - (AB - B^2 - B^3) = 0$$

$$A = \frac{aP}{R^2T^2} \qquad B = \frac{bP}{RT}$$

The Peng-Robinson equation is a great improvement over Van der Waals equation and it's accuracy makes it often useful for engineering calculations. It is also useful for mixtures which will be considered later in this book.

Another useful equation is the Redlich-Kwong equation [4]:

$$P = \frac{RT}{v-b} - \frac{a}{\sqrt{T}v(v+b)} \qquad 3.25$$

$$a = \frac{0.42748R^2 T_c^{2.5}}{P_c} \qquad b = \frac{0.08664RT_c}{P_c}$$

Because the equation showed very good properties when it was first published, many researchers have worked to improve the equation with various amounts of success. Tatakad et. al. [5] reviewed the accuracy of many of the proposed improvements and have made recommendations on which modifications work best for different scenarios including predicting the behavior of mixtures. The polynomial form of the Redlich-Kwong equation is generally written:

$$v^3 - \frac{RT}{P}v^2 + \frac{1}{P}\left(\frac{a}{\sqrt{T}} - bRT - Pb^2\right)v - \frac{ab}{P\sqrt{T}} = 0 \qquad 3.26$$

A compressibility factor form of the Redlich-Kwong equation is:

$$z^3 - z^2 + (A - B - B^2)z - AB = 0 \qquad 3.27$$

$$A = \frac{aP}{R^2 T^{2.5}} \qquad B = \frac{bP}{RT}$$

As with the Peng-Robinson equation, the Redlich-Kwong equation along with improvements made to the equation have proven utility for engineering calculations.

3.e.v Higher Order Equations of State

Two higher order equations should be mentioned, but these equations are complex and given a lack of suitable constants for a wide range of components, these equations are not used often. An early complex equation was given by Beattie and Bridgeman:

$$P = \frac{RT}{v^2}\left(1 - \frac{c}{vT^3}\right)(v+B) - \frac{A}{v^2} \qquad 3.28$$

$$A = A_0\left(1 - \frac{a}{v}\right) \qquad B = B_0\left(1 - \frac{b}{v}\right)$$

This equation has 5 adjustable constants A_0, B_0, a, b, and c, and is fourth order in v. The Benedict-Web-Rubin (BWR) equation was developed from the Beattie-Bridgeman equation:

$$P = \frac{RT}{v} + \left(B_0 RT - A_0 - \frac{C_0}{T^2}\right)\frac{1}{v^2} + \frac{(bRT - a)}{v^3} + \frac{a\alpha}{v^6} + \frac{c}{T^2 v^3}\left(1 + \frac{\gamma}{v^2}\right)\exp\left(\frac{-\gamma}{v^2}\right) \qquad 3.29$$

Eight adjustable constants, A_0, B_0, C_0, a, b, c, α, and γ are contained in the equation and it is sixth order in v. These two higher order EOS's are largely empirical and have a limited number of gases for which constants have been determined. However, once a good set of constants have been determined, these equations do a very good job of simulating the P-V-T behavior of the gas over a wide range. Many modifications exist and descriptions can be found in the open literature.

3.e.vi   Corresponding States and Generalized Correlations

One of the most important engineering tools for correlating P-V-T relationships begins with the observation that all pure substances have very similar P-V-T surfaces when they are normalized to P_r-T_r surfaces. This aligns the critical points for the surface of each substance at $P_r = T_r = 1$. Once this is considered, then we need to measure only one P_r-T_r surface and we can apply the results to all substances because they will all have similar z's when compared at the same T_r and P_r. This principle is known as the *theorem of corresponding states*. In it's most fundamental form, the theorem of corresponding states is:

$$z = f(P_r, T_r)$$

But as we have already discussed, critical compressibilities vary among different substances, and this can be readily seen given data in Appendix C. So a P_r-T_r surface for a single substance obviously cannot accurately give z_c's for all substances. Hougen, Watson and Ragatz [8] presented tables based upon a three parameter corresponding states correlation:

$$z = f(P_r, T_r, z_c) \qquad 3.30$$

Tables for $z_c = 0.27$ were presented along with corrections based upon deviations from this central value of z_c.

More recently, Pitzer and coworkers [9] introduced the three parameter correlation:

$$z = f(P_r, T_r, \omega) \qquad 3.31$$

We have already encountered the acentric factor, ω, which is directly related to vapor pressure of a substance and therefore indicative of how strongly molecules attract other molecules. This third parameter in the correlation corrects for such interactions. The three parameter corresponding states correlation using ω as in equation 3.31 has found wide acceptance for engineering calculations.

The three parameter correlation has two important forms. The first is in tabular form as given in Appendix D which is known as the Lee-Kesler tables [10]. To use this table, compute P_r and T_r for the substance and look up z^0 and z^1 in the table for the computed values of P_r and T_r. The compressibility factor is then given by:

$$z = z^0 + \omega z^1 \qquad 3.32$$

The z^0 part has been determined from behaviors of simple fluids such as argon. The second term in equation 3.32 then corrects for fluids which are which are not simple and may be polar. The compressibility factor obtained for most substances that are non-polar or only sightly polar is generally within 2 - 3% of the correct value. Highly polar substances generally give results to 5% of the correct value, but when applied to substances which associate, the resulting errors can be large. However, for the vast majority of engineering calculations which do not require extreme accuracy, this correlation is excellent and simple to use provided the critical properties and acentric factor are known.

The second form of the three parameter corresponding states correlation is derived from the Lee-Kesler tables and puts the correlation into equation form. The equation is based on the Pitzer Generalized Virial Coefficient correlation which is based on the truncated form of the virial equation (equation 3.12):

$$z = 1 + \frac{BP}{RT} = 1 + B_r \frac{P_r}{T_r} \qquad 3.33$$

B_r is the reduced second virial coefficient:

$$B_r = \frac{BP_c}{RT_c} \qquad 3.34$$

To obtain values for B_r:

$$B_r = B^0 + \omega B^1 \qquad 3.35$$

and:

$$B^0 = 0.083 - \frac{0.422}{T_r^{1.6}} \qquad B^1 = 0.139 - \frac{0.172}{T_r^{4.2}} \qquad 3.36 \text{ and } 3.37$$

Equations 3.36 and 3.37 are fitted from the Lee-Kesler tables and represent the behavior of the function in the table very well in the region $0.8 < T_r < 4.0$ and $P_r < 2.0$. These equations are much easier to use than reading values from a table, and the results are basically the same for the limited T_r, P_r region.

3.f Correlations for Liquid Volumes

Though the cubic equations of state give liquid volumes, the results are often less accurate than needed. The Lee-Kesler table contains values for z in the liquid region ($T_r < 1$ and $P_r > 1$), and these values represent reasonably accurate values for many engineering applications and the correlation requires only the critical constants and the acentric factor. However, this correlation is not suitable for highly polar substances.

Other generalized correlations focusing only on the liquid region are available. One is the Rackett equation [11] which gives molar volumes of saturated liquids:

$$v^{sat} = v_c z_c^{(1-T_r)^{2/7}} \qquad 3.38$$

Results from this equation are generally accurate to 2% or better, and application of this equation is simple. More complex correlations such as the COSTALD equation developed by Thomson, Brobst, and Hankinson [12] exist and give very accurate results for both pure components and mixtures.

3.g The Steam Tables

Steam tables have been placed in tabular form in Appendix G. Note that the term "steam" generally refers to pure water vapor in the strictest sense. However, the steam table also includes entries for liquid, and we will use the term "steam" interchangeably with "water" in most instances. Two sets of steam tables are given in this book; one in SI units and the second in Field Units. Each of the tables has three parts and diagrams representing depictions are also included for steam. The fir table is a saturated table which gives values of saturated liquid and saturated vapor on each side of the saturation dome. These tables run up to the critical temperature of steam (water) where the saturation done ends. Tables with even increments of pressure and even increments of temperature are presented in the saturation table. These two forms follow exactly the same saturation curves but since these data represent two-phase equilibrium data, the phase rule tells us that only one independent variable, usually either temperature or pressure, may be specified. Note that columns for both the liquid properties (v^L, h^L, etc.) and vapors (v^V, h^V, etc.) are given in these tables.

The second part of each table are values for superheated steam. To be precise, some of the entries are actually supercritical steam whenever conditions are above the critical temperature and pressure. Since we have a single phase, we must specify two independent variables, so the entries are arranged in pressure blocks with varying temperature. Note that next to the pressure for any given pressure block there is a temperature given in parentheses (though only when the saturation temperature at that pressure is below the critical temperature). For example, in the steam table, one heading for a block is "P = 50 kPa (81.317 °C)". The temperature given in parentheses is the saturation temperature at that pressure. Students may check these numbers against the numbers in the saturation table. Also note that the first line of data below the block heading is labeled "(sat)" to the left rather than having a temperature corresponding to that line of data. Obviously, this designates that the row of data is for the saturated state, but it is for the vapor phase only. Finally, note that sometimes there is a missing row of data such as in the 50 kPa block in the steam table. This is due to the fact that the saturation temperature at that pressure is above the temperature corresponding to the row where the data is missing.

The third part of each table is a subcooled liquid table. As with the superheated steam table, two independent variables are required, and the table is arranged in pressure blocks with varying temperature.

Volumes are one of the entries in the tables. These values have been derived from experimental data, but over many decades they have been checked, corrected, and correlated extensively. As such, the values are not dependent upon an approximate representation with an equation of state and are only approximate in that they represent the volumetric data only to the number of significant digits presented. Enthalpies and internal energies are now also familiar quantities to students and values for these energy functions are also given in the tables. Another quantity, s known as entropy, has not yet been introduced in this book, but is in the steam table and can be retrieved from the table just as the other variables. We will discuss the meaning of entropy in future chapters of this book.

If a desired quantity is between values in the table, use interpolation to get the desired quantity. For example, the following entries exist in the SI saturated steam table with the even temperature increment table:

T (°C)	P (kPa)	v^V (m³/kg)
115	169.18	1.0358
120	198.67	0.89119

To find the entry for the vapor volume at 117.3°C set up a linear ratio of the differences:

$$\frac{117.3 - 115}{120 - 115} = \frac{v^V\big|_{117.3} - 1.0358}{0.89119 - 1.0358}$$

which yields:

$$v^V\big|_{117.3} = 0.96928 \text{ m}^3/\text{kg}$$

Other tables for other pure components which comprise Appendices H - P are basically set up the same as the steam tables and may be treated similarly, but the data for the other components are not as extensive as the steam table.

3.h Simple Processes with Ideal Gases

As we discussed in section 3.c, the nature of an ideal gas leads us to find that the internal energy and enthalpy of ideal gases are independent of pressure and volume. We can arrive at this conclusion from a qualitative viewpoint, since by definition of an ideal gas there are no energetic interactions of molecules, the volume of an ideal gas should make no difference in its internal energy because changing the volume doesn't affect any energetic property of the gas. This also applies to pressure since pressure and volume are intimately linked through the ideal gas law. We will later prove this rigorously, but for now, we will simply accept this and move forward with some calculations for simple processes using an ideal gas.

If we reconsider equations 2.24 and 2.25 in light of the knowledge that h and u are independent of P and V for an ideal gas, we get:

$$\Delta u = \int_{T_1}^{T_2} C_V^{IG} dT \quad \text{and} \quad \Delta h = \int_{T_1}^{T_2} C_P^{IG} dT \qquad \text{(ideal gas) 3.39 and 3.40}$$

These equations are now written without pressure or volume restrictions and apply to *all* processes for ideal gases. Recall the temperature functionality for heat capacity that was introduced in Chapter 2. Now that we understand the concept of an ideal gas, we recall equation 2.26 to note that constants in Appendix B for the heat capacity equation are for the ideal gas state:

$$C_p^{IG}/R = A + BT + CT^2 + DT^3 + E/T^2 \quad \text{where T is in K} \qquad 2.26$$

and equation 2.27, the integrated equation, is:

$$\int_{T_1}^{T_2} \frac{C_p^{IG}}{R} dT = A(T_2 - T_1) + \frac{B}{2}(T_2^2 - T_1^2) + \frac{C}{3}(T_2^3 - T_1^3) + \frac{D}{4}(T_2^4 - T_1^4) - E\left(\frac{1}{T_2} - \frac{1}{T_1}\right) \qquad 2.27$$

Equations 2.26 and 2.27 are important to remember in the context of evaluating equation 3.40 (especially) and several other equations to be introduced later.

Consider some simple processes with ideal gases. For each of these situations, we will assume we have an ideal gas in a frictionless piston and cylinder arrangement and we carry out heating, cooling, expansion, compression, etc. using the device.

3.h.1 Constant Temperature (Isothermal) Process

According to equations 3.39 and 3.40, Δu and Δh are zero for an isothermal process. Thus, the First Law states:

$$\Delta u = 0 = q - w \qquad \text{so} \qquad q = w \qquad 3.41$$

We can expand or compress the gas in the piston and cylinder holding the temperature constant by placing the apparatus in a constant temperature bath then expanding/compressing slowly so heat flows in or out to maintain the temperature. Assuming the process is reversible:

$$w = \int_{v_1}^{v_2} P dv = \int_{v_1}^{v_2} \frac{RT}{v} dv \qquad \text{(reversible) } 3.42$$

Since T is constant, it can come out of the integral resulting in:

$$q = w = RT \ln\left(\frac{v_2}{v_1}\right) \qquad \text{(reversible) } 3.43$$

If $v_2 > v_1$, the process is an expansion and both q an w are positive. Heat must be added to the system to maintain the temperature as the gas expands, and positive work is done on the surroundings pushing the piston. For a compression, heat must be removed and work must be done on the system making both q and w negative. Using Pv=RT and substituting into equation 3.43, we find equivalently:

$$q = w = -RT \ln\left(\frac{P_2}{P_1}\right) \qquad \text{(reversible) } 3.44$$

3.h.2 Constant Pressure (Isobaric) Process

$$\Delta h = q = \int_{T_1}^{T_2} C_p^{IG} dT \qquad 3.45$$

$$w = \int_{v_1}^{v_2} P dv = P(v_2 - v_1) \qquad \text{(reversible) } 3.46$$

$$\Delta u = q - w = \int_{T_1}^{T_2} C_p^{IG} dT - P(v_2 - v_1) = \int_{T_1}^{T_2} C_p^{IG} dT - R(T_2 - T_1) = \int_{T_1}^{T_2} C_v^{IG} dT \qquad 3.47$$

In equation 3.47 we have used $C_p^{IG} - R = C_v^{IG}$ for an ideal gas.

3.h.3 Constant Volume (Isochoric) Process

For constant volume, w = 0.

$$\Delta u = q = \int_{T_1}^{T_2} C_v^{IG} dT \qquad 3.48$$

$$\Delta h = \int_{T_1}^{T_2} C_p^{IG} dT \qquad 3.49$$

3.h.4 Adiabatic Process

An adiabatic process is a process where no heat is added or removed from the system. This is easily accomplished by insulating the system, and expansion/compression can then occur. Upon adiabatic expansion, the temperature of the system will fall and will rise upon adiabatic compression. We still have:

$$\Delta u = \int_{T_1}^{T_2} C_v^{IG} dT \text{ and } \Delta h = \int_{T_1}^{T_2} C_p^{IG} dT \qquad 3.39 \text{ and } 3.40$$

However, the change in temperature that accompanies the process is not obvious, and it is needed to evaluate equations 3.39 and 3.40. Writing and solving the First Law:

$$\Delta u = q - w \qquad 3.50$$

$$\int_{T_1}^{T_2} C_v^{IG} dT = 0 - \int_{v_1}^{v_2} P dv \qquad 3.51$$

Differentiate equation 3.51, substitute the ideal gas law, and rearrange:

$$C_v^{IG} dT = -P dv = -\frac{RT}{v} dv \qquad 3.52$$

$$C_v^{IG} \frac{dT}{T} = -R \frac{dv}{v} \qquad 3.53$$

Equation 3.53 can be integrated approximately for the case where C_v^{IG} is constant or by using an appropriate averaged value of C_v^{IG}. This yields:

$$C_v^{IG} \ln\left(\frac{T_2}{T_1}\right) = -R \ln\left(\frac{v_2}{v_1}\right) \qquad 3.54$$

$$\left(\frac{T_2}{T_1}\right)^{C_v^{IG}} = \left(\frac{v_2}{v_1}\right)^{-R} \text{ or } \left(\frac{T_2}{T_1}\right) = \left(\frac{v_2}{v_1}\right)^{\frac{-R}{C_v^{IG}}} = \left(\frac{v_2}{v_1}\right)^{\frac{C_v^{IG} - C_p^{IG}}{C_v^{IG}}} \qquad 3.55$$

Using $C_p^{IG}/C_v^{IG} = \gamma$:

$$\frac{T_2}{T_1} = \left(\frac{V_2}{V_1}\right)^{1-\gamma} \qquad 3.56$$

If we know the initial temperature and the starting and ending volumes, we can compute T_2 using equation 3.56 then we can substitute into equations 3.39 and 3.40 to get Δu and Δh. Starting and ending pressures may be obtained from the ideal gas law.

If we know the initial temperature and starting and ending pressures, we can substitute the ideal gas law into equation 3.56 to obtain:

$$\frac{T_2}{T_1} = \left(\frac{P_2}{P_1}\right)^{\frac{\gamma-1}{\gamma}} \qquad 3.57$$

3.i   Chapter 3 Example Problems

3.1   A 10 liter solid vessel is divided exactly in half by a thin membrane. On one side of the membrane is an ideal gas at 1 atm and 50°C and on the other side is the same gas at 100 atm and 50°C. The membrane is then ruptured and the contents mix and reach steady state. Assuming the vessel is fully insulated, a) Calculate ΔU, Q and W. b) Calculate the final temperature of the gas.

Solution:

a) Take the entire vessel contents initially including the membrane as the system. $\Delta U = Q - W$ where $Q = 0$ because the system is insulated and $W = 0$ because the vessel doesn't change volume. Therefore $\Delta U = 0$.

b) Since $\Delta U = \int C_v dT$, $dT = 0$ so $T_2 = T_1 = 50°C = 323K$.

$$n_{side\ 1} = \frac{(1\,\text{atm})(5\,\text{L})}{\left(0.08206\,\frac{\text{L}\cdot\text{atm}}{\text{mole}\cdot\text{K}}\right)(323\,\text{K})} = 0.1886 \text{ moles}$$

$$n_{side\ 2} = \frac{(100\,\text{atm})(5\,\text{L})}{\left(0.08206\,\frac{\text{L}\cdot\text{atm}}{\text{mole}\cdot\text{K}}\right)(323\,\text{K})} = 18.86 \text{ moles}$$

$n_{total} = 19.05$ moles.

$$19.05 = \frac{(P)(10\,L)}{\left(0.08206\dfrac{L\cdot atm}{mole\cdot K}\right)(323K)}$$

which yields P_{final} = **50.5 atm**. T_{final} = **50°C**.

3.2 One mole of steam initially at 260°C and 800 kPa is cooled at constant volume until it is a saturated vapor. Calculate ΔU, Q, and W for the process.

Solution:

Interpolating in the steam table:
v_1(260°C, 800 kPa) = 0.29947 m³/kg
u_1(260°C, 800 kPa) = 2732.3 kJ/kg
$v_2 = v_1$ so
u_2(v = 0.29947 m³/kg, saturated vapor) = 2568.6 kJ/kg
Δu = 2568.6 - 2732.3 = -163.7 kJ/kg = -163.7 J/g
ΔU = -163.7 J/g x 18.02 g/mole x 1 mole = **-2949.9 J**
W = 0 (constant volume)
Q = ΔU = -2949.9 J

3.3 1 mole of n-butane gas is contained in a 2.0 L vessel at 400 K. Use the Pitzer generalized virial coefficient correlation to find the pressure.

Solution:

This problem is more difficult than problems where we know the temperature and pressure. Most problems are direct when we know T and P, but since we know T and V instead, we will generally have to iterate to find P. These are the kinds of problems that engineers should become adept at solving. It should be easy to see that we can always just guess a pressure, then determining the volume becomes direct, and all we need to do is guess the right pressure to get the specified volume. The part that we need to think about is how to efficiently do the guessing.

For n-butane, Appendix C yields T_c = 425.1 K, P_c = 3796 kPa = 37.96 bar, and ω = 0.206. Then T_r = 0.941. We can't compute P_r initially, so let's guess a pressure. Let's guess the pressure we expect if n-butane was an ideal gas and this should at least get us started.

P_{guess1} = (1 mole)(0.08314 L-bar/mole-K)(400K)/2 L = 16.628 bar

Then, P_r = 0.438. Let's use this guess to determine a value for z, then compute P for the second guess, then continue. From equations 3.36 and 3.37:

B^0 = -0.382 B^1 = -0.083

Equation 3.35 gives:

$B_r = B^0 + \omega B^1 = -0.399$

B_r will not change as long as T does not change and this is true for this particular problem.

From equation 3.33:

$z = 1 + B_r P_r/T_r = 1 + (-0.399)(0.438)/0.941 = 0.814$

Using PV = znRT, we find V = 1.6283 L. We need a lower pressure to get V = 2 L. Try:

P_{guess2} = 15 bar P_r = 0.395   z = 0.832   V = 1.8453 L
Try P_{guess3} = 14 bar P_r = 0.369   z = 0.844   V = 2.0037 L

which could be considered close enough for many calculations or if we want to get closer, we need a very slightly higher P. Eventually we can get to:

P_{guess4} = **14.0227 bar** P_r = 0.369   z = 0.843   V = 2.0000 L

3.4 Use Van der Waals equation to calculate the work done when 1 mole of ethylene expands reversibly from 1 liter to 20 liters at a constant temperature of 300 K. What is the initial pressure?

Solution:

For ethylene, T_c = 282.4 K, P_c = 5042 kPa (Appendix C) = 50.42 bar. Using equations 3.20 and 3.21:

$$a = \frac{(27)\left(83.14 \frac{cm^3 \cdot bar}{mole \cdot K}\right)(282.4^2 K^2)}{(64)(50.4 bar)} = 4.614 \times 10^6 \frac{cm^6 \cdot bar}{mole^2}$$

$$b = \frac{83.14 \cdot 282.4}{8 \cdot 50.4} = 58.23 \frac{cm^3}{mole}$$

using VDW equation:

$$w = \int_{1000 cm^3}^{20000 cm^3} \left(\frac{RT}{v-b} - \frac{a}{v^2}\right) dv = \left[RT \ln\left(\frac{v_2 - b}{v_1 - b}\right) + a\left(\frac{1}{v_2} - \frac{1}{v_1}\right)\right]_{1000 cm^3}^{20000 cm^3}$$

giving w = 71760 cm³-bar/mole and W = w · 1 mole = 71760 cm³-bar. Since we are dealing with 1 mole of gas, to compute the initial pressure:

$$P_1 = \frac{\left(83.14 \frac{cm^3 \cdot bar}{mole \cdot K}\right) \cdot 300K}{(1000 - 58.23) cm^3/mole} - \frac{4.614 \times 10^6 \frac{cm^6 \cdot bar}{mole^2}}{(1000 cm^3/mole)^2} = \mathbf{21.87 bar}$$

3.5 For methanol at 440 K and 25 bar, find z (the compressibility factor) for the vapor using the Peng-Robinson equation. To make sure you find the largest real root, find all three z roots of Peng-Robinson equation at this temperature and pressure.

Solution:

First, let's find parameters. From Appendix C for methanol, $T_c = 513.4$ K, $P_c = 82.20$ bar and $\omega = 0.565$. From equation 3.23 and equations for the parameters:

a = 11955369 cm^6-bar/mole^2 b = 40.40 cm^3/mole

Plugging these into equation 3.23 along with P = 25 bar and T = 440 K, we then are looking for v that satisfies the equation. Students may verify that there are three solutions corresponding to 62.48, 241.42, and 1119.01 cm^3/mole using whatever means to solve the cubic equation. These are equivalent to the three values of **z corresponding to 0.0427, 0.1650, and 0.7647**. The largest real root, **z = 0.7647**, corresponds to the compressibility factor for the vapor.

3.6 A frictionless piston and cylinder arrangement initially contains $V_1 = 1.0$ m^3 of pure, saturated vapor water at $P_1 = 500$ kPa (0.5 Mpa) at state I. Assume all processes are reversible.
a. How many kg of water does the piston and cylinder arrangement contain?
b. What are the total enthalpy (H_1) and internal energy (U_1) of the water in kJ?
The piston is then locked in place so that the volume is constant at $V_2 = V_1$ while the piston and cylinder arrangement is put on a hot plate and heated until $T_2 = 300°C$ at state II.
c. Estimate the pressure (P_2) in kPa using the ideal gas law for P-V-T behavior of the water vapor.
d. Using the steam tables, what are the internal energy (U_2) in kJ, total enthalpy (H_2) in kJ, and pressure (P_2) in kPa after the heating process from state I → II ?

Solution:

a. From the steam table, saturated vapor at 500 kPa, $v_1 = 0.37481$ m^3/kg, $h_1 = 2748.1$ kJ/kg and $u_1 = 2560.7$ kJ/kg.

$$m_1 = \frac{1 \text{ m}^3}{0.37481 \frac{m^3}{kg}} = \mathbf{2.668 \text{ kg}}$$

b. $U_2 = 2560.7$ kJ/kg * 2.668 kg = **6831.9 kJ**. $H_2 = 2748.1$ kJ/kg * 2.668 kg = **7331.9 kJ**.

c. $v_2 = v_1 = 0.37481$ m³/kg and $T_2 = 300°C = 573$ K.

For the ideal gas law, we need the volume to be molar volume rather than specific volume. $v_2 = 0.37481$ m³/kg = 0.37481 L/g × 18.02 g/mole = 6.754 L/mole.

$$P(6.754 \text{ L/mole}) = 0.08314 \frac{\text{L} \cdot \text{bar}}{\text{mole} \cdot \text{K}} 573\text{K}$$

P (by ideal gas law) = 7.053 bar = **705.3 kPa**

d. Look in the steam tables for $v_2 = 0.37481$ m³/kg and $T_2 = 300°C$, interpolating yields P_2 = **693.8 kPa** (this is a better value for the pressure than the value given by the ideal gas law, but the two values are within 2%), u_2 = 2799.6 kJ/kg, and h_2 = 3059.6 kJ/kg. U_2 = **7469.3 kJ** and H_2 = **8163.0 kJ**.

3.7 Find the temperature rise when ethylene is compressed adiabatically from 400 K and 1 bar to 10 bar. Assume ethylene is an ideal gas for this calculation. How much work is required per mole of ethylene to accomplish this compression?

Solution:

We will use equation 3.57 for this calculation. We need $\gamma = C_p/C_v$ for this. Since this is an ideal gas, we can use heat capacity values from Appendix B. In Appendix B, the heat capacities are given as functions of temperature, so we should probably chose some average temperature to evaluate γ. Let's choose 500 K as an approximate average temperature. For ethylene:

$$\frac{C_P^{IG}}{R} = 1.9993 + 0.013916 \cdot 500\text{K} - 5.3561 \times 10^{-6}(500\text{K})^2$$
$$+ 7.3142 \times 10^{-10}(500\text{K})^3 - 42295/(500\text{K})^2$$

with R = 8.314 J/mole-K, we get C_P^{IG} = 53.97 J/mole-K. With $C_p^{IG} = C_v^{IG} + R$, C_v^{IG} = 45.66 J/mole-K and γ = 1.182. Plugging into equation 3.57, we get T_2/T_1 = 1.43 so **T_2 = 570 K**.

Our guess of an average temperature of 500 K is not too bad, and given that the heat capacity is not a strong function of temperature, we will stick with our values. We could continue to iterate choosing a new average temperature and repeating, but we would probably find little difference. This calculation would still be approximate, but we will have to cover the Second Law of Thermodynamics before we can get a more exact answer.

With q = 0 (adiabatic), $\Delta u = \int C_v \, dT = -w$ and using equation 3.10 with the heat capacity equation for C_p^{IG} for ethylene:

$$\Delta u = \int_{400}^{570} (C_p^{IG} - R) dT$$

$$\Delta u = \int_{400}^{570} [R(1.9993 + 0.013916 \cdot T - 5.3561 \times 10^{-6} T^2 + 7.3142 \times 10^{-10} T^3$$
$$- 42295/T^2) - R] dT$$

integrating and evaluating yields $\Delta u = -w = 9012$ J/mole. Work is negative because we have to do work *on* the system for the compression.

3.8 One lbm of water vapor is contained in a frictionless piston/cylinder initially at 240°F and 10 psia. The water is then compressed reversibly and isothermally until the pressure is 300 psia. a) What are the initial and final phases of the water? b) Determine Δu and Δh for the process.

Solution:

a) There is nothing more important than the phase, and students should learn that they can often figure out the correct phase by logical reasoning and some rudimentary knowledge of the components of interest. The normal boiling point of water is 212°F, meaning it's vapor pressure at 212°F is 1 atm (14.7 psia). We are at a higher T (240°F) and initially at a lower P (10 psia). Therefore, point 1 is undoubtedly a vapor. Now, point 2 is at the same T not too far above it's normal boiling point, but at 300 psia which is well above 14.7 psia. Estimating, it appears point 2 will be a liquid, but this is somewhat in doubt. Let's go to the saturation curve for water which is in Appendix G.1. The vapor pressure (saturation pressure) of water at 240°F is 24.986 psia and we are at a higher P than that, so it is a liquid at these conditions.

b) State 1 - superheated vapor:
$T_1 = 240°F$
$P_1 = 10$ psia
Appendix G3:
$v_1 = 41.326$ ft³/lbm
$u_1 = 1089.0$ BTU/lbm
$h_1 = 1165.5$ BTU/lbm

State 2 - compressed liquid:
$T_2 = 240°F$
$P_2 = 300$ psia
Appendix G4:
$v_2 = 0.016914$ ft³/lbm
$u_2 = 208.29$ BTU/lbm
$h_2 = 208.76$ BTU/lbm

$\Delta u = 208.29 - 1089.0 = -880.71$ BTU/lbm $\Delta h = 208.76 - 1165.5 = -965.74$ BTU/lbm

These are negative because energy must be taken out to condense a vapor to a liquid.

3.9 One mole of propane is contained in a 1 liter vessel at 400 K. Find the pressure using Van der Waals equation and using the Pitzer generalized virial coefficient correlation.

Solution:

It is easier to do the calculation using Van der Waals equation, because Van der Walls equation is explicit in pressure meaning that no iteration is required. The molar volume is

1 L/1 mole =1 L/mole. For Van der Waals equation using the critical properties given in Appendix C:

$$a = \frac{27\left(0.08314 \frac{\text{L·bar}}{\text{mole·K}}\right)^2 (369.9\text{K})^2}{64(42.51\text{bar})} = 9.386 \frac{\text{L}^2 \cdot \text{bar}}{\text{mole}^2}$$

$$b = \frac{0.08314 \frac{\text{L·bar}}{\text{mole·K}} \cdot 369.9\text{K}}{8(42.51\text{bar})} = 0.09043 \frac{\text{L}}{\text{mole}}$$

$$P = \frac{0.08314 \frac{\text{L·bar}}{\text{mole·K}} \cdot 400\text{K}}{1 \frac{\text{L}}{\text{mole}} - 0.09043 \frac{\text{L}}{\text{mole}}} - \frac{9.386 \frac{\text{L}^2 \cdot \text{bar}}{\text{mole}^2}}{1 \frac{\text{L}^2}{\text{mole}^2}} = \mathbf{27.18\ bar}$$

Now, since we need the pressure, not the volume, to do the Pitzer generalized correction (i.e; we need $P_r = P/P_c$) so it is an iterative process. We could use the pressure given by the ideal gas law to start the iteration, but we have the pressure that Van der Waals equation gives which should be somewhat better than the ideal gas law, so let's use P = 27.18 bar for the first pressure we guess.

$P_{r,\text{guess1}}$ = 27.18 bar/42.51 bar = 0.6394 T_r = 400 K/369.9 K = 1.081

From equations 3.36 and 3.37:

B^0 = 0.083 - 0.422/$1.081^{1.6}$ = -0.2896 B^1 = 0.139 - 0.172/$1.081^{4.2}$ = 0.01499

From equation 3.35:

B_r = -0.2896 + 0.157 · 0.01499 = -0.2872

and from equation 3.34:

-0.2872 = B · 42.51/(0.08314 · 369.9)   B = -0.2078

From equation 3.33:

z = 1 + -0.2078 · 27.18/(0.08314 · 400) = 0.8302

v = zRT/P = 0.8302 · 0.08314 · 400/27.18 = 1.016 L/mole

With our first guess of P_{guess1} = 27.18 bar, we got v = 1.016 L/mole instead of the desired v = 1 L/mole. We need a slightly higher pressure, so let's try P_{guess2} = 27.5 bar.

B is still the same because it is only a function of T and the critical properties.

$P_{guess2} = 27.5$ bar $z = 0.8282$ $v = 1.002$ L/mole

Thus, **P = 27.5 bar** is well within 1% of the exact number and close enough.

3.10 **Use the Peng-Robinson Equation of State and data in Appendix C to determine the compressibility factors of pure benzene and pure toluene at 240°C and 10 bar total pressure.**

Solution:

From Appendix C we find:

	Benzene	Toluene
T_c (K)	562.0	591.7
P_c (bar)	49.06	41.26
ω	0.217	0.273

using equation 3.23 and equations for reduced constants for Peng-Robinson:

	Benzene	Toluene	
T (K)	513.16	513.16	
P (bar)	10	10	
T_r	0.9131	0.8673	
P_r	0.2038	0.2424	
α	1.0629	1.1094	
a (bar·L²/mole²)	21.6267	29.7540	
b (L/mole)	0.0741	0.0928	
A	0.1188	0.1635	
B	0.0174	0.0217	
a1	1	1	these are
a2	-0.9826	-0.9783	coefficients for
a3	0.08317	0.11856	the polynomial
a4	-0.001757	-0.003071	form in z
z	**0.8915**	**0.8418**	largest root
other root	0.0557	0.1000	lowest
other root	0.0350	0.0365	middle

the largest real root is the vapor root which is what was being sought.

3.11 **Using the Redlich-Kwong equation, plot P vs. v for the 400 K isotherm for methanol and estimate the three possible volume roots to the equation if the pressure is 25 bar.**

Solution:

This plot is shown in Figure 3.i.1. At 25 bar, we see intersections at v's of **about 70, 280, and 950 cm³/mole**. These are rough. Solve the equation for better values.

3.12 Compare the pressures given by the Redlich-Kwong equation, the Peng-Robinson equation, and the Lee-Kesler Tables when 100 g benzene is contained in a 1 L vessel at 350°C.

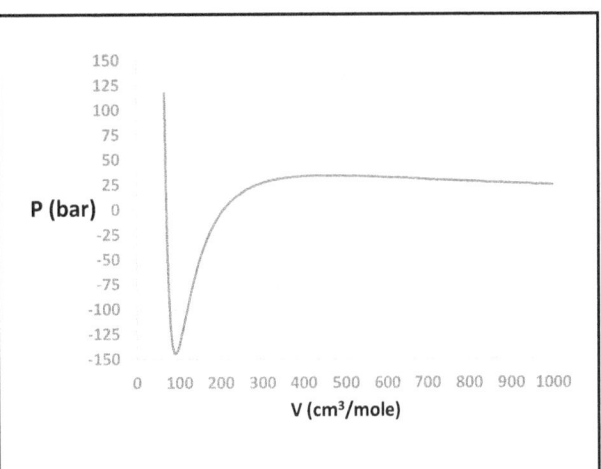

Figure 3.i.1. Redlich-Kwong equation P-v plot for example problem 3.11.

Solution:

Redlich-Kwong parameters using critical properties for benzene from Appendix C:

$$a = \frac{0.42748 \cdot \left(0.08314 \tfrac{L \cdot bar}{mole \cdot K}\right)^2 \cdot (562.0K)^{2.5}}{49.06 \, bar} = 450.97 \frac{L^2 \cdot bar \cdot K^{0.5}}{mole^2}$$

$$b = \frac{0.08664 \cdot 0.08314 \tfrac{L \cdot bar}{mole \cdot K} \cdot 562.0K}{49.06 \, bar} = 0.08252 \frac{L}{mole}$$

Since we have 100 g benzene, that is 100 g/78.114 g/mole = 1.282 moles, so v = 1 L/1.282 moles = 0.7811 L/mole. Plug into the Redlich-Kwong equation:

$$P = \frac{0.08314 \cdot (350 + 273)}{0.7811 - 0.08252} - \frac{450.97}{(350 + 273)^{0.5} \cdot 0.7811 \cdot (0.7811 + 0.08252)} = \mathbf{47.36 \, bar}$$

Students should carefully check the units. Peng-Robinson parameters:

$$\alpha = \left[1 + \left(0.37464 + 1.54226 \cdot 0.217 - 0.26992 \cdot 0.217^2\right)\left(1 - \left(\frac{623}{562.0}\right)^{0.5}\right)\right]^2 = 0.9277$$

$$a = \frac{0.45724 \cdot \left(0.08314 \tfrac{L \cdot bar}{mole \cdot K}\right)^2 \cdot (562.0K)^2 \cdot 0.9277}{49.06 \, bar} = 18.88 \frac{L^2 \cdot bar}{mole^2}$$

$$b = \frac{0.08664 \cdot 0.08314 \tfrac{L \cdot bar}{mole \cdot K} \cdot 562.0K}{49.06 \, bar} = 0.0741 \frac{L}{mole}$$

$$P = \frac{0.08314 \cdot (350+273)}{0.7811 - 0.0741} - \frac{18.88}{0.7811(0.7811+0.0741) + 0.0741(0.7811-0.0741)} = \mathbf{47.06\ bar}$$

Again, check the units to confirm. For the Lee Kesler tables, we need $T_r = 623/562.0 = 1.11$, and we need P_r but since we are looking for P, we will need to do an iterative process. We will use $P_{guess1} = 47$ bar (from results above, otherwise we might use the ideal gas equation to get started), then $P_{r,guess1} = 47/49.06 = 0.958$. This is a 2D interpolation in Table D.1. To do this, we first interpolate in one dimension (create "new" points at $T_r = 1.11$), then the next dimension (interpolate between "new" points at $T_r = 1.11$ to find $P_r = 0.958$). Here are the numbers from the table and the results of the interpolation:

z^0 - T_r/P_r	0.80	0.958	1.00		z^1 - T_r/P_r	0.80	0.958	1.00
1.10	0.7649		0.6880		1.10	0.0236		0.0476
1.11	0.7726	0.7147	0.6993		1.11	0.0268	0.0456	0.0506
1.15	0.8032		0.7443		1.15	0.0396		0.0625

$z^0 = 0.7147 \quad z^1 = 0.0456$.
$z = z^0 + \omega z^1 = 0.7147 + 0.217 \cdot 0.0456 = 0.7246$
$v = zRT/P = 0.7246 \cdot 0.08314 \cdot (350+273)/47 = 0.7985$ L/mole

However, we want $v = 0.7811$ L/mole, so we need a slightly higher pressure. Take:

$P_{guess2} = 48$ bar $\quad P_{r,\ guess2} = 0.978 \quad z^0 = 0.7073 \quad z^1 = 0.0480 \quad z = 0.7177$
$v = 0.7745$ L/mole

One more loop with $P_{guess3} = 47.7$ bar:

$P_{guess3} = 47.7$ bar $\quad P_{r,\ guess3} = 0.972 \quad z^0 = 0.7094 \quad z^1 = 0.0473 \quad z = 0.7197$
$v = 0.7815$ L/mole which is close enough. Therefore **P = 47.7 bar**.

The three results from Redlich-Kwong, Peng-Robinson, and Lee-Kesler Tables are quite close to each other.

3.13 **Steam initially at 550°C and 400 kPa is cooled at constant pressure until it is a saturated vapor. Calculate Δu and Δh for the process. Assuming the process is reversible, calculate q and w. If the saturated vapor at 400 kPa is cooled reversibly at constant T until it is a saturated liquid, what is w?**

Solution:

Using the steam table:

h_1 (550°C, 400 kPa) = 3485.5 kJ/kg

h_2 (sat. vapor, 400 kPa) = 2738.1 kJ/kg
u_1 (550°C, 400 kPa) = 3129.8 kJ/kg

However, u_2 (sat. vapor, 400 kPa) is not in the table, but we can compute it using $h = u + Pv$. v_2 (sat. vapor, 400 kPa) = 0.46238 m³/kg so $Pv = 400 \cdot 0.46238 = 185.0$ m³·kPa/kg = 185.0 kJ/kg.

u_2 (sat. vapor, 400 kPa) = 2738.1 kJ/kg - 185.0 kJ/kg = 2553.1 kJ/kg

Δu = 2553.1 - 3129.8 = -576.7 kJ/kg   **Δh = 2738.1 - 3485.5 = -747.4 kJ/kg**

Since the process is reversible and constant P, **q = Δh = -747.4 kJ/kg.**

Using $\Delta u = q - w$:

w = 576.7 + -747.4 = -170.7 kJ/kg

Now, the process of cooling the saturated vapor at 400 kPa to a saturated liquid is a constant T and P process. v_1 (sat. vapor, 400 kPa) = 0.46238 m³/kg and v_2 (sat. liquid, 400 kPa) = 0.001084 m³/kg.

w = $\int Pdv$ = 400 kPa (0.001084 - 0.46238) m³/kg = -184.5 m³·kPa/kg = -184.5 kJ/kg

3.14 Use the Pitzer generalized virial coefficient correlation (z = 1 + BP/RT) to compute the work done when one mole of isobutane is compressed from it's critical point to 50 bar reversibly and isothermally.

Solution:

From Appendix C for isobutane (i-butane), T_c = 407.8 K, P_c = 36.29 bar, and ω = 0.189, so the starting state is at 407.8 K and 36.29 bar. The final state will be 407.8 K and 50 bar. The generalized virial coefficient will be a constant since T is constant.

T_r = 407.8 K/407.8 K = 1.0. From equations 3.36 and 3.37:

$B^0 = 0.083 - 0.422/1.0^{1.6} = -0.339$ $B^1 = 0.139 - 0.172/1.0^{4.2} = -0.033$

From equation 3.35:

B_r = -0.339 - 0.033 · 0.189 = -0.378

and from equation 3.34:

-0.378 = B · 36.29 bar /(407.8 K · 0.08314 L·bar/mole·K)   then   B = -0.353 L/mole

$z = Pv/RT = 1 + BP/RT$ so $Pv = RT + BP$ and $P(v - B) = RT$

$P = RT/(v - B)$ is an equivalent arrangement of $z = 1 + BP/RT$.

Since the process is reversible, $w = \int P dv$ and substituting for P:

$$w = \int_{v_1}^{v_2} \frac{RT}{v - B} dv = RT \ln\left(\frac{v_2 - B}{v_1 - B}\right)$$

We need v_1 and v_2.

$$36.29 \text{ bar} = \frac{0.08314 \frac{L \cdot bar}{mole \cdot K} \cdot 407.8 \text{ K}}{v_1 + 0.353 \frac{L}{mole}} \qquad 50 \text{ bar} = \frac{0.08314 \frac{L \cdot bar}{mole \cdot K} \cdot 407.8 \text{ K}}{v_2 + 0.353 \frac{L}{mole}}$$

$v_1 = 0.5813$ L/mole   and $v_2 = 0.3251$ L/mole

$$w = 0.008314 \frac{kJ}{mole \cdot K} \cdot 407.8 \text{K} \cdot \ln\left(\frac{0.3251 + 0.0353}{0.05813 + 0.353}\right) = \mathbf{-7.11 \text{ kJ/kg}}$$

3.15 Ammonia gas is at 160°C at a pressure such that it's specific volume is 0.3465 m³/kg. Estimate the pressure using the ideal gas law and the ammonia tables in Appendix J.2. Compare the two values and say which value you consider to be more accurate.

Solution:

The molecular weight (molar mass) of ammonia is 17.03 g/mole. Therefore the molar volume is 0.3465 m³/kg · 17.03 g/mole · 1 kg/1000 g = 0.005901 m³/mole. For an ideal gas:

P · 0.005901 m³/mole = 0.008314 m³·kPa/mole·K · (160 + 273) K

P = 610 kPa

Going to Table J.2 for superheated ammonia at 160°C, we see that the volume lies between 600 and 800 kPa, so we interpolate between v = 0.34697 m³/kg at 600 kPa and v = 0.25884 m³/kg at 800 kPa. The result is **P = 601 kPa**. The result from the ammonia table is the more accurate.

Student Problems:

1. Use the steam table to determine each property

a. P = 1000 kPa   v = 0.37389 m³/kg
b. P = 150 kPa   h = 2323.5 kJ/kg

c. T = 600°C   h = 3688.0 kJ/kg
d. T = 400°C   s = 7.3989 kJ/kg-K (even though you don't know what "s" means, you can look it up)
e. P = 50 kPa   s = 6.8778 kJ/kg-K

Answers:
a. s = 7.8765 kJ/kg-K   h = 3571.5 kJ/kg   T = 542.1°C
b. s = 6.2618 kJ/kg-K   u = 2178.4 kJ/kg   T = 111.35°C
c. s = 7.6306 kJ/kg-K   v = 0.17171 m^3/kg   P = 2346.2 kPa
d. h = 3262.0 kJ/kg   u = 2956.0 kJ/kg   P = 1155.3 kPa
e. h = 2391.7 kJ/kg   u = 2247.5 kJ/kg   T = 81.32°C

2. Compare the pressures given by the Redlich-Kwong equation, the Peng-Robinson equation, and the Lee-Kesler Tables when 80 g 2-propanol is contained in a 1 L vessel at 375°C.

3. Find the temperature drop when ammonia is expanded adiabatically from 400 K and 20 bar to 2 bar. Assume ammonia is an ideal gas for this calculation. How much work is done per mole of ammonia to accomplish this compression?

4. Using the Peng-Robinson equation, plot P vs. v for the 200°C isotherm for ethylbenzene and estimate the three possible volume roots to the equation if the pressure is 10 bar. Answer: 150, 950 and 2700 cm^3/mole.

5. Use the Pitzer generalized virial coefficient correlation ($z = 1 + BP/RT$) to compute the work done when one mole of acetone is compressed from 500 K, 10 bar to 30 bar reversibly and isothermally.

3.j References

1. Van der Waals, J.D.; "The equation of state for gases and liquids", Nobel Prize Lectures, Physics 1910.
2. Walas, S. M.; "Phase Equilibrium in Chemical Engineering", Butterworth Publishers, Boston (1985).
3. Peng, D.Y. and Robinson, D.B.; Ind. Eng. Chem. Fundamentals 15, 59 (1976).
4. Redlich, O, and Kwong, J.N.S.; "On the Thermodynamics of Solutions. V. An Equation of State. Fugacities of Gaseous Solutions.", Chem. Rev. 44, 233 (1949).
5. Tatakad, R.R., Spencer, C.F., and Adler, S.B.; "A Comparison of Eight Equations of State to Predict Gas-Phase Density and Fugacity", I&EC Proc. Res. Dev. 18, 726 (1979).
6. Beattie, J.A. and Bridgeman, O.C.; "A New Equation of State for Fluids. I. Application to Gaseous Ethyl Ether and Carbon Dioxide", J. Amer. Chem. Soc. 49, 1665 (1927).
7. Benedict, M., Webb, G.B., Rubin, L.C.; "An Empirical Equation for Thermodynamic Properties of Light Hydrocarbons and Their Mixtures: I. Methane, Ethane, Propane, and n-Butane", J. Chem. Phys. 8, 334 (1940).
8. Hougen, O.A, Watson, K.M., and Ragatz, R.A.; *Chemical Process Principles, Part II, Thermodynamics, 2nd ed.*, Wiley, New York (1959).
9. see Pitzer, K.S.; *Thermodynamics, 3rd ed.*, McGraw-Hill, New York (1995).

10. Lee, B.I. and Kesler, M.G.; "A Generalized Thermodynamic Correlation Based on Three-Parameter Corresponding States", <u>AIChE Journal 21</u>, 510-527 (1975).
11. Rackett, H.G; "Equation of State for Saturated Liquids", <u>J. Chem. Eng. Data 15</u>, 514 (1970).
12. Thomson, G.H., Brobst, K.R. and Hankinson, R.W.; "An improved Correlation for Densities of compressed Liquids and Liquid Mixtures", <u>AIChE Journal 28</u>, 671 (1982).

Chapter 4 - Thermochemistry

Many processes involve heating or cooling substances or may include chemical reactions which have associated heats of reaction. Often, chemical reactions and heating/cooling happen simultaneously. Generally, we refer to the heat which increases or decreases the internal energy of a substance, resulting in a temperature change *sensible heat*. Heats of reaction can be determined from tabulated data. We take each of these processes separately then consider how to handle calculations when both happen simultaneously.

4.a Sensible Heat Effects

As we have already alluded to, the heat capacity of most chemical species is a function of temperature. This principle can be derived from statistical thermodynamics, but the details of the calculations are beyond the scope of this book. However, the final results of the statistical thermodynamic considerations are useful. A summary is as follows.

For an ideal gas, the constant volume heat capacity can be computed as the sum of kinetic (ke), rotational (rot), vibrational (vib) and electronic (el) contributions at the molecular level. Thus:

$$C_v^{IG} = C_v^{ke} + C_v^{rot} + C_v^{vib} + C_v^{el} \qquad 4.1$$

To be clear, we are referring to the kinetic, rotational, vibrational, and electronic energy of the molecules, not the system as a whole. When heat is added to a substance, depending upon the chemistry of the substance, the heat may divide into more than one of these components. An energy increase in the kinetic energy component results in an observed temperature increase, so if energy goes to other components, more energy is required to get a particular temperature increase than in the case where all the energy goes to the kinetic energy component. Thus, the heat capacity is higher. Another way to think of this is that each of the components in equation 4.1 may have capacity to hold energy, so the heat capacity is higher when several terms in equation 4.1 are active.

From statistical thermodynamic arguments, the kinetic energy contribution is 3/2R and the rotational contribution, for diatomic (molecules composed of two atoms) and polyatomic molecules (molecules composed of at least 3 atoms) is R. The electronic energy component usually does not affect the heat capacity as most molecules are in their ground states. The vibrational component, also for diatomic and polyatomic molecules, relates to bond vibrational energy and this is the component that carries the majority of the temperature dependence. Vibrational energies increase with temperature, and the more atoms the molecule is composed of, the more vibrational modes the molecule has, leading to stronger temperature dependence. Thus, for monoatomic substances such as helium, neon, argon, etc. that have no rotational or vibrational components:

$$C_v^{IG} = \frac{3}{2}R$$

and for diatomic and linear polyatomic molecules:

$$C_v^{IG} = \frac{5}{2}R + C_v^{vib}$$

for non-linear polyatomic molecules:

$$C_v^{IG} = \frac{6}{2}R + C_v^{vib}$$

At room temperature, the bond vibrational energy can be close to zero, and since for an ideal gas, $C_p^{IG} = C_v^{IG} + R$ we have:

$$C_p^{IG} = \frac{5}{2}R \qquad \text{(monoatomic, low T, approximate) 4.2}$$

$$C_p^{IG} = \frac{7}{2}R \text{ (diatomic or linear polyatomic)}$$

$$\text{or } C_p^{IG} = \frac{8}{2}R \text{ (non-linear polyatomic)} \qquad \text{(low T, approximate) 4.3}$$

In Appendix B, one column is devoted to the heat capacity of a variety of substances at room temperature (298.15K). The values of the heat capacities at 298.15K of the monoatomic gases He and Ar are reflected well by equation 4.2 in Appendix B, and the values are virtually constant with T. He and Ar do not have any bond vibration components. Diatomic gases such as N_2 and O_2 have C_p^{IG} values near 7/2 R at 298.15K, but the values increase as T increases, consistent with the hypothesis that the bond vibrational energies increase with temperature. Non-linear polyatomic molecules such as n-hexane (which has 57 normal modes of bond vibration) have heat capacities at 298.15K well above 8/2R, reflecting the fact that due to the many bond vibrational modes in the molecule, some are already active and holding energy even at room temperature.

C_p^{IG} for methane more than doubles when T is increased from room temperature to about 1500K. Other molecules exhibit even higher temperature dependencies. Obviously, this temperature dependence cannot be ignored except over small temperature ranges. Therefore, the temperature dependence of the heat capacities for substances are generally correlated with temperature using some type of polynomial equation. Many different forms have been used, and in this book, we have provided coefficients for the polynomial:

$$\frac{C_p^{IG}}{R} = A + BT + CT^2 + DT^3 + E/T^2 \qquad 4.4$$

The equation has more adjustable constants than many of the equations used previously in engineering texts, but many other correlations fit this one simply by making some of the constants in the other correlation zero. Computations are still very simple. In fact, using equation 2.25 to apply to an ideal gas:

$$\Delta h^{IG} = R \int_{T_1}^{T_2} \frac{C_p^{IG}}{R} dT = R \left[\int_{T_1}^{T_2} \left(A + BT + CT^2 + DT^3 + E/T^2 \right) dT \right]$$

$$\frac{\Delta h^{IG}}{R} = A(T_2 - T_1) + \frac{B}{2}(T_2^2 - T_1^2) + \frac{C}{3}(T_2^3 - T_1^3) + \frac{D}{4}(T_2^4 - T_1^4) - E\left(\frac{1}{T_2} - \frac{1}{T_1}\right) \qquad 4.5$$

Using equation 4.5 with constants from Appendix B, we can compute enthalpy changes for ideal gases for many engineering applications. Be sure to note that K must be used for the temperature scale and note the applicable temperature ranges for the various components given in Appendix B. Using equation 4.4 outside those ranges is risky. Turn to the published literature for thermodynamic data in such instances.

4.b Heats of Formation, Reaction and Combustion

4.b.i Generalized Stoichiometry

For engineering purposes, we use shorthand to simplify equations and calculations. This leads to the concept of *generalized stoichiometry*, which puts all chemical reactions into a standard form that translates to use in equations easily. Let's take a symbolic chemical reaction:

$$aA + bB + cC + \ldots \rightarrow rR + sS + tT + \ldots \qquad 4.6$$

Uppercase letters represent chemical species and lowercase letters represent the stoichiometric coefficients. The left hand side of equation 4.6 by convention is termed "reactants", and the right hand side is termed "products". Students should see that all stoichiometric chemical reactions can be placed in this symbolic form.

Now, let's rewrite equation 4.6 in more of an algebraic form, thinking of the species A, B, C, R, S, T… like variables with coefficients and the chemical reaction as an algebraic expression:

$$0 = -aA - bB - cC \ldots + rR + sS + tT + \ldots \qquad 4.7$$

We have purposely arranged equation 4.7 so that reactants have negative coefficients and products have positive coefficients. We emphasize this by placing 0 on the left hand side of the equation where the reactants were originally written by convention. The negative coefficients for the reactants translates to the fact that reactants are being used or going away. Products, on the other hand have positive coefficients because they are being formed.

Equation 4.7 can be more compactly written in the generalized stoichiometry form:

$$0 = \sum v_i A_i \qquad 4.8$$

In equation 4.8, v_i's are defined as being negative for reactants and positive for products. A_i's are the individual species. Any stoichiometric chemical equation can be written in this form. For example:

$$2CO(g) + O_2(g) \rightarrow 2CO_2(g)$$

In this equation, $v_{CO} = -2$, $v_{O_2} = -1$, and $v_{CO_2} = +2$. Be careful when using this concept, especially remembering that by definition, the v_i's for reactants are negative. Note that v_i's are strictly unitless, as they are coefficients, not moles. However, we often translate v_i's to number of moles when dealing with actual calculations, so also be prepared to see when v_i's are translated to moles and be sure to use the correct units.

4.b.ii Enthalpies of Formation

In order to determine enthalpy changes when reactions take place, we need to set a baseline by defining h = 0 for particular species and conditions. We set the enthalpy of the elements in their *standard state of aggregation* at 298.15K and 1 bar to have zero enthalpy. If the element is gaseous, it must be in the ideal gas state or corrected to the ideal gas state at 298.15K and 1 bar. For standard state of aggregation, we mean in what form the element naturally exists at 298.15K and 1 bar. Good examples are elemental gases such as N_2, O_2, H_2 and others that naturally exist in the diatomic form. The diatomic forms are considered the standard state of aggregation rather than N, O, and H as atomic species. Solids often exist in crystalline form and may naturally exist as different crystal forms at 298.15K and 1 bar. An important example in this vein is carbon which commonly exists as graphite and diamond, but graphite has a lower enthalpy than diamond so graphite is chosen as the standard state of aggregation. Other solid elements at 298.15K and 1 bar get similar consideration and an exact designation should be written in the tables where data resides.

Consider a chemical reaction where a compound is made from pure chemical elements. For example:

$$C(s) + 2H_2(g) \rightarrow CH_4(g) \qquad 4.9$$

If we specify that for this reaction, a stoichiometric mixture of elements under standard conditions forms exactly one mole of methane under standard conditions (1 bar, 298.15K, ideal gas), equation 4.9 is known as the *standard reaction of formation* of methane. Clearly, we could write similar stoichiometric equations for the formation of any chemical species. The enthalpy change associated with any standard reaction of formation is known as the *standard enthalpy of formation* of the product species and it is the difference in enthalpy between the products and reactants. The use of the standard conditions 298.15K and 1 bar is arbitrary but conventional, and a broad range of standard enthalpies of formation can be found in the open literature with these standard conditions. The standard state pressure was 1 atm in years past, but the new standard of 1 bar is now popular and the differences in enthalpies between the two standards are very small. Standard enthalpies of formation are tabulated in Appendix F. Note than enthalpies of formation for elemental species are not listed because by definition, their enthalpies of formation are zero.

Take a system starting with one mole of carbon in it's standard state and 2 moles of hydrogen gas in its ideal gas state at 298.15K and 1 bar as in equation 4.9. Now, allow the carbon and hydrogen to combine at constant T and P until one mole of methane is formed. For this process, a first law consideration yields:

$$\Delta h^o_{f,CH_4,298} = q \qquad 4.10$$

In equation 4.10, we have truncated "298.15" to "298" for brevity and will do that often in this chapter (and book). The subscript temperature reiterates the temperature at which the enthalpy is determined. The "o" superscript indicates standard enthalpies of reaction or formation, and the superscript specifies the usual conditions required for standard states. In order to keep the reaction mixture at 298.15K, q must be added and $\Delta h^o_{f,CH_4,298}$ represents this heat. If the heat added is positive, the reaction requires heat and is termed *endothermic*. If heat added is negative (i.e., the reaction produces heat), the reaction is *exothermic*. Equation 4.10 indicates why enthalpies of formation and reaction are sometimes termed heats of formation or reaction and these terms are often used interchangeably.

4.b.iii Enthalpies of Reaction

The *standard enthalpy of reaction* extends to all stoichiometric reactions, not just those combining elements to form a product. A standard enthalpy of formation is a special type of a standard enthalpy of reaction. All species participating in a reaction, both reactants and products, must have a specified state of aggregation (usually the standard state of aggregation) at 298.15K and 1 bar and gases must be in the ideal gas state (or corrected to ideal gas conditions). The major utility of standard enthalpies of formation is that they can be used to compute the standard enthalpy of reaction for any reaction that consists of species for which we know the standard enthalpy of formation:

$$\Delta h^o_{r,298} = \sum v_i \Delta h^o_{f,i,298} \qquad 4.11$$

However, in practice it is not often that we want to carry out the reaction at 298.15K. How can the enthalpy change be determined for other temperatures? Consider a gaseous reaction occurring at arbitrary temperature T, then let's set up the following process. Start with the reactants at T then 1) cool (or heat) the reactants to 298.15K, 2) let the reaction proceed to products at 298.15K, and 3) heat (or cool) the products back to temperature T. This process is depicted in Figure 4.b.1. Since enthalpy is a state function independent of the process path, we can sum the changes over steps 1-3 and equate the results to the overall process.

Steps 1 and 3 are simply sensible heat processes. The process is at constant P so for step 1 we have:

$$\Delta h_1 = \int_T^{298} C_{p,mix}^{IG} dT \qquad 4.12$$

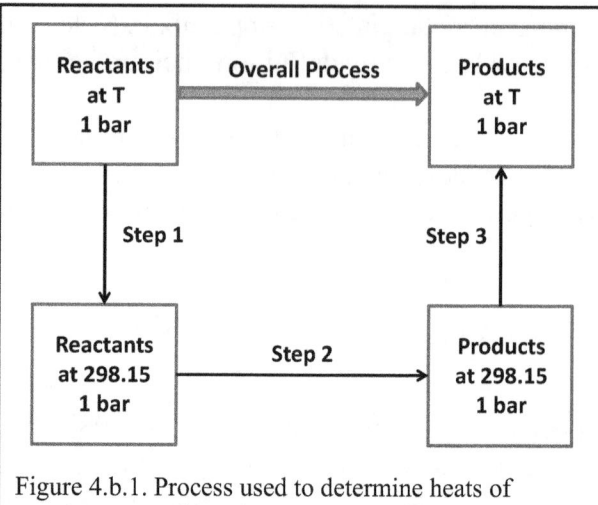

Figure 4.b.1. Process used to determine heats of reaction at an arbitrary temperature.

In equation 4.12, $C_{p,mix}^{IG}$ is that associated with a stoichiometric mixture of just the reactants. Since it is stoichiometric, the v_i's tell us exactly the number of moles of each reacting species. However, we need to be careful of the signs and the units, recalling that v_i's are unitless. Reactant v_i's are defined as negative, so we put a negative sign in front to indicate the relative amounts of each reactant species which are positive numbers. Thus:

$$C_{p,mix}^{IG} = \sum_{reactants}(-v_i)C_{p,i}^{IG} \qquad 4.13$$

Substituting equation 4.13 into 4.12, we will invert the integral limits, compensating by again changing the sign on the v_i's yielding:

$$\Delta h_1 = \int_{298}^{T} \sum_{reactants} v_i C_{p,i}^{IG} dT \qquad 4.14$$

Step 2 is a standard enthalpy of reaction at 298.15K and 1 bar. Step 3 is the opposite direction of Step 1 using products instead of reactants. This time the sign of the v_i's is the same sign as the relative number of moles, so:

$$\Delta h_3 = \int_{298}^{T} \sum_{products} v_i C_{p,i}^{IG} dT \qquad 4.15$$

Now we can combine 4.14 and 4.15 noting the summation is over all species (reactants and products) when combined. Adding all three steps we get:

$$\Delta h_{r,T}^{o} = \Delta h_{r,298}^{o} + \int_{298}^{T} \sum_{all} v_i C_{p,i}^{IG} dT = \sum v_i \Delta h_{f,i,298}^{o} + \int_{298}^{T} \sum v_i C_{p,i}^{IG} dT \qquad 4.16$$

As long as we have heat capacity and standard heat of formation data for each species in the reaction, we can evaluate equation 4.16. Using the heat capacity data in Appendix B, we can write the heat capacity to substitute into equation 4.16 as:

$$\sum \frac{v_i C_{p,i}^{IG}}{R} = \sum v_i \left(A_i + B_i T + C_i T^2 + D_i T^3 + E_i T^{-2} \right)$$

$$\sum v_i C_{p,i}^{IG} = \left(\sum v_i A_i + \sum v_i B_i T + \sum v_i C_i T^2 + \sum v_i D_i T^3 + \sum v_i E_i T^{-2} \right) R \qquad 4.17$$

The sums in equation 4.17 are easily done, and if we substitute equation 4.17 into 4.16, we can integrate easily as in equation 4.5. The result is:

$$\frac{\Delta h_{r,T}^\circ}{R} = \frac{\Delta h_{r,298}^\circ}{R} + \left(\sum v_i A_i \right)(T - 298) + \left(\frac{\sum v_i B_i}{2} \right)(T^2 - 298^2)$$

$$+ \left(\frac{\sum v_i C_i}{3} \right)(T^3 - 298^3) + \left(\frac{\sum v_i D_i}{4} \right)(T^4 - 298^4) - \left(\sum v_i E_i \right)\left(\frac{1}{T} - \frac{1}{298} \right) \qquad 4.18$$

Students should note that it is not required to use the heat capacity functionality tabulated in this book. Any heat capacity equation can be substituted into equation 4.16 and integrated. The temperature functionalities of the heat capacities may even differ among species, but attention should be paid to consistency in units when taking data from different sources.

4.b.iv Heats of Combustion

The *standard heat of combustion* is defined as the enthalpy change when one mole of a compound is burned completely using oxygen. There are various possibilities for products of the reaction, but in the case that the compound is a hydrocarbon consisting of hydrogen, carbon, and oxygen only, the products are generally defined to be H_2O and CO_2. With a sulfur-containing compound, SO_2 is usually considered to be the product of the reaction but sometimes an SO_3 product is assumed. If the compound contains nitrogen, we assume an N_2 product. However, one complication is the phase condition of the product water which can either be liquid or vapor. Usually, liquid water is assumed, but the difference between the heat of combustion for water as a liquid versus water as a gas is simply the heat of vaporization of water, and either value is equally useful.

With these assumptions, for a compound that contains only carbon and hydrogen:

$$C_x H_y + \left(x + \frac{y}{4} \right) O_2 = x CO_2 + \frac{y}{2} H_2 O \qquad 4.19$$

and for a compound that contains only C, H, O, S, and N the general stoichiometric equation is:

$$C_x H_y O_z S_t N_u + \left(x + \frac{y}{4} - \frac{z}{2} + t \right) O_2 \to x CO_2 + \left(\frac{y}{2} \right) H_2 O + t SO_2 + \frac{u}{2} N_2 \qquad 4.20$$

The main reason that heats of combustion are important is that they may be measured fairly easily and directly by carrying out the combustion in either a batch or flow calorimeter. In either process, the compound for which the heat of combustion is sought is burned, usually with a large excess of oxygen to assure complete combustion, and the products of combustion are cooled back to near their initial temperature, generally near room temperature, using a water bath or flowing water. A precise measure of the temperature change of the water then yields the amount of heat transferred to the water and the result is a raw heat of combustion. Corrections can be made for small differences in the initial and final temperature of combustion process, for any partial combustion which may have occurred, and for the partial condensation of water at the end of the process. Once an accurate value for the heat of combustion is obtained, the heat of formation of the compound of interest can be extracted, assuming that the heats of formation of the products are known. Heats of combustion thus were widely used in years past to construct an enthalpy of formation table for a wide range of compounds. Because the table of enthalpies of formation are more fundamental and may even be used to compute heats of combustion, heats of combustion are not often tabulated anymore. However, students may find tables in the literature.

4.c   Adiabatic Flame Temperature

An example process where both heats of reaction and sensible heat effects are observed simultaneously is in the calculation of adiabatic flame temperature. Consider a fuel being ignited in oxygen or air in a device such as a Bunsen burner or welding torch. Upon combustion, the fuel releases a large amount of energy as an exothermic combustion reaction, but unlike processes such as a table of enthalpies of formation where we start and end at the same temperature, the enthalpy change for the exothermic reaction transfers to the combustion products thus raising the temperature of the products of the reaction. In the limit where no heat is lost (adiabatic), the temperature of the combustion products is the adiabatic flame temperature. In practice, actual flame temperatures are lower than the adiabatic flame temperature because there may be significant heat lost from the combustion process especially by radiation, and also the combustion process might not give complete combustion products CO_2 and H_2O. The adiabatic flame temperature is therefore an upper limit.

Let's set up a process that we can use to calculate the adiabatic flame temperature. Such a process is given in Figure 4.c.1 where we have assumed that the process starts and ends at 1 bar, and we further assume that all the gases and mixtures of gases are in the ideal gas state. Since we deal exclusively with ideal gases, the actual pressure doesn't matter and we could just as well assume the pressure is 1 atm. The fuel starts as a mixture with oxygen or air, and we need a minimum of stoichiometric oxygen to burn the fuel completely to CO_2 and H_2O. If we use air as the oxidant, N_2 (and other minor

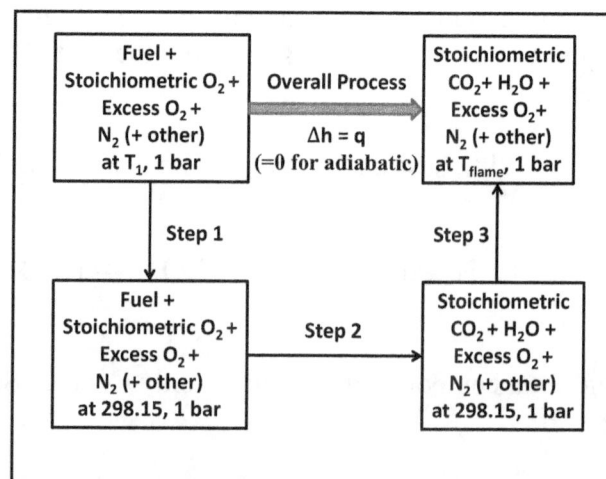

Figure 4.c.1. Process used to determine an adiabatic flame temperature.

components we will neglect) will come with the oxygen in the ratio of about 79/21 moles of N_2 for every mole of O_2. We can further have excess O_2 and further N_2 will come with the excess O_2 if we use air as the oxidant. In Figure 4.c.1, we have also included the possibility of "other" substances that do not react also being mixed with the reactant initially. Synthetic air mixtures with helium or argon come to mind, but such a case would be unusual.

Typically, the fuel and oxidant would be close to room temperature to start. If the starting temperature is 298.15K, $\Delta h_1 = 0$ in Step 1, otherwise, we need to determine the sensible heat effect to heat/cool the fuel plus oxidant from T_1 to 298K. This can be written:

$$\Delta H_1 = R \int_{T_1}^{298} \sum_{\text{reactants}} n_i \frac{C_{p,i}^{IG}}{R} dT \qquad 4.21$$

In equation 4.21, "reactants" includes all the species in the reacting mixture including fuel, stoichiometric oxygen, excess oxygen, nitrogen, and other. If we take 1 mole of fuel as a basis, then $n_{\text{fuel}} = 1$ mole, and the moles of other species can be determined or specified. Note that we now have an uppercase "H" for enthalpy meaning the number is in BTU or kJ or an energy equivalent instead of lowercase "h" indicating per mole because we have multiplied by an actual number of moles. For a fuel that is purely a hydrocarbon with molecular formula C_xH_y as in equation 4.19, stoichiometric O_2 would be $(x+y/4)$ moles. Other types of fuel are possible, and the stoichiometric amount of oxygen should be computed. Excess oxygen is generally expressed as a percentage above the stoichiometrically required amount, so that 25% excess O_2 would mean 1.25 times the amount of stoichiometric O_2 is mixed with the fuel. If air is the oxidant, then 79/21 moles of N_2 for every mole of O_2 (both stoichiometric and excess) is included in the reaction mixture. Other non-reacting species can be included as specified.

In Step 2, the fuel is burned completely. Stoichiometric products are formed, and the standard heat of reaction is released since Step 2 occurs at constant T and P:

$$\Delta H_2 = Q = n_{\text{fuel}} \Delta h_{r,298}^o \qquad 4.22$$

All of the components in the reaction mixture that do not participate in the reaction (excess O_2, N_2, and "other") do not contribute energy effects since they follow through Step 2 at constant T. We do need to compute the moles of each species in the combustion product carefully. There will be stoichiometric CO_2 and H_2O formed as a result of the reaction, then excess O_2 (if present) will be part of the products, and the amount of N_2 (if air is used at the oxidant) that originally was part of the reaction mixture is still part of the combustion product. There could also be other stoichiometric species if the fuel is composed of elements other than carbon and hydrogen, and if other non-reacting species are present initially, those will be part of the combustion product. However, there should be no fuel because we assume it to be burned completely. Moles of each species in the combustion product must be computed.

Finally, in Step 3, we need to compute the sensible heat effect in raising the temperature of the combustion product to T_{flame}:

$$\Delta H_3 = R \int_{298}^{T_{flame}} \sum_{products} n_i \frac{C_{p,i}^{IG}}{R} dT \qquad 4.23$$

Since the overall process is adiabatic when we are computing an adiabatic flame temperature and constant pressure, Q = 0 and:

$$\Delta H = \Delta H_1 + \Delta H_2 + \Delta H_3 = Q = 0 =$$

$$R \int_{T_1}^{298} \sum_{reactants} n_i \frac{C_{p,i}^{IG}}{R} dT + n_{fuel} \Delta h_{r,298}^\circ + R \int_{298}^{T_{flame}} \sum_{products} n_i \frac{C_{p,i}^{IG}}{R} dT \qquad 4.24$$

In equation 4.24, the only unknown is T_{flame} so we can solve. This same method of determining the flame temperature may be applied to processes other than flame temperatures.

4.d Chapter 4 Example Problems

4.1 Compute the adiabatic flame temperature for methane (CH_4) burned in 10% excess air starting at room temperature (298K).

For this case, $\Delta H_1 = 0$ for equation 4.24 since reactants start at 298K. Consider then the moles of each species in the reacting mixture taking n_{fuel} = 1 mole of CH_4. $CH_4 = C_1H_4$ so moles of stoichiometric O_2 via equation 4.19 are (1+ 4/4) = 2 moles. With 10% excess O_2:

$$n_{O_2} = 1.1 \times 2 = 2.2 \text{ moles}$$

For every mole of O_2, there are 79/21 moles of N_2 for air. Therefore:

$$n_{N_2} = \frac{79 \times 2.2}{21} = 8.276 \text{ moles}$$

Products of the combustion in moles are:

$$n_{CH_4} = 0 \quad n_{O_2} = 0.2 \quad n_{N_2} = 8.276 \quad n_{CO_2} = 1 \quad n_{H_2O} = 2$$

To evaluate equation 4.24, we will use these numbers for moles of each species in the product, and we will use equation 4.4 for the heat capacities of the product gases and get the constants for the heat capacity equation from Appendix B. This yields:

$$\Delta H_3 = R \int_{298.15}^{T_{flame}} \sum_{products} n_i \left(A_i + B_i T + C_i T^2 + D_i T^3 + E_i T^{-2} \right) dT \qquad 4.25$$

Using moles of product from above, and pulling the numbers for the first term of the integral from Appendix B, we get:

$$\sum_{products} n_i A_i = 0.2(3.0643) + 8.276(2.8185) + 3.8803 + 2(3.1950) = 34.1371 \quad 4.26$$

Similarly, we get:

$$\sum_{products} n_i B_i = 0.0214278 \qquad \sum_{products} n_i C_i = -6.6793 \times 10^{-6}$$

$$\sum_{products} n_i D_i = 7.6492 \times 10^{-10} \qquad \sum_{products} n_i E_i = 192683.8$$

Plug all back into equation 4.24 with $\Delta H_1 = 0$:

$$-(1\,\text{mole})\Delta h^\circ_{r,298} = 8.314\,\text{J/mole}\cdot\text{K}[(34.1371)(T_{flame} - 298)$$
$$+ 0.0214278(T_{flame}^2 - 298^2)/2 - 6.6793 \times 10^{-6}(T_{flame}^3 - 298^3)/3 \quad\quad 4.27$$
$$+ 7.6492 \times 10^{-10}(T_{flame}^4 - 298^4)/4 - 192683.8(T_{flame}^{-1} - 298^{-1})]$$

From Appendix F, we find for the combustion of methane:

$$\Delta h^\circ_{r,298} = \Delta h^\circ_{f,CO_2,298} + 2\Delta h^\circ_{f,H_2O,298} - \Delta h^\circ_{f,CH_4} =$$
$$-393509\,\text{J/mole} - 2\cdot 241818\,\text{J/mole} - (-74500)\,\text{J/mole} = \quad 4.28$$
$$-802645\,\text{J/mole}$$

Plugging into 4.27, the only unknown is T_{flame} which can be solved to yield:

$$T_{flame} = 2188\,K$$

Note that in equation 4.25 the largest term is the term representing N_2. Much of the heat that is generated by the combustion of methane in a device such as a Bunsen burner goes to increasing the temperature of the N_2 that is in air, comes into the process as part of the mixture of reactants, and leaves in the hot flame. In the case of a welding torch, pure O_2 is used as the oxidant so there is no N_2 entering with the reactants, and the N_2 term would be eliminated in equation 4.25. The product gases would consist only of excess O_2, CO_2 and H_2O, resulting in a much higher adiabatic flame temperature. This is the reason that O_2 is substituted for air in such devices. For such a case, T_{flame} might be so high that the heat capacity equation does not have the range to properly simulate the heat capacities of the product gases.

4.2 Propane and 10% excess air at 100°C are mixed and burned adiabatically at atmospheric pressure. Assuming complete combustion to CO_2 and H_2O, what is the final temperature of the combustion products?

This is an adiabatic flame temperature problem, but this time, ΔH_1 of equation 4.21 is not zero. Let's begin with balancing out the moles entering and leaving the overall process. The stoichiometric equation is:

$$C_3H_8 + 5O_2 \rightarrow 3CO_2 + 4H_2O$$

With 10% excess air and on the basis of one mole of C_3H_8, entering the process is:

$$n_{C_3H_8} = 1 \quad n_{O_2} = 5.5 \quad n_{N_2} = 20.7 \quad n_{CO_2} = 0 \quad n_{H_2O} = 0$$

and leaving is:

$$n_{C_3H_8} = 0 \quad n_{O_2} = 0.5 \quad n_{N_2} = 20.7 \quad n_{CO_2} = 3 \quad n_{H_2O} = 4$$

First, get the heat capacity equation coefficients to do the calculations to cool the reactants to 298 K:

$$\sum n_i A_i = 1 \cdot (-1.7622) + 5.5 \cdot (3.0643) + 20.7 \cdot (2.8185) = 73.4344$$

Similarly:

$$\sum n_i B_i = 0.0814417 \quad \sum n_i C_i = -3.6813 \times 10^{-5}$$
$$\sum n_i D_i = 6.7429 \times 10^{-9} \quad \sum n_i E_i = 525052.5$$

plug into $\Delta H = \int C_p dT$:

$$\frac{\Delta H_1}{R} = \left(\sum n_i A_i\right)(298 - 398) + \left(\frac{\sum n_i B_i}{2}\right)(298^2 - 398^2)$$
$$+ \left(\frac{\sum n_i C_i}{3}\right)(298^3 - 398^3) + \left(\frac{\sum n_i D_i}{4}\right)(298^4 - 398^4) - \left(\sum n_i E_i\right)\left(\frac{1}{298} - \frac{1}{398}\right)$$

which yields $\Delta H_1 = -84806$ J.

ΔH_2 is the combustion step for one mole of propane. Looking up the heats of formation for the stoichiometric reaction, we find:

$$\Delta H_2 = -1 \cdot (-104.0) + 3 \cdot (-393.3) + 4 \cdot (-241.8) = -2043.1 \text{ kJ} = -2043100 \text{ J}$$

ΔH_3 steps from T = 298 to the flame temperature. We use new sums for the heat capacity coefficients because now we have products of the reaction.

$$\sum n_i A_i = 0.5 \cdot (3.0643) + 20.7 \cdot (2.8185) + 4 \cdot (3.1590) + 3 \cdot (3.8803) = 84.1520$$

similarly:

$$\sum n_i B_i = 0.0536250 \quad \sum n_i C_i = -1.7335 \times 10^{-5}$$
$$\sum n_i D_i = 2.0585 \times 10^{-9} \quad \sum n_i E_i = 435234.4$$

then:

$$\frac{\Delta H_3}{R} = \left(\sum n_i A_i\right)(T_{flame} - 298) + \left(\frac{\sum n_i B_i}{2}\right)(T_{flame}^2 - 298^2)$$
$$+ \left(\frac{\sum n_i C_i}{3}\right)(T_{flame}^3 - 298^3) + \left(\frac{\sum n_i D_i}{4}\right)(T_{flame}^4 - 298^4) - \left(\sum n_i E_i\right)\left(\frac{1}{T_{flame}} - \frac{1}{298}\right)$$

4.29

Summing over the whole process, $\Delta H_1 + \Delta H_2 + \Delta H_3 = 0 = -84806$ J $- 2043100$ J $+ \Delta H_3$ (where ΔH_3 is given in the equation above) yields an equation with the only variable being T_{flame} which can be solved iteratively for T_{flame} to give **T_{flame} = 2318 K**.

4.3 Compute the theoretical flame temperature for n-butane gas burned in a synthetic air mixture which contains 10 mole% O_2 in 90 mole% Ar. Assume the starting temperature is room temperature (298 K) and assume the combustion mixture is stoichiometric.

The term "theoretical flame temperature" is often used interchangeably with "adiabatic flame temperature". From the problem description, ΔH_1 in equation 4.24 = 0. We are burning n-butane so the stoichiometric equation is:

$$C_4H_{10} + 6\tfrac{1}{2} O_2 \rightarrow 4 CO_2 + 5 H_2O$$

In equation 4.24, using Appendix F for heats of formation, $\Delta H_2 = 4 \cdot (-393300) + 5 \cdot (-241800) - 1 \cdot (-127200) = -2655000$ J.

flowing in: $n_{C_4H_{10}} = 1 \quad n_{O_2} = 6.5 \quad n_{Ar} = 58.5$

flowing out: $n_{CO_2} = 4 \quad n_{H_2O} = 5 \quad n_{Ar} = 58.5 \quad (n_{O_2} = 0 \quad n_{C_4H_{10}} = 0)$

Flowing out, both butane and oxygen are zero because the combustion mixture is stoichiometric. Getting the heat capacity constants from Appendix B and summing using moles of the species flowing out:

$\sum n_i A_i = 177.6247$ $\sum n_i B_i = 0.0268209$ $\sum n_i C_i = -8.0429 \times 10^{-6}$

$\sum n_i D_i = 9.1796 \times 10^{-10}$ $\sum n_i E_i = -47220.7$

ΔH_3 in equation 4.24 is exactly the same form as equation 4.29 in the previous problem but we substitute the heat capacity constants computed in this problem. Then, $\Delta H_1 + \Delta H_2 + \Delta H_3 = 0 = -2655000 \text{ J} + \Delta H_3$. Solve for $T_{flame} = $ **1915 K**.

4.4 One mole of SO_2 (g) and one mole of O_2 (g) are contained in a frictionless piston and cylinder arrangement initially at 1 bar and 25°C. The SO_2 is then allowed to react at constant P until all of it has been converted to SO_3 (g). During the process, 104600 J of heat are transferred to the surroundings. What is the final temperature of the reaction products?

This problem is the same as an adiabatic flame temperature calculation, but it is not adiabatic as heat is transferred to the surroundings. Since the heat is transferred to the surroundings, the heat is negative by our formal sign convention. Thus, setting up the problem using Figure 4.c.1 as a guide, $\Delta H_1 + \Delta H_2 + \Delta H_3 = Q = -104600$ J. Since the process starts at 298 K, $\Delta H_1 = 0$.

For ΔH_2, the reaction stoichiometry is:

$$SO_2 + \tfrac{1}{2}O_2 \rightarrow SO_3$$

Using heats of formation in Appendix F, $\Delta H_2 = 1 \cdot (-395400) - 1 \cdot (-296200) = -99200$ J. The products of the combustion are:

$$n_{SO_3} = 1 \quad n_{O_2} = \tfrac{1}{2}$$

For Step 3:

$\sum n_i A_i = 7.5280$ $\sum n_i B_i = 0.0062146$ $\sum n_i C_i = -2.7351 \times 10^{-6}$

$\sum n_i D_i = 4.1849 \times 10^{-10}$ $\sum n_i E_i = -107255.4$

$$\frac{\Delta H_3}{R} = \left(\sum n_i A_i\right)(T_3 - 298) + \left(\frac{\sum n_i B_i}{2}\right)(T_3^2 - 298^2)$$

$$+ \left(\frac{\sum n_i C_i}{3}\right)(T_3^3 - 298^3) + \left(\frac{\sum n_i D_i}{4}\right)(T_3^4 - 298^4) - \left(\sum n_i E_i\right)\left(\frac{1}{T_3} - \frac{1}{298}\right)$$

Summing, $\Delta H_1 + \Delta H_2 + \Delta H_3 = 0 - 99200$ J $+ \Delta H_3 = Q = -104600$ J or $\Delta H_3 = -5400$ J. Substituting the expression in temperature for ΔH_3 and solving for T_3, $T_3 = $ **208 K**. More heat

4.5 Calculate the standard heat of reaction for cyclohexane dehydrogenating to give benzene and hydrogen at 100°C.

The reaction stoichiometry is:

$$C_6H_{12} \rightarrow C_6H_6 + 3H_2$$

We will evaluate equation 4.18 with T = 100°C = 373 K. Accordingly, $\Delta h^\circ_{r,298}$ = 1·(82800) J/mole - 1·(-123000) J/mole = 205800 J/mole. For the heat capacity coefficients:

$\sum v_i A_i = 4.8261$ $\sum v_i B_i = 0.0013406$ $\sum v_i C_i = -1.3271 \times 10^{-5}$

$\sum v_i D_i = 5.7292 \times 10^{-9}$ $\sum v_i E_i = 279529.0$

Students should recall equation 4.18 uses the stoichiometric coefficients, v_i's, which are positive for products and negative for reactants, rather than numbers of moles as we did in the previous four problems where we were heating or cooling an actual mixture of components. Equation 4.18 becomes:

$$\frac{\Delta h^\circ_{r,373}}{R} = \frac{205800 \text{ J/mole}}{R} + (4.8261)(373 - 298) + \left(\frac{0.0013406}{2}\right)(373^2 - 298^2)$$

$$+ \left(\frac{-1.3271 \times 10^{-5}}{3}\right)(373^3 - 298^3) + \left(\frac{5.7292 \times 10^{-9}}{4}\right)(373^4 - 298^4) - (279529.0)\left(\frac{1}{373} - \frac{1}{298}\right)$$

$$\Delta h^\circ_{r,373} = 205800 \text{ J/mole} + 488.22 \times 8.314 = \mathbf{209859 \text{ J/mole}}$$

4.6 Gaseous methanol flowing at 1.0 mole/hr is mixed with stoichiometric air and burned completely to CO_2 and H_2O. Even though the burner is not designed to transfer heat out, it loses 100000 J/hr to the surroundings by radiation. Estimate the temperature of the flame.

Solution:

We will assume that the air and methanol are initially at 1 bar and 298 K. We will take a basis of 1 hr so $n_{methanol}$ = 1 mole and base all our calculations on that. Setting up the problem using Figure 4.c.1 as a guide, $\Delta H_1 + \Delta H_2 + \Delta H_3 = Q = -100000$ J. Heat is transferred to the surroundings so it is negative by our formal sign convention.

Since we assume the reactants start at 298 K, $\Delta H_1 = 0$. ΔH_2 is the heat of reaction for:

$$CH_3OH + 1\tfrac{1}{2}O_2 \rightarrow CO_2 + 2H_2O$$

$$\Delta H_2 = \Delta H^\circ_{r,298} = 1(-393300J) + 2(-241800J) - 1(-201200J)) = -675700J$$

Entering, in moles:

$$n_{CH_3OH} = 1 \quad n_{O_2} = 1.5 \quad n_{N_2} = 5.643 \quad n_{H_2O} = 0 \quad n_{CO_2} = 0$$

Leaving, in moles:

$$n_{CH_3OH} = 0 \quad n_{O_2} = 0 \quad n_{N_2} = 5.643 \quad n_{H_2O} = 2 \quad n_{CO_2} = 1$$

For Step 3:

$\sum n_i A_i = 26.1031 \qquad \sum n_i B_i = 0.0169999 \qquad \sum n_i C_i = -5.1522 \times 10^{-6}$

$\sum n_i D_i = 5.7552 \times 10^{-10} \quad \sum n_i E_i = 133289.9$

$$\frac{\Delta H_3}{R} = \left(\sum n_i A_i\right)(T_3 - 298) + \left(\frac{\sum n_i B_i}{2}\right)\left(T_3^2 - 298^2\right)$$

$$+ \left(\frac{\sum n_i C_i}{3}\right)\left(T_3^3 - 298^3\right) + \left(\frac{\sum n_i D_i}{4}\right)\left(T_3^4 - 298^4\right) - \left(\sum n_i E_i\right)\left(\frac{1}{T_3} - \frac{1}{298}\right)$$

Summing, $Q = -100000$ J $= \Delta H_1 + \Delta H_2 + \Delta H_3 = 0 - 675700$ J $+ \Delta H_3$. This gives $\Delta H_3 = 575700$ J. Plugging in the expression in temperature for ΔH_3 and solving for T_3, we get **$T_3 = 2058$ K**.

4.7 Find the standard heat of heat of reaction (Δh_r°) for the formation of ethylene oxide from ethylene at 500 K. The reaction is:

$$C_2H_4 + \tfrac{1}{2}O_2 \rightarrow C_2H_4O$$

Solution:

We will evaluate equation 4.18 with T = 500 K. Accordingly, $\Delta h_{r,298}^\circ = 1 \cdot (-52600)$ J/mole - $1 \cdot (52500)$ J/mole = -105100 J/mole. For the heat capacity coefficients:

$\sum v_i A_i = 1 \cdot (-1.9316) - 1 \cdot (1.9993) - \tfrac{1}{2} \cdot (3.0643) = -5.4631$

similarly:

$\sum v_i B_i = 0.0149905 \qquad \sum v_i C_i = -1.2598 \times 10^{-5} \quad \sum v_i D_i = 3.5187 \times 10^{-9} \quad \sum v_i E_i = 73357.6$

Equation 4.18 becomes:

$$\frac{\Delta h^\circ_{r,500}}{R} = \frac{-105100 \text{ J/mole}}{R} + (-5.4631)(500-298) + \left(\frac{0.0149905}{2}\right)(500^2 - 298^2)$$
$$+ \left(\frac{-1.2598 \times 10^{-5}}{3}\right)(500^3 - 298^3) + \left(\frac{3.5187 \times 10^{-9}}{4}\right)(500^4 - 298^4) - (73357.6)\left(\frac{1}{500} - \frac{1}{298}\right)$$

$$\Delta h^\circ_{r,500} = -105100 \text{ J/mole} + 8.314 \cdot (-161.64) = \textbf{-106444 J/mole}$$

4.8 Calculate the standard heat of reaction for the methane/steam reforming reaction at 600°C. The reaction is:

$$CH_4 + 2H_2O \to CO_2 + 4H_2$$

Solution:

We will again evaluate equation 4.18 with T = 273 + 600 = 873 K. Then:
$\Delta h^\circ_{r,298} = [1 \cdot (-393300) + 4 \cdot (0) - 1 \cdot (-74900) - 2 \cdot (-241800)] \cdot$ J/mole = 165200 J/mole. For the heat capacity coefficients:

$\sum v_i A_i = 11.5612$ $\sum v_i B_i = -0.0131845$ $\sum v_i C_i = 4.7840 \times 10^{-6}$

$\sum v_i D_i = -5.9207 \times 10^{-10}$ $\sum v_i E_i = -181705.5$

Equation 4.18 becomes:

$$\frac{\Delta h^\circ_{r,873}}{R} = \frac{165200 \text{ J/mole}}{R} + (11.5612)(873-298) + \left(\frac{-0.0131845}{2}\right)(873^2 - 298^2)$$
$$+ \left(\frac{4.7840 \times 10^{-6}}{3}\right)(873^3 - 298^3) + \left(\frac{-5.9207 \times 10^{-10}}{4}\right)(873^4 - 298^4) - (-181705.5)\left(\frac{1}{873} - \frac{1}{298}\right)$$

$$\Delta h^\circ_{r,873} = 165200 + 2741.3 \times 8.314 = \textbf{187991 J/mole}$$

Student Problems:

1. Repeat problem 4.3 but assume the starting temperature is 100°C and assume the combustion mixture is 10% excess O_2.

2. Calculate the standard heat of reaction for the ammonia synthesis reaction at 500°C. The reaction is $N_2 + 3 H_2 \to 2 NH_3$.

3. n-Butane and 25% excess air at 100°C are mixed and burned adiabatically at atmospheric pressure. Assuming complete combustion to CO_2 and H_2O, what is the final temperature of the combustion products?

Chapter 5 - The Second Law of Thermodynamics

5.a The Nature of Entropy and the Second Law

The First Law of Thermodynamics places restrictions on processes, which simply put means that energy is conserved in any operation. This concept is not at all hard to grasp, and we can even observe this principle with everyday processes as simple as heating a pan of water on a stove, or freezing an ice cube in the freezer, or tossing a ball in the air and watching it come back down. We note that energy moves from one place to another in these processes and though we don't have a quantitative measure in everyday life, we can see qualitatively that the conservation of energy principle is reasonable.

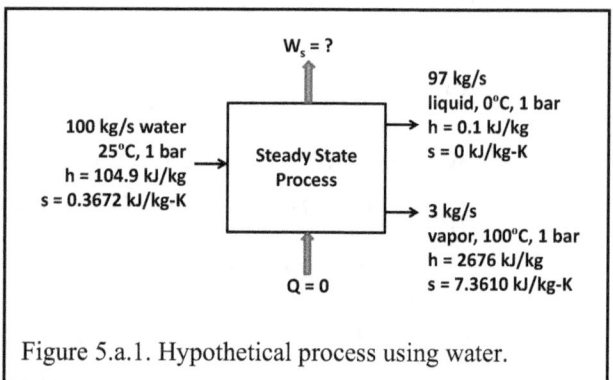

Figure 5.a.1. Hypothetical process using water.

The Second Law of Thermodynamics is not so easily grasped, but it is much simpler than the thought that it is "scary and impossible" to understand that many beginning engineering students seem to have. So we begin this chapter by making some relatively simple observations that go directly to the necessary concepts. In Figure 5.a.1, we have setup a simple hypothetical process that takes room temperature and pressure water, runs it through some process, and produces cold liquid water and low pressure steam. We have specified this process to be adiabatic (Q = 0) and steady state. Apply the First Law:

$$\dot{m}\Delta h = \left(97\frac{kg}{s}\cdot 0.1\frac{kJ}{kg} + 3\frac{kg}{s}\cdot 2676\frac{kJ}{kg}\right) - \left(100\frac{kg}{s}\cdot 104.9\frac{kJ}{kg}\right) = \dot{Q} - \dot{W}_s \qquad 5.1$$

With $\dot{Q}=0$, we get $\dot{W}_s = 2452\,kW$ so by the First Law calculation, the process produces power. The First Law is fully satisfied with the process, and this process is the greatest! It takes plain tap water, makes cold water we could use for air-conditioning or refrigeration around our house, steam to heat or make hot water for the shower, and produces work that we could convert to electricity to power our lights. Furthermore, we don't have to use any fossil fuels because the process requires no heat, so there are no greenhouse gases produced. But it satisfies the First Law so it must be possible? Have we just solved the world's problems?

This process is counterintuitive because we have never seen room temperature water spontaneously split into hot and cold water streams and produce work at the same time. Unfortunately, the answer is that the process is impossible, and as we unfold the Second Law of Thermodynamics, we will find it violates this law. In the same steam table we used to look up enthalpies, we can also look up the quantity called "entropy", denoted "s". Let's use the values for s from the steam table to compute $\Delta\dot{S}$ for this process:

$$\Delta \dot{S} = \left(97\frac{kg}{s} \cdot 0\frac{kJ}{kg \cdot K} + 3\frac{kg}{s} \cdot 7.361\frac{kJ}{kg \cdot K}\right) - \left(100\frac{kg}{s} \cdot 0.3672\frac{kJ}{kg \cdot K}\right) = -14.64\frac{kJ}{s \cdot K} \qquad 5.2$$

If such a process was possible, it would lead to a decrease in entropy for the universe.

As we continue this chapter, we will relate this function entropy to heat added in a process, but for a moment let's consider entropy as a measure of disorder. In fact, statistical thermodynamic calculations for an ideal gas relate entropy directly to this concept. So a process in which entropy of the universe declines means that the process reduces disorder or adds order to the universe. Again, this is counterintuitive. For example, take a bottle and order black and white marbles in several layers in the bottle, then shake, stop, and wait for the marbles to settle then observe the results. The marbles mix and take on random positions in layers in the bottle. Keep shaking then observing. Do the marbles ever reorder themselves in black and white layers? Perhaps, but in fact, the probability that will happen is extremely small. Shaking the bottle to reorder the marbles in layers by color is similar to trying to find a process for which entropy declines, like the one shown in Figure 5.a.1.

When energy is added to a system such as a mole of an ideal gas, the energy distributes everywhere among all the molecules of the system. Some goes to increase the kinetic energy of the molecules and distributes among low, medium and high kinetic energy molecules. Some goes to rotational and vibrational components in a similar fashion. The energy goes everywhere in the system like throwing a cup of dust in the wind. We can even consider the energy on the microscopic scale in quantum packets almost like individual marbles. These packets distribute rather randomly by bumping up quantum states of random molecules in the system. So the process of getting the energy back out cannot be accomplished exactly in reverse because the energy would have to order itself in doing so.

Consider the process depicted in Figure 5.a.2. The overall process is an arbitrary change from T_1, V_1, P_1 to T_2, V_2, P_2 for one mole of an ideal gas. Consider the gas as the system, and consider all the processes as reversible. For the overall process $\Delta u = q - w$ and since we are working with an ideal gas, $\Delta u = \int_{T_1}^{T_2} C_v^{IG} dT$. However, because q and w depend upon how we accomplish the process we don't know the individual values of q and w, we just know that the difference is Δu.

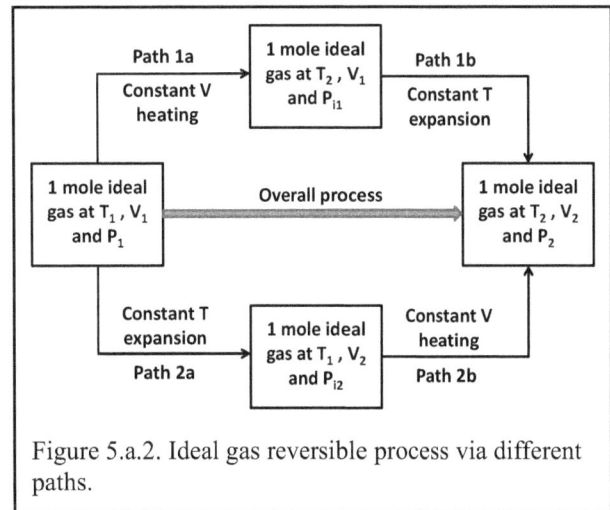

Figure 5.a.2. Ideal gas reversible process via different paths.

Consider then the separate Paths 1 and 2 shown on the figure. We can accomplish either of these paths by placing the gas in a piston and cylinder arrangement and performing operations in sequence. In Path 1, we first hold $V = V_1$ constant via locking the piston in place while heating, increasing the temperature from T_1 to T_2 (Step a), then we hold temperature constant and allow the gas to expand to get to V_2, T_2, and P_2. In Path 2, the separate steps are reversed so we expand at constant temperature first, then heat at constant volume. Now, apply the First Law:

Path 1:
Step a:
$$w_a = 0 \qquad q_a = \Delta u_a = \int_{T_1}^{T_2} C_v^{IG} dT$$

Step b:
$$\Delta u_b = 0 \qquad q_b = w_b = \int P dv = \int_{v_1}^{v_2} \frac{RT_2}{v} dv = RT_2 \ln\left(\frac{v_2}{v_1}\right)$$

Overall:
$$\Delta u = \int_{T_1}^{T_2} C_v^{IG} dT \qquad q = \int_{T_1}^{T_2} C_v^{IG} dT + RT_2 \ln\left(\frac{v_2}{v_1}\right) \qquad w = RT_2 \ln\left(\frac{v_2}{v_1}\right)$$

Similarly for Path 2, we will skip the individual steps (left to the student to do) and present the final result:

Path 2:
Overall:
$$\Delta u = \int_{T_1}^{T_2} C_v^{IG} dT \qquad q = \int_{T_1}^{T_2} C_v^{IG} dT + RT_1 \ln\left(\frac{v_2}{v_1}\right) \qquad w = RT_1 \ln\left(\frac{v_2}{v_1}\right)$$

There are several things to note from this analysis. First, regardless of whether we follow Path 1 or 2, Δu is the same, which emphasizes that Δu is a state function and is independent of path. Similarly, q and w are different for both paths illustrating that they are not state functions. However, what is the difference between q and w for Path 1 and 2? The difference is simply the temperature at which heat is added during the constant temperature expansion process. In Path 1, the constant T expansion occurs at T_2 while in Path 2, the constant T expansion occurs at T_1. Suppose we were to compute $\int \frac{dq}{T}$ for both paths:

Path 1:
Overall:
$$\int \frac{dq}{T} = \int_{T_1}^{T_2} \frac{C_v^{IG}}{T} dT + R \ln\left(\frac{v_2}{v_1}\right)$$

Path 2:
Overall:

$$\int \frac{dq}{T} = \int_{T_1}^{T_2} \frac{C_v^{IG}}{T} dT + R \ln\left(\frac{v_2}{v_1}\right)$$

This begs a question. Since $\int \frac{dq}{T}$ is the same for both paths, is it a state function? Let's test this question using mathematical tools we have available. First, let's test q to see how the math works. This should remind us it is not a state function.

$$\Delta u = q - w \quad \text{or} \quad du = \delta q - \delta w \quad \text{or} \quad \delta q = \Delta u + \delta w \qquad 5.3$$

In equation 5.3, we have use the inexact differential representation δ for q and w differentials since they are not state functions, but this is only for mathematical language that is exact. Let's assume we have a reversible process with an ideal gas, then:

$$\delta q = C_v^{IG} dT + P dv \qquad 5.4$$

Recall that for an exact function z = f(x,y) with differential written dz = Mdx + Ndy:

$$\left.\frac{\partial M}{\partial y}\right)_x = \left.\frac{\partial N}{\partial x}\right)_y \qquad 5.5$$

Apply to equation 5.4:

$$\left.\frac{\partial C_v^{IG}}{\partial v}\right)_T \overset{?}{=} \left.\frac{\partial P}{\partial T}\right)_v \qquad 5.6$$

Take each side of equation 5.6 separately:

$$\frac{\partial C_v^{IG}}{\partial v} = \frac{\partial}{\partial v}\left(\frac{\partial u}{\partial T}\right) = \frac{\partial}{\partial T}\left(\frac{\partial u}{\partial v}\right) = 0 \qquad 5.7$$

In equation 5.7 we have used the definition of C_v, then reversed the order of differentiation, then noted that for an ideal gas, u is only a function of T only, not v.

For the other side of equation 5.6, using the ideal gas Pv = RT:

$$\left.\frac{\partial P}{\partial T}\right)_v = \frac{R}{v}\left.\frac{\partial T}{\partial T}\right)_v = \frac{R}{v} \qquad 5.8$$

Since the result of equation 5.7 does not equal the result of 5.8, equations 5.5 and 5.6 are not satisfied so q is not an exact function. This is what we expected.

Now, let's try the function ds = δq/T. Dividing equation 5.4 by T:

$$ds = \frac{\delta q}{T} = \frac{C_v^{IG}}{T} dT + \frac{P}{T} dv \qquad 5.9$$

Apply equation 5.5:

$$\frac{\partial \left(\frac{C_v^{IG}}{T}\right)}{\partial v}\bigg)_T \stackrel{?}{=} \frac{\partial \left(\frac{P}{T}\right)}{\partial T}\bigg)_v \qquad 5.10$$

Take each side of equation 5.10 individually starting with the left-hand side:

$$\frac{\partial \left(\frac{C_v^{IG}}{T}\right)}{\partial v}\bigg)_T = \frac{1}{T}\frac{\partial C_v^{IG}}{\partial v}\bigg)_T = 0 \qquad 5.11$$

This follows from equation 5.7. Now the right-hand side of equation 5.10:

$$\frac{\partial \left(\frac{P}{T}\right)}{\partial T}\bigg)_v = \frac{\partial \left(\frac{RT}{vT}\right)}{\partial T}\bigg)_v = \frac{1}{v}\frac{\partial R}{\partial T}\bigg)_v = 0 \qquad 5.12$$

Therefore, equations 5.11 and 5.12 are equal, equation 5.10 is satisfied along with equation 5.5, and equation 5.9 is exact. ds is exact for an ideal gas undergoing a reversible process. We will continue with the assumption that ds is exact for all processes and for all substances, but more development of the function is warranted.

5.b Heat Engines and Statement of the Second Law of Thermodynamics

In order to continue our development of the Second Law, we need to define *heat engines* and *thermal reservoirs*. A heat engine is a device which produces work from heat. It absorbs heat at a high temperature, converts some of that heat to work, and rejects any remaining heat at a lower temperature. We will consider these as "black boxes" because we generally do not know the inner workings of the device, though we will develop a working idea of a heat engine. We also consider heat engines as cyclic in nature. Even though one or more devices may be operating within the engine, these devices must all come back to their original layout and working fluids must cyclically return to the same initial state. This means that the energy content of the heat engine (as a system) has not changed when it completes a cycle.

Generally, the heat to operate a heat engine is absorbed from a heat reservoir, which is defined as a device that can accept or reject any quantity of heat without changing temperature. A very close approximation to a thermal reservoir is a lake, a steam boiler, or the earth's atmosphere. At this time, we will refer to the temperature of a thermal reservoir as θ which is a temperature scale defined by thermal arguments. We will later relate θ to T (the ideal gas temperature scale). The only thing we initially know about θ is that when θ of one thermal reservoir is greater than θ for another thermal reservoir, heat can flow spontaneously from the higher thermal temperature to the lower and not the other way around.

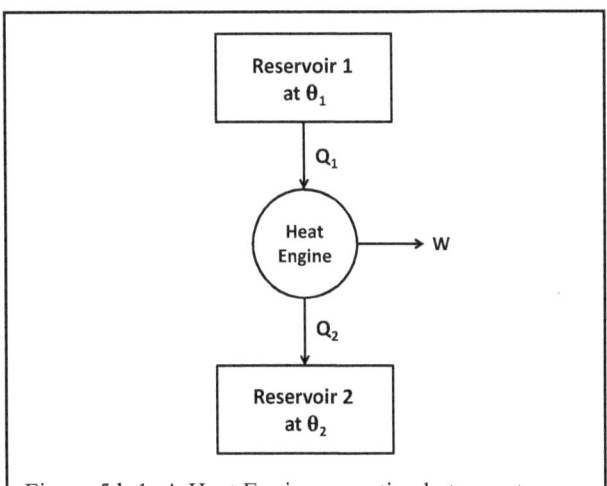

Figure 5.b.1. A Heat Engine operating between two thermal reservoirs.

Figure 5.b.1 shows a typical process for our discussion on heat engines. The engine takes an amount of heat Q_1 from a thermal reservoir at θ_1, does work W and rejects an amount of heat Q_2 to a reservoir at θ_2. Consider the heat engine as a system, and note that we will temporarily abandon our formal sign convention in which we treat heat added as positive and we will simply draw arrows indicating the expected direction the heat or work is flowing. If we compute a negative heat or work, we will know we drew the arrow the wrong direction. A First Law calculation for the engine as the system yields:

$$\Delta U = Q_1 - Q_2 - W \qquad 5.13$$

Since the engine is cyclic, when it returns to it's initial state, $\Delta U = 0$, so:

$$Q_1 - Q_2 = W \qquad 5.14$$

We are now ready to give a statement of the Second Law of Thermodynamics: *It is impossible to convert heat absorbed completely into work in a cyclic process.* Though there are a number of different ways to state the Second Law, all can be shown to be equivalent, and we will use this statement. We present this law without proof and the reader is referred to the open literature for discussion of this topic. This law has never been disproven either by theory or observation. From the Second Law, we can immediately see that it requires Q_2 to be greater than zero. Also, we define the thermal efficiency, η, of an engine according to:

$$\eta = \frac{W}{Q_1} = \frac{W_{net}}{Q_{abs}} \qquad 5.15$$

The Second Law also tells us that η cannot be equal to 1. Note that we distinguish heat absorbed (in this case Q_1) from heat rejected (in this case Q_2) and that in the more general sense, Q_{abs} is the correct nomenclature. Also, work could possibly be both used and produced within the black box, so the work in equation 5.15 is the net amount of work done by the process.

Now consider Corollary 1: *The efficiency of a reversible engine is always greater than the efficiency of an irreversible engine when they are operated between the same temperature reservoirs.* To prove this, consider Figure 5.b.2 where we have an irreversible engine and reversible engine operating between the same two thermal reservoirs. We will use *reductio ad absurdum* logic to prove this so we begin by assuming the opposite, so we assume that the irreversible engine has a higher efficiency than the reversible engine. We scale the size of the two engines such that Q_2 is the same for both engines. Since via our logic process, the irreversible engine has a higher efficiency than the reversible engine, the irreversible engine has to make more work than the reversible engine, and we write the work as $W + \varepsilon$ where ε is a positive number and is the excess work the irreversible engine makes. A First Law calculation around the irreversible engine then requires that $Q_1 + \varepsilon$ be absorbed from the high temperature reservoir since we have fixed the size of Q_2.

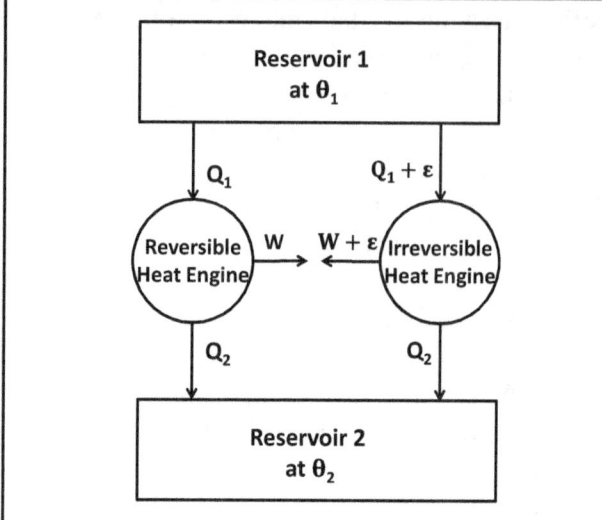

Figure 5.b.2. A reversible and irreversible heat engine operating between the same two thermal reservoirs.

Recall that the term "reversible" means just that, so let's run the reversible engine in reverse. See Figure 5.b.3. The reversible engine would require W to run it in reverse, and we will use the irreversible engine to supply that work. ε is therefore left over work and is positive. Now perform a First Law balance considering the process inside the dotted ellipse in Figure 5.b.3 as the system. Q_1, Q_2, and W are each of opposite sign for the reversible and irreversible engines so they cancel out in the First Law balance leaving ε net heat absorbed and ε net work done. Therefore ε heat absorbed is converted completely into ε work in a cyclic engine. This is a direct violation of our statement of the Second Law. Via *reductio ad absurdum*, Corollary 1 is proven true.

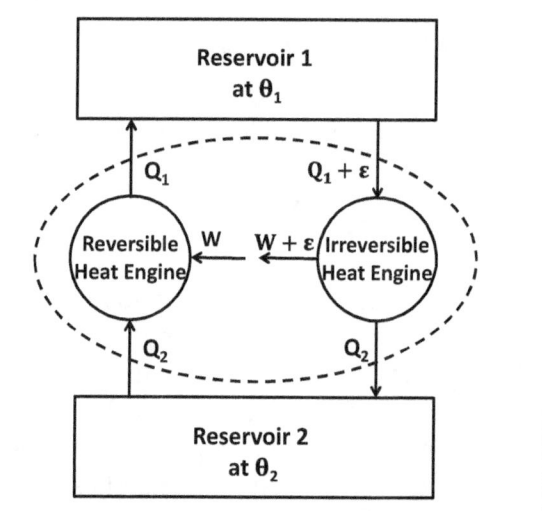

Figure 5.b.3. A reversible heat engine operating in reverse and irreversible engine operating between the same two thermal reservoirs.

Corollary 2: *The efficiencies of two reversible engines operating between the same temperature reservoirs are the same.* To prove this corollary, assume one reversible engine has an efficiency greater than the other, then change the title "Irreversible Heat Engine" in Figure 5.b.3 to "Reversible Heat Engine of Higher Efficiency". Follow exactly the same logic as for Corollary 1. Corollary 2 is proven.

Now let's see what we can deduce regarding the thermal temperature scale, θ. Let's take a reversible engine and duplicate it three times so we have E1, E2, and E3. Arrange the duplicates as in Figure 5.b.4. Here, we insert a thermal reservoir at θ_3 with temperature somewhere between θ_1 and θ_2 between E2 and E3. If the dotted rectangle is taken as the system, we see that that system operates exactly as E1, taking in Q_1 and rejecting Q_2. Thus, using Corollary 2 we recognize that $W_1 = W_2 + W_3$. We could vary θ_3 anywhere between θ_1 and θ_2 and together, engines E2 and E3 would have to yield the same total work as E1. Suppose we allowed θ_3 to approach θ_1, then the efficiency of E3 would approach the efficiency of E1 and the efficiency of E2 would approach zero. This observation would hold vice versa for θ_3 approaching θ_2. The efficiencies of E2 and E3 would vary widely if we started θ_3 slightly below θ_1 and varied it almost to θ_2. Apparently, the efficiency of a reversible engine is dependent only upon the temperatures of the reservoirs used to run the engines as this is the only variable we have to adjust the efficiency. Now:

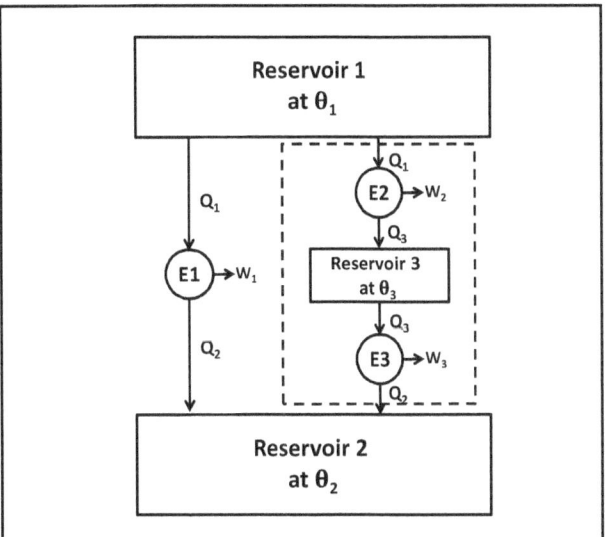

Figure 5.b.4. Three reversible heat engines operating between thermal reservoirs.

$$\eta_{E1} = \frac{W}{Q_1} = \frac{Q_1 - Q_2}{Q_1} = 1 - \frac{Q_2}{Q_1} \qquad 5.16$$

Since η_{E1} is a function of the reservoir temperatures only, equation 5.16 also results in:

$$\frac{Q_2}{Q_1} = f(\theta_1, \theta_2) \text{ only} \qquad 5.17$$

Similarly for E2 and E3:

$$\frac{Q_3}{Q_1} = f(\theta_1, \theta_3) \text{ and } \frac{Q_2}{Q_3} = f(\theta_3, \theta_2) \qquad 5.18 \text{ and } 5.19$$

If we multiply equations 5.18 and 5.19 together, Q_3 divides out, then equating to 5.17, we get:

$$\frac{Q_3}{Q_1}\frac{Q_2}{Q_3} = f(\theta_1,\theta_3) \cdot f(\theta_3,\theta_2) = f(\theta_1,\theta_2) = \frac{Q_2}{Q_1} \qquad 5.20$$

We need to ponder this a bit and come up with a function for $f(\theta_i, \theta_j)$ that allows elimination of θ_3 when applied like in equation 5.20. It's not difficult to note that dividing out a numerator and denominator to eliminate θ_3 works if applied correctly. Try:

$$f(\theta_i,\theta_j) = \frac{\theta_j}{\theta_i} \qquad 5.21$$

Then:

$$f(\theta_1,\theta_3) \cdot f(\theta_3,\theta_2) = \frac{\theta_3}{\theta_1}\frac{\theta_2}{\theta_3} = \frac{\theta_2}{\theta_1} = f(\theta_1,\theta_2) \qquad 5.22$$

So, relating back to Q_1 and Q_2 and using the function given in equation 5.21:

$$\frac{Q_2}{Q_1} = \frac{\theta_2}{\theta_1} \qquad 5.23$$

$$\boxed{\eta = \frac{W_{net}}{Q_{abs}} = 1 - \frac{Q_2}{Q_1} = 1 - \frac{\theta_2}{\theta_1} = \frac{\theta_1 - \theta_2}{\theta_1}} \qquad 5.24$$

Equation 5.24 gives the thermal efficiency for *all* reversible engines. This is an important equation as it relates the efficiency of a reversible heat engine to just the temperatures of the thermal reservoirs.

5.c Carnot Cycle

In order to relate the thermal temperature scale, θ to the ideal gas temperature scale, T, we will create a heat engine using an ideal gas as the working fluid. Consider a frictionless piston and cylinder arrangement holding one mole of an ideal gas as the system. The piston/cylinder arrangement we will use is perfectly insulated except for the bottom of the cylinder where heat may be added or removed through a conductive bottom. The whole device may be shifted back and forth from an insulating block, to a thermal reservoir at temperature T_1, or to another thermal

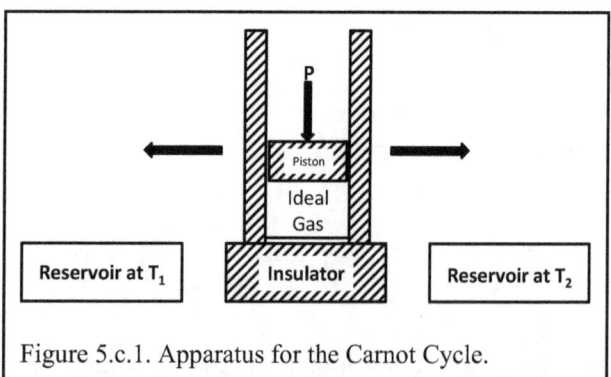

Figure 5.c.1. Apparatus for the Carnot Cycle.

reservoir at temperature T_2 which is lower than T_1. When the device sits on the insulating block, the device is fully insulated. See Figure 5.c.1.

The process, depicted in Figure 5.c.2, begins with the device sitting on the insulating block with the system at state P_A, V_A, and T_1. The device is then moved to the thermal reservoir at T_1. An isothermal expansion is performed until the system reaches P_B, V_B, and T_1. Heat Q_1 is absorbed during this step in the process. Note that heat must be absorbed during expansion or the temperature of the ideal gas would drop. The device is then moved back to the insulating block, and expansion continues adiabatically until the state reaches P_C, V_C, and T_2. T_2 is lower than T_1 because adiabatic expansion results in a falling temperature. Next, we slide the device to the

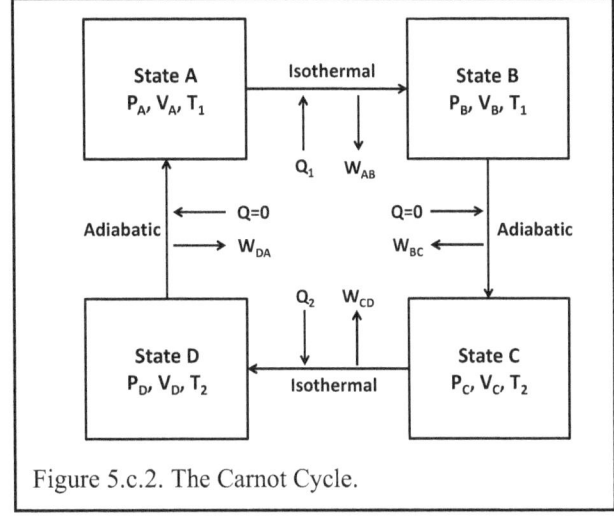

Figure 5.c.2. The Carnot Cycle.

reservoir at T_2 where we perform an isothermal compression to P_C, V_C, and T_2. Heat Q_2 is rejected to the thermal reservoir during this isothermal compression step. The device goes back to the insulating block and adiabatic compression is performed until we get back to the initial state at P_A, V_A, and T_1. This completes one cycle.

Note that this process, known as the Carnot Cycle, is a type of heat engine. It absorbs heat from a high temperature reservoir at T_1 and rejects heat to a lower temperature reservoir at T_2. Next, we need to determine the net amount of work done by the process, then we can compute the efficiency of the cycle. Since the Carnot Cycle is a reversible cycle, the efficiency determined for this cycle will be the same as all other reversible cycles and will represent the maximum efficiency any heat engine can achieve since a reversible engine always has a higher thermal efficiency than an irreversible engine.

Refer to section 3.h where we have already done the analysis of the individual steps in the Carnot Cycle. Let's begin with an analysis of the two adiabatic steps going from State B to C and State D to A. In both steps, since they are adiabatic, $Q = 0$, and the work for these steps can be computed from $\Delta u = -w$. Therefore:

$$\Delta u_{BC} = -w_{BC} = \int_{T_1}^{T_2} C_v^{IG} dT \text{ and } \Delta u_{DA} = -w_{DA} = \int_{T_2}^{T_1} C_v^{IG} dT \qquad \text{5.25 and 5.26}$$

The integral limits in equations 5.25 and 5.26 are reversed so:

$$w_{BC} + w_{DA} = 0 \qquad \text{5.27}$$

For the two isothermal processes, $\Delta u = 0$ so $q = w$:

$$q_{AB} = w_{AB} = RT_1 \ln\left(\frac{P_A}{P_B}\right) \quad \text{and} \quad q_{CD} = w_{CD} = RT_2 \ln\left(\frac{P_C}{P_D}\right) \qquad 5.28 \text{ and } 5.29$$

Summing all the work terms, and using equation 5.27:

$$w_{net} = w_{AB} + w_{BC} + w_{CD} + w_{DA} = RT_1 \ln\left(\frac{P_A}{P_B}\right) + RT_2 \ln\left(\frac{P_C}{P_D}\right) \qquad 5.30$$

Heat is absorbed in step A to B, so $q_{abs} = q_{AB}$. Then:

$$\eta = \frac{w_{net}}{q_{abs}} = \frac{RT_1 \ln\left(\frac{P_A}{P_B}\right) + RT_2 \ln\left(\frac{P_C}{P_D}\right)}{RT_1 \ln\left(\frac{P_A}{P_B}\right)} = \frac{T_1 \ln\left(\frac{P_A}{P_B}\right) + T_2 \ln\left(\frac{P_C}{P_D}\right)}{T_1 \ln\left(\frac{P_A}{P_B}\right)} \qquad 5.31$$

Let's consider the ratios P_A/P_B and P_C/P_D. Steps D to A and B to C are adiabatic steps, so recalling equation 3.57 in chapter 3 and applying:

$$\frac{T_1}{T_2} = \left(\frac{P_A}{P_D}\right)^{\frac{\gamma-1}{\gamma}} \quad \text{and} \quad \frac{T_1}{T_2} = \left(\frac{P_B}{P_C}\right)^{\frac{\gamma-1}{\gamma}} \qquad 5.32 \text{ and } 5.33$$

eliminating temperatures by equating equations 5.32 and 5.33 and solving for the pressure ratios:

$$\frac{P_A}{P_D} = \frac{P_B}{P_C} \quad \text{or} \quad \frac{P_C}{P_D} = \frac{P_B}{P_A} \qquad 5.34 \text{ and } 5.35$$

$$\ln\left(\frac{P_C}{P_D}\right) = \ln\left(\frac{P_B}{P_A}\right) = -\ln\left(\frac{P_A}{P_B}\right) \qquad 5.36$$

Substituting equation 5.36 into 5.31:

$$\eta = \frac{w_{net}}{q_{abs}} = \frac{T_1 \ln\left(\frac{P_A}{P_B}\right) - T_2 \ln\left(\frac{P_A}{P_B}\right)}{T_1 \ln\left(\frac{P_A}{P_B}\right)} = \frac{T_1 - T_2}{T_1} \qquad 5.37$$

which is the final result we are looking for. Just as in the general case for the heat engine, the efficiency, η, of this specific process is dependent only upon the temperatures of the thermal reservoirs. Comparing equation 5.37 with 5.24, we see that θ (the thermal temperature scale) is proportional to T (the ideal gas temperature scale) and the simplest way to treat the proportionality is to simply make the two scales the same. $\theta = T$.

5.d Definition of Entropy

Going back to equation 5.37 and rearranging:

$$\eta = \frac{w_{net}}{q_{abs}} = \frac{q_1 - q_2}{q_1} = 1 - \frac{q_2}{q_1} = \frac{T_1 - T_2}{T_1} = 1 - \frac{T_2}{T_1} \qquad 5.38$$

which yields:

$$\frac{q_2}{q_1} = \frac{T_2}{T_1} \quad \text{or} \quad \frac{q_1}{T_1} = \frac{q_2}{T_2} \qquad 5.39$$

Recall that q_2 was removed from the engine, so if we return to our formal sign convention where heat added to a process is positive, we need to substitute $-q_2$ for this quantity of heat. Equation 5.39 can then be rearranged and written using our formal sign convention:

$$\frac{q_1}{T_1} = \frac{-q_2}{T_2} \quad \text{or} \quad \frac{q_1}{T_1} + \frac{q_2}{T_2} = 0 \qquad 5.40$$

differentiating and noting that thermal reservoirs are constant in temperature:

$$\frac{dq_1}{T_1} + \frac{dq_2}{T_2} = 0 \qquad 5.41$$

This is the result for a full cycle and the result goes to zero. Also, we are working with a reversible process when this was developed so this could just as well have been written:

$$\oint \frac{dq_{rev}}{T} = 0 \qquad 5.42$$

The integral sign with the circle indicates an integral over a closed path or full cycle. For a non-closed path, the integral will not return to zero, so we define the function for the partial path as entropy, s:

$$ds = \frac{dq_{rev}}{T}$$
$$\Delta s = \int \frac{dq_{rev}}{T} \qquad 5.43$$

Note that just like $\Delta u = 0$ for the system used in a cyclic process such as the heat engine, $\Delta s = 0$ (equation 5.42). This is indicative of a state function.

In the event we have an irreversible heat addition, the entropy will increase above that of a reversible process, and:

$$\Delta s \geq \int \frac{\delta q}{T} \quad \text{or} \quad ds \geq \frac{\delta q}{T} \qquad 5.44$$

Equation 5.44 is known as the *inequality of Clausius*.

In equations 5.43 and 5.44 we see that the higher the temperature associated with available heat, the lower its entropy. In the concept of entropy as a measure of disorder, this means that higher temperature energy has the equivalent of lower disorder than low temperature energy. We also see in equation 5.37 that this translates to a higher fraction of the energy available to do work.

5.e Third Law of Thermodynamics

The Third Law of Thermodynamics relates to entropy and provides an absolute baseline for measuring and computing entropy. *The Third Law of Thermodynamics states that the entropy of a perfect crystal at absolute zero (0 K) is zero.* A perfect crystal is a solid which has perfect order, repeating the spacing and orientation of molecules in a regular pattern. This part of the Third Law equates directly to the concept of entropy as a measure of disorder. If a solid is not perfectly ordered, then the disorder it possesses is random in nature and the entropy associated with that disorder results in a non-zero measure even at absolute zero. Using the Third Law, it is possible to compute absolute entropies of substances.

5.f Entropy Change for an Ideal Gas

Consider an ideal gas that undergoes an arbitrary process and changes state from P_1, T_1, V_1 to P_2, T_2, V_2. Let's compute Δs for such a process. We will setup a 2 step process to do this which is shown in Figure 5.f.1. For Step a, the process is isothermal so $\Delta u = 0$. Then:

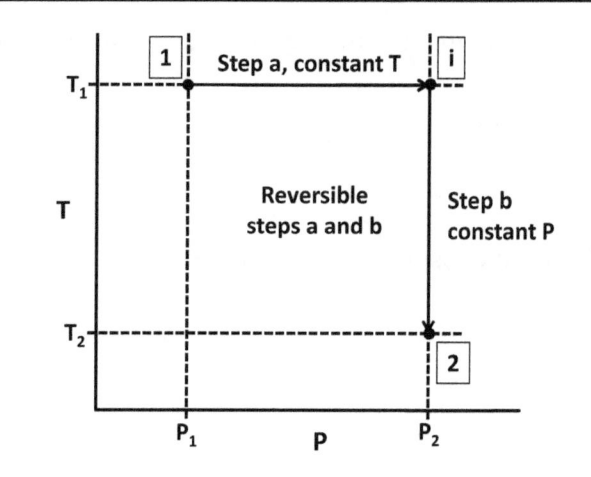

Figure 5.f.1. Process used for determining the entropy change of an ideal gas.

$$q_{rev} = w_{rev} = -RT_1 \ln\left(\frac{P_2}{P_1}\right) \quad \text{reversible process}$$

$$\Delta s_a = \int \frac{dq_{rev}}{T_1} = \frac{1}{T_1}\int dq_{rev} = \frac{q_{rev}}{T_1} = \frac{-RT_1 \ln\left(\frac{P_2}{P_1}\right)}{T_1}$$

$$\Delta s_a = -R \ln\left(\frac{P_2}{P_1}\right) \qquad 5.45$$

For Step b:

$$dq_{rev} = C_p^{IG} dT \quad \text{constant pressure, reversible}$$

$$\Delta s_b = \int_{T_1}^{T_2} \frac{C_p^{IG}}{T} dT \qquad 5.46$$

summing:

$$\boxed{\Delta s = \int_{T_1}^{T_2} \frac{C_p^{IG}}{T} dT - R \ln\left(\frac{P_2}{P_1}\right)} \qquad \text{(ideal gas) } 5.47$$

algebraic manipulation using the ideal gas law and $C_p^{IG} = C_v^{IG} + R$ shows this is equivalent to:

$$\boxed{\Delta s = \int_{T_1}^{T_2} \frac{C_v^{IG}}{T} dT + R \ln\left(\frac{V_2}{V_1}\right)} \qquad \text{(ideal gas) } 5.48$$

Equations 5.47 an 5.48 should be remembered for future use along with results from chapter 3:

$$\Delta u = \int_{T_1}^{T_2} C_v^{IG} dT \quad \text{and} \quad \Delta h = \int_{T_1}^{T_2} C_p^{IG} dT \qquad \text{(ideal gas) 3.39 and 3.40}$$

These equations, 5.47 - 5.48, and 3.39 - 3.40 may be used for *all* processes with an ideal gas regardless of whether they are reversible or not since the equations consist exclusively of state functions. Also, we have already evaluated the integral in equation 3.40 for the form of the heat capacity equation we use in this book in conjunction with Appendix B yielding equation 2.27:

$$\int_{T_1}^{T_2} \frac{C_p^{IG}}{R} dT = A(T_2 - T_1) + \frac{B}{2}(T_2^2 - T_1^2) + \frac{C}{3}(T_2^3 - T_1^3) + \frac{D}{4}(T_2^4 - T_1^4) - E\left(\frac{1}{T_2} - \frac{1}{T_1}\right) \qquad 2.27$$

Similarly, the integral in equation 5.47 can be evaluated:

$$\int_{T_1}^{T_2} \frac{C_p^{IG}}{R} \frac{dT}{T} = A \ln(T_2/T_1) + B(T_2 - T_1) + \frac{C}{2}(T_2^2 - T_1^2) + \frac{D}{3}(T_2^3 - T_1^3) + \frac{E}{-2}\left(\frac{1}{T_2^2} - \frac{1}{T_1^2}\right) \qquad 5.49$$

5.g Lost Work

Consider the entropy change of a thermal reservoir. See Figure 5.g.1, focusing first on the apparatus designated A). Here, we allow a transfer of heat of magnitude Q_1 from one reservoir to another. Now, let's take Reservoir 1 as the system and consider its entropy change. The calculation is:

$$\Delta S = \int \frac{dQ_{rev}}{T} \qquad 5.50$$

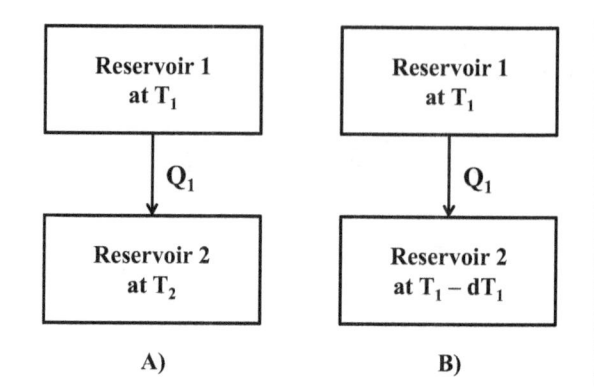

Figure 5.g.1. Diagram for computation of the entropy change for a thermal reservoir.

Unfortunately, since the temperature difference that is shown in A) is not differentially small, this process is irreversible whereas equation 5.50 requires a reversible heat exchange to evaluate. Therefore, we need to invent a reversible process that accomplishes the same process. This is what we need to do in many instances in order to compute thermodynamic functions correctly. In B) in Figure 5.g.1, we have proposed a reversible process. Now the temperature difference is dT, but Reservoir 1 really doesn't observe the temperature of Reservoir 2 to which the heat Q_1 is transferred. Therefore, this process appears the same as the process in A) to Reservoir 1 and at the same time is reversible. Clearly, since heat leaves Reservoir 1, by our formal sign convention:

$$\Delta S_{R1} = \frac{-Q_1}{T_1} \qquad 5.51$$

since T_1 is constant. This is the general expression for the entropy change for a thermal reservoir, and applies to all processes whether they are reversible or not. However we do need to be careful of the sign of Q for the reservoir. Q can either be + or - depending on whether heat is gained or lost by the reservoir.

Now, consider the entropy change of the universe for the irreversible process shown in A). As before we get the same entropy change for Reservoir 1 as in equation 5.51, but since the heat, Q_1, is transferred out of Reservoir 1 to Reservoir 2, Q_1 is positive for Reservoir 2:

$$\Delta S_{R2} = \frac{Q_1}{T_2} \qquad 5.52$$

so, for the universe:

$$\Delta S_{total} = Q_1 \left(\frac{1}{T_2} - \frac{1}{T_1} \right) \qquad 5.53$$

For $T_1 > T_2$, ΔS_{total} is greater than zero as expected. Note that the process where $T_2 > T_1$ with heat flowing from Reservoir 1 to Reservoir 2 would give ΔS_{total} less than zero which is impossible. When the flow is spontaneous, heat must flow from high T to low T.

Now, recall that we have had heat engines working between two thermal reservoirs, and if we had used a reversible heat engine to take in Q_1, we would have done work. So we can look at the process of transferring heat via the irreversible process in A) in Figure 5.g.1 as a process which loses work because we could have done work which would have been the maximum amount if we used it in a reversible engine. Let's put some numbers to the process to do some calculations. See Figure 5.g.2 where we have redrawn Q_2 to give our formal sign convention when we work on the engine as our system. We expect to compute Q_2 to be negative.

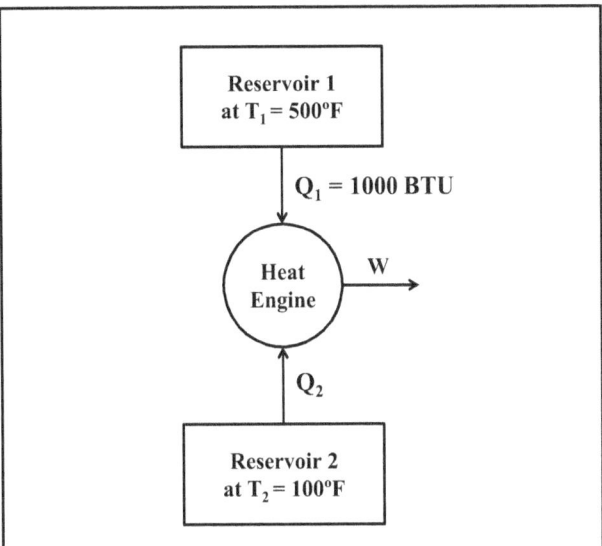

Figure 5.g.2. A Heat Engine operating between two thermal reservoirs.

For the engine as the system:

$$\Delta U = 0 = Q_1 + Q_2 - W$$

$$\eta = \frac{W}{Q_1} = \frac{T_1 - T_2}{T_1} \qquad 5.54$$

$$W = Q_1 \left(\frac{T_1 - T_2}{T_1}\right) = (1000 \text{ BTU})\left(\frac{960°R - 560°R}{960°R}\right) = 417 \text{ BTU}$$

$$417 = 1000 + Q_2 \quad \text{or} \quad Q_2 = -583 \text{ BTU}$$

Using Reservoir 1 as the system:

$$\Delta S_{R1} = \frac{-1000 \text{ BTU}}{960°R} = -1.04 \frac{\text{BTU}}{°R} \qquad 5.55$$

Using Reservoir 2 as the system:

$$\Delta S_{R2} = \frac{583 \text{ BTU}}{560°R} = 1.04 \frac{\text{BTU}}{°R} \qquad 5.56$$

ΔS for the engine is zero, and we see that for the universe, $\Delta S_{total} = \Delta S_{R1} + \Delta S_{R2} = 0$ indicating (as expected) a reversible process overall.

Suppose we didn't have an engine and just transferred 1000 BTU from Reservoir 1 to 2. Then equation 5.53 is applicable, and:

$$\Delta S_{total} = 1000 \text{ BTU}\left(\frac{1}{560°R} - \frac{1}{960°R}\right) = 0.744 \text{ BTU/°R} \qquad 5.57$$

In this case, we would have "lost" 417 BTU of work (noted W_{lost}) that the reversible engine could have done and we would have a corresponding increase in the entropy of the universe.

We can put these results into more general equations. For the irreversible case with no work done, equation 5.53 applies:

$$\Delta S_{total} = Q_1\left(\frac{1}{T_2} - \frac{1}{T_1}\right) \qquad 5.53$$

And for the reversible case:

$$\Delta S_{total} = 0 \text{ and } W = T_2 Q_1\left(\frac{1}{T_2} - \frac{1}{T_1}\right) \qquad 5.58$$

From equations 5.53 and 5.58, we can see that we could have computed the lost work as:

$$W_{lost} = T_2 \Delta S_{total} \qquad 5.59$$

This is a general principle that we can use for other cases. If we can compute ΔS_{total} for any irreversible process, multiplying that entropy change by the temperature of the reservoir where rejected heat is traveling will result in the work which could have been done if the system could have been operated in a reversible manner. Generally, the reservoir where rejected heat is traveling is the temperature of the atmosphere or the surroundings.

5.h   Chapter 5 Example Problems

5.1   Compute Δs and Δh for CO (carbon monoxide) in it's ideal gas state when it changes from 25°C and 1 bar to 200°C and 12 bar.

Solution:

For Δs, equation 5.47 can be applied since it is specified to be an ideal gas.

$$\Delta s = \int_{298}^{473} \frac{R(2.7963 + 0.0017182T - 6.3511 \times 10^{-7}T^2 + 8.2914 \times 10^{-11}T^3 + 21075.2T^{-2})}{T} dT$$

$$- R \cdot \ln\left(\frac{12}{1}\right)$$

This integral form has already been evaluated and is given in equation 5.49. **Δs = -7.1618 J/mole·K**

$$\Delta h = \int_{298}^{473} R(2.7963 + 0.0017182T - 6.3511 \times 10^{-7}T^2 + 8.2914 \times 10^{-11}T^3 + 21075.2T^{-2}) dT$$

Δh = 5117.3 J/mole

5.2   A turbine operating in an environment at 25°C receives air at 1500 K and 1000 kPa then expands it to 951 K and 100 kPa. If the turbine is adiabatic, find the lost work.

Solution:

Equation 5.59 tells us that $w_{lost} = T\Delta s$ where T is the temperature of surroundings = 298 K. We will assume air acts as an ideal gas and we will approximate it's composition as 21% O_2, 79% N_2. For the process:

$$\Delta s = \int_{1500}^{951} (0.79 C_{p,N_2}^{IG} + 0.21 C_{p,O_2}^{IG}) \frac{dT}{T} - R \ln\left(\frac{100}{1000}\right)$$

Getting constants for N_2 and O_2 from Appendix B, integrating as in equation 5.49, we get:

for N_2:
$$\int_{T_1}^{T_2} \frac{C_p^{IG}}{R} \frac{dT}{T} = 2.8185 \ln(951/1500) + 0.0015534(951 - 1500) + \frac{-5.2779 \times 10^{-7}}{2}(951^2 - 1500^2)$$
$$+ \frac{6.3840 \times 10^{-11}}{3}(951^3 - 1500^3) + \frac{22375.7}{-2}\left(\frac{1}{951^2} - \frac{1}{1500^2}\right)$$

for O_2:
$$\int_{T_1}^{T_2} \frac{C_p^{IG}}{R} \frac{dT}{T} = 3.0643 \ln(951/1500) + 0.0016886(951 - 1500) + \frac{-6.8712 \times 10^{-7}}{2}(951^2 - 1500^2)$$
$$+ \frac{1.0656 \times 10^{-10}}{3}(951^3 - 1500^3) + \frac{2393.3}{-2}\left(\frac{1}{951^2} - \frac{1}{1500^2}\right)$$

evaluating and plugging into equation 5.47:
Δs = 0.79·(-15.3231) + 0.21·(-16.2230) - 8.314·(-2.3026)
Δs = 3.6316 J/mole·K

therefore:
w_{lost} = 298 K · 3.6316 J/mole·K = 1082 J/mole

5.3 A cyclic heat engine employs an ideal gas with a constant heat capacity to perform the following cycle:
a. adiabatic compression from P_1-v_1-T_1 to P_2-v_2-T_2
b. isobaric expansion from P_2-v_2-T_2 to P_3-v_3-T_3
c. adiabatic expansion from P_3-v_3-T_3 to P_4-v_4-T_4
d. isobaric compression from P_4-v_4-T_4 to P_1-v_1-T_1

Sketch the process on the P-v plane and find the thermal efficiency of the process.

Solution:

A sketch of the process is given in Figure 5.h.1. The thermal efficiency of the process is given by:

$$\eta = \frac{w_{net}}{q_{abs}}$$

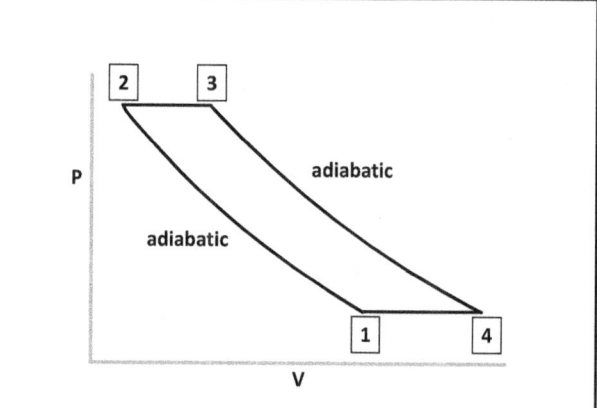

Figure 5.h.1. P-v diagram for the process given in Chapter 5 example problem 5.3.

since the process is cyclic, $\Delta u_{overall} = 0$ so $w_{net} = q_{net}$ for a whole cycle. Heat must be added in step b because volume is going up while holding P constant. Therefore, $q_b = q_{abs} = C_p(T_3-T_2)$. Likewise, heat must be rejected in step d to get the volume to go down while holding P constant. $q_d = q_{rej} = C_p(T_1-T_4)$.

No heat is added or subtracted in step a or step c, so $q_{net} = q_b + q_d = w_{net}$. Substituting:

$$\eta = \frac{C_p(T_3 - T_2) + C_p(T_1 - T_4)}{C_p(T_3 - T_2)} = 1 - \frac{T_1\left(\frac{T_4}{T_1} - 1\right)}{T_2\left(\frac{T_3}{T_2} - 1\right)}$$

with $P_2 = P_3$ and $P_1 = P_4$:

$$\frac{P_2}{P_1} = \frac{P_3}{P_4} = \left(\frac{T_2}{T_1}\right)^{\frac{\gamma}{\gamma-1}} = \left(\frac{T_3}{T_4}\right)^{\frac{\gamma}{\gamma-1}} \quad \text{so} \quad \frac{T_2}{T_1} = \frac{T_3}{T_4} \quad \text{and} \quad \frac{T_3}{T_2} = \frac{T_4}{T_1}$$

plugging in above:

$$\eta = 1 - \frac{T_1}{T_2} = 1 - \left(\frac{P_1}{P_2}\right)^{\frac{\gamma-1}{\gamma}}$$

5.4 **Neopentane is initially at 100°C and 500 bar. It then is subjected to a process and the**

entropy goes up so that $\Delta s = 30$ J/mole·K. If the process is isothermal, what is the new pressure? If the process is isobaric, what is the new temperature? Assume neopentane acts as an ideal gas.

Solution:

For the isothermal case, $\Delta s = 30$ J/mole·K $= -8.314$ J/mole·K·ln(P_2/500 bar). This gives **P_2 = 13.55 bar.**

For the isobaric case:

$$\frac{\Delta s}{R} = -0.0020\ln\left(\frac{T_2}{373}\right) + 0.0568490(T_2 - 373) + \frac{-2.5753\times10^{-5}}{2}\left(T_2^2 - 373^2\right)$$

$$+ \frac{4.7545\times10^{-9}}{3}\left(T_2^3 - 373^3\right) + \frac{-22821.8}{-2}\left(\frac{1}{T_2^2} - \frac{1}{373^2}\right) - \ln\left(\frac{500}{500}\right)$$

This equation is solved iteratively for T_2 and result is **T_2 = 450.2 K.**

5.5 Diethylether initially at 200°C and 1 bar is compressed adiabatically and reversibly until it reaches 20 bar. As we will learn, this process takes place at constant entropy because q = 0 and reversible means no frictional losses. Assuming diethylether is an ideal gas and taking $\Delta s = 0$ for this process, find T_2.

Solution:

Using constants from Appendix B, plugging into equation 5.47 and integrating:

$$\frac{\Delta s}{R} = -0.8695\ln\left(\frac{T_2}{473}\right) + 0.0539234(T_2 - 473) + \frac{-3.0436\times10^{-5}}{2}\left(T_2^2 - 473^2\right)$$

$$+ \frac{6.7814\times10^{-9}}{3}\left(T_2^3 - 473^3\right) + \frac{148717.4}{-2}\left(\frac{1}{T_2^2} - \frac{1}{473^2}\right) - \ln\left(\frac{20}{1}\right)$$

This equation is solved iteratively for T_2 and result is **T_2 = 548.7 K.**

5.6 A furnace burning natural gas acts as a thermal reservoir at 1200°C, and transfers heat to a heat exchanger. See Figure 5.h.2 for a schematic of the flow diagram. Using the steam table, find Δh, Δs, and $\dot{Q}$. Find Δs for the reservoir, compute the entropy

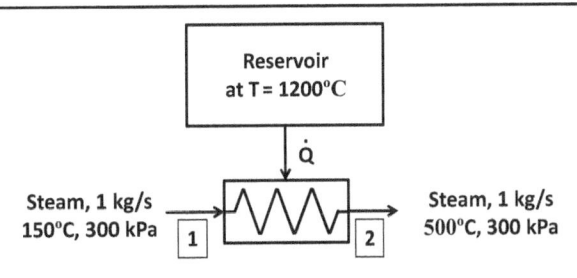

Figure 5.h.2. Flow diagram for the process given in Chapter 5 example problem 5.5.

change of the universe, and find the lost work for the surroundings temperature of 300K.

Solution:

From the steam table, h_1 = 2761.2 kJ/kg, s_1 = 7.0791 kJ/kg·K, h_2 = 3486.6 kJ/kg, and s_2 = 8.3271 kJ/kg·K. **Δh = 725.4 kJ/kg and Δs = 1.248 kJ/kg·K**.

Using the First Law applied to the heat exchanger:

$$\dot{m}\Delta h = \dot{Q} - \dot{W}_s \quad 1\,kg/s \cdot 725.4\,kJ/kg = \dot{Q} - 0$$

$$\dot{Q} = 725.4 \text{ kJ/s}$$

for the reservoir, the heat added is negative, so $\Delta \dot{s}_{res}$ = -725.4 kJ/s/1473 K = **-0.4945 kJ/s·K**.

$$\Delta \dot{s}_{univ} = -0.4945 \tfrac{kJ}{s \cdot K} + 1 \tfrac{kg}{s} \cdot 1.248 \tfrac{kJ}{kg \cdot K} = \mathbf{0.7555 \tfrac{kJ}{s \cdot K}}$$

$$\dot{W}_{lost} = 300K \cdot 0.7555 \tfrac{kJ}{s \cdot K} = \mathbf{226.7 \tfrac{kJ}{s}}$$

Student Problems:

1. Compute Δs and Δh for the following substances in their ideal gas state:

a. N_2, 300 K, 1 bar to 600K, 12 bar
b. CO_2, 500 K, 1 bar to 400 K, 8 bar
c. m-xylene, 150°C, 50 kPa to 300°C, 80 kPa
d. carbon disulfide, 600 K, 10 bar to 400 K, 5 bar

2. Methylchloride undergoes a constant entropy process (Δs = 0) starting at 100°C, 150 kPa and ending at 450 kPa. What is the final temperature, the enthalpy change, and the work done assuming the process is reversible and adiabatic?

3. A compressor operating in an environment at 25°C receives air at 50°C and 100 kPa compressing it to 350°C and 750 kPa. If the compressor is adiabatic, find the lost work.

Chapter 6 - Relationships Among Thermodynamic Properties for Pure Fluids

6.a Fundamental Relationships

Beginning with the First Law for a reversible process in differential form, $du = dq - dw$ (exact differentials for q and w since the process is reversible are appropriate), we can use $dw = Pdv$. But we have two possibilities for dq:

1. $dq = C_p dT$

2. $dq_{rev} = Tds$

Let's use the latter and substitute into the First Law:

$$du = Tds - Pdv \qquad 6.1$$

or multiplying through by moles:

$$\boxed{dU = TdS - PdV} \qquad 6.2$$

Even though equations 6.1 and 6.2 were developed for a reversible process, they are applicable to *any* process since they consist solely of state functions. Let's dissect the underlying meaning of equation 6.2:

$$\boxed{\text{Change in Energy Content of the System}} = \boxed{\text{Thermal Potential}} \times \boxed{\text{Thermal Change}} + \boxed{\text{Mechanical Potential}} \times \boxed{\text{Mechanical Change}}$$

Equations 6.1 and 6.2 should be seen as a combination of the First and Second Laws of Thermodynamics. Now, using equation 2.8 and differentiating:

$$\boxed{H = U + PV} \qquad 2.8$$

$$dH = dU + PdV + VdP \qquad 6.3$$

adding equations 6.2 and 6.3:

$$\boxed{dH = TdS + VdP} \qquad 6.4$$

Similar to the way we defined enthalpy, we define the free energy functions:

$$A = U - TS \quad \text{Helmholtz Free Energy} \tag{6.5}$$

$$G = H - TS \quad \text{Gibbs Free Energy} \tag{6.6}$$

Differentiating and substituting equation 6.5 or 6.6 as appropriate:

$$dA = -SdT - PdV \tag{6.7}$$

$$dG = -SdT + VdP \tag{6.8}$$

Equations 6.2, 6.3, 6.7, and 6.8 relate the most important of the thermodynamic functions, U, H, A, G, and S to each other and to temperature and pressure. Dividing these equations through by either moles or mass yields the molar or specific property relations. These equations are very useful and will be referred to throughout the remainder of this book.

6.b   Maxwell Relations

Recall that for an exact function $z = f(x,y)$ with differential written $dz = Mdx + Ndy$:

$$\left(\frac{\partial M}{\partial y}\right)_x = \left(\frac{\partial N}{\partial x}\right)_y \tag{5.5}$$

Let's apply equation 5.5 to equations 6.2, 6.3, 6.7, and 6.8, respectively:

$$\left(\frac{\partial T}{\partial V}\right)_S = -\left(\frac{\partial P}{\partial S}\right)_V \tag{6.9}$$

$$\left(\frac{\partial T}{\partial P}\right)_S = -\left(\frac{\partial V}{\partial S}\right)_P \tag{6.10}$$

$$\left(\frac{\partial P}{\partial T}\right)_V = \left(\frac{\partial S}{\partial V}\right)_T \tag{6.11}$$

$$\boxed{\left.\frac{\partial V}{\partial T}\right)_P = -\left.\frac{\partial S}{\partial P}\right)_T} \qquad 6.12$$

Equations 6.9 - 6.12 are known as the *Maxwell Relations*. These relations allow us to transform difficult derivatives to other derivatives. For example, in equation 6.11, the entropy derivative on the right hand side seems of mathematical interest only. However, that can be transformed to the change in pressure with temperature holding volume constant. Such a derivative could be imagined by simply putting a gas in a constant volume container, raising the temperature of the container and determining how the pressure changes.

6.c   T, P, and V Explicit Forms of the Energy Functions

In this section, we will find expressions for the energy functions H and S as functions of T, P, V and derivatives thereof. From H and S, we can extract U, A, and G using equations 2.8, 6.5 and 6.6. From these relationships, we can extract all the energy functions from an equation of state. Let's start with H, beginning with equation 6.4:

$$dH = TdS + VdP \qquad 6.4$$

If we divide this equation by dP and restrict to constant T, we get:

$$\left.\frac{\partial H}{\partial P}\right)_T = T\left.\frac{\partial S}{\partial P}\right)_T + V \qquad 6.13$$

Using equation 6.12 (one of the Maxwell Relations), we can substitute into equation 6.13 to get:

$$\left.\frac{\partial H}{\partial P}\right)_T = V - T\left.\frac{\partial V}{\partial T}\right)_P \quad \text{or} \quad \left.\frac{\partial h}{\partial P}\right)_T = v - T\left.\frac{\partial v}{\partial T}\right)_P \qquad 6.14$$

This equation is extremely useful because the enthalpy of any pure substance can be extracted from an equation of state for that substance. So, let's try this for an ideal gas:

$$V = n\frac{RT}{P} \qquad 6.15$$

for constant n (moles):

$$\left.\frac{\partial V}{\partial T}\right)_P = n\frac{R}{P} \qquad 6.16$$

substituting in equation 6.14:

$$\left.\frac{\partial H}{\partial P}\right)_T = n\frac{RT}{P} - Tn\frac{R}{P} = 0 \qquad 6.17$$

Equation 6.17 tells us that enthalpy does not change with pressure holding temperature constant for an ideal gas. *Enthalpy is independent of pressure for an ideal gas.* We have used this fact without proof in previous chapters in this book, and we have given qualitative arguments why it should be true, but we now have proved this. Using equation 2.8 with 6.17, since $Pv = RT$, we also see that U is not a function of pressure for an ideal gas.

If we take $H = f(T,P)$, then the total derivative of H is given by:

$$dH = \left.\frac{\partial H}{\partial T}\right)_P dT + \left.\frac{\partial H}{\partial P}\right)_T dP \qquad 6.18$$

We quickly recognize the two partial derivatives in equation 6.18 (see equations 6.14 and 2.23) and we can write:

$$\boxed{dH = nC_P dT + \left[V - T\left.\frac{\partial V}{\partial T}\right)_P\right]dP \quad \text{or} \quad dh = C_P dT + \left[v - T\left.\frac{\partial v}{\partial T}\right)_P\right]dP} \qquad 6.20$$

For entropy, $S = f(T,P)$:

$$dS = \left.\frac{\partial S}{\partial T}\right)_P dT + \left.\frac{\partial S}{\partial P}\right)_T dP \qquad 6.21$$

If we go back to equation 6.4, divide by dT and restrict to constant P:

$$\left.\frac{\partial H}{\partial T}\right)_P = T\left.\frac{\partial S}{\partial T}\right)_P = nC_P \qquad 6.22$$

The other partial derivative in equation 6.21 that we want is one of the Maxwell Relations, equation 6.12. Substituting into equation 6.21 we get:

$$\boxed{dS = \frac{nC_P}{T}dT - \left.\frac{\partial V}{\partial T}\right)_P dP \quad \text{or} \quad ds = \frac{C_P}{T}dT - \left.\frac{\partial v}{\partial T}\right)_P} \qquad 6.23$$

From equations 6.20 and 6.23 we can derive any of the energy functions as functions of temperature and pressure if we know the heat capacity and an equation of state for the substance.

6.d Residual Properties

We have already developed the energy equations H and S as functions of temperature and pressure for ideal gases; see equations 3.40 and 5.47:

$$\Delta h = \int_{T_1}^{T_2} C_P^{IG} dT \qquad \text{(ideal gas) 3.40}$$

$$\Delta s = \int_{T_1}^{T_2} \frac{C_P^{IG}}{T} dT - R \ln\left(\frac{P_2}{P_1}\right) \qquad \text{(ideal gas) 5.47}$$

These functions form a baseline for the ideal gas that can be used in the analysis of the energy functions for real gases. In fact, we only have available heat capacities for the ideal gas state (equivalent to $P \to 0$), and we would need tables of C_P as a function of pressure if we wanted to apply equations such as 6.23 without using the ideal gas state as a baseline. To setup the analysis using the ideal gas as the baseline, we define a *residual property*, M^R such that:

$$M^R = M - M^{IG} \qquad 6.24$$

M can be any function such as V, H or S.

Let's first find an expression for H^R. We begin starting at zero pressure, then $H^R = 0$ because the substance is in it's ideal gas state. Consider then one particular temperature then integrate equation 6.14 over pressure to get:

$$\boxed{H^R = \int_0^P \left[V - T\left(\frac{\partial V}{\partial T}\right)_P\right] dP \quad \text{or} \quad h^R = \int_0^P \left[v - T\left(\frac{\partial v}{\partial T}\right)_P\right] dP} \qquad \text{(constant T) 6.25}$$

The note "constant T" on equation 6.25 reminds us that the integral is to be done at constant temperature. However, this does not mean that partial derivatives under the integral are done at constant T. The partial derivatives should treat T as any other variable in the equation from which the derivative is being derived. Treating S in exactly the same manner, we get:

$$\boxed{S^R = -\int_0^P \left[\left(\frac{\partial V}{\partial T}\right)_P - \frac{nR}{P}\right] dP \quad \text{or} \quad s^R = -\int_0^P \left[\left(\frac{\partial v}{\partial T}\right)_P - \frac{R}{P}\right] dP} \qquad \text{(constant T) 6.26}$$

Assuming we can evaluate the residual properties, let's see how to use equations 6.25 - 6.26.

Consider the process given in Figure 6.d.1. For the gas at 1 bar pressure, it must either be an ideal gas or corrected to ideal gas conditions at 1 bar just as the heat capacities in Appendix B have been corrected. For enthalpy, Step 1 is $H^{IG} - H = -H^R$ for State 1. Step 2 is $H - H^{IG} = H^R$ for State 2. Summing up the three steps, we get:

$$\Delta H = \int_{T_1}^{T_2} nC_p^{IG} dT + H_2^R - H_1^R$$

or

$$\Delta h = \int_{T_1}^{T_2} C_p^{IG} dT + h_2^R - h_1^R \qquad 6.27$$

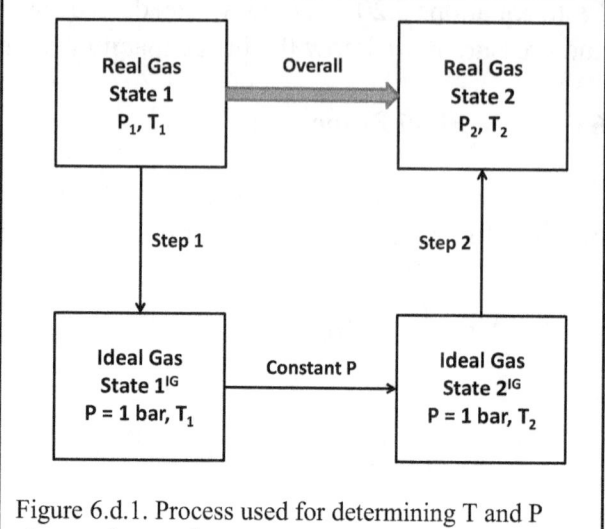

Figure 6.d.1. Process used for determining T and P changes for a real gas.

Similarly for entropy, we get:

$$\Delta S = \int_{T_1}^{T_2} \frac{nC_p^{IG}}{T} dT - nR\ln\left(\frac{P_2}{P_1}\right) + S_2^R - S_1^R \quad \text{or} \quad \Delta s = \int_{T_1}^{T_2} \frac{C_p^{IG}}{T} dT - R\ln\left(\frac{P_2}{P_1}\right) + s_2^R - s_1^R \qquad 6.28$$

All that is left is to evaluate the residual properties and we can evaluate the enthalpy and entropy changes for real gases.

6.d.i Residual Properties from Equations of State

Let's begin the analysis of residual properties from equations of state beginning with going back to the compressibility factor:

$$Pv = zRT \quad \text{or} \quad PV = znRT \qquad 3.5$$

We will treat n as constant. Then from equation 3.5:

$$\left(\frac{\partial V}{\partial T}\right)_P = \frac{nR}{P}\left[z + T\left(\frac{\partial z}{\partial T}\right)_P\right] \qquad 6.29$$

and plugging into equation 6.25:

$$H^R = \int_0^P \left[V - \frac{nRT}{P}z - \frac{nRT^2}{P}\left(\frac{\partial z}{\partial T}\right)_P\right] dP \qquad \text{(constant T) } 6.30$$

The first two terms in the bracket sum to zero in equation 6.30, then rearranging the remainder:

$$\boxed{\frac{H^R}{nRT}=-T\int_0^P \left(\frac{\partial z}{\partial T}\right)_P \frac{dP}{P} \quad \text{or} \quad \frac{h^R}{RT}=-T\int_0^P \left(\frac{\partial z}{\partial T}\right)_P \frac{dP}{P}}$$ (constant T) 6.31

similarly, evaluating S^R:

$$\boxed{\frac{S^R}{nR}=-T\int_0^P \left(\frac{\partial z}{\partial T}\right)_P \frac{dP}{P}-\int_0^P (z-1)\frac{dP}{P} \quad \text{or} \quad \frac{s^R}{R}=-T\int_0^P \left(\frac{\partial z}{\partial T}\right)_P \frac{dP}{P}-\int_0^P (z-1)\frac{dP}{P}}$$ (constant T) 6.32

Equations 6.31 and 6.32 are equivalent to equations 6.25 and 6.26 but in compressibility factor format instead of volume format.

Let's evaluate S^R and H^R using Van der Waals equation beginning with S^R. First, since Van der Waal's equation is explicit in pressure, it is simpler to evaluate partial derivatives of P rather than V which appears in equation 6.26. So, we can transform equation 6.26 to use derivatives of pressure beginning with the chain rule for partial derivatives applied to volume, temperature and pressure:

$$\left(\frac{\partial V}{\partial T}\right)_P \left(\frac{\partial T}{\partial P}\right)_V \left(\frac{\partial P}{\partial V}\right)_T = -1 \qquad 6.33$$

rearrange, then multiply by dV and restrict the whole equation to constant T:

$$\left(\frac{\partial V}{\partial T}\right)_P dP = -\left(\frac{\partial P}{\partial T}\right)_V dV \qquad \text{(constant T) 6.34}$$

With this substitution, equation 6.26 can be written:

$$S^R = \int_\infty^V \left(\frac{\partial P}{\partial T}\right)_V dV + nR\int_0^P \frac{dP}{P} \qquad \text{(constant T) 6.35}$$

In equation 6.35, since we have changed the left side of the expression for S^R from an integral in pressure to an integral in volume, the limits of the integral now reflect the appropriate volume limits. Now for Van der Waal's equation, equation 3.9, and multiplying through by n or n^2 to get V's instead of v's:

$$P=\frac{RT}{v-b}-\frac{a}{v^2}=\frac{nRT}{V-nb}-\frac{n^2 a}{V^2} \qquad 3.9$$

$$\left.\frac{\partial P}{\partial T}\right)_V = \frac{nR}{V-nb} = \frac{R}{v-b} \qquad 6.36$$

$$S^R = \int_\infty^V \frac{nR}{V-nb}dV + nR\int_0^P \frac{dP}{P} \qquad 6.37$$

The ∞ and 0 limits are going to give some problems with respect to the integrals being finite. Looking ahead, we will find that the two infinities will offset each other. To handle this problem, we will replace the 0 and ∞ limits temporarily with P^* and V^*, respectively, perform the integrations, collect terms, then apply the limiting conditions $P^* \to 0$ and $V^* \to \infty$. As such:

$$S^R = \int_{V^*}^V \frac{nR}{V-nb}dV + nR\int_{P^*}^P \frac{dP}{P} \qquad 6.38$$

$$S^R = nR\ln\left(\frac{V-nb}{V^*-nb}\right) + nR\ln\left(\frac{P}{P^*}\right) \qquad 6.39$$

$$S^R = nR\ln\left[\frac{P(V-nb)}{P^*(V^*-nb)}\right] = nR\ln\frac{P(V-nb)}{P^*V^* - P^*nb} \qquad 6.40$$

Now, take the limit as $P^* \to 0$ and $V^* \to \infty$. $P^*nb \to 0$ and $P^*V^* \to nRT$ since it will approach ideal gas behavior as $P \to 0$. The final result is:

$$S^R = nR\ln\left[\frac{P(V-nb)}{nRT}\right] \quad \text{or} \quad s^R = R\ln\left[\frac{P(v-b)}{RT}\right] \qquad 6.41$$

Working in a similar fashion using equation 6.25, we find:

$$H^R = PV - nRT - \frac{n^2 a}{V} \quad \text{or} \quad h^R = Pv - RT - \frac{a}{v} \qquad 6.42$$

A procedure for using equations 6.41 and 6.42 is to find values of a and b from critical constants, then Van Der Waal's equation must be solved for the volume at the desired temperature and pressure. As a reminder, the equation is cubic, and it will generally yield 3 roots, the largest of which is generally the desired root when working with gases that are not near the saturation dome. See Chapter 14, section c for an exact determination of when to use the upper real root as a vapor root. The volume along with the constants a and b then go into the equations for the residuals.

Other cubic equations of state such as Peng-Robinson and Redlich-Kwong and higher order equations of state such as Beattie-Bridgeman and Benedict-Web-Rubin can be evaluated in a similar fashion, though the mathematics is more extensive. The reader is referred to the open literature for results of these calculations. Walas [1] is a good source for such results.

6.d.ii   Residual Properties from Corresponding States

In chapter 3, we introduced the Lee-Kesler tables to use with the compressibility factor correlation:

$$z = z^0 + \omega z^1 \qquad 3.27$$

This equation can be applied in equations 6.31 and 6.32 to yield equations for use with the Lee-Kesler tables. To evaluate equations 6.31 and 6.32, we need the derivative:

$$\left.\frac{\partial z}{\partial T}\right)_P = \left.\frac{\partial z^0}{\partial T}\right)_P + \omega \left.\frac{\partial z^1}{\partial T}\right)_P \qquad 6.43$$

From here, we need take appropriate derivatives of the data in the Lee-Kesler table to substitute into equation 6.43, then we need to evaluate the integrals in equations 6.31 and 6.32 with the differentiated data to get h^R and s^R. This work was done in the original article by Lee and Kesler [2] and values for the dimensionless quantities similar to s^{R0}/R, s^{R1}/R, h^{R0}/RT, and h^{R1}/RT were tabulated. These values have been adapted to nomenclature used in this book and are included Appendix D. These can be used with:

$$\frac{h^R}{RT_c} = \frac{h^{R0}}{RT_c} + \omega \frac{h^{R1}}{RT_c} \qquad 6.44 \qquad\qquad \frac{s^R}{R} = \frac{s^{R0}}{R} + \omega \frac{s^{R1}}{R} \qquad 6.45$$

Knowing reduced temperature and pressure, values for h^R and s^R may be obtained from the table and used in equations 6.27 and 6.28 to evaluate Δh and Δs for real gases. This procedure is generally of sufficient accuracy for many engineering calculations and can be applied to a wide variety of substances for which we have P_c, T_c, and ω.

The Pitzer Generalized virial coefficient and supporting equations in equations 3.33 - 3.37 have also already been presented:

$$z = 1 + \frac{BP}{RT} = 1 + B_r \frac{P_r}{T_r} \qquad 3.33 \qquad\qquad B_r = \frac{BP_c}{RT_c} = B^0 + \omega B^1 \qquad 3.34\text{-}3.35$$

$$B^0 = 0.083 - \frac{0.422}{T_r^{1.6}} \qquad 3.36 \qquad B^1 = 0.139 - \frac{0.172}{T_r^{4.2}} \qquad 3.37$$

This correlation can also be adapted easily to find h^R and s^R, and in fact, students are encouraged to try using equation 3.33 in equations 6.31 and 6.32 to extract h^R and s^R. Temperature derivatives of B^0 and B^1 from equations 3.36 and 3.37 are required, which yields:

$$\frac{dB^0}{dT_r} = \frac{0.675}{T_r^{2.6}} \qquad 6.45 \qquad \frac{dB^1}{dT_r} = \frac{0.722}{T_r^{5.2}} \qquad 6.46$$

these are used in conjunction with:

$$\frac{h^R}{RT_c} = P_r \left[B^0 + \omega B^1 - T_r \left(\frac{dB^0}{dT_r} + \omega \frac{dB^1}{dT_r} \right) \right] \qquad 6.47$$

$$\frac{s^R}{R} = -P_r \left(\frac{dB^0}{dT_r} + \omega \frac{dB^1}{dT_r} \right) \qquad 6.48$$

6.e   Two Phase Systems and the Clapeyron Equation

Consider a system consisting of a pure component with two phases, α and β, in equilibrium with each other. Suppose we hold the temperature of the system constant by enclosing the system in a frictionless piston/cylinder arrangement and immersing it in a thermostated bath. Since the α phase can convert to the β phase and vice versa, we can change the volume while the pressure also remains constant. If we add or remove heat (which comes from the thermostated bath) at constant T and P, we will convert some of phase α to β or vice versa. Since T and P are constant, equation 6.8 tells us that this process of converting phase α to β while maintaining phase equilibrium occurs at constant Gibbs free energy, g. Thus, $g^\alpha = g^\beta$. This must be true or g for the whole system would change when converting one phase into the other. If we consider a differential amount of phase α converted to β, then $dg^\alpha = dg^\beta$. Applying equation 6.8:

$$v^\alpha dP^{sat} - s^\alpha dT = v^\beta dP^{sat} - s^\beta dT \qquad 6.49$$

Here, P^{sat} is the saturation pressure at which phase equilibrium occurs. Equation 6.49 tells us that if we change T, we must change P^{sat} in order to maintain phase equilibrium, and solving equation 6.49, we see:

$$\frac{dP^{sat}}{dT} = \frac{\Delta s^{\alpha\beta}}{\Delta v^{\alpha\beta}} \qquad 6.50$$

Since we can imagine a reversible process of adding heat at constant temperature and pressure and converting one phase to the other, $\Delta s^{\alpha\beta} = q^{rev}/T$ and $q^{rev} = \Delta h^{\alpha\beta}$ which is the enthalpy change associated with the phase change. Equation 6.50 can therefore be written:

$$\boxed{\frac{dP^{sat}}{dT} = \frac{\Delta h^{\alpha\beta}}{T\Delta v^{\alpha\beta}}} \qquad 6.51$$

Equation 6.51 is known as the Clapeyron equation.

Let suppose that we want to apply equation 6.51 to vapor-liquid equilibrium (VLE). Then, P^{sat} is also known as the vapor pressure of the substance, $\Delta h^{\alpha\beta} = \Delta h^{vap}$ which is the heat of vaporization of the liquid, and $\Delta v^{\alpha\beta}$ is the difference between the volume of the vapor and the liquid phases. Suppose we make three approximations:

1. $v^{vap} \gg v^{liq}$
2. $v^{vap} = RT/P$ (ideal gas above the liquid)
3. Δh^{vap} is constant with temperature

With these approximations, equation 6.51 can be written:

$$\frac{dP^{sat}}{P^{sat}} = \frac{\Delta h^{vap}}{RT^2} dT \qquad 6.52$$

Integrating equation 6.52 using our third approximation, we arrive at the Clausius-Clapeyron equation:

$$\boxed{\ln\left(\frac{P_2^{sat}}{P_1^{sat}}\right) = -\frac{\Delta h^{vap}}{R}\left(\frac{1}{T_2} - \frac{1}{T_1}\right)} \qquad \text{(approximate) } 6.53$$

Note that equation 6.51 is an exact equation governing all types of phase equilibrium, whereas equation 6.53 is approximate and applies only to vapor-liquid equilibrium.

Equation 6.53 has a number of applications. First, if we were to measure two different saturation (vapor) pressures at two different temperatures for a pure component, then we could extract an approximation for the heat of vaporization for the substance. Also, equation 6.53 says that $\ln P^{sat}$ is linear in $1/T$, and thus, a linear correlation for vapor pressure as a function of temperature could be extracted. However, a somewhat more accurate correlation known as the Antoine equation has been derived from the Clausius-Clapeyron equation and is used often to correlate vapor pressures:

$$\ln P^{sat} = A - \frac{B}{T+C} \qquad 6.54$$

In equation 6.54, A, B, and C are constants and Appendix E has these constants tabulated for a number of pure components.

6.f Chapter 6 Example Problems

6.1 Find Δh and Δs for the following components using the Lee-Kesler Table (Appendix D) for the EOS:
a.   1-Butanol with $T_1 = 500$ K, $P_1 = 5$ bar, $T_2 = 600$ K, $P_2 = 50$ bar.
b.   Ammonia with $T_1 = 300°C$, $P_1 = 100$ bar, $T_2 = 25°C$, $P_2 = 1$ bar.
c.   Fluoromethane with $T_1 = 150$ K, $P_1 = 200$ bar, $T_2 = 600$ K, $P_2 = 200$ bar.

Solution:

This problem is a straightforward calculation since we know T and P:

	a. 1-Butanol		b. Ammonia		c. Fluoromethane	
T_c (K)	561.9		405.5		317.28	
P_c (bar)	44.2		113.59		59.06	
ω	0.597		0.261		0.207	
	State 1	State 2	State 1	State 2	State 1	State 2
T (K)	500	600	573	298	150	600
P (bar)	5	50	100	1	200	200
T_r	0.890	1.068	1.413	0.735	0.473	1.891
P_r	0.113	1.131	0.880	0.009	3.386	3.386
values of h^{R0}, h^{R1}, s^{R0}, and s^{R1} from Appendix D via interpolation						
h^{R0}/RT_c	-0.143	-1.62	-0.506	-0.018	-5.377	-0.976
h^{R1}/RT_c	-0.169	-0.788	-0.099	-0.03	-9.308	0.289
s^{R0}/R	-0.108	-1.164	-0.259	-0.016	-4.774	-0.42
s^{R1}/R	-0.166	-0.743	-0.139	-0.033	-11.825	-0.105
h^{R0} (J/mole)	-669.07	-7577.39	-1705.89	-60.68	-14183.81	-2574.56
h^{R1} (J/mole)	-789.51	-3681.25	-333.76	-101.14	-24553.26	762.34
s^{R0} (J/mole·K)	-0.9009	-9.6775	-2.1533	-0.1330	-39.6910	-3.4919
s^{R1} (J/mole·K)	-1.3801	-6.1773	-1.1556	-0.2744	-98.3131	-0.8730
h^R (J/mole)	-1140.73	-9776.61	-1793.14	-87.12	-19267.83	-2416.71

s^R (J/mole·K)	-1.7254	-13.3679	-2.4554	-0.2047	-60.0478	-3.6726

Ideal Gas Result, equations 3.40 and 5.47

Δh^{IG} (J/mole)	17554.48		-10980.31	19346.37
Δs^{IG} (J/mole·K)	12.81		12.50	54.80

Real Gas Result, equations 6.27 and 6.28

Δh (J/mole)	8918.60		-9274.29	36197.50
Δs (J/mole·K)	1.16		14.75	111.17

6.2 The Joule-Thompson coefficient is defined as:

$$\mu = \left.\frac{\partial T}{\partial P}\right)_h$$

Show that an equivalent expression is:

$$\mu = \frac{RT^2}{C_P}\left.\frac{\partial z}{\partial T}\right)_P$$

Solution:

Begin by writing out the chain rule for μ from it's definition:

$$\left.\frac{\partial T}{\partial P}\right)_h \left.\frac{\partial P}{\partial h}\right)_T \left.\frac{\partial h}{\partial T}\right)_P = -1$$

The first partial in this expression is μ, and we recognize the third partial derivative as C_P. Going back to equation 6.14, we find an expression for the reciprocal of the partial derivative in the middle above:

$$\left.\frac{\partial h}{\partial P}\right)_T = v - T\left.\frac{\partial v}{\partial T}\right)_P \qquad 6.14$$

using $Pv = zRT$, $v = zRT/P$, and:

$$\left.\frac{\partial v}{\partial T}\right)_P = \frac{RT}{P}\left.\frac{\partial z}{\partial T}\right)_P + \frac{zR}{P}$$

Plugging back into equation 6.14:

$$\left.\frac{\partial h}{\partial P}\right)_T = \frac{zRT}{P} - \frac{RT^2}{P}\left.\frac{\partial z}{\partial T}\right)_P - \frac{zRT}{P} = -\frac{RT^2}{P}\left.\frac{\partial z}{\partial T}\right)_P$$

Plugging all the partials back into the chain rule and rearranging, we find:

$$\eta = \frac{RT^2}{C_P}\left.\frac{\partial z}{\partial T}\right)_P \qquad \text{Q.E.D.}$$

6.3 **Given that an expression for the Joule-Thompson coefficient is:**

$$\mu = \frac{RT^2}{C_P}\left.\frac{\partial z}{\partial T}\right)_P$$

Find an expression for μ in terms of B, the generalized virial coefficient of Pitzer. Then, calculate a value for μ for O_2 gas at 300 K and 40 atm using the Pitzer generalized virial coefficient as the EOS for oxygen.

Solution:

The generalized virial coefficient of Pitzer is B which comes from:

$$z = 1 + \frac{BP}{RT}$$

Accordingly:

$$\left.\frac{\partial z}{\partial T}\right)_P = \frac{P}{R}\left(-\frac{B}{T^2} + \frac{1}{T}\frac{dB}{dT}\right) = \frac{P}{RT}\left(\frac{dB}{dT} - \frac{B}{T}\right)$$

With this expression and using:

$$\eta = \frac{RT^2}{C_P}\left.\frac{\partial z}{\partial T}\right)_P$$

we find:

$$\eta = \frac{T}{C_P}\left(\frac{dB}{dT} - \frac{B}{T}\right)$$

Using data from Appendix C, for O_2 at 300 K and 40 atm = 40.53 bar, T_r = 300 K/ 154.6 K = 1.94 and P_r = 40.53 bar/50.46 bar = 0.803 and ω = 0.026. We also have previously:

$$B_r = \frac{BP_c}{RT_c} = B^0 + \omega B^1 \qquad 3.30$$

$$B^0 = 0.083 - \frac{0.422}{T_r^{1.6}} \qquad 3.31$$

$$B^1 = 0.139 - \frac{0.172}{T_r^{4.2}} \qquad 3.32$$

$$\frac{dB^0}{dT_r} = \frac{0.675}{T_r^{2.6}} \qquad 6.45$$

$$\frac{dB^1}{dT_r} = \frac{0.722}{T_r^{5.2}} \qquad 6.46$$

Plugging in the reduced properties and ω for O_2, we find B = -15.4 cm³/mole and dB/dT = 0.199 cm³/mole·K. Using C_p/R = 3.5364 given at 298 K for O_2 in Appendix B, we have:

$$\eta = \frac{300K}{25.47 \frac{J}{mole \cdot K}} \left(0.199 \frac{cm^3}{mole \cdot K} - \frac{-15.4 \frac{cm^3}{mole}}{300K} \right) \cdot \frac{0.1J}{cm^3 \cdot bar} = \mathbf{0.295 \ K/bar}$$

6.4 Given the following data for VLE of pure chlorine:

T (°F)	P^{sat} (psia)	v^{liq} (ft³/lbm)	v^{vap} (ft³/lbm)
40	61.276	0.01101	1.163
60	85.606	0.01126	0.8503
80	116.46	0.01154	0.6357
100	154.8	0.01185	0.4839
120	201.65	0.01218	0.3739

Find Δh^{vap}, Δs^{vap}, and Δg^{vap} at 80°F.

Solution:

There are two ways to approach this problem based on the Clapeyron equation, equation 6.51 and on the Clausius-Clapeyron equation, equation 6.53, derived from the Clapeyron equation. In order to use the Clapeyron equation, we need the derivative dP^{sat}/dT which we can obtain roughly as a derivative of the data about the desired temperature, 80°F:

$$\frac{dP^{sat}}{dT} \approx \frac{\Delta P^{sat}}{\Delta T} = \frac{85.606 - 154.80}{60 - 100} \frac{\text{psia}}{°R} = 1.73 \frac{\text{psia}}{°R}$$

Plugging into equation 6.51 and supplying the other data as necessary:

$$1.73 \frac{\text{psia}}{°R} = \frac{\Delta h^{vap}}{(80+460)° R (0.6357 - 0.01154) \frac{\text{ft}^3}{\text{lbm}}}$$

$$\Delta h^{vap} = 583 \frac{\text{psia} \cdot \text{ft}^3}{\text{lbm}} \cdot \frac{1.986 \frac{\text{BTU}}{\text{lbmole} \cdot °R}}{10.7317 \frac{\text{psi} \cdot \text{ft}^3}{\text{lbmole} \cdot °R}} = 108 \text{ BTU/lbm} = 7558 \text{ BTU/lbmole}$$

Since $\Delta h = q$ and $\Delta s = q/T$ for the constant T and P VLE process, $\Delta s^{vap} = \Delta h^{vap}/T = 108/(80 + 460) = 0.200$ BTU/lbm·°R. $\Delta g^{vap} = 108 - (80 + 460) \cdot 0.200 = 0$ as expected.

The other method uses the Clausius-Clapeyron equation, 6.53:

$$\ln\left(\frac{P_2^{sat}}{P_1^{sat}}\right) = -\frac{\Delta h^{vap}}{R}\left(\frac{1}{T_2} - \frac{1}{T_1}\right) = \ln\left(\frac{85.606}{154.80}\right) = -\frac{\Delta h^{vap}}{1.986 \frac{\text{BTU}}{\text{lbmole} \cdot °R}}\left(\frac{1}{520° R} - \frac{1}{560° R}\right)$$

This gives Δh^{vap} = 7999 BTU/lbmole, Δs^{vap} = 14.8 BTU/lbmole·°R = 0.212 BTU/lbm·°R, and Δg^{vap} = 0. Both results are similar, but both are approximate. To get a better value, a better numerical derivative of the data could be obtained and used in the Clapeyron equation.

6.5 Carbon monoxide (CO) is subjected to a compression process which takes place at constant entropy ($\Delta s = 0$). The initial state is 1 bar and 140 K and the final state is 10 bar. Find the final temperature and the enthalpy change for this process using the Lee-Kesler table for the EOS.

Solution:

The solution method for this process is iterative because we don't know the final temperature to evaluate the residual properties for s. So, we begin by assuming CO is an ideal gas to get an approximate final temperature to begin the iterative process. Using equation 5.47 for an ideal gas:

$$\Delta s = 0 = \int_{T_1}^{T_2} \frac{C_p^{IG}}{T} dT - R \ln\left(\frac{P_2}{P_1}\right) = \int_{140K}^{T_2} \frac{C_p^{IG}}{T} dT - R \ln\left(\frac{10 \text{bar}}{1 \text{bar}}\right)$$

Using constants from Appendix B we can assume T_2, calculate Δs, and iterate until we get

$\Delta s = 0$. As an example, try $T_2 = 200$ K:

$$\Delta s = R \int_{140K}^{200K} \frac{2.7963 + 0.0017182T - 6.3511 \times 10^{-7} + 8.2914 \times 10^{-11} + 21075.2/T^2}{T} dT$$

$$- R \ln\left(\frac{10 \text{bar}}{1 \text{bar}}\right) = -7.7675 \frac{J}{\text{mole} \cdot K}$$

To make $\Delta s = 0$, we need to try a higher T. Continuing with a new higher T then continuing and finishing the iteration, we find $\Delta s = 0$ when $T_2 = 259.9$ K. This is the ideal gas result. Now that we have a trial value for T_2, we can evaluate s^R_1 and s^R_2.

s^R_1 is evaluated at $T_r = 140$ K/132.9 K $= 1.053$ and $P_r = 1.0$ bar/34.98 bar $= 0.029$. Interpolating in the Lee-Kesler table we find:

$$s^R_1 = s^{R0} + \omega s^{R1} = R(-0.017 + 0.0533 \cdot -0.020) = -0.1502 \text{ J/mole} \cdot K$$

This value of s^R_1 is constant for the iteration on T_2.

s^R_2 is evaluated at $T_r = 259.9$ K/132.9 K $= 1.956$ and $P_r = 10$ bar/34.98 bar $= 0.286$. Interpolating in the Lee-Kesler table we find:

$$s^R_2 = s^{R0} + \omega s^{R1} = R(-0.033 + 0.0533 \cdot -0.012) = -0.2828 \text{ J/mole} \cdot K$$

Then:

Trial 1 with $T_2 = 259.9$ K: $\quad \Delta s = 0 + [-0.1502 - (-0.2828)]$ J/mole·K $= -0.1326$ J/mole·K

We now see that the residuals are quite small (near ideal gas at both the start and finish) and the two values s^R_1 and s^R_2 partially offset each other so that with $T_2 = 259.9$ K, Δs is nearly zero with the Lee-Kesler table corrections. A second iteration changes the trial temperature by only 0.2 K, so our result is good enough because this very small change in temperature would change s^R_1 and s^R_2 very little so **$T_2 = 261.1$ K**.

Now, Δh can be evaluated from equation 6.27. We need h^R_1 ($T_r = 1.053$, $P_r = 0.029$) and h^R_2 ($T_r = 1.965$, $P_r = 0.286$). Interpolating in the Lee-Kesler table and combining the results according to equation 6.44, we find $h^R_1 = -29.96$ J/mole and $h^R_2 = -90.53$ J/mole.

$$\Delta h = R \int_{140K}^{259.9K} \left(2.7963 + 0.0017182T - 6.3511 \times 10^{-7} + 8.2914 \times 10^{-11} + 21075.2/T^2\right) dT$$
$$+ (-90.53) - (-29.96)$$

giving **$\Delta h = 3621.48$ J/mole**.

6.6 **Given:**

$$\left.\frac{\partial u}{\partial v}\right)_T = T\left.\frac{\partial P}{\partial T}\right)_v - P \qquad 6.55$$

show that u is a function of T only for a gas which obeys the EOS:

P(v-b) = RT   where b is a constant

Solution:

Rearranging the EOS:

$$P = \frac{RT}{v-b}$$

then:

$$\left.\frac{\partial P}{\partial T}\right)_v = \frac{R}{v-b}$$

Plugging back in equation 6.55:

$$\left.\frac{\partial u}{\partial v}\right)_T = T\frac{R}{v-b} - P = P - P = 0$$

since u does not change with volume for this EOS, it must be a function of T only.

6.7   Find an expression for s^R for the Joule-Thompson EOS, $z = 1 + \alpha P/RT^3$ where α is a constant. Using this result, find Δs for argon when it is compressed isothermally from 1 bar to 100 bar at 200 K given $\alpha = -1.9 \times 10^6$ cm³·K²/mole at 200 K for argon.

Solution:

An equation which relates s^R to compressibility factor is;

$$\frac{s^R}{R} = -T\int_0^P \left.\frac{\partial z}{\partial T}\right)_P \frac{dP}{P} - \int_0^P (z-1)\frac{dP}{P} \qquad \text{(constant T) 6.32}$$

evaluating the partial using the Joule-Thompson EOS:

$$\left(\frac{\partial z}{\partial T}\right)_P = \frac{-3\alpha P}{RT^4}$$

Plug in to equation 6.32:

$$s^R = \frac{3\alpha P}{T^3} - \frac{\alpha P}{T^3} = \frac{2\alpha P}{T^3}$$

To evaluate Δs, since this process is isothermal, $\int C_P\, dT/T = 0$ leaving:

$$\Delta s = -R \ln\left(\frac{100\,\text{bar}}{1\,\text{bar}}\right) + s_2^R - s_1^R$$

Accordingly:

$$s_2^R = \frac{2\left(-1.9\times 10^6\, \frac{\text{cm}^3\cdot\text{K}^2}{\text{mole}}\right)(100\,\text{bar})}{200^3\,\text{K}^3} \cdot \frac{1\,\text{J}}{10\,\text{cm}^3\cdot\text{bar}} = -4.75\,\frac{\text{J}}{\text{mole}}$$

Similarly, $s_1^R = -0.0475$ J/mole. Then:

$\Delta s = -38.39 + (-4.75) - (-0.0475) = -42.99$ J/mole

6.8 Steam at 4000 kPa and 300°C is expanded isothermally to 2500 kPa. Compute Δh and Δs for this process using a) the steam table, and b) the Pitzer generalized virial coefficient as the EOS. Compare the results. Which result do you think is more accurate?

Solution:

a) Looking in the steam table, Δh = 3009.6 - 2961.7 = 47.9 kJ/kg and Δs = 6.6459 - 6.3639 = 0.2820 kJ/kg·K.

b) Using the heat capacity data for water in Appendix B, $\Delta h^{IG} = 0$ and Δs^{IG} = 3.9076 J/mole·K = 0.2171 kJ/kg·K. We now need to correct these for non-ideal gas behavior using the Pitzer generalized virial coefficient EOS. Accordingly, using equations 6.45 - 6.48 and equations for the generalized virial coefficient from chapter 3:

		State 1	State 2
T_c (K)	647.10		
P_c (kPa)	22064		
ω	0.351		
T(K)		573	573

P (kPa)	4000	2500
T_r	0.885	0.885
P_r	0.181	0.113
B^0	-0.4296	-0.4296
B^1	-0.1476	-0.1476
dB^0/dT_r	0.926	0.926
dB^1/dT_r	1.359	1.359
h^R/RT_c	-0.3125	-0.1953
s^R/R	-0.2543	-0.1590
h^R (J/mole)	-1681.29	-1050.81
s^R (J/mole·K)	-2.1147	-1.3217
Δh (J/mole)	630.48	
Δs (J/mole·K)	4.7006	

Or, in terms of units used in part a), **Δh = 35.03 kJ/kg and Δs = 0.2611 kJ/kg·K**. The comparison with the steam table values is okay for Δs but is quite rough for Δh, though we are dealing with relatively small changes in the properties because T is constant. But the EOS values are certainly less accurate than we would often want to achieve as we expect for the steam table to give the more accurate result because it is based upon actual data which has been checked, corrected, and corroborated over time.

6.9 **For methane gas at 350 K and 50 bar, use Van der Waals equation to calculate a) the molar volume, b) the residual enthalpy, h^R, and c) the residual entropy, s^R.**

Solution:

For methane, T_c = 190.6 K, P_c = 45.99 bar. Then from equations 3.20 and 3.21, the parameters for Van der Waals equation are:

$$a = 2.3035 \times 10^6 \frac{cm^6 \cdot bar}{mole^2} \quad b = 43.07 \frac{cm^3}{mole}$$

Plugging into Van der Waals equation:

$$50\,bar = \frac{83.14 \frac{cm^3 \cdot bar}{mole \cdot K} \cdot 350K}{v - 43.07 \frac{cm^3}{mole}} - \frac{2.3035 \times 10^6 \frac{cm^6 \cdot bar}{mole^2}}{v^2}$$

this equation is cubic in volume if it is multiplied out, and a quick way to solve it for v is iteratively, often looking for the largest real root if there are three real roots. However, in this case, it is clear T is above the critical T, so liquification is not possible and there should be only one real root. Iterating, we find **v = 547.5 cm³/mole**.

b) To find h^R, we use equation 6.42. Plugging in values:

$$h^R = 50\,\text{bar} \cdot 547.5\frac{\text{cm}^3}{\text{mole}} - 83.14\frac{\text{cm}^3\cdot\text{bar}}{\text{mole}\cdot\text{K}} \cdot 350\text{K} - \frac{2.3035\times 10^6\,\frac{\text{cm}^6\cdot\text{bar}}{\text{mole}^2}}{547.5\,\frac{\text{cm}^3}{\text{mole}}}$$

h^R = -5930 cm³·bar/mole = -593.0 J/mole

c) To find s^R, we use equation 6.41. Plugging in values:

$$s^R = 8.314\frac{\text{J}}{\text{mole}\cdot\text{K}}\ln\left(\frac{50\,\text{bar}\cdot(v-43.07\,\frac{\text{cm}^3}{\text{mole}})}{83.14\,\frac{\text{cm}^3\cdot\text{bar}}{\text{mole}\cdot\text{K}}\cdot 350\text{K}}\right)$$

s^R = -1.189 J/mole·K

6.10 Helium enters a compressor where it is compressed isothermally at 540°R from 12 atm to 150 atm. Calculate Δh and Δs assuming a) the gas behaves as an ideal gas and b) the gas obeys the EOS:

$$Pv = RT - a\frac{P}{T} + bP$$

given: a = 11.13 ft³·R/lbmole and b = 0.2445 ft³/lbmole for helium.

Solution:

a) For an ideal gas, h is a function of T only and since $T_2 = T_1$, **Δh = 0**. Δs = -Rln(P_2/P_1) for isothermal process with an ideal gas, so **Δs = 21.00 J/mole·K**.

b) The details of determining h^R and s^R from the EOS is left to the student as Student Problem 4 below. Results of the calculations show that:

$$h^R = -\frac{2aP}{T} + bP \quad \text{and} \quad s^R = -\frac{aP}{T^2}$$

So:

$$h_1^R = -\frac{2\cdot 11.13\,\frac{\text{ft}^3\cdot°\text{R}}{\text{lbmole}}}{540°\,\text{R}} + 0.2245\frac{\text{ft}^3}{\text{lbmole}}\cdot 12\,\text{atm} = 2.4393\frac{\text{ft}^3\cdot\text{atm}}{\text{lbmole}}$$

similarly, $h_2^R = 30.4917$ ft³·atm/lbmole.

$$s_2^R = -\frac{11.13 \frac{ft^3 \cdot °R}{lbmole} \cdot 12 atm}{540^2 (°R)^2} = -4.580 \times 10^{-4} \frac{ft^3 \cdot atm}{lbmole \cdot °R}$$

and $s_2^R = -5.725 \times 10^{-3}$ ft³·atm/lbmole·°R.

By equation 6.27, since the $\int C_P dT = 0$ because the process is isothermal, $\Delta h = 30.4917 - 2.4393 = 28.0523$ ft³·atm/lbmole = **177.44 J/mole.**

By equation 6.28, $\Delta s = -0.73024 \cdot \ln(150/12) + (-5.725 \times 10^{-3}) - (-4.580 \times 10^{-4}) = -1.8497$ ft³·atm/lbmole·°R = **-21.06 J/mole·K.**

6.11 Find Δh and Δs for n-pentane when it is heated and compressed from 400 K and 1 bar to 493 K and 27 bar using a) the ideal gas EOS, and b) the Lee-Kesler table for the EOS.

Solution:

a) For an ideal gas:

$$\Delta h = \int_{T_1}^{T_2} C_P^{IG} dT \qquad \text{(ideal gas) 3.40}$$

$$\Delta s = \int_{T_1}^{T_2} \frac{C_P^{IG}}{T} dT - R \ln\left(\frac{P_2}{P_1}\right) \qquad \text{(ideal gas) 5.47}$$

Using constants from Appendix B for n-pentane, we find **$\Delta h = 15507.24$ J/mole** and **$\Delta s = 7.3549$ J/mole·K** for the ideal gas.

b) Using the Lee-Kesler table, we find $\Delta h_1^R = -211.00$ J/mole, $\Delta h_2^R = -4442.77$ J/mole, $\Delta s_1^R = -0.3758$ J/mole·K, $\Delta s_2^R = -6.8625$ J/mole·K. **$\Delta h = 15507.24 + (-4442.77) - (-211.00) = 11275.47$ J/mole** and **$\Delta s = 7.3549 + (-6.8625) - (-0.3758) = 0.8682$ J/mole·K.**

Student Problems:

1. Provide the details for determining h^R from Van der Waals equation which results in equation 6.42.

2. Find an expression for h^R using the Joule-Thompson EOS as given in problem 6.7.

3. Find Δh and Δs for methanol (CH_3OH) for a process beginning at 1 bar, 100°C and ending

at 10 bar, 200°C using the Joule-Thompson EOS with $\alpha = -7.5 \times 10^7$ cm^3·K^2/mole for methanol. Ans: $\Delta s = -7.4087$ J/mole·K and $\Delta h = 4587.11$ J/mole.

4. Determine expressions for h^R and s^R given the EOS:

$$Pv = RT - a\frac{P}{T} + bP$$

and b in the EOS are constants.

5. The Joule-Thompson coefficient is defined as:

$$\mu = \left(\frac{\partial T}{\partial P}\right)_h$$

Show that $\mu = 0$ for an ideal gas, and determine μ for a gas that obeys the equation of state $P(v-b) = RT$ where b is a constant.

6.g. References

1. Walas, S. M.; "Phase Equilibrium in Chemical Engineering", Butterworth Publishers, Boston (1985).
2. Lee, B.I. and Kesler, M.G.; "A Generalized Thermodynamic Correlation Based on Three-Parameter Corresponding States", AIChE Journal 21, 510-527 (1975).

Chapter 7 - Macroscopic Balances for Process Equipment

7.a Conservation of Mass

Consider the generalized process system shown in Figure 7.a.1. We have an arbitrary number of flows entering the system and an arbitrary number of flows exiting the system. We assume that the flows are constant in composition and that they are each steady flows. Initially, there may be mass m_0 in the system. Consider the system as a black box so there can be any number of devices that produce or require work (pumps, turbines, etc.) and $\Sigma \dot{W}_s$ represents all of the various shaft work requirements for those devices. For the general case, we consider the possibility that

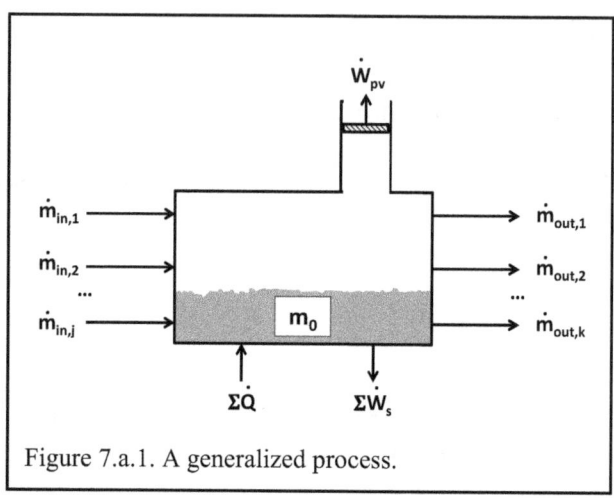

Figure 7.a.1. A generalized process.

the system could change volume, and a possible change in volume is depicted in the figure as a piston/cylinder arrangement with associated work $\dot{W}_{pv}$. An analogous way to treat this would be to treat it like a balloon that can inflate or deflate. In either case, work is done on the surroundings when the system expands. Similar to the various shaft work terms, $\Sigma \dot{Q}$ represents all of the various amounts of heat which could be added to various pieces of processing equipment such as boilers and heat exchangers.

For many considerations, we will assume a steady-state process, but that is not always the case. If the sum of inlet flows is different from the sum of outlet flows, the mass in the system will be changing with time. So, we write an unsteady-state mass balance as:

$$\frac{dm_{sys}}{dt} = \sum_{i=1}^{j} \dot{m}_{in,i} - \sum_{i=1}^{k} \dot{m}_{out,k} \qquad 7.1$$

Usually, we deal with flowrates that are constant with time as is assumed in equation 7.1. Suppose that the sum of mass that flows in is greater than the sum of mass that flows out, then equation 7.1 tells us, as expected, that the total mass in the system is increasing with time.

For steady-state, the mass in the system does not change with time which yields:

$$\sum_{i=1}^{j} \dot{m}_{in,i} = \sum_{i=1}^{k} \dot{m}_{out,k} \qquad \text{(steady-state) } 7.2$$

The same amount of mass must be flowing out as flows in or the system is not at steady-state. Equations 7.1 and 7.2 are known as *continuity equations*.

7.b   First Law Balance

In order to simplify the number of terms in equations that we are preparing to set up, we define two new terms:

$$H^{tot} = H + KE + PE \quad \text{and} \quad U^{tot} = U + KE + PE \qquad \text{7.3 and 7.4}$$

The associated equations given by dividing through by mass or moles also hold for equations 7.3 and 7.4.

Let's then consider the energy that exists in the system and the flow of energy in and out. Each of the flows, when multiplied by their respective h^{tot} represents flow of energy into or out of the system. The enthalpies automatically include the flow work associated with each flow stream, so we don't have to consider separate terms for flow work. Also, the system initially contains energy associated with the initial mass m_0. This is not a term involving flow, so we use u^{tot} for the energy associated with mass enclosed in the system. In general, we get for the unsteady-state balance:

$$\boxed{\frac{dE_{sys}}{dt} = \sum_{i=1}^{j} \dot{m}_{in,i} h^{tot}_{in,i} - \sum_{i=1}^{k} \dot{m}_{out,i} h^{tot}_{out,i} + \sum \dot{Q} - \sum \dot{W}_s - \dot{W}_{pv}} \qquad 7.5$$

Now, consider some simplifications. First, for the steady-state process, the volume of the system must be constant which makes $\dot{W}_{pv} = 0$. Clearly, the overall energy content of the system must not be changing with time, so that derivative can be set to zero. This yields what we often call the enthalpy balance:

$$\boxed{\sum_{i=1}^{k} \dot{m}_{out,i} h^{tot}_{out,i} - \sum_{i=1}^{j} \dot{m}_{in,i} h^{tot}_{in,i} = \sum \dot{Q} - \sum \dot{W}_s} \qquad \text{(steady-state) 7.6}$$

The left-hand side of equation 7.5 is a more general way of writing Δh for multiple flows going in and out of the system, and the right-hand side includes the possibility of multiple devices where heat is added and work is done. This equation should be very understandable in terms of the simpler equations developed in chapter 2.

Another very useful simplification is the single-inlet/single-outlet system. For steady-state, in this case, $\dot{m}_{in} = \dot{m}_{out}$ and:

$$\boxed{\dot{m}\Delta h^{tot} = \dot{Q} - \dot{W}_s}$$ (steady-state) 7.7

Next, let's consider an unsteady-state process such as filling a tank or emptying a compressed gas cylinder. We will assume that the contents of the system are uniform in composition, temperature, and pressure throughout (or spatially constant), though they may be changing with time. Let's begin by integrating the continuity equation over time for steady flows ($\dot{m}_i$ = constant):

$$\boxed{m - m_0 = \sum_{i=1}^{j} \int \dot{m}_{in,i} dt - \sum_{i=1}^{k} \int \dot{m}_{out,i} dt = \sum_{i=1}^{j} \dot{m}_{in,i} \Delta t - \sum_{i=1}^{k} \dot{m}_{out,i} \Delta t}$$ 7.8

The integrals, $\int \dot{m} dt$, could be evaluated if the flows are changing with time, and the result of the integration is simply total mass which has been flowed over the time period Δt just like the result $\dot{m}\Delta t$ is a total mass flowing in the time period Δt.

Let's dissect equation 7.8 by taking a simple example. Suppose we are filling a tank by adding water at 10 lbm/min while at the same time, 1 lbm/min is leaking out because we forgot to close a valve at the bottom of the tank. The unsteady-state continuity equation (equation 7.1) yields $dm_{sys}/dt = 10 - 1 = 9$ lbm/min. If we watch the filling process for a total of 5 minutes, equation 7.8 yields $m - m_0 = 10$ lbm/min x 5 min - 1 lbm/min x 5 min = 45 lbm. Over the time period of 5 minutes, we added a net of 45 lbm of water to the system.

Now, integrating equation 7.5 over a period of time 0 to t:

$$E_{sys} - E_{sys,0} = \sum_{i=1}^{j} \dot{m}_{in,i} h_{in,i}^{tot} \Delta t - \sum_{i=1}^{k} \dot{m}_{out,i} h_{out,i}^{tot} \Delta t + \sum \int_0^t \dot{Q} dt - \sum \int_0^t \dot{W}_s dt - \int_0^t \dot{W}_{pv} dt$$ 7.9

In this equation, we have again assumed steady flows which is typical, but consider $\dot{Q}$ and $\dot{W}_s$. For an example, if a thermal reservoir is adding heat to the system, as the temperature of the system approaches that of the thermal reservoir, the rate of flow of heat may slow as the difference in temperature driving force drops. Likewise, if a compressor is compressing gas to fill a tank, the work requirement for the compressor will likely increase with time as the pressure of the tank increases causing an increasing pressure at the exit of the compressor with time. Therefore, these terms have been left in integral form. However, note that each term in equation 7.9 carries units of energy such as kJ. The integral expressions may be represented by an average power applied over a period of time. That is, $\int_0^t \dot{Q} dt = Q$ which is a total amount of energy added over the time period. The same concept applies to W_s.

Next, consider the left-hand side of equation 7.9. The energy content of the system is the internal energy times the number of moles or mass. This gives the final result:

$$\boxed{\begin{aligned}mu_{sys} - m_o u_{sys,o} &= \sum_{i=1}^{j} \dot{m}_{in,i} h_{in,i}^{tot} \Delta t - \sum_{i=1}^{k} \dot{m}_{out,i} h_{out,i}^{tot} \Delta t + \sum \int_0^t \dot{Q} dt \\ &- \sum \int_0^t \dot{W}_s dt - \int_0^t \dot{W}_{pv} dt\end{aligned}} \qquad 7.10$$

Note that in equation 7.10, the terms $\dot{m}_i \Delta t$ could be replaced with m_i which is the total mass added to or exiting from the system over the time period. This replacement could also be done if the flow is not steady in which case m_i is the total mass added (or removed) over the time period Δt.

7.c Second Law Analysis of Processes

Going back to chapter 5 to the inequality of Clausius:

$$\Delta s \geq \int \frac{\delta q}{T} \quad \text{or} \quad ds \geq \frac{\delta q}{T} \qquad 5.44$$

for a system, we could rewrite this inequality as:

$$\Delta s_{sys} = \int \frac{\delta q}{T} + s_{gen} \qquad 7.11$$

where s_{gen} is a positive number and represents the entropy increase that accompanies an irreversible process. In this form, if $s_{gen} = 0$ then the process must be reversible, so the δq can be replace with dq. Written in the differential form, equation 7.11 yields:

$$ds_{sys} = \frac{dq}{T} + ds_{gen} \qquad 7.12$$

divide by dt to get:

$$\frac{ds_{sys}}{dt} = \frac{1}{T}\frac{dq}{dt} + \frac{ds_{gen}}{dt} \qquad 7.13$$

As in Figure 7.a.1, there may be multiple places in the system where heat is added, so anticipating application to that process, let's get equation 7.13 to reflect that possibility, and let's write the two derivatives on the right side as rates:

$$\frac{ds_{sys}}{dt} = \sum \frac{\dot{q}}{T} + \dot{s}_{gen} \quad \text{or} \quad \frac{dS_{sys}}{dt} = \sum \frac{\dot{Q}}{T} + \dot{S}_{gen} \qquad 7.14$$

Also we have not yet considered that there are flows bringing entropy in an out if we are going to apply it to a general process such as in Figure 7.a.1, so let's include those in equation 7.14 to give the final form of the unsteady-state entropy balance for a process:

$$\boxed{\frac{dS_{sys}}{dt} = \sum_{i=1}^{j} \dot{m}_{in,i} s_{in,i} - \sum_{i=1}^{k} \dot{m}_{out,i} s_{out,i} + \sum \frac{\dot{Q}}{T} + \dot{S}_{gen}} \qquad 7.15$$

Suppose we have a state-state process:

$$\boxed{\sum_{i=1}^{k} \dot{m}_{out,i} s_{out,i} - \sum_{i=1}^{j} \dot{m}_{in,i} s_{in,i} = \sum \frac{\dot{Q}}{T} + \dot{S}_{gen}} \qquad \text{(steady-state) } 7.16$$

and if there is a single inlet and single outlet:

$$\boxed{\dot{m}(s_{out} - s_{in}) = \sum \frac{\dot{Q}}{T} + \dot{S}_{gen}} \qquad \text{(steady-state) } 7.17$$

For the transient process taking place over a time period Δt, we need to include the entropy of the mass in the system at both the beginning and end of the time period:

$$\boxed{m s_{sys} - m_o s_{sys,o} = \sum_{i=1}^{j} \dot{m}_{in,i} s_{in,i} \Delta t - \sum_{i=1}^{k} \dot{m}_{out,i} s_{out,i} \Delta t + \sum \int_0^t \dot{Q} dt + \dot{S}_{gen}} \qquad 7.18$$

Comments applying to equations 7.9 and 7.10 regarding evaluating the terms in these equations also apply for equation 7.18.

7.d Irreversibility, Exergy, and Availability

In this section, we will determine expressions for some thermodynamic terms which are often applied to sustainability considerations. We begin by working with the term *irreversibility* which is defined as:

$$i = w_{s,rev} - w_s \qquad 7.19$$

which is clearly the amount of work which could have been done but was not because of irreversibilities. For the simple heat engine, this quantity is the same as the lost work which was developed in section 5.g, but here we wish to extend the definition to more general flow systems. Figure 7.d.1 shows the equivalent of a heat engine converted such that it is included in a flow process. We assume that heat is extracted from the high temperature reservoir and rejected to a low temperature reservoir just as the heat engine we studied in chapter 5, and we assume this is a steady-state process. We have simply added reference to a flow into and out of the device. We would first like to determine the maximum amount of work that can be done for this process, so let's suppose this is a reversible process.

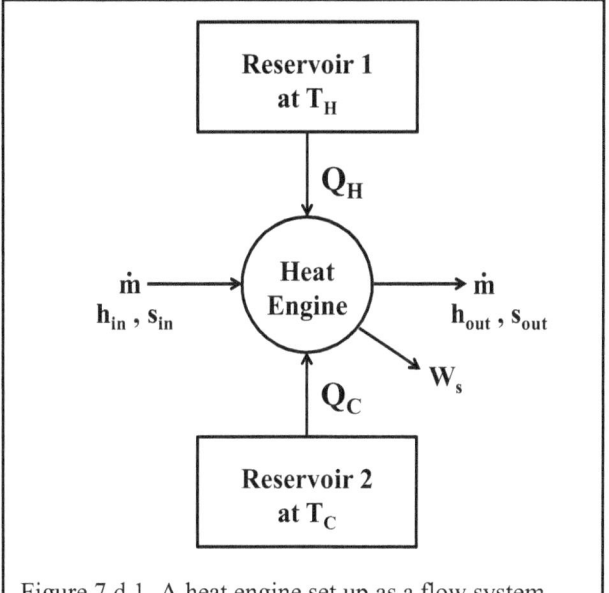

Figure 7.d.1. A heat engine set up as a flow system.

If we apply equation 7.17 to the process in Figure 7.d.1 assuming that the process is reversible ($\dot{S}_{gen} = 0$), we find:

$$\dot{m}(s_{out} - s_{in}) = \frac{\dot{Q}_H}{T_H} + \frac{\dot{Q}_C}{T_C} \qquad 7.19$$

after dividing through by $\dot{m}$, this can be rearranged to:

$$-q_C = q \frac{T_C}{T_H} - T_C(s_{out} - s_{in}) \qquad 7.20$$

Equation 7.6 (the First Law) gives:

$$\dot{m}\left(h_{out}^{tot} - h_{in}^{tot}\right) = \dot{Q}_H + \dot{Q}_C - \dot{W}_s \qquad 7.21$$

and dividing through by $\dot{m}$ and rearranging gives:

$$w_s = q_H + q_C - \left(h_{out}^{tot} - h_{in}^{tot}\right) \qquad 7.22$$

Substituting equation 7.20 into 7.22 gives the reversible work:

$$w_{s,rev} = -\left(h_{out}^{tot} - h_{in}^{tot}\right) + T_c\left(s_{out} - s_{in}\right) + q_H\left(1 - \frac{T_C}{T_H}\right) \qquad 7.23$$

In order to find an appropriate expression for the irreversible case, we note that we want to compare the reversible case to the case where q_H, s_{in}, h_{in}, s_{out}, and h_{out} are exactly the same as the reversible case. We find that in order to make this direct comparison, q_C must be zero. Thus:

$$w_s = q_H - \left(h_{out}^{tot} - h_{in}^{tot}\right) \qquad 7.24$$

Substituting equations 7.23 and 7.24 into equation 7.19 (the definition of irreversibility):

$$\boxed{i = w_{s,rev} - w_s = T_C\left(s_{out} - s_{in}\right) - q_H \frac{T_C}{T_H} = T_C\left(s_{out} - s_{in} - \frac{q_H}{T_H}\right)} \qquad 7.25$$

Alternately, rewriting equation 7.19 to include $\dot{S}_{gen}$ for the irreversible case, dividing by $\dot{m}$, and setting q_C zero:

$$\left(s_{out} - s_{in}\right) = \frac{q_H}{T_H} + s_{gen} \qquad 7.26$$

Comparing with equation 7.25, we find:

$$\boxed{i = T_C s_{gen}} \qquad 7.27$$

This is the same result as the lost work given by equation 5.58. Note that for the preponderance of cases, T_C is the temperature of the surroundings which is approximately 300K. If we can compute s_{gen} for any process, multiplying that value by the temperature of the surroundings gives the irreversibility and this is an upper limit for the energy savings that could be realized if process efficiency was improved to reduce i.

Exergy is also known as flow availability and is quickly derived from equation 7.23 with the following considerations. The quantity we are looking for in exergy is the maximum amount of energy available to do work in a flowing stream. The reversible work, $w_{s,rev}$ is not the correct quantity because some of that work derives from the heat absorbed in the process, whereas we want to associate exergy to the amount of available work that the flowing stream has without the provision of additional high temperature heat. Therefore, in equation 7.23 we will use $q_h = 0$.

Consider the other terms in equation 7.23. Clearly, if we wish to get the maximum amount of work out of a system that has both KE and PE, we want to convert all the KE to work so that the final state has zero KE or zero velocity. We also want to convert as much PE as possible to work, so the final state would leave the system at the reference height z_0. Making these substitutions in equation 7.23, we get the maximum potential for a flow stream to do work which is known as exergy, ψ, as:

$$\psi = \left(h_{in} - T_C s_{in} + \frac{v^2}{2g_c} + \frac{g}{g_c}z \right) - \left(h_{out} - T_C s_{out} + \frac{g}{g_c}z_0 \right) \qquad 7.28$$

When the exiting conditions are considered to be atmospheric temperature and pressure and h_{out} and s_{out} are determined at these conditions, the last term in parentheses in equation 7.28 is known as the dead state. A flow stream in this condition has no more potential to do work which is equivalent to $\psi = 0$.

7.e Steady-State Flow Devices

Let's consider some common pieces of equipment used for steady-state flow processes. In this section, we use "u" for velocity to distinguish velocity from molar or specific volume, and as such, "u" should not be confused with internal energy. We begin with the development of Bernoulli's equation which is a very important equation in fluid dynamics.

7.e.i Bernoulli's Equation

Recall equation 6.4:

$$dh = Tds + vdP \qquad 6.4$$

Apply this to a flowing system, and consider a reversible process:

$$dh = dq_{rev} + vdP \qquad 7.29$$

Integrating:

$$\Delta h = q + \int_{P_1}^{P_2} vdP \qquad 7.30$$

But, dividing equation 7.7 through by $\dot{m}$, we also have from the First Law for a flow process:

$$\Delta h + \Delta \left(\frac{u^2}{2g_c} \right) + \frac{g\Delta z}{g_c} = q - w_s \qquad 7.31$$

Substituting 7.30 into 7.31, eliminating q which appears on both sides of the equation, and solving for w_s:

$$-w_s = \int_{P_1}^{P_2} vdP + \Delta\left(\frac{u^2}{2g_c}\right) + \frac{g\Delta z}{g_c} \qquad 7.32$$

But equation 7.32 is for a reversible process. Generally in fluid dynamics, the irreversibilities, especially the frictional losses due to viscous dissipation, are accumulated into a single term called F for frictional loss:

$$\boxed{-w_s = \int_{P_1}^{P_2} vdP + \Delta\left(\frac{u^2}{2g_c}\right) + \frac{g\Delta z}{g_c} + F} \qquad 7.33$$

This is a common form for Bernoulli's equation. Often, this equation is simplified for flow of a liquid in a section of piping where there is no pump or other work device. Then, $w_s = 0$, and since we are dealing with a liquid, v can be considered constant so the integral may be evaluated. Using $v = 1/\rho$ where ρ is the density:

$$\boxed{\frac{\Delta P}{\rho} + \Delta\left(\frac{u^2}{2g_c}\right) + \frac{g\Delta z}{g_c} + F = 0} \qquad 7.34$$

Equation 7.34 is also a common form of Bernoulli's equation for flow of a liquid in a pipe. In addition to defining frictional losses in flow, it relates the pressure drop in the pipe to kinetic energy in an expanding or contracting cross-sectional area of the duct and to potential energy changes such as flow in a vertical pipe.

7.e.ii Maximum Velocity in a Pipe

Consider a gas flowing steadily in a pipe of constant cross-sectional area. For this situation:

$$\Delta h + \Delta\left(\frac{u^2}{2g_c}\right) = 0 \qquad 7.32$$

So, if we un-throttle the flow, we can convert enthalpy into kinetic energy and increase the velocity. However, there is a limit as to the amount of enthalpy which can be converted based upon Second Law restrictions. Differentiating 7.32:

$$dh = -\frac{udu}{g_c} \qquad 7.33$$

Also, the continuity equation governs how velocity increases as the molar volume decreases. For a duct (or pipe) with cross-sectional area A:

$$\dot{m} = \frac{uA}{v} \qquad 7.34$$

for $\dot{m}$ and A constant, differentiation of equation 7.34 yields:

$$du = \frac{u\,dv}{v} \qquad 7.35$$

Equation 7.35 relates a change in velocity to a change in molar volume. Substitution of equation 7.35 in 7.33 gives:

$$dh = -\frac{u^2}{g_c}\frac{dv}{v} \qquad 7.36$$

then, substitution of equation 7.36 into equation 6.4 provides a relationship to entropy:

$$Tds = -\frac{u^2}{g_c}\frac{dv}{v} - vdP \qquad 7.37$$

If the process is reversible, ds = 0 and the irreversible case gives ds > 0. Irreversibility corresponds to frictional losses, which could be large for high velocities, but evidently, the maximum velocity which can be achieved corresponds to ds = 0. For ds = 0:

$$\frac{u_{max}^2}{g_c}\frac{dv}{v} + vdP = 0 \qquad \text{(constant entropy) } 7.38$$

Rearrangement results in:

$$u_{max}^2 = -g_c v^2 \left(\frac{\partial P}{\partial v}\right)_s \qquad 7.39$$

This is the sonic velocity.

7.e.iii Valves

Consider a throttling device, which could be a partially closed valve or other flow restriction, in a flow process, see Figure 7.e.1. Equation 7.7 is applicable since there is one inlet and one outlet and we will assume steady-state. $\dot{Q}=\dot{W}_s=0$ since a flow restrictor neither takes in heat nor does any work on the surroundings. This leaves $\Delta h^{tot} = 0$. Though there will be a velocity change across the restriction due to a pressure drop if the flow is gaseous, for many applications the kinetic energy change across the restriction is small enough to be neglected. A horizontal valve will have $\Delta PE = 0$, and even if the valve is vertical, the potential energy change is small since the change in elevation is small. *With these caveats, a restrictor (or valve) can generally be characterized by $\Delta h = 0$.* The enthalpy entering the device is the same as the enthalpy leaving. So, knowing the inlet T and P, h_{in} can be determined, then $h_{out} = h_{in}$.

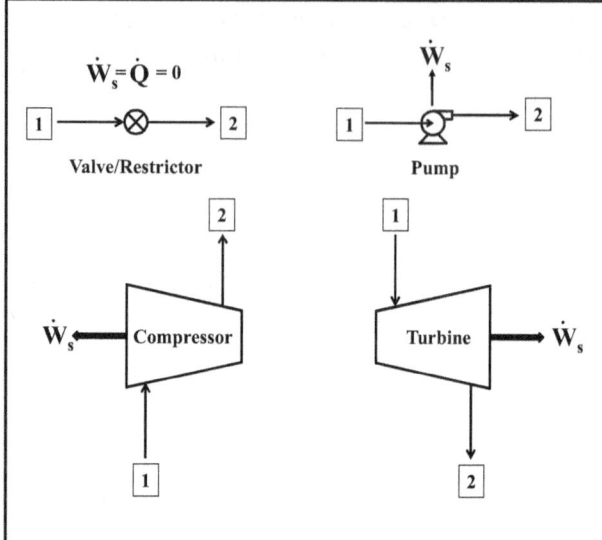

Figure 7.e.1. Four types of flow devices, generally assumed to be steady-state.

To establish the state of the system leaving the valve, we need to know just one other property. Often we know the exiting pressure which establishes the state of the system. Otherwise, we need to know one other property in addition to the enthalpy.

In some cases, the kinetic energy change cannot be neglected. Consider how to compute the velocity in a piping system so that velocity changes and therefore the kinetic energy effect can be computed. For a duct of any shape, the cross-sectional area of the duct is A, and we have:

$$\boxed{\dot{m}=\frac{u_1 A_1}{v_1}=\frac{u_2 A_2}{v_2}} \qquad 7.40$$

Equation 7.40 allows us to compute velocities and therefore velocity changes across a valve or other single-inlet/single-outlet piping device if we know the cross-sectional areas of the pipes leading to and leaving the system. For circular pipes:

$$\dot{m}=\frac{\pi r_1^2 u_1}{v_1}=\frac{\pi r_2^2 u_2}{v_2} \qquad 7.41$$

$$\frac{u_2}{u_1}=\frac{v_2}{v_1}\left(\frac{r_1}{r_2}\right)^2 \qquad 7.42$$

Equation 7.42 simplifies for some common situations. For liquids, v is independent of pressure so

v_1 and v_2 divide out. Also, we often have the same size inlet pipe as the outlet pipe. In that case, r_1 and r_2 divide out. Equations 7:40 - 7.42 allow calculation of KE differences between the inlet and outlet and they are applicable to other single-inlet/single-outlet processes.

7.e.iv Heat Exchangers

There are many configurations of heat exchangers and this book does not treat design and operation of these devices but we will consider the heat balances. Figure 7.e.2 shows two basic ways we can treat the process of exchanging heat. In the single stream arrangement, we treat the heat load, q, as coming from an external source. This may be radiation, a flame, cooling water, a heat reservoir, tracing on a pipe, air flow, or many other possibilities. But in this single stream configuration, we don't consider changes in whatever entity is used as the source of heat. We simply treat the heat transfer as q. Accordingly, this is a single-inlet/single-outlet

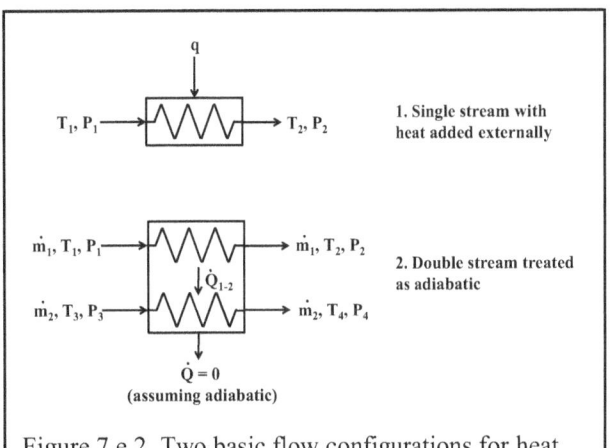

Figure 7.e.2. Two basic flow configurations for heat exchangers.

device with q added. If we assume steady-state, equation 7.7 applies with $\dot{W}_s = 0$. Generally, since the inlet is connected to the outlet with an open pipe, we assume that the pressure drop is zero (or negligible) so that $P_2 = P_1$. This not always the case but is the general assumption we usually make in the absence of other information.

Consider the other configuration shown in Figure 7.e.2 with a double stream. The internals of the exchanger are designed so that heat can flow easily from stream 1 to stream 2 and is shown in the figure as $\dot{Q}_{1-2}$ which is referred to as the heat load or duty for the heat exchanger. We could take either stream in the double stream configuration as the system, and the result would be exactly like the single stream case. Note that $\dot{Q}_{1-2}$ is positive for one stream and negative for the other since heat is added to one stream and lost from the other. However, we generally treat the whole device as a multi-stream steady-state device. In this case, equation 7.6 applies assuming steady-state operation. With $\dot{W}_s = 0$ and assuming no heat lost from the device as a whole, $\dot{Q} = 0$, so we have:

$$\dot{m}_1 \left(h_2^{tot} - h_1^{tot} \right) + \dot{m}_2 \left(h_4^{tot} - h_3^{tot} \right) = 0 \qquad 7.43$$

Often, we are trying to find one of the four temperatures given the flowrates and three temperatures, or we might be interested in one of the flowrates given all the temperatures. In most instances we neglect KE and PE differences, then the enthalpies can be evaluated at their respective temperatures and pressures. Note that either stream may be used to evaluate the heat load:

$$\left|\dot{Q}_{1-2}\right| = \left|\dot{m}_1(h_2 - h_1)\right| = \left|\dot{m}_2(h_4 - h_3)\right| \qquad 7.44$$

We have used the absolute value because the specific situation will determine which way the heat is flowing. Again, as in the case of the single stream configuration, we often take pressure drop as negligible for each stream so $P_2 = P_1$ and $P_4 = P_3$.

7.e.v Pumps

The term "pump" is usually applied to a device which increases the pressure of a flowing liquid stream, see Figure 7.e.1. For steady-state, single-inlet/single-outlet:

$$\dot{m}\Delta h^{tot} = \dot{Q} - \dot{W}_s \qquad 7.45$$

Pumps are not designed to add heat to a flowing stream, so usually we take the process to be adiabatic then $\dot{Q}$ is zero. Neglecting KE and PE effects, we get:

$$\Delta h = -w_s \qquad 7.46$$

Note that by our formal sign convention for heat and work, we expect the pump work to be negative, so Δh by equation 7.46 is expected to be positive. Going back to equation 7.33 and applying the equation for a liquid (v is constant with P) and neglecting KE, PE and F (frictional losses), we get:

$$-w_s = \int_{P_1}^{P_2} v\,dP = \Delta h = v(P_2 - P_1) \qquad 7.47$$

Equation 7.47 can be used to determine the work requirement for a pump, and it is also useful to consider how to correct enthalpies of liquids for the effects of pressure. Subcooled liquid water tables are sparse with data when they are available. In most cases where high accuracy is not necessary, when the enthalpy of a subcooled liquid is needed, it is recommended that the *enthalpy of the saturated liquid at the same temperature (but lower pressure) be substituted.*

Let's see what error such a substitution would impart by taking data from the steam tables. Looking up the enthalpy of the saturated liquid at 25°C, we have h = 104.83 kJ/kg at P^{sat} = 3.1699 kPa and v^{liq} = 0.001003 m³/kg (which we assume constant with pressure). Below is a table of values for T = 25°C at other pressures calculated via equation 7.47:

P (kPa)	vΔP (kJ/kg)	h (kJ/kg)
3.1699	0.000	104.83
100	0.097	104.93
500	0.498	105.33

P (kPa)	vΔP (kJ/kg)	h (kJ/kg)
1000	1.000	105.83

The calculated enthalpies given in the table agree to within 0.1 kJ/kg of tabulated values in the open literature. The results give a good representation regarding when a pressure correction needs to be applied when subcooled liquid data is needed. Engineering judgement and the required accuracy of calculations play into these considerations, but we can see that for pressures up to a few bar, the correction is minimal.

7.e.vi Compressors

A compressor is quite similar to a pump in the operation it accomplishes, but the term "compressor" applies to gases instead of liquids, see Figure 7.e.1. Beginning with equation 7.45, we first want to discuss $\dot{Q}$ in the context of minimizing work requirements for the compression process. Figure 7.e.3 shows a P-v diagram for an ideal gas. Isotherms are drawn in solid lines, while adiabats (long dashes) come from a combination of equations 3.56 and 3.57:

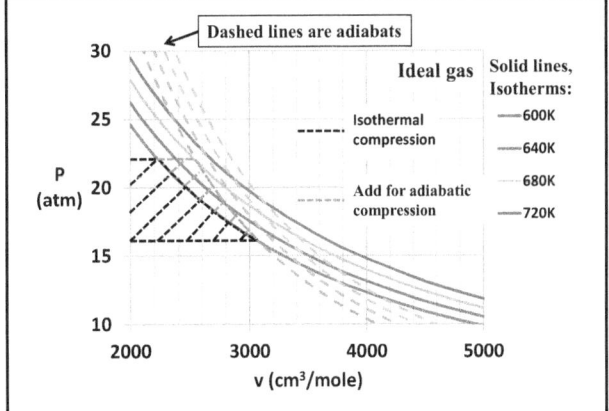

Figure 7.e.3. P-v Diagram for an ideal gas for consideration of isothermal vs. adiabatic compression.

$$\left(\frac{P_2}{P_1}\right)^{\frac{\gamma-1}{\gamma}} = \left(\frac{v_2}{v_1}\right)^{\gamma-1} \qquad 7.48$$

Now, consider the work requirement for isothermal compression. For example, suppose we want to compress an ideal gas from 16 to 22 atm isothermally at 600K. From equation 7.33 neglecting KE and PE effects, the work requirement is:

$$-w_s = \int_{P_1}^{P_2} v dP \qquad 7.49$$

This integral is depicted in the area highlighted by the black short dash region of Figure 7.e.3. If we do the compression adiabatically, the area determined for the isothermal compression *plus* the area represented by the red short dash region represents the integral. If we analyze the situation more closely, we can see that the adiabatic work requires energy (in the form of work) to both raise the pressure and raise the temperature of the gas. We can see from Figure 7.e.3 that for the adiabatic compression from 16 to 22 atm, the temperature of the gas would go from 600K to about 680K. Therefore for a smaller work requirement, an isothermal compression is better than an adiabatic compression. But an isothermal compression must reject heat to the surroundings to maintain the temperature at a constant value as the gas compresses.

Unfortunately, the mechanical process of compressing a gas is much faster than the process by which heat can be removed from a gas. Therefore, in order to do the compression isothermally, we would generally have to slow the compressor to a painfully slow mechanical process in order to reject the heat quick enough to maintain isothermality. As a result, compressors generally operate more closely to adiabatic operation rather than isothermal operation even though more work is required.

In this light, consider adiabatic reversible operation of a compressor. Since q = 0 and the process is reversible, $\Delta s = 0$ or $s_{out} = s_{in}$. Reversible, adiabatic operation is also termed *isentropic* operation. Typically we know T and P entering the compressor, so that establishes the state of the system and s_{in} is determined. s_{out} is then known, and we need only one more property to establish the state of the system, and this is usually P_2 though any other property could also be known. Alternatively, if we know how much power the compressor is consuming and the inlet conditions, the outlet conditions can be determined. Consider equation 6.28 as a computational tool for the situation for any gas for which we know C_P, T_c, P_c, ω, and P_1, T_1, and P_2:

$$\Delta S = \int_{T_1}^{T_2} \frac{nC_p^{IG}}{T} dT - nR\ln\left(\frac{P_2}{P_1}\right) + S_2^R - S_1^R \quad \text{or} \quad \Delta s = \int_{T_1}^{T_2} \frac{C_p^{IG}}{T} dT - R\ln\left(\frac{P_2}{P_1}\right) + s_2^R - s_1^R \qquad 6.28$$

We want $\Delta s = 0$ knowing P_1, T_1, and P_2 so:

$$0 = \int_{T_1}^{T_2} \frac{C_p^{IG}}{T} dT - R\ln\left(\frac{P_2}{P_1}\right) + s_2^R - s_1^R \qquad 7.50$$

Knowing P_1 and T_1 we can directly compute s_1^R. We then begin an iterative loop by guessing T_2 so we can compute a rough value for s_2^R. Now, the only thing we need to solve equation 7.50 is T_2. Once we find a value for T_2 which makes equation 7.50 zero, we have a better estimate of T_2 to evaluate a new value for s_2^R, then we can iterate again on T_2 to make equation 7.50 zero and we continue until the value we guessed for T_2 to evaluate s_2^R is the same value that solves equation 7.50. We now know P_1, T_1, P_2 and T_2. Equation 6.27 can then be solved for Δh:

$$\Delta h = \int_{T_1}^{T_2} C_p^{IG} dT + h_2^R - h_1^R \qquad 6.27$$

$\Delta h = -w_s$ completes the calculation of the work requirement for the reversible case.

If we have a table of T, P, h, and s values (such as the steam table), we look up s_1 and h_1 then we establish $s_2 = s_1$ for the reversible case. We search through the table at P_2 looking for s_2. Often, we must interpolate to get this value, but we now have T_2 and h_2 from the table at that same point. $h_1 - h_2$ gives the work requirement for the compressor.

7.e.vii Multi-Stage Compression Processes

In Figure 7.e.4 we show a two stage compression with interstage cooling process. Compressor 1 compresses the gas from P_1 to P_2. Assuming adiabatic reversible operation of the compressor, $T_2 > T_1$ leaving the compressor. Then the gas goes into the interstage cooler where heat is removed such that the temperature is returned back to the original inlet temperature, T_1. The gas enters a second compressor and leaves the process at T_3 and P_3. This process is shown on the ideal gas Pv plane in Figure 7.e.5. In the figure, we have taken $T_1 = 680K$, $P_1 = 16$ atm. Compressor 1 compresses adiabatically to about 19.5 atm, then the gas enters the interstage cooler and is cooled at *constant pressure* back to $T_1 = 680K$. Compressor 2 compresses adiabatically from 680K, 19.5 atm to 22 atm. Notice that the area enclosed in the black solid lines represents the work required for this compression sequence.

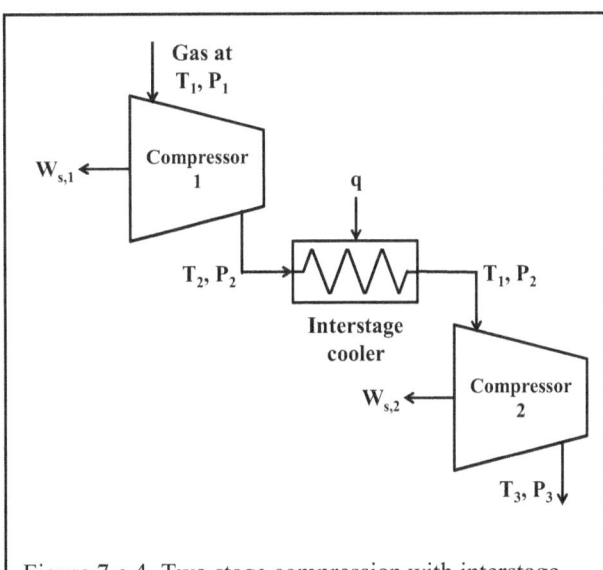

Figure 7.e.4. Two stage compression with interstage cooling.

Now, suppose we had accomplished the entire compression from $T_1 = 680K$, $P_1 = 16$ atm to 22 atm with a single adiabatic compressor. The blue area represents *additional* work that the single compressor would do compared to the two stage compression process with interstage cooling. The key to this savings is the interstage cooling. Notice that the interstage cooling step brings the two stage compression process closer to a single isothermal process in work requirement. We compared the work requirement for an adiabatic versus an isothermal compressor in reference to Figure 7.e.3, and an isothermal compression requires less work than the adiabatic compression

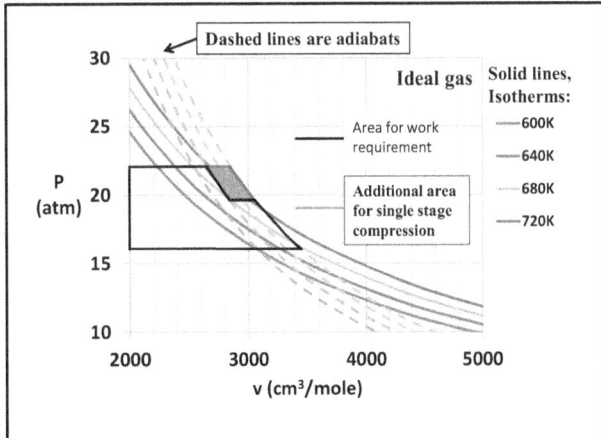

Figure 7.e.5. P-v Diagram for an ideal gas for the two stage compression process of Figure 7.e.4.

operating from the same initial temperature and pressure to the final pressure. It should be simple to see that dividing the adiabatic compression process into more and more adiabatic compression steps with interstage cooling between each compressor approaches the same work requirement as the single stage isothermal compression.

Now, let's ask the question why the interstage pressure in the example shown in Figure 7.e.5 was chosen to be ~19.5 atm. This interstage pressure was basically chosen by eye and as we shall see, this interstage pressure is not optimum (minimum) with respect to the least amount of work required. Let's frame this optimization problem by asking the question: "What interstage pressure gives the minimum total work required?" Let's set up the total work required as a function of interstage pressure then look for a minimum. For compressor 1, we take KE = PE = 0, q = 0 (adiabatic), reversible, and we assume that heat capacity is approximately constant giving:

$$-w_{s,1} = \Delta h \approx C_p^{IG}(T_2 - T_1) \qquad 7.51$$

and for compressor 2, we have an equivalent expression. Then summing the two work requirements:

$$-w_{s,tot} = -w_{s,1} - w_{s,2} = C_p^{IG}(T_2 - T_1 + T_3 - T_1) = T_1 C_p^{IG}\left(\frac{T_2}{T_1} + \frac{T_3}{T_1} - 2\right) \qquad 7.52$$

But we would like the work requirement in terms of pressure not temperature, so using equation 3.57 which relates temperature to pressure for a reversible adiabatic process with an ideal gas:

$$\frac{T_2}{T_1} = \left(\frac{P_2}{P_1}\right)^{\frac{\gamma-1}{\gamma}} \qquad 3.54$$

and substituting into equation 7.52, we get:

$$-w_{s,tot} = T_1 C_p^{IG}\left[\left(\frac{P_2}{P_1}\right)^{\frac{\gamma-1}{\gamma}} + \left(\frac{P_3}{P_2}\right)^{\frac{\gamma-1}{\gamma}} - 2\right] \qquad 7.53$$

We want the minimum work with respect to P_2 holding P_1 and P_3 constant:

$$\frac{dw_{s,tot}}{dP_2} = 0 \qquad 7.54$$

Taking the derivative, setting equal to zero and rearranging:

$$P_2^{\frac{2(\gamma-1)}{\gamma}} = P_1^{\frac{(\gamma-1)}{\gamma}} P_3^{\frac{(\gamma-1)}{\gamma}} \qquad 7.55$$

which results in:

$$P_2 = \sqrt{P_1 P_3} \qquad 7.56$$

Therefore, the correct intermediate pressure to minimize total work requirements is the *geometric mean of the inlet and outlet pressures* for the overall process. This assumes may things including that the interstage cooler cools the exhaust of the first compressor back to the same temperature as the temperature of the gas as it enters the process. Notice that in the case the intermediate pressure is given by equation 7.56, P_{out}/P_{in} for both compressors is the same and since T_{in} is the same for both compressors, that is equivalent to V_{out}/V_{in} (the compression ratio) is the same for both compressors. Also, equation 7.53 tells us the work requirement in each compressor is the same. These two compressors are basically the same compressor, though compressor 2 experiences higher pressures than compressor 1.

All of these arguments can be extended to multi-stage compression processes where the number of compressors is greater than 2. A three stage sequence would require two intermediate pressures with interstage cooling between compressors 1 and 2, and 2 and 3. The optimum intermediate pressures would derive from the cube root of the product of the inlet and outlet pressures for the whole process.

7.e.viii   Turbines

Turbines are basically governed by the same equations and same assumptions as compressors, but the work is positive for turbines and negative for compressors, see Figure 7.e.1. Nonetheless, a base case for turbines is that they are steady-state, adiabatic and reversible so that equation 7.46 is applicable. The calculations for turbines parallels calculations for compressors in section 7.e.vi in the according to basic assumptions, $s_{out} = s_{in}$.

7.e.ix   Thermal Efficiencies of Pumps, Compressors, and Turbines

In sections dealing with pumps, compressors, and turbines, we have dealt with reversible pieces of equipment up to this point. What do we do when the equipment is irreversible? To handle this question, we define the thermal efficiency, η, of these devices according to:

1.   Devices that require work (pumps and compressors):

$$\eta = \frac{w_s|_{\Delta s=0}}{w_s|_{actual}} \qquad 7.57$$

2.   Devices that produce work (turbines):

$$\eta = \frac{w_s|_{actual}}{w_s|_{\Delta s=0}} \qquad 7.58$$

For all devices, each work requirement in equation 7.57 and 7.58 is determined based upon the reversible and irreversible device *having the same inlet and outlet pressure*. Notice that for devices that require work, the actual work requirement for the actual, irreversible device is greater than the isentropic work requirement. On the other hand, for devices that produce work, the isentropic device produces more work than the actual, irreversible device. You can remember this by noting that these thermal efficiencies are defined such that $\eta \leq 1$, where $\eta = 1$ translates to the reversible device.

To solve for outlet properties for irreversible devices if we are given the efficiency, we solve the isentropic base case first, then use the thermal efficiency to determine the actual work. Since $-w_s = \Delta h$ we can determine the actual enthalpy change and therefore the exiting enthalpy. Given the exiting enthalpy and exiting pressure (thermal efficiencies are based upon the same inlet and out pressures), the exiting state is determined. If we know the actual work requirement and want the thermal efficiency, we solve the isentropic base case and use that result along with the actual work requirement in either equation 7.57 or 7.58.

7.f Chapter 7 Example Problems

7.1 A turbine like that shown in Figure 7.e.1 has a thermal efficiency of 0.75. Steam enters the turbine at $P_1 = 8000$ kPa, $T_1 = 600°C$ and exhausts at $P_2 = 400$ kPa. Find h_2, s_2, and T_2 for the exhaust steam and the (actual) work produced.

Solution:

We will make typical assumptions for this turbine: steady-state, $q = 0$ and $\Delta KE = \Delta PE = 0$. These typical assumptions will always be made unless there is information provided that allows us to include the situation in calculations.

The typical method used to solve a process such as this which has a thermal efficiency for a device including a pump, compressor, or turbine is to *solve the isentropic base case first*. The isentropic base case is for the adiabatic, reversible device so that $\Delta s = 0$ and this is why it is termed isentropic. For the isentropic base case, we use a prime, ', to embellish the quantities to distinguish them from the actual properties. So, let's begin by looking up the properties of the steam entering the turbine using the superheated steam table (Appendix G.3). $h_1(600°C, 8000\text{ kPa}) = 3642.4$ kJ/kg and $s_1(600°C, 8000\text{ kPa}) = 7.0221$ kJ/kg·K.

Now, for the isentropic base case, $s_2' = s_1 = 7.0221$ kJ/kg·K and $P_2 (= P_2') = 400$ kPa. Note that as stated previously, the thermal efficiency is based upon the same inlet and outlet pressure for the actual and isentropic cases. We are looking in the steam table in the block for $P = 400$ kPa and we search down the column for s looking for 7.0221 kJ/kg·K. We find this entry would be between $T = 150°C$ and $200°C$ where $s = 6.9306$ and 7.1723 kJ/kg·K, respectively. Interpolating:

s (kJ/kg·K)	T°C	h (kJ/kg)
6.9306	150	2752.8

s (kJ/kg·K)	T °C	h (kJ/kg)
7.0221	168.9	2793.7
7.1723	200	2860.9

So, $T_2' = 168.9°C$ and $h_2' = 2793.7$ kJ/kg. $\Delta h' = 2793.7 - 3642.4 = -848.7$ kJ/kg $= -w_s'$. A turbine that operated adiabatically and reversibly would produce 848.7 kJ/kg of work. This isentropic case corresponds to $\eta_{turbine} = 1$.

Now, the actual turbine will produce less work than the isentropic case, and we use the thermal efficiency of the turbine to find the actual work. **$w_s = 0.75 \times 848.7 = 636.5$ kJ/kg.** Note that the ' is gone - this is the actual work the turbine produces.

With $\Delta h = -w_s = -636.5$ kJ/kg, $h_2 - 3642.4 = -636.5$ kJ/kg, we find **$h_2 = 3005.9$ kJ/kg**. This is the actual enthalpy leaving the turbine rather than h_2'. To find T_2 and s_2, we go back to the steam table, again looking in the 400 kPa block in the superheated steam table, but now following down the column for h looking for the entry 3005.9 kJ/kg. We find that this entry is between 250 and 300°C. Interpolating:

h (kJ/kg)	T°C	s (kJ/kg·K)
2964.5	250	7.3804
3005.9	270.2	7.4560
3067.1	300	7.5677

So, **$T_2 = 270.2°C$ and $s_2 = 7.4560$ kJ/kg·K**. Note that s has increased for the actual turbine which is indicative of an irreversible process. When the thermal efficiency of a turbine is less than 1, we must have the entropy increase when q = 0.

7.2 **R-134a refrigerant is compressed from 10°C, 2 bar to 20 bar in a compressor that has a thermal efficiency of 0.8. The flowrate of R-134a is 0.2 kg/min. Find the actual compressor power requirement $\dot{W}_s$ and the properties of the refrigerant leaving the compressor, T_2, s_2, and h_2.**

Solution:

With $T_1 = 10°C$ and $P_1 = 2$ bar (200 kPa), we find in the R-134a table that $h_1 = 409.73$ kJ/kg and $s_1 = 1.7961$ kJ/kg·K. Next, solve the isentropic base case.

$s_2' = s_1 = 1.7961$ kJ/kg·K and with $P_2 = 20$ bar (2000 kPa), we see this point is between 90 and 100 °C. Interpolating, we find $h_2' = 463.10$ kJ/kg and $T_2' = 93.9°C$. $\Delta h' = 463.10 - 409.73 = 53.37$ kJ/kg $= -w_s'$.

Now we note a fundamental difference between a compressor and turbine. The compressor *requires* work which is correct by our calculation according to our formal sign convention, so the actual work requirement is derived by *dividing* the isentropic work requirement by η rather than multiplying as in the case for the turbine. The actual work requirement for a compressor is therefore greater than the isentropic work requirement. This yields $-w_s' = 53.37/0.8 = 66.71$ kJ/kg. $\dot{W}_s = 66.71$ **kJ/kg · 0.2 kg/min = 13.34 kJ/min = 0.222 kW**.

To find the final properties, $\Delta h = \Delta h'/0.8 = 53.37/0.8 = 66.71$ kJ/kg. $h_2 = \Delta h + h_1 = 409.73 + 66.71 = $ **476.44 kJ/kg**. We now have $h_2 = 476.44$ kJ/kg and $P_2 = 20$ bar (2000 kPa), so we look in the R-134a table and find this between 100 and 110°C. Interpolating, $T_2 = $ **105.1°C** and $s_2 = $ **1.8318 kJ/kg·K**.

7.3 **Carbon dioxide gas flows across a valve, entering at 30°C and 5000 kPa, and leaving at 600 kPa. Neglecting KE effects, use the CO_2 tables (Appendix M) to determine the exiting temperature and the entropy change for the process.**

Solution:

Entering properties are $h_1 = 451.44$ kJ/kg and $s_1 = 1.8691$ kJ/kg·K. For $\Delta KE = 0$, $\Delta h = 0$ across a valve, so $h_2 = h_1 = 451.44$ kJ/kg and we also know $P_2 = 600$ kPa. Looking in the 600 kPa block for CO_2, we see T is between -30 and -40°C. Interpolating, we find $T_2 = $ **-31.3°C** and $s_2 = 2.2050$ kJ/kg·K. $\Delta s = $ **0.3359 kJ/kg·K**.

7.4 **Repeat problem 7.3 but include a ΔKE calculation. Data for the calculation is that the valve has a 1" schedule 40 pipe leading into and out of it and the velocity of CO_2 entering the valve is 1 m/s. T_1, P_1, and P_2 are the same as in problem 7.3.**

Solution:

Now we wish to solve $\Delta h + \Delta ke = 0$. We have the inlet velocity, but we need the outlet velocity to compute Δke. In order to do this, let's use equation 7.34 to compute the mass flowrate through the device. We need the cross-sectional area of the pipe, so we look in the open literature and find that the inside diameter of a 1" schedule 40 pipe is 1.05". This translates to a 0.494 in² cross-sectional area which can be converted to 0.0003187 m². We also need the specific volume which can be found in the CO_2 table at 30°C and 5000 kPa giving $v_1 = 0.00806$ m³/kg. Plugging into equation 7.34:

$$\dot{m} = \frac{u \cdot A}{v} = \frac{1\frac{m}{s} \cdot 0.0003187\, m^2}{0.00806\, \frac{m^3}{kg}} = 0.0395 \frac{kg}{s}$$

The conservation of mass equation (or the continuity equation), equation 7.2, tells us that this is also the mass flowrate leaving the valve for steady-state operation. We can apply equation 7.34 again on the downstream side of the valve, but the temperature and pressure have changed. We have T_2 from the calculation we did in problem 7.3, but that calculation assumed $\Delta h = 0$ and we want $\Delta h + \Delta ke = 0$. However, we need at least an estimate of T_2 to get a value of v_2 to plug in to find the exiting velocity, so we will start an iterative process using T_2 we found in problem 7.3, $T_{2,\,guess1}$ = -31.3°C, and we know P_2 = 600 kPa. Looking in the table, we find v_2 = 0.07138 m³/kg. Pugging into equation 7.34:

$$\dot{m} = 0.0395 \frac{kg}{s} = \frac{u_2 \cdot 0.0003187\, m^2}{0.07138 \frac{m^3}{kg}}$$

Solving, u_2 = 8.847 m/s (remember that u in this context is velocity, not internal energy). Now we can do a Δke calculation:

$$\Delta ke = \frac{\left(8.847^2 - 1^2\right)\frac{m^2}{s^2}}{2 \cdot 1 \frac{kg \cdot m}{N \cdot s^2}} = 38.63 \frac{N \cdot m}{kg} = 38.63 \frac{J}{kg} = 0.03863 \frac{kJ}{kg}$$

Rather than going back into the table with a new h_2, which we could get from $\Delta h + \Delta ke$ = 0, we can estimate what temperature change this suggests using 0.03863 kJ/kg = 0.03863 J/g ≈ $C_p \Delta T$ ≈ 35 J/mole·K x 1 mole/44g x ΔT. This gives ΔT ≈ 0.048°C, which means T_2 would need to adjusted by 0.048°C and we could perform another loop. However, we note that v_2 will change very little with this small T correction, and we conclude **the kinetic energy effect is small enough to be neglected**.

7.5 Repeat problem 7.3 but use the heat capacity correlations of Appendix B in conjunction with the Lee-Kesler tables, Appendix D, for the EOS to find T_2 and s_2. Neglect the KE effect. Compare your answer to the result from problem 7.3.

Solution:

We still have Δh = 0, and since Δh is a function of T only for an ideal gas, ΔT = 0. So we will use an ideal gas guess, $T_2 = T_1$ = 30°C to start an iterative process. From Appendix C we find: T_c = 304.1 K, P_c = 73.77 bar, and ω = 0.272. Then:

	State 1	State 2
T(K)	303	303
P (bar)	50	6
T_r	0.996	0.996
P_r	0.678	0.081
h^{R0}/RT_c	-0.921	-0.086
h^{R1}/RT_c	-0.845	-0.080
s^{R0}/R	-0.669	-0.056
s^{R1}/R	-0.797	-0.071
h^{R0} (J/mole)	-2327.71	-216.21
h^{R1} (J/mole)	-2137.10	-201.46
s^{R0} (J/mole·K)	-5.5637	-0.4685
s^{R1} (J/mole·K)	-6.6274	-0.5868

h^R (J/mole)	-2908.92	-271.00
s^R (J/mole·K)	-7.3661	-0.6281

So, if T is constant across the valve, we plug into equation 6.27 and we get $\Delta h = 0 + (-271.00) - (-2908.92) = 2637.92$ J/mole. However, we are looking for $\Delta h = 0$. So, we need a lower T_2.

Try $T_2 = 0°C$. h^R_1 is the same because the inlet T doesn't change, so with $T_2 = 0°C$ and $P_2 = 6$ bar. This gives $h^R_2 = -338.82$ J/mole. $\int C_p dT = -1101.73$ J/mole. So $\Delta h = -1101.73 + (-338.82) - (-2908.92) = 1468.37$ J/mole.

We need a still lower T. Try $T_2 = -40°C$. This results in $h^R_2 = -511.90$ J/mole and $\int C_p dT = -2480.42$ J/mole. $\Delta h = -2480.42 + (-511.90) - (-2908.92) = -83.40$ J/mole. Now T_2 is a little high.

Try $T_2 = -38°C$. This results in $h^R_2 = -500.04$ J/mole and $\int C_p dT = -2414.37$ J/mole. $\Delta h = -2414.37 + (-500.04) - (-2908.92) = -5.49$ J/mole. We could do another iteration, but a change of 2°C resulted in a change of almost 80 J/mole in Δh, so the next guess would be fractions of a degree.

So, **$T_2 = -38°C$** and with that temperature, $s^R_2 = -1.5346$ J/mole·K for use in equation 6.28, which yields **$\Delta s = 14.460$ J/mole·K**. $T_2 = -31.3°C$ and $\Delta s = 0.3359$ kJ/kg·K = 14.780 J/mole·K is the result from problem 7.3 using the CO_2 table for comparison purposes. The comparison is reasonably good.

7.6 A desuperheater is a device that mixes subcooled liquid water with superheated steam to yield a saturated vapor. A schematic of such a device is shown in Figure 7.f.1. A flow of superheated vapor at 0.25 kg/s at 1800 kPa and 400°C is mixed with a subcooled liquid at 2000 kPa and 30°C to yield the saturated vapor. There is a small pressure drop in the apparatus so that the saturated vapor leaves at 1750 kPa. Find the flowrate of subcooled liquid, $\dot{m}_2$, and the flowrate of the saturated vapor leaving the device. Assume the device operates adiabatically.

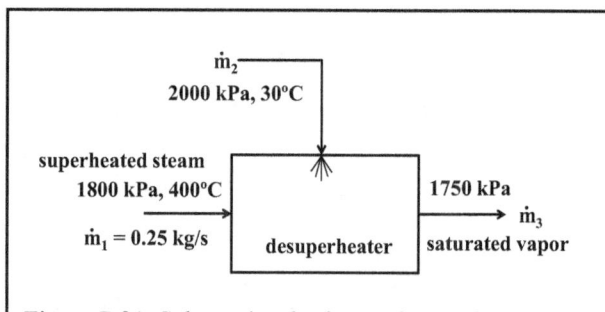

Figure 7.f.1. Schematic of a desuperheater for use with problem 7.6.

Solution:

Let's start with a material balance and apply equation 7.2 assuming state-state operation:

$$\sum \dot{m}_{in} = \dot{m}_1 + \dot{m}_2 = \sum \dot{m}_{out} = \dot{m}_3$$

Next, apply equation 7.6, the enthalpy balance, with $\dot{Q} = \dot{W} = 0$:

$$\dot{m}_3 h_3 - \left(0.25 \tfrac{kg}{s} h_1 + \dot{m}_2 h_2\right) = 0$$

We interpolate in the steam table to find h_3 (1750 kPa, sat. v) = 2793.0 kJ/kg and we find h_1 (1800 kPa, 400°C) = 3254.9 kJ/kg and h_2(2000 kPa, 30°C) = 127.55 kJ/kg. Plugging what we know back in to the enthalpy balance and eliminating $\dot{m}_3$ using the material balance:

$$\left(0.25 \tfrac{kg}{s} + \dot{m}_2\right) \cdot 2793.0 \tfrac{kJ}{kg} - \left(0.25 \tfrac{kg}{s} \cdot 3254.9 \tfrac{kJ}{kg} + \dot{m}_2 \cdot 127.55 \tfrac{kJ}{kg}\right) = 0$$

Solving, we find $\dot{m}_2$ = **0.0433 kg/s.** Plugging back into the material balance yields $\dot{m}_3$ = **0.2933 kg/s.**

7.7 **It is desired to cool a 1 kg/s stream of ammonia gas from 450°C and 500 kPa to 200°C. To have the energy available for other processes, we want to use steam for the cooling medium, converting saturated liquid steam at 190°C to saturated vapor. For this process, we will use a heat exchanger shown in Figure 7.f.2. Find the flowrate of steam for this process and the heat exchanger duty. Make a list of assumptions which you would typically use.**

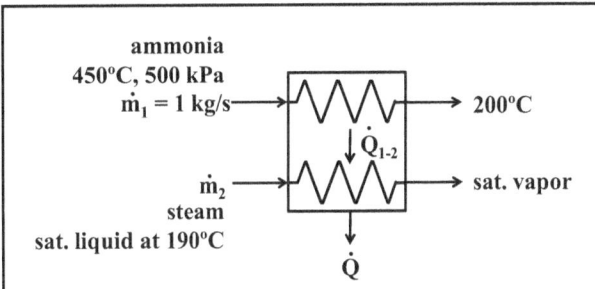

Figure 7.f.2. Schematic of a heat exchanger for use with problem 7.7.

Solution:

Assumptions: 1) Steady state operation, 2) $\dot{Q} = 0$ (adiabatic operation). We don't want to lose heat from this process so the device will be insulated, 3) the two fluid streams are physically separated by the piping, 4) each fluid passes through at approximately constant pressure. There is not other information to go on, and this is not a bad assumption in most instances. This assumption results in ammonia leaving at 500 kPa and saturated steam leaving at it's saturation pressure at 190°C, and 5) KE and PE effects are negligible.

Let's put the enthalpy balance, equation 7.6, into the nomenclature for this problem, grouping enthalpies with flowrates:

$$\dot{m}_1 \left(h_{out}^{ammonia} - h_{in}^{ammonia}\right) + \dot{m}_2 \left(h_{out}^{steam} - h_{in}^{steam}\right) = 0$$

For ammonia, h_{in} (450°C, 500 kPa) = 2748.4 kJ/kg and h_{out} (200°C, 500 kPa) = 2083.3 kJ/kg.
For steam, h_{in} (190°C, sat. liq.) = 807.43 kJ/kg and h_{out} (190°C, sat. vap.) = 2785.3 kJ/kg.

Plugging in the actual values that we know:

$$1\tfrac{kg}{s}(2083.3 - 2748.4)\tfrac{kJ}{kg} + \dot{m}_2(2785.3 - 807.43)\tfrac{kJ}{kg} = 0$$

which yields $\dot{m}_2 = 0.3363$ kg/s. The heat exchanger duty is $|\dot{Q}_{1-2}|$ in Figure 7.f.2. $\dot{Q}_{1-2}$ is negative for ammonia (heat leaves this stream) and is positive for steam. Each side of the exchanger gives the same duty (in the absolute value sense) when the exchanger operates adiabatically. $|\dot{Q}_{1-2}| = 0.3363$ kg/s · (2785.3 - 807.43) kJ/kg = **665.16 kW**. Students may verify that the other side gives the same result.

7.8 A line flowing with N_2 gas at 30°C and 40 bar has taps to take off N_2 to pressurize tanks. Tanks are initially empty when attached, then the valve is opened to let gas in until the pressure in the tank is 30 bar. Figure 7.f.3 has a schematic of the process with a 48 L tank attached. Find the final temperature of the tank and the amount of N_2 that has been delivered to the tank.

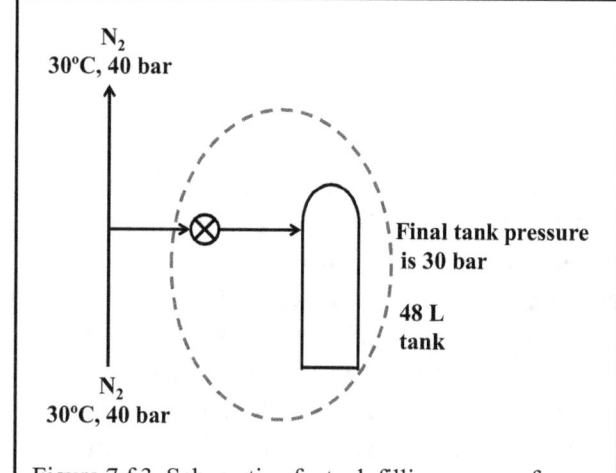

Figure 7.f.3. Schematic of a tank filling process for use with problem 7.8.

Solution:

This is an unsteady-state process because the mass of N_2 in the tank changes with time. Therefore, we choose the unsteady-state material balance and energy balance, equations 7.1 and 7.10, respectively. We consider a period of time over which the tank is filled, and we take the system to be inside the dotted line in Figure 7.f.3. A key point is that choosing the system boundary to include the inlet valve means the flow of N_2 entering the system is at constant T and P. If we chose just the tank with the line leading into it, the flow of N_2 going into the tank has a changing T and P with time which doesn't work well with equations that have already been integrated over time.

Looking at equation 7.1, we see that there is only one stream flowing in and integrating over a period of time, m_{in} flows into the tank and $m_{in} = m - m_o$. Since the tank is initially empty, $m_o = 0$, so the continuity equation for this case simply tells us that what flows into the system over the time period while filling the tank is the mass in the system at the end of the filling process.

Equation 7.10 has been integrated over a period of time for the situation where contents of the system are uniform in composition, temperature, and pressure throughout as we assume in this case. Removing the sums in equation 7.10 because we have only a single stream to deal with results in:

$$mu_{sys} - m_o u_{sys,o} = \dot{m}_{in} h_{in}^{tot} \Delta t - \dot{m}_{out} h_{out}^{tot} \Delta t + \int_0^t \dot{Q} dt - \int_0^t \dot{W}_s dt - \int_0^t \dot{W}_{pv} dt$$

Consider a number of simplifications for this problem. First, $m_o = 0$ because the tank is initially empty. With the continuity equation considerations given above, $m = \dot{m}_{in} \Delta t = m_{in}$. There are no significant KE or PE effects, so $h_{in}^{tot} = h_{in}$ which is the enthalpy of the stream flowing into the system and is at constant T and P. $\dot{m}_{out} h_{out}^{tot} \Delta t = 0$ because there is no outlet flow. There is no heat or work done and the system is not expanding in size so $\dot{W}_{pv} = 0$. This results in a relatively simple equation: $mu_{sys} = m_{in} h_{in}$ or $u_{sys} = h_{in}$. The internal energy of the system at the end of the filling process is equal to the enthalpy of the fluid entering the system.

Looking in the N_2 table given in Appendix K, h_{in} (30°C, 40 bar) = 306.42 kJ/kg. Now we want T (u = 306.42 kJ/kg, 30 bar). Interpolating between 140 and 150°C in the 30 bar block in the table, **T_{final} = 144.8°C**. To find the mass of N_2 delivered to the tank, the N_2 table gives v (144.8°C, 30 bar) = 0.04177 m³/kg = 41.77 L/kg. **m = 48 L/41.77 L/kg = 1.15 kg**.

7.9 A turbine receives steam at 6000 kPa and 500°C and discharges the steam at 60 kPa as saturated vapor. Find the thermal efficiency of the turbine assuming that the turbine operates adiabatically.

Solution:

Inlet conditions, 6000 kPa, 550°C: s_1 = 7.0307 kJ/kg·K and h_1 = 3541.3 kJ/kg

Outlet conditions, 60 kPa, sat. v: s_2 = 7.5311 kJ/kg·K and h_2 = 2652.9 kJ/kg

Δh = 2652.9 - 3541.3 = -888.4 kJ/kg = $-w_s$ for adiabatic operation.

Isentropic base case: s_2' = s_1 = 7.0307 kJ/kg·K and P_2 = 60 kPa. This results in wet steam with s_2' between the saturated vapor and saturated liquid entropy at 60 kPa. x_2' = 0.9216 and h_2' = 2473.21 kJ/kg. $\Delta h'$ = 2473.21 - 3541.3 = -1068.09 kJ/kg = $-w_s'$.

η = w_s'/w_s = 888.4/1068.09 = 0.8318 or 83.18%.

7.10 A compressor compresses fluoromethane (CH_3F) from 1 bar, 30°C to 12.5 bar, 250°C. Find the thermal efficiency of the compressor. Use the Pitzer generalized viral coefficient correlation for the EOS and assume adiabatic operation.

Solution:

For fluoromethane, $T_c = 317.28$ K, $P_c = 59.06$ bar, $\omega = 0.207$. For the isentropic base case, we solve equation 6.28 so that $\Delta s = 0$. The results of this calculation are in the table below in the column for State 2'. The actual case is given in the column for State 2.

	State 1	State 2'	State 2
T(K)	303	490.5	523
P (bar)	1.00	12.50	12.50
T_r	0.955	1.55	1.65
P_r	0.0169	0.2116	0.2116
B^0	-0.371	-0.127	-0.107
B^1	-0.070	0.111	0.118
dB^0/dT_r	0.761	0.217	0.184
dB^1/dT_r	0.917	0.075	0.054
h^R/RT_c	-0.022	-0.098	-0.086
s^R/R	-0.0161	-0.0493	-0.0413
h^R (J/mole)	-57.78	-259.24	-225.54
s^R (J/mole-K)	-0.134	-0.410	-0.343
Δh^{IG} (J/mole)		8378.96	10072.43
Δs^{IG} (J/mole-K)		0.277	3.619
Δh (J/mole)		8177.50	9904.67
Δs (J/mole-K)		0.000	3.409

With results from the table, $\eta = w_s/w_s' = -\Delta h/-\Delta h' = 8177.50/9904.67 = 0.826$ or 82.6%.

Student Problems:

1. Repeat problem 7.10 but use ammonia as the working fluid and use the ammonia tables in Appendix J instead of an EOS. Assume adiabatic operation.

2. R-134a refrigerant is expanded in an adiabatic turbine that has a thermal efficiency of 0.8 from 200°C, 20 bar to 5 bar. The flowrate of R-134a is 1 kg/s. Find $\dot{W}_s$ (the actual work done by the turbine) and the properties of the refrigerant leaving the turbine, T_2, s_2, and h_2.

3. A radiator is used to cool a 1 kg/s stream of CO_2 from 100°C and 1000 kPa to 30°C. Air flows over the radiator as the cooling medium. Find the heat load on the radiator, and the entropy and enthalpy change for the CO_2.

4. Steam enters an adiabatic turbine at $P_1 = 2500$ psia, $T_1 = 1500$°F and exhausts at $T_2 = 440$°F, $P_2 = 40$ psia. Find the efficiency of the turbine.

Chapter 8 - Processes for the Production of Power

The production of power is a requirement for modern human existence. In this chapter we discuss rudimentary thermodynamic aspects of important processes that are used for bulk production of power.

8.a The Steam Rankine Cycle

A schematic of a steam Rankine cycle is given in Figure 8.a.1. In the most general process, water begins as a liquid at point 1 at a relatively low temperature and pressure, often very roughly near ambient conditions. The liquid water is then pumped up to a relatively high pressure at point 2 where it enters a boiler. Heat, often generated by burning a fossil fuel or from a nuclear reaction, is added in the boiler bringing the water to a high pressure and temperature vapor at point 3. The steam then enters a turbine where is does work as it expands in the device, often driving a

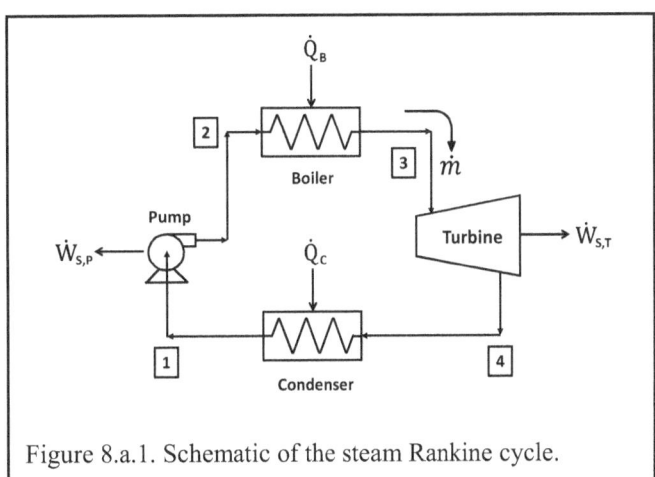

Figure 8.a.1. Schematic of the steam Rankine cycle.

generator which produces electricity. As the steam exits the turbine, it's temperature and pressure have dropped considerably so that it is a vapor at a relatively low temperature and pressure as it exits at point 4. Removing heat from the vapor in the condenser condenses the steam to liquid water at point 1 which completes the cycle.

Overall, this process converts a fraction of the heat added to the boiler, $\dot{Q}_B$, into useful work. Though the work required to operate the pump is often negligible, the net amount of work produced by the process is $\dot{W}_{S,net} = \dot{W}_{S,T} + \dot{W}_{S,P}$. Note that $\dot{W}_{S,P}$ is negative by our formal sign convention for heat and work because the pump requires work to operate. Therefore, the thermal efficiency of the process is given by:

$$\eta = \frac{\dot{W}_{S,net}}{\dot{Q}_B} \qquad 8.1$$

since $\dot{Q}_B$ is the heat absorbed in the process. We can compare this efficiency to the efficiency of the Carnot cycle given in equation 5.37:

$$\eta = \frac{w_{net}}{q_{abs}} = \frac{T_1 - T_2}{T_1} \qquad 5.37$$

Imagining that we run the Carnot cycle using the steam leaving the boiler as the high temperature thermal reservoir and ambient atmospheric conditions as the low temperature reservoir, we can get the maximum efficiency of the equivalent Carnot cycle. T_1 would be the temperature of the steam leaving the boiler and T_2 would be in the vicinity of 300 K. Clearly, the efficiency of the equivalent Carnot cycle goes up with increasing temperature leaving the boiler, and this is also observed in the steam Rankine cycle. In practice, we would like to generate as high a temperature and pressure as possible leaving the boiler to maximize the efficiency, but this will be limited by the materials of construction of the boiler, piping, and turbine.

Consider more details of what the conditions around the process shown in Figure 8.a.1 would be. We know that a high temperature and pressure leaving the boiler at point 3 is desired, and any heat lost from the turbine is heat which could have been used to produce some useful work. Therefore, it is generally desirable to operate the turbine adiabatically. We could then apply the base case of a reversible, adiabatic turbine and use the thermal efficiency of the turbine to determine the actual conditions leaving the turbine at point 4. This calculation was discussed previously in section 7.e.ix.

The pressure leaving the turbine often is below atmospheric pressure, because the condenser will use a cooling medium near ambient conditions (~300 K) and steam will condense well below atmospheric pressure at that temperature. Often, the calculation for adiabatic, reversible (isentropic) operation of the turbine will result in the hypothetical turbine exhaust being wet. However, the actual case where the turbine efficiency is less than 1 may not be wet even if the hypothetical isentropic case yields a wet exhaust. In practice, turbines are generally designed and operated such that steam condensate never exists inside the turbine. Droplets of steam in the turbine could result in damage to the turbine blades.

The condenser can often be treated as approximately constant pressure for the steam flow, as only pressure drop due to flow in open pipes will be observed. Steam leaving the condenser should be completely liquid as it will next enter the suction side of the pump. The steam (water) leaving the condenser might be near a saturated liquid, but subcooled is better for the pump operation.

The pump simply raises the pressure of the liquid leaving the condenser, and so it may be treated as constant temperature if the volume of the liquid is constant. The enthalpy change and the work requirement across the pump is close to $v^L \Delta P$ under this same assumption though pump efficiency could also be considered yielding $\Delta h = v^L \Delta P / \eta_{pump}$. The boiler can also be treated as roughly constant pressure. Entering the boiler is a high pressure liquid and leaving is a superheated vapor.

A quick calculation can be done on the process with a P-H diagram for steam. First, we note the desired pressure for the steam leaving the boiler and draw a horizontal line extending from the liquid region across the saturation dome (if this pressure is below the critical pressure which might possibly not be the case) and up to the desired temperature of the steam leaving the boiler in the superheated region. The desired temperature and pressure of the steam leaving the boiler establishes point 3 on Figure 8.a.1.

Similarly, a horizontal line can be drawn at the expected pressure of the condenser. This will correspond to the saturation temperature of steam 5 or 10°C above the temperature of the cooling media used to condense the steam. For example, if the cooling media is water taken from a lake or river at about 20°C, then the condenser temperature should be 25 or 30°C so that heat can be transferred from the condensing steam to the cooling medium. A saturated liquid or slightly subcooled liquid at the condenser temperature establishes point 1 on Figure 8.a.1. Point 4' (for the isentropic base case for the turbine exhaust) can be established by following a line of constant entropy from point 3 down to the pressure established for the condenser. Point 4 (the actual case for the non-isentropic turbine exhaust) can be established by finding the actual enthalpy change across the turbine using the isentropic base case enthalpy change and multiplying by the turbine thermal efficiency. Finally, point 2 may be established by following a line of constant temperature from point 1 up to the pressure of the boiler. If this line isn't shown on the P-H diagram, then Δh across the pump may be established using $\Delta h = v^L \Delta P / \eta_{pump}$ and h_2 can be established at the pressure of the boiler. Once all points (1, 2, 3, and 4) have been established, enthalpies, entropies, temperatures and pressures can be determined. The cycle efficiency can be established, heat loads for the condenser and boiler may be determined, and the turbine and pump works can be computed.

The discussion applies to the common conditions for the steam Rankine cycle. Many other possibilities exist. The most useful alternative is for sources of heat that are at a much lower temperature than those generated by burning a fossil fuel such as solar or geothermal sources. In such a case, it can be desirable to replace the steam working fluid with a fluid that exhibits vapor/liquid phase transitions at much lower temperatures than steam such as a refrigerant like R-134a or NH_3.

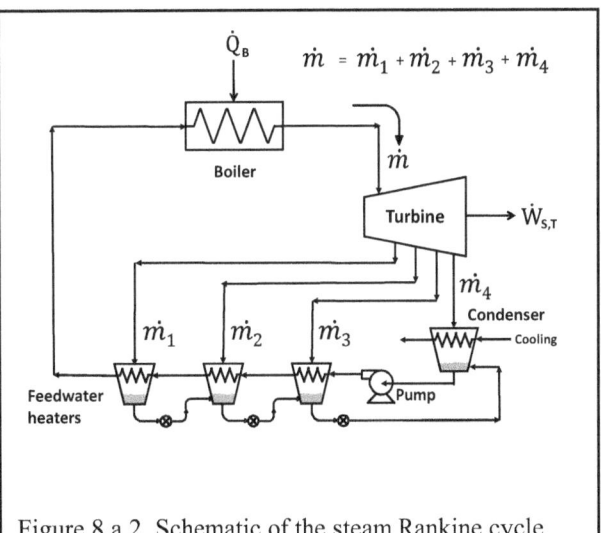

Figure 8.a.2. Schematic of the steam Rankine cycle with feedwater heaters.

The efficiency of the Rankine process can often be improved with the use of feedwater heaters - see Figure 8.a.2. Steam at successively lower temperatures and pressures are withdrawn from the turbine and used to preheat the liquid headed to the boiler. Three feedwater heaters are shown in Figure 8.a.2, but more or less could be used. The overall effect of the feedwater heaters is to reduce temperature differences around the process, but especially in the boiler. As has been noted previously, reversible processes operate with only differentially different temperature differences, and the boiler especially has a high actual temperature difference. Preheating the water before it enters the boiler reduces this temperature difference and generally improves the thermal efficiency of the process.

8.b   The Otto Engine

A schematic of the Otto cycle, also often referred to as the gasoline engine, is shown in Figure 8.b.1. This is often also called a four-stroke engine which refers to the four parts of the process: intake, compression, power, and exhaust. Two-stroke engines also exist but are much less common and beyond the scope of this discussion. Consider each of the four strokes in the four stroke cycle in sequence.

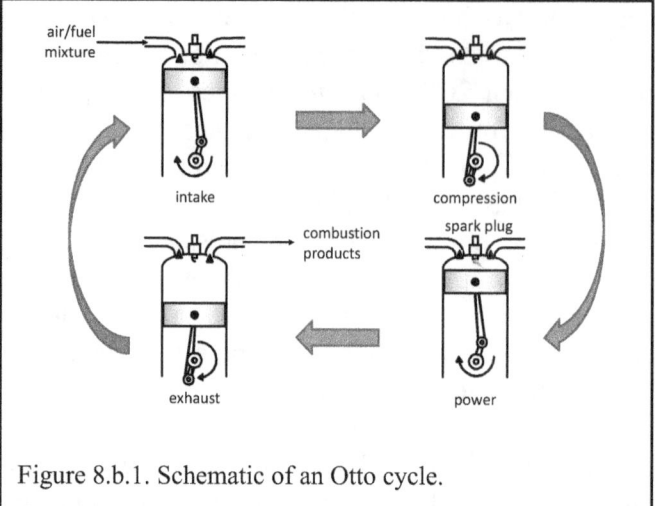

Figure 8.b.1. Schematic of an Otto cycle.

In the intake stroke, the intake valve is opened and the exhaust valve is closed as the piston begins moving downward and a mixture of fuel and air are drawn into the cylinder. Fuel could also be injected into the air as the air is drawn in rather than the mixture being drawn in, but in either case, the cylinder contains a mixture of fuel and air at the end of the stroke. The final mixture is generally near a stoichiometric mixture of air/fuel meaning that there is exactly enough oxygen in the air to completely burn the fuel to carbon dioxide and water with no oxygen left over. However, the mixture is often slightly lean meaning there is more air than required. A rich mixture, one that has more fuel than the stoichiometric mixture, is generally undesired as this leaves unburned fuel at the end of the process. The air/fuel ratio has strong ramifications on the ultimate engine emissions which will not be discussed here.

When the intake stroke is complete, both the intake and exhaust valves are closed and as the piston begins it's travel back up, the air/fuel mixture is compressed in the compression stroke. The compression ratio is limited by the fact that as the mixture is compressed, it's temperature rises rapidly (close to adiabatic compression temperature rise), and too much compression will ignite the air/fuel mixture (auto-ignition) prior to the piston reaching the end of the stroke. The amount of compression which can be achieved without auto-ignition is a function of the fuel properties. Compression is also limited by regulation to reduce engine emissions, but this discussion is beyond the scope of this book.

At the end of compression, a spark generated by the engine ignites the air/fuel mixture. This results in a rapid increase in temperature and pressure in the cylinder, and the power stroke ensues with both valves still closed. At the end of this stroke, products of the combustion process are contained in the cylinder.

In the exhaust stroke, the exhaust valve is opened and the piston rises in the cylinder pushing the combustion products out of the cylinder. The engine is then ready to begin the intake stroke which completes the cycle.

This cycle can be plotted on the P (or log P) versus the volume contained within the cylinder as in Figure 8.b.2 as depicted in the graph labeled "Actual Cycle". Starting again at the intake stroke with the intake valve open, the piston moves down increasing the volume in the cylinder. This occurs at almost constant pressure assuming there is little pressure drop across the intake valve.

Once the intake process is complete, with both valves closed, compression occurs ultimately reducing the cylinder volume back to its minimum (at "top dead center" in automotive jargon) and as the cylinder volume decreases, the pressure rises. This step is close to adiabatic compression, so the temperature rises as well. When the spark ignites the air/fuel mixture, the temperature and pressure rise very rapidly and not far from being at constant volume. However, it does take some time for the flame caused by the spark to travel, and during this small time period, the piston has moved a small amount thus increasing the volume. Therefore, the combustion step is not fully constant volume and this fact is reflected in the actual cycle graph given in Figure 8.b.2. Also contributing to the fact that the combustion step is not truly at constant volume is "spark advance" in automotive jargon. The spark to ignite the combustion is actually advanced to slightly before top dead center so that the piston arrives at top dead center at the same time that the shock wave from the combustion arrives at the piston.

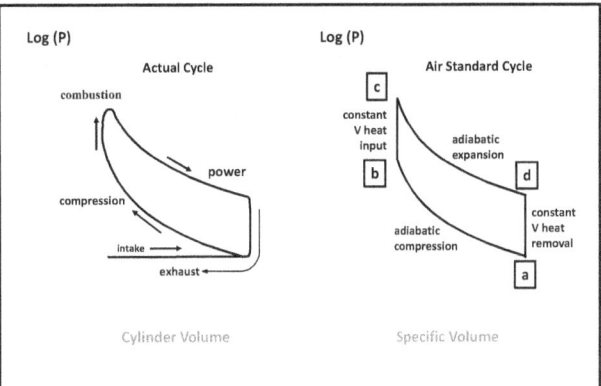

Figure 8.b.2. P-V diagram for the Otto cycle – actual vs. air standard (idealized) cycle.

The power stroke ensues and the step is almost adiabatic as the volume in the cylinder increases, resulting in lower temperature and pressure at the end of the stroke. When the exhaust valve opens, the contents of the cylinder are at a higher pressure than ambient, so a rapid loss of pressure occurs, and this is often heard as a bang which is generally throttled by a muffler system. The exhaust stoke then expels the remainder of the exhaust gas as the piston moves back to top dead center. This completes the cycle.

Because of many complexities of this actual system, it is useful for us to invent a system which is easier to analyze thermodynamically but that also has the salient features of the actual process. This process is shown in Figure 8.b.2 and is labeled the "Air Standard Cycle". Imagine that we use only air to approximate the contents of the cylinder for the entire cycle. Clearly, this is not correct since we have the products of combustion after the combustion step which is not pure air, and the intake step brings in vaporized fuel as well as air. However, air is almost 80% nitrogen which does not participate in the combustion reaction to any appreciable extent, so a preponderance of the gas in the cylinder does not change in the process, so this is a reasonable simplification. We will no longer have a change in the cylinder volume to draw in the air fuel mixture nor will we have the cylinder volume change to expel the remnants of the exhaust because we will use the same air over and over again in the process. Therefore, we can use the specific (or molar) volume of the air instead of the cylinder volume in analyzing the process and this becomes the abscissa of the graph. The air standard cycle begins at point "a" and adiabatic compression of the air ensues which is process a - b in Figure 8.b.2 and we further assume reversible compression for a - b.

Now, consider the combustion step. We assume that a reasonable approximation to this step is an instantaneous injection of heat. The combustion process gives essentially that result, though the actual process takes place over a small volume change and results in combustion products. We will neglect this volume change and we have already discussed substitution of air for the combustion products. This is step b - c. Next is the power stroke which we take as being adiabatic, reversible expansion of the air. This is step c - d.

In the actual cycle, the exhaust stroke begins with the exhaust valve opening which results in a virtually instantaneous loss of pressure. In the actual process, hot combustion gases are expelled almost instantaneously to the exhaust manifold in this step and a cooler air fuel mixture eventually replaces the hot exhaust gas. Instead of expelling the hot gas in the air standard cycle, we simply imagine an instantaneous cooling of the air roughly equivalent to the heat that is lost by rapidly expelling the hot exhaust gas in the actual cycle. The air standard cycle assumes an instantaneous withdrawal of heat (which occurs at constant volume). This is step d - a in Figure 8.b.2.

Our next task is to find the thermal efficiency of the air standard cycle. Consider each step in order:

$$a \rightarrow b: \quad q_{ab} = 0, \quad -w_{ab} = \Delta h_{ab} \qquad 8.2$$

$$b \rightarrow c: \quad \Delta h = q_{bc}, \quad -w_{bc} = 0 \qquad 8.3$$

$$c \rightarrow d: \quad q_{cd} = 0, \quad -w_{cd} = \Delta h_{cd} \qquad 8.4$$

$$d \rightarrow a: \quad \Delta h_{da} = q_{da}, \quad -w_{da} = 0 \qquad 8.5$$

The thermal efficiency is the fraction of heat absorbed converted to useful work. The heat in this process is absorbed in the combustion step and is q_{bc}. The net work, $w_{net} = w_{ab} + w_{cd}$ but keep in mind that the work done in the compression step is negative since work is done on the system in that step. Now an overall first law balance for the entire cycle gives:

$$\Delta h_{overall} = 0 = \Delta h_{ab} + \Delta h_{bc} + \Delta h_{cd} + \Delta h_{da} \qquad 8.6$$

$$0 = -w_{ab} + q_{bc} - w_{cd} + q_{da} \qquad 8.7$$

$$q_{bc} + q_{da} = w_{ab} + w_{cd} = w_{net} \qquad 8.8$$

Therefore, the thermal efficiency is given by:

$$\eta = \frac{w_{net}}{q_{abs}} = \frac{q_{bc} + q_{da}}{q_{bc}} \qquad 8.9$$

Both q_{bc} and q_{da} are added at constant volume, so assuming air is an ideal gas and that all the steps are reversible, $q_{da} = C_v (T_a - T_d)$ and $q_{bc} = C_v (T_c - T_b)$. Of course, these are approximate but the accuracy could be improved with a proper choice for an average value of C_v. Substituting into equation 8.9 and simplifying:

$$\eta = 1 - \frac{T_d - T_a}{T_c - T_b} \qquad 8.10$$

Since we have an ideal gas, for each point in the cycle:

$$T_i = \frac{P_i v_i}{R} \qquad 8.11$$

Substituting equation 8.11 into 8.10 for each temperature and noting that $v_b = v_c$ and $v_a = v_d$, we can reduce equation 8.10 to:

$$\eta = 1 - \frac{v_a}{v_b}\left(\frac{P_c - P_a}{P_c - P_b}\right) \qquad 8.12$$

The compression ratio, r, is defined as the volume in the cylinder at point "a" divided by the volume at point "b". Thus:

$$\eta = 1 - r\left(\frac{P_c - P_a}{P_c - P_b}\right) \qquad 8.13$$

Since points "a" and "b" and points "c" and "d" are connected through adiabatic, reversible processes, $P_a v_a^\gamma = P_b v_b^\gamma$ and $P_c v_c^\gamma = P_d v_d^\gamma$. Furthermore, since $v_b = v_c$ and $v_a = v_d$, $P_a v_a^\gamma = P_b v_b^\gamma = P_c v_c^\gamma = P_d v_d^\gamma$. Using these expressions in equation 8.13 and preforming more algebraic manipulation, we come to the final result:

$$\eta = 1 - \left(\frac{1}{r}\right)^{\gamma - 1} \qquad 8.14$$

With a typical compression ratio of about 8 and $\gamma = 1.4$ for air, this results in $\eta = 0.44$, Thus, 44% of the heating value of the fuel can be converted to useful work in the process.

8.c The Diesel Engine

A schematic of a Diesel cycle is shown in Figure 8.c.1. This cycle is similar to the Otto cycle but with some important differences. In the intake stroke, only air is brought into the cylinder and compression ensues also without any fuel. Compression ratios are much higher in Diesel than Otto cycles, so the temperature of the air in the cylinder at the end of the compression stroke is high enough to immediately ignite fuel. Thus, when fuel is injected starting at the beginning of the power stroke, it immediately burns.

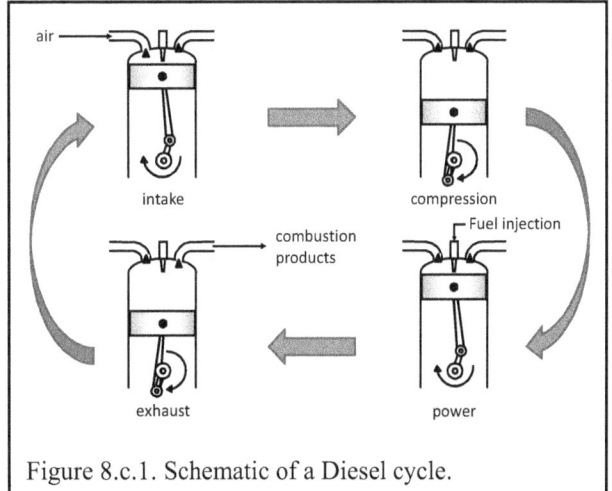

Figure 8.c.1. Schematic of a Diesel cycle.

However, the fuel is not injected instantaneously, so the volume in the cylinder increases during a short but significant time frame over which fuel is injected. The fuel addition ends before the volume of the cylinder has increased to it's maximum at the end of the power cycle, then the exhaust step takes place which is identical to the Otto cycle exhaust stroke.

As we did in the case for the Otto cycle, we invent an air standard cycle that mimics this process, and this is shown in Figure 8.c.2. The only difference here is that the step shown as b - c on the Diesel diagram is at constant pressure rather than constant volume. We can rationalize why b - c is approximately constant pressure by considering what is happening there. As fuel is added, the combustion process tends to raise the temperature which would give a corresponding pressure rise, but the volume in the cylinder is simultaneously increasing which would result in the temperature falling due to adiabatic expansion. We assume that these two competing physical situations result in the pressure remaining approximately constant

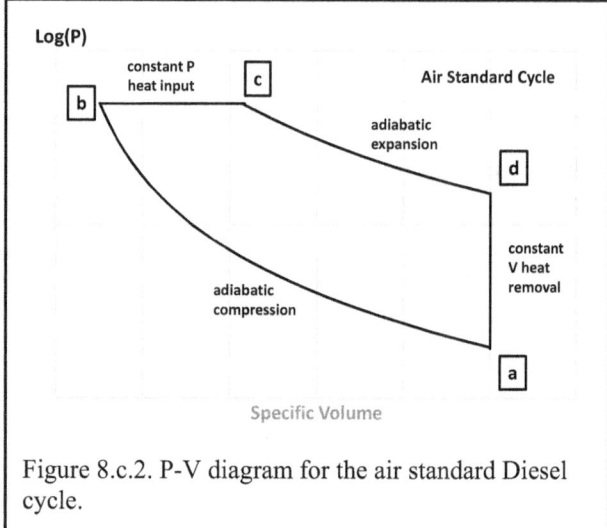

Figure 8.c.2. P-V diagram for the air standard Diesel cycle.

during the timeframe over which fuel is added. At point "c", fuel addition is complete and adiabatic expansion ensues. As in the case of the Otto cycle, we mimic the exhaust step, d - a in Figure 8.c.2, as an instantaneous removal of heat resulting in a constant volume process.

Note that we now have two volume ratios to deal with. These are the compression ratio, $r_c = v_a/v_b$ which is the same ratio as we used in the Otto cycle, and the volume ratio change after the fuel addition finishes, known as the expansion ratio or cut-off ratio, $r_e = v_d/v_c$. Analysis of the process to determine the thermal efficiency is similar to the Otto cycle analysis and we get:

$$\eta = 1 - \frac{C_v(T_d - T_a)}{C_p(T_c - T_b)} \qquad 8.15$$

Equation 8.15 is similar to equation 8.10, but note that C_p rather than C_v appears in the denominator because the process is at constant pressure in the Diesel cycle rather than constant volume. Further analysis is similar to the Otto cycle but considerably more algebraically messy, so we present here just the final result:

$$\eta = 1 - \frac{1}{\gamma}\frac{(1/r_e)^\gamma - (1/r_c)^\gamma}{(1/r_e) - (1/r_c)} \qquad 8.16$$

8.d The Brayton Cycle

The Brayton cycle is a power production cycle using a compressor and turbine built on a single shaft with energy provided by a combustion process. Several varieties exist but these engines are often known as gas turbine engines. A schematic of such a process is given in Figure 8.d.1. Since the compressor is driven by work produced in the turbine, the net work done by the system is the work done by the turbine in excess of the work done on the compressor.

Figure 8.d.1. Schematic of the Brayton cycle.

In the Brayton cycle, air enters a compressor where the pressure is raised before it enters a combustion chamber. In the combustion chamber, fuel is added and burned, then the hot, high pressure gases enter a turbine. Adiabatic expansion in the turbine provides both the power to run the compressor and additional work which can be used for a variety of applications including generation of electricity, propulsion of an airplane by turning a propeller blade, or running machinery such as 1967 - 1969 Indianapolis 500 race cars powered by gas turbine engines.

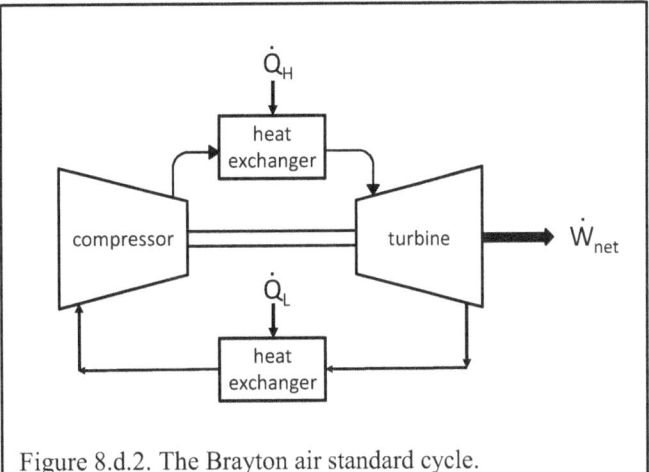

Figure 8.d.2. The Brayton air standard cycle.

The Brayton air standard cycle is shown in Figure 8.d.2, where air is being circulated in this hypothetical process. The combustion of the fuel is simulated with the addition of heat using a heat exchanger, and the exhaust/intake gas difference is simulated with rapid removal of heat in a heat exchanger. The Brayton air standard cycle is shown on the T-S diagram in Figure 8.d.3 with the assumption that each step is reversible. Consider the thermodynamic analysis of this cycle.

Noting a First Law balance around the entire process, we find the thermal efficiency is given by:

$$\eta = \frac{\dot{W}_{net}}{\dot{Q}_{abs}} = \frac{\dot{Q}_H + \dot{Q}_L}{\dot{Q}_H} \qquad 8.17$$

then, since heat is added at constant pressure in the heat exchangers and assuming an appropriate average heat capacity is used:

$$\eta = 1 - \frac{C_P(T_4 - T_1)}{C_P(T_4 - T_2)} = 1 - \frac{T_1(T_4/T_1 - 1)}{T_2(T_3/T_2 - 1)} \qquad 8.18$$

Now, since by our assumptions $P_1 = P_4$ and $P_2 = P_3$, and noting that points 1 and 2 in addition to points 3 and 4 are connected by adiabatic, reversible operations:

$$\frac{P_2}{P_1} = \left(\frac{T_2}{T_1}\right)^{\frac{\gamma-1}{\gamma}} = \frac{P_3}{P_4} = \left(\frac{T_3}{T_4}\right)^{\frac{\gamma-1}{\gamma}} \qquad 8.19$$

so:

$$\frac{T_3}{T_4} = \frac{T_2}{T_1} \quad \text{yielding} \quad \frac{T_3}{T_2} = \frac{T_4}{T_1} \qquad 8.20$$

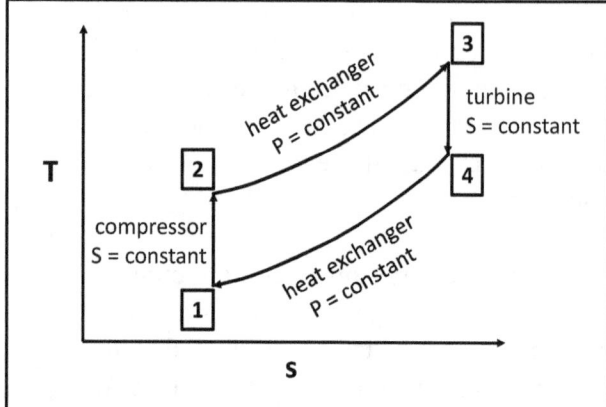

Figure 8.d.3. T-s diagram for the Brayton air standard cycle.

plugging back into equation 8.18:

$$\eta = 1 - \frac{T_1}{T_2} \qquad 8.21$$

Equation 8.21 can be written in terms of pressure as:

$$\eta = 1 - \frac{1}{\left(P_2/P_1\right)^{(\gamma-1)/\gamma}} \qquad 8.22$$

8.e The Jet Engine

A schematic of a jet engine is shown in Figure 8.e.1. This cycle has many similarities to the Brayton in that a compressor and turbine built on a single shaft drive the unit with fuel being added between the compressor and turbine. The difference is that the jet engine is designed to maximize exhaust gas velocity which produces thrust rather than producing work to drive some external process.

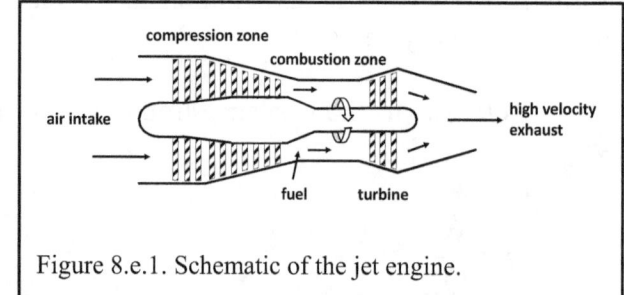

Figure 8.e.1. Schematic of the jet engine.

8.e Chapter 8 Example Problems

8.1 A steam Rankine cycle operates according to the schematic in Figure 8.a.1. The following data is available:

Point 1: $P_1 = 75$ kPa, sat. liquid
Point 2: $P_2 = 1000$ kPa
Point 3: $T_3 = 600°C$, $P_3 = 1000$ kPa
Point 4: $T_4 = 350°C$, $P_4 = 75$ kPa

Find η_T, the thermal efficiency of the turbine, and η_{cycle}, the thermal efficiency of this Rankine cycle.

Solution:

Properties at each point that we need are as follows (interpolating as needed):

Point 1: 75 kPa, sat. liquid, $T_1 = 91.7°C$, $h_1 = 384.2$ kJ/kg, $s_1 = 1.213$ kJ/kg·K
Point 2: $P_2 = 1000$ kPa, $T_2 \approx 91.7°C$, $h_2 = 384.2 + w_{s,pump}$ kJ/kg, $s_1 \approx 1.213$ kJ/kg·K
Point 3: $T_3 = 600°C$, $P_3 = 1000$ kPa, $h_3 = 3698.6$ kJ/kg, $s_3 = 8.031$ kJ/kg·K
Point 4': $P_4' = 75$ kPa, $s_4' = s_3 = 8.031$ kJ/kg·K, $h_3' = 2893.4$ kJ/kg
Point 4: $T_4 = 350°C$, $P_4 = 75$ kPa, $h_4 = 3176.3$ kJ/kg, $s_4 = 8.547$ kJ/kg·K

$w_{s,pump} = v^L \cdot (P_2 - P_1) = 0.001037$ m³/kg·(1000 - 75) kPa = 0.96 m³·kPa/kg = 0.96 kJ/kg
$w_{s,T} = -(3176.3 - 3698.6)$ kJ/kg = 522.3 kJ/kg
$w_{s,T}' = -(2893.4 - 3698.6)$ kJ/kg = 805.2 kJ/kg
$\eta_T = 522.3/805.2 = 0.649$

$w_{s,net} = -(3176.3 - 3698.6)$ kJ/kg - 0.96 kJ/kg = 521.3 kJ/kg
$Q_{abs} = h_3 - h_2 = 3698.6 - (384.2 + 0.96)$ kJ/kg = 3315.4 kJ/kg
$\eta_{cycle} = 521.3/3315.4 = 0.157$

8.2 A steam Rankine cycle operates according to the schematic in Figure 8.a.1. The following data is available, but if information is missing, make a reasonable assumption:

Point 1: $P_1 = 10$ kPa
Point 2: $P_2 = 2000$ kPa
Point 3: $T_3 = 500°C$, $P_3 = 2000$ kPa
Point 4: $T_4 = 100°C$, $P_4 = 10$ kPa

Find η_T, the thermal efficiency of the turbine, and η_{cycle}, the thermal efficiency of this Rankine cycle.

Solution:

For the liquid leaving the condenser (Point 1), with no other information available, we will assume a saturated liquid. Properties at each point that we need are as follows (interpolating as needed):

Point 1: $P_1 = 10$ kPa, sat. liquid, $T_1 = 45.81°C$, $h_1 = 191.81$ kJ/kg
Point 2: $P_2 = 2000$ kPa, $T_2 \approx 45.81°C$, $h_2 = 191.81 + w_{s,pump}$ kJ/kg
Point 3: $T_3 = 500°C$, $P_3 = 2000$ kPa, $h_3 = 3468.2$ kJ/kg, $s_3 = 7.4337$ kJ/kg·K
Point 4': $P_4' = 10$ kPa, $s_4' = s_3 = 7.4337$ kJ/kg·K, $x_3' = 0.9046$ (steam is wet), $h_3' = 2355.8$ kJ/kg
Point 4: $T_4 = 100°C$, $P_4 = 10$ kPa, $h_4 = 2687.5$ kJ/kg

$w_{s,T} = (3468.2 - 2687.5)$ kJ/kg $= 780.7$ kJ/kg
$w_{s,T}' = (3468.2 - 2355.8)$ kJ/kg $= 1112.4$ kJ/kg
$\eta_T = 780.7/1112.4 = 0.702$

$w_{s,pump} = v^L \cdot (P_2 - P_1) = 0.001010$ m³/kg·$(2000 - 10)$ kPa $= 2.01$ m³·kPa/kg $= 2.01$ kJ/kg
$w_{s,net} = 780.7$ kJ/kg $- 2.01$ kJ/kg $= 778.7$ kJ/kg
$Q_{abs} = h_3 - h_2 = 3468.2 - (191.81 + 2.01)$ kJ/kg $= 3274.4$ kJ/kg
$\eta_{cycle} = 778.7/3274.4 = 0.238$

8.3 A power station is to be installed in the Indian Ocean to supply power needs for an offshore oil rig. A deep water trough in the ocean floor near the oil rig can be used as a cooling water source and solar panels will be installed to serve as a thermal supply for the boiler. Instead of steam, since we will working at much lower temperatures than combustion of a fossil fuel can provide, we will use ammonia as a working fluid. Assume that the condenser operates at 20°F and the maximum achievable temperature in the boiler is 150°F (heated by solar panels). Use the property numbering system in Figure 8.a.1. Assume that the turbine is adiabatic and reversible. Make reasonable assumptions where you are missing data. Find the thermal efficiency of the cycle and the flowrate of ammonia required to produce 100 hp.

Solution:

Make the following assumptions consistent with the problem statement:

Point 1 leaving the condenser: $T_1 = 20°F$, sat. liquid
Point 2 leaving the pump: $P_2 = P_3$, $h_2 = h_1 + w_{s,pump}$
Point 3 leaving the boiler: $T_3 = 150°F$, $s_3 = s_4$
Point 4 leaving the turbine: $T_4 = 20°F$, sat vapor

Thus, we can find everything we need at Points 1 and 4, then we can work backward from Point 4 to Point 3, then Point 3 to Point 2. Thus:

Point 1: $T_1 = 20°F$, $P_1 = 48.181$ psia, $h_1 = 135.51$ BTU/lbm, $v_1 = 0.024735$ ft³/lbm
Point 4: $T_4 = 20°F$, $P_4 = 48.181$ psia, $h_4 = 688.31$ BTU/lbm, $s_4 = 1.4800$ BTU/lbm·°F
Point 3: $T_3 = 150°F$, $s_3 = s_4 = 1.4800$ BTU/lbm·°F. This point is between 125 and 150 psia at 150°F. Interpolate to find $P_3 = 138.14$ psia, $h_3 =$ CBTU/lbm

Point 2: $P_2 = P_3 = 138.14$ psia, $h_2 = h_1 + 0.024735$ ft³/lbm·(138.14 - 48.181) psia = h_1 + 2.23 psia·ft³/lbm = h_1 + 0.41 BTU/lbm = 135.51 + 0.41 = 135.93 BTU/lbm

$-w_{s,pump} = 0.024735$ ft³/lbm·(138.14 - 48.181) psia = 2.23 psia·ft³/lbm = 0.41 BTU/lbm
$-w_{s,T} = h_4 - h_3 = 688.31 - 750.67 = -62.36$ BTU/lbm
$-w_{s,net} = -62.36 + 0.41 = -61.95$ BTU/lbm
$Q_{abs} = h_3 - h_2 = 750.67 - 135.93 = 614.74$ BTU/lbm
$\eta_{cycle} = 61.95/614.74 = 0.101$
100 hp = $-\dot{W}_{s,net} = 61.95$ BTU/lbm · $\dot{m}$ · 1.34 hp/0.948 BTU/s

$\dot{m} = 1.14$ lbm/s

8.4 A steam Rankine cycle with one feedwater heater takes steam at **5000 kPa** and **800°C** into a turbine. The turbine has a tap that delivers steam to the feedwater heater at 400 kPa where it condenses at that pressure, heating the feed going to the boiler to within 10°C of the temperature at which it is condensing. The final exhaust from the turbine is at 50 kPa, and it is condensed at that pressure before going into the pump. Find $\dot{m}_1$ and $\dot{m}_2$ for $\dot{m}$ = 1 kg/s, the power output of the turbine, and the thermal efficiency of the cycle.

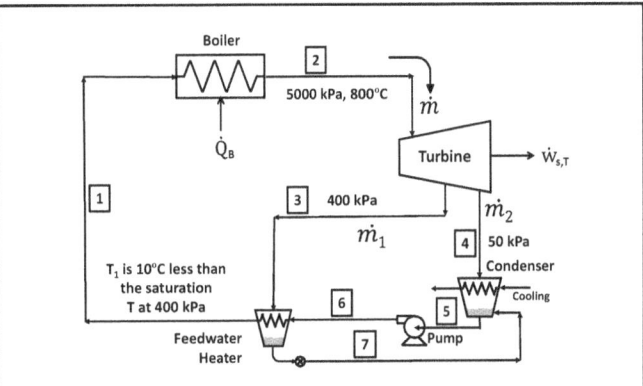

Figure 8.e.1. Schematic of the steam Rankine cycle with one feedwater heater for Problem 8.4

Solution:

The way to approach a complicated flowpath such as that shown in this problem is to first work around the cycle and determine all the enthalpies using whatever information is available and making reasonable assumptions where data is missing. Let's start where we know everything at Point 2, and interpolate in the steam table for points as necessary as we work around the schematic:

Point 2: at 5000 kPa, 800°C, $h_2 = 4137.7$ kJ/kg, $s_2 = 7.7458$ kJ/kg·K
Point 3: $s_3 = s_2 = 7.7458$ kJ/kg·K, $P_3 = 400$ kPa, so $h_3 = 3173.8$ kJ/kg, $T_3 = 351.8$°C
Point 4: $s_4 = s_2 = 7.7458$ kJ/kg·K, $P_4 = 50$ kPa, so $h_3 = 2702.5$ kJ/kg, $T_3 = 110.3$°C
Point 5: $P_5 = 50$ kPa, assume sat. liquid, so $h_5 = 340.54$ kJ/kg, $T_5 = 81.32$°C
Point 6: $P_6 = 5000$ kPa, $T_6 \approx 81.32$°C, so $h_6 = 340.54$ kJ/kg + $w_{s,pump}$

For Point 1, we need to consider that the steam at 400 kPa, 351.8°C at Point 3 condenses in the feedwater heater, and we assume that there is little pressure drop so the liquid leaving the feedwater heater is saturated liquid at 400 kPa. Therefore, the feedwater heater temperature is 143.61°C. The feedwater is heated to within 10°C of that temperature so:

Point 1: $P_1 = 5000$ kPa, $T_1 = 143.61 - 10 = 133.61°C$, so $h_1 = 565.0$ kJ/kg

For point 7, the liquid leaving the feedwater heater is saturated liquid (at 400 kPa), so it's enthalpy is 604.65 kJ/kg. It flows across a valve, which is a constant enthalpy device, before getting to point 7. Therefore:

Point 7: $h_7 = 604.65$ kJ/kg, $P_7 = P_5 = 50$ kPa

Now, let's solve for the flows, work, and efficiency. Consider an enthalpy balance around the feedwater heater. The flow at Points 3 and 7 is $\dot{m}_1$ but that flow recombines with $\dot{m}_2$ in the condenser so the flow at Points 5, 6, and 7 is $\dot{m} = 1$ kg/s. The enthalpy balance is:

$$\dot{m}_1 h_7 + \dot{m} h_1 = \dot{m}_1 h_3 + \dot{m} h_6 \quad \text{or} \quad 604.65\dot{m}_1 + 565.0 \cdot 1\text{kg/s} = 3173.8\dot{m}_1 + 340.54\dot{m}$$

This gives $\dot{m}_1 = 0.08737$ kg/s. $\dot{m}_2 = 0.91263$ kg/s.

For the work output of the turbine, there are two sections. The first section operating from 5000 kPa and 800°C has a flowrate of 1 kg/s. $\dot{m}\Delta h = -\dot{W}_{s,T1} = 1$ kg/s·(3173.8 - 4137.7) kJ/kg = -963.9 kW. The second section has a flowrate of 0.91263 operating between Points 3 and 4. $\dot{m}_2 \Delta h = \dot{W}_{s,T2} = 0.91263$ kg/s·(2702.5 - 3173.8) kJ/kg = -430.1 kW. $\dot{W}_{s,T}$ = **1394.0 kW**.

$$\dot{Q}_B = \dot{m}(h_2 - h_1) = 1 \text{ kg/s} \cdot (4137.7 - 565.0) \text{ kJ/kg} = 3572.7 \text{ kW. } \eta_{cycle} = 1394.0/3572.7 = 0.3902.$$

8.5 An Otto engine (Figure 8.b.2) has a compression ratio of 8.5. The air/fuel mixture taken into the engine has a heating value of 1200 kJ/kg on the basis of per kg of the air/fuel mixture (which is treated as air in the air standard cycle). Assuming the intake step is at 1 bar and 300 K, estimate the temperature and pressure in the cylinder at the end of the compression process (Point b on Figure 8.b.2) and the temperature and pressure in the cylinder at the end of the combustion step (Point c on Figure 8.b.2). What is the thermal efficiency of the process?

Solution:

With a compression ratio of 8.5, starting at 1 bar and 300 K and compressing adiabatically, we can use equation 3.56 for an adiabatic process with an ideal gas. Referring to points a, b, c, and d on Figure 8.b.2, this gives:

$$\frac{T_b}{T_a} = \frac{T_b}{300\text{K}} = \left(\frac{V_b}{V_a}\right)^{1-\gamma} = \left(\frac{1}{8.5}\right)^{1-1.4}$$

Solving, **T_b = 706.4 K**. From equation 3.57 using the points given in Figure 8.b.2:

$$\frac{T_b}{T_a} = \frac{706.1}{300} = \left(\frac{P_b}{P_a}\right)^{\frac{\gamma-1}{\gamma}} = \left(\frac{P_b}{1\,\text{bar}}\right)^{\frac{1.4-1}{1.4}}$$

P_b = 20.00 bar.

To find T_c and P_c, we use 1200 kJ/kg = $\int C_v^{IG} dT$ since the process is at constant volume. To use the C_p^{IG} equation in Appendix B, $\int C_v^{IG} dT = \int (C_p^{IG} - R)\, dT = \int C_p^{IG} dT - R(T_c - T_b)$. However, we need to watch out units. The heating value is 1200 kJ/kg but values of C_p^{IG} in Appendix B are in J/mole·K which is the same as kJ/kmole·K. We need to multiply 1200 kJ/kg by 29.89 kg/kmole (the approximate molar mass of air) to give 35868 kJ/kmole to have consistent units with the heat capacities in Appendix B. With T_b = 706.4 K, we can iterate on T_c to give **T_c = 2032 K**. We could also do an approximate calculation by taking C_v^{IG} = constant but with lesser accuracy. At ≈ 1500 K, a rough average temperature between T_b and T_c, $C_v^{IG} \approx 25$ kJ/kmole·K so ΔT = 35868 kJ/kmole/25 kJ/kmole·K = 1438 K. Tc = 706.4 + 1438 = 2144 K which is not a bad comparison with 2032 K considering the approximate nature.

P_c can be found with the ideal gas law noting $v_b = v_c$. $P_b/T_b = P_c/T_c$ so **P_c = 20.00 bar ·2032/706.4 = 57.53 bar**. η comes from equation 8.14:

$$\eta = 1 - \left(\frac{1}{r}\right)^{g-1} = 1 - \left(\frac{1}{8.5}\right)^{0.4} = 0.575$$

Student Problems:

1. Repeat Example Problem 8.4, but this time, find the pressure tapped off at Point 3 which gives equal work in each of the two sections of the turbine. In all cases, treat the temperature of the water at Point 1 as being 10°C below the saturation pressure of the steam that is tapped off at Point 3. This will be an iterative process.

2. Create a plot of η for an automotive engine vs. the compression ratio for the range 6 < η < 12.

3. A power plant operates on thermal gradients in the ocean. The schematic is exactly the same as the Rankine cycle in Figure 8.a.1 which uses steam, but this cycle uses R134a as the working fluid since the temperatures are much lower. Warm seawater at 30°C on the top of the ocean is used in the boiler, and there needs to be a driving force of 5°C difference in temperature for the ocean's heat to flow into the R-134a in the boiler, so assume the boiler produces R134a at 25°C and 0.575 MPa. The condenser is placed in deep ocean water where the ocean's temperature is about 0°C, so assume the condenser operates at 5°C. Use good engineering practice and appropriate assumptions. $\eta_{turbine} = 0.75$ and $\eta_{pump} = 1.0$.

a) Fill in the following table:

Point	T (°C)	P (MPa)	h (kJ/kg)	s (kJ/kg-K)
1(sat. liquid)	5			
2				
3	25	0.575		
4' (isentropic base case)				
4 (actual)				

b) Find the pump work assuming the pump (η_{pump}) has a thermal efficiency of 1.0 and that the specific volume liquid is constant at 0.0008 m³/kg.
c) Find the turbine work assuming the turbine ($\eta_{turbine}$) has a thermal efficiency of 0.75
d) Find the flowrate of the R134a if the actual net power output is to be 1 MW.

4. Find η, the thermal efficiency of a diesel engine that has $r_c = 20$ and $r_e = 2$. Prepare a plot of η vs r_c for $18 < r_c < 22$ with curves for constant values of $1.5 < r_e < 3.0$. How is high efficiency related to each variable?

Chapter 9 - Refrigeration and Liquification Cycles

From home air conditioners, refrigerators, and freezers, to air separation, industrial cooling and pharmaceutical applications requiring strict temperature control, various kinds of refrigeration and liquification processes are needed to support modern society. Fresh fruits, vegetables, meats, and beverages rely on refrigerated transportation processes. This chapter considers some of the thermodynamic aspects of typical processes used to accomplish refrigeration and also the industrially important liquification process.

9.a   Vapor Compression Refrigeration

The Carnot cycle run in reverse is an example of a refrigeration process. Referring to Figures 5.c.1 and 5.c.2, we would reverse the direction of the cycle depicted in Figure 5.c.2 using the same apparatus that is depicted in Figure 5.c.1. Recall from the discussion of heat engines and the Carnot cycle that the (forward) Carnot cycle results in high temperature heat at T_1 being removed from the high temperature thermal reservoir, a fraction of that heat converted to useful work, and the remainder of the heat is sent to the low temperature reservoir at T_2. Recall also that the Carnot cycle is defined such that all processes are reversible, meaning that the processes can go backward resulting in an exact restoration of the universe. Therefore, the reverse Carnot cycle would take heat out of the low temperature reservoir at T_2 and, in addition to the net amount of work required, send it to the high temperature reservoir at T_1. In this process, the low temperature reservoir has been refrigerated. Compare this to say a refrigerator in your kitchen. Since the kitchen is at a higher temperature than the refrigerator compartment, heat is continuously transferring to the inside of the refrigerator thereby slowly raising its temperature. At some point, the refrigerator detects the compartment temperature is too hot, and the mechanical equipment turns on. The result over a short period of time is that heat is removed from the refrigerator compartment and is sent to the kitchen, which is at a higher temperature. This is exactly the result of the reverse Carnot cycle.

The parameter that we are interested in when dealing with refrigeration is known as the *coefficient of performance*, ω. It is defined as the ratio of the amount of refrigeration, or refrigeration capacity, to the amount of work required to accomplish the refrigeration. This calculation is relatively simple by following the Carnot cycle in reverse and results in:

$$\omega = \frac{T_{cold}}{T_{hot} - T_{cold}} = \frac{\text{refrigeration capacity}}{\text{work required}} \qquad 9.1$$

Unlike thermal efficiencies (η) which are defined such that they are always less than one, ω does not have such a bound, and we often find that many units of heat can be transferred up the thermal hill for every unit of work required to run the process. Note that the Carnot refrigerator would have the maximum ω which is possible. Consider the kitchen refrigerator for a quick calculation. The refrigerator compartment is at about 35°F, and the kitchen is at about 72°F. For this case:

$$\omega = \frac{(35+460)°R}{(72-35)°R} = 13.4 \qquad 9.2$$

So more than 13 units of heat can be transferred from the refrigerator compartment to the surrounding kitchen for every unit of work done by the refrigeration process when using the reverse Carnot cycle. No other process could do better.

Vapor compression cycles are the most common type of refrigeration processes used in some of the most well known devices such as home refrigerators, freezers, and air conditioners. Industry and transportation also rely on this process. A basic cycle is shown in Figure 9.a.1.

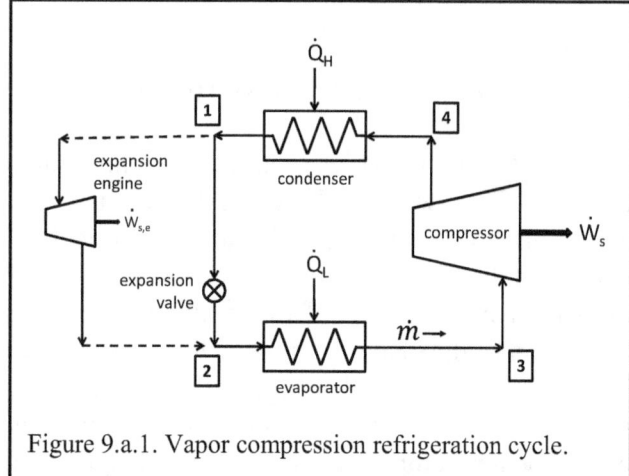

Figure 9.a.1. Vapor compression refrigeration cycle.

First, consider the process shown that has an expansion valve. In the most common process, a high pressure liquid refrigerant not far above the temperature of the surroundings where heat is being rejected exists at point 1. When it flows through the expansion valve to a low pressure at point 2, it partially vaporizes and the temperature drops to below the temperature of the compartment being cooled. $\dot{Q}_L$ is added to the flowing refrigerant (thus, removed from the compartment being cooled) in the evaporator resulting in a saturated or slightly superheated vapor at point 3. When the vapor is then compressed in the compressor ($\dot{W}_S$ is negative) a high temperature, high pressure vapor results at point 4. This high temperature vapor then flows through the condenser where high temperature heat is rejected to the surroundings. Note that $\dot{Q}_H$ will be negative by our formal sign convention.

A possible modification to this process uses an expansion engine which is shown with dotted lines in Figure 9.a.1. The expansion engine would replace the expansion valve. Work would be derived in the expansion when the high pressure liquid at point 1 expands to the vapor/liquid mixture at point 2. The work produced by the expansion engine would reduce the net amount of work required to operate the process, resulting in a higher ω, but the possibility of using an expansion engine is mostly theoretical because the cost associated with the installation and operation of the expansion engine is prohibitive.

Figure 9.a.2 gives an outline of how this process might look on the P-H plane for R-134a. Suppose we have a home refrigerator operating with R-134a as a refrigerant. We begin by assuming that the high pressure liquid leaving the condenser is at about 100°F ≈ 40°C which is bit above the temperature of the kitchen where the heat from the condenser is rejected. This establishes point 1 marked in Figure 9.a.2. Expansion across the expansion valve is very close to constant enthalpy which sends us directly down on the P-H diagram to a lower temperature. This lower temperature needs to be below the temperature inside the refrigerator so we choose 0°C = 32°F. This establishes point 2. In the evaporator, the process uses heat from the refrigerator to heat the refrigerant. We assume this is accomplished at roughly constant pressure and we assume we heat at least until all the liquid is gone and we have a saturated or slightly superheated vapor. This establishes point 3 on the diagram. Finally, the compressor operates at constant entropy if it is reversible and adiabatic, or the entropy will increase to account for irreversiblites. Following a line of constant entropy from point 3 to point 4 and connecting up with the same pressure as point 1 establishes point 4. Point 1 and point 4 are at about the same pressure because there is only pressure drop in the condenser due to flow which is generally negligible.

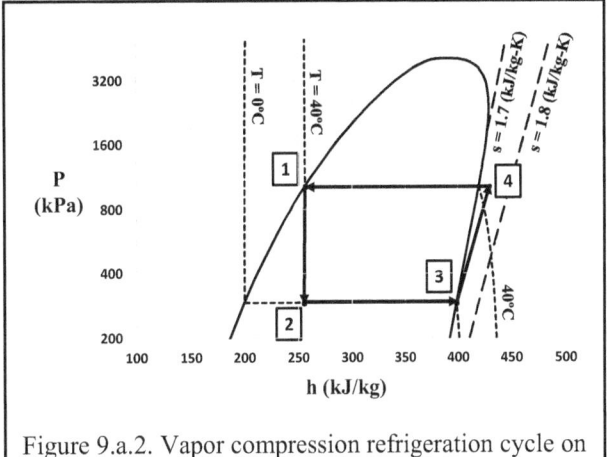

Figure 9.a.2. Vapor compression refrigeration cycle on the R-134a P-H diagram.

In the absence of other information, choose point 1 on Figure 9.a.1 to be a saturated liquid, point 3 to be a saturated vapor, and for the condenser and evaporator to operate at constant pressure; i.e., $P_3 = P_2$ and $P_1 = P_4$. Design based upon the temperatures that will be encountered. At what temperature is the compartment to operate and what is the temperature of the space where the heat is being rejected?

Knowing the location of each point on the P-H diagram establishes h_1, h_2, h_3, and h_4. The refrigeration capacity is $\dot{m}(h_3 - h_2)$, the net work requirement (recalling $\Delta h = -w_s$) is $\dot{m}(h_4 - h_3)$, and the heat load on the condenser is $\dot{m}(h_1 - h_4)$. The coefficient of performance is:

$$\omega = \frac{h_3 - h_2}{h_4 - h_3} \qquad 9.3$$

Figure 9.a.3 shows a cascaded refrigeration system which can be used to reach very low temperatures. The high temperature (upper) loop, points 1 - 4, is identical to the loop given in Figure 9.a.1 except that the evaporator is used to remove heat from the condenser of the low temperature (lower) loop, points 5 - 8. By arranging the cascade like this, the low temperature loop can reject heat to a lower temperature, thus making the possible temperature range for the evaporator of the low temperature loop much lower. You can think of this as being like a small refrigerator is placed inside a large refrigerator so that the small refrigerator rejects heat to the temperature of the compartment of the larger refrigerator. In practice, the two refrigeration loops generally use two different refrigerants.

This leads to a discussion of the choice of a refrigerant. The first properties that a refrigerant must possess are physical properties that are appropriate for the temperature range. The refrigerant should have a normal boiling point below the temperature of the refrigerated compartment so that the vapor/liquid equilibrium mixture in the evaporator is above atmospheric pressure. In this case, if there are any tiny leaks, the system will not tend to leak air in. Also, neither the condenser nor evaporator should operate at too high a pressure so that high pressure piping is not required. These two requirements give a rough bound to the vapor pressure curve for the refrigerant. Also, the freezing point of the refrigerant should be well below any temperatures encountered in the refrigeration system. Other physical properties such as viscosity can also be important.

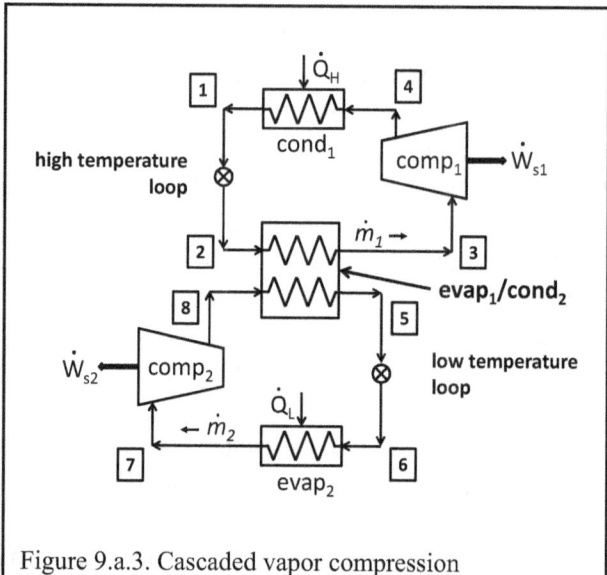

Figure 9.a.3. Cascaded vapor compression refrigeration cycle to reach low temperatures.

Chemical properties also have numerous requirements. The refrigerant must have long term stability in the system, and must not be corrosive to inexpensive piping. Though a refrigeration system could be built out of chemically inert materials such as stainless steel, this could easily add costs to the system which are prohibitive. In recent years, atmospheric stability has become an issue. If a refrigerant contains chlorine, and many do, there is concern that it could affect the ozone layer so refrigerants should decompose at least slowly in the atmosphere so they are depleted before reaching the upper levels of the atmosphere. Also, global warming potential has become an issue as well. Finally, for most installations, refrigerants should be non-toxic. Ammonia is an excellent refrigerant for many purposes, but its toxicity is a problem for consumer use and this property often limits it's application to industrial installations.

Finally, we define a heat pump which can be a vapor compression cycle. Instead of the vapor compression cycle of Figure 9.a.1 where the evaporator withdraws heat from a cold compartment and rejects it to high temperature surroundings, imagine that we desire to heat a compartment such as a house. The condenser of a vapor compression cycle could be inserted in the compartment (house) and heat is then rejected to the compartment. Some of the heat comes from the low temperature surroundings and the net effect is that heat from the low temperature surroundings + net work done by the cycle is sent to the high temperature compartment. This is a form of heating and is called a heat pump.

9.b Absorption Refrigeration

Another type of refrigeration system is an absorption refrigeration system depicted in Figure 9.b.1. These are sometimes called natural gas air-conditioners in which case $\dot{Q}_G$ is provided by burning natural gas. They are occasionally installed for home use and are generally very inexpensive to operate because of the low cost of natural gas. However, because of high installation costs, consumers often choose vapor compression systems over absorption refrigeration systems.

Points 1 - 4 in Figure 9.b.1 are identical to a vapor compression cycle. High pressure vapor refrigerant at point 4 is delivered to a condenser where it is condensed (point 1) while

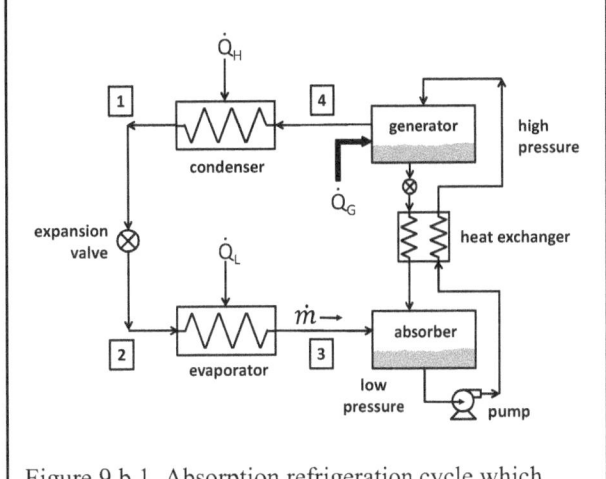

Figure 9.b.1. Absorption refrigeration cycle which generally uses an ammonia refrigerant.

rejecting heat to the surroundings then sent to an expansion valve where the pressure drops (point 2) then evaporated in the evaporator to point 3 which provides the refrigeration. However, in order for this process to work, we must have a refrigerant that has a widely different solubility in a solute at high and low temperatures. The refrigerant used is ammonia which absorbs (dissolves) easily in water around ambient temperature and desorbs readily at higher temperatures.

Low pressure ammonia at point 3 absorbs in water in the absorber at a temperature a little below ambient temperature. This strong ammonia solution is pumped through a heat exchanger where it picks up heat, thus freeing some of the ammonia from solution, then it travels to the generator. Heat is added by combustion of natural gas to the generator which is maintained at a high pressure by further desorption of ammonia. The gas phase in the generator is almost pure ammonia at a high pressure which supplies the ammonia at point 4 to the condenser. The solution of ammonia in the generator is weak in ammonia. The hot, weak ammonia solution in the generator heads back through the heat exchanger to preheat the strong ammonia solution coming from the absorber, then goes into the absorber at a low temperature ready to pick up ammonia again.

Together, the generator, absorber, pump and heat exchanger act like a compressor, bringing ammonia from the evaporator at point 3 at a low pressure and delivering it at a high pressure to the condenser at point 4.

9.c Joule-Thompson Liquification

Consider the process of liquefying light gases such as oxygen or nitrogen. Oxygen has a critical temperature of 154.2 K and normal boiling point of 90.2 K (see Appendix C). Nitrogen has a critical temperature of 126.2 K and normal boiling point of 77.4 K. These species will not liquify above their critical temperatures regardless of how high the pressure might be, and these temperatures are not easily achievable even with refrigeration. Distillation of liquid air to get pure nitrogen and pure oxygen (among other components) is a common route to purifying these species and the pure components are quite important. Therefore processes have been designed to liquify these species.

One of the most common processes is the Linde liquification process which is based upon the Joule-Thompson effect. When ideal gases expand in going from high to low pressure through a valve or restriction (constant enthalpy), there is no change in temperature since the enthalpy of an ideal gas is not dependent upon pressure. However, real gases do experience a temperature change.

Refer to the P-H diagram for ammonia in Appendix J. Consider first an expansion process at constant enthalpy (across a valve or restriction) for ammonia in it's ideal gas region which is at low pressure and high temperatures. Let's take 160°C and 100 kPa where h = 2000 kJ/kg, then consider dropping the pressure at constant enthalpy (vertical line on the P-H diagram). At 10 kPa and h = 2000 kJ/kg, we see that the temperature would still be very close to 160°C. This is the ideal gas region.

Now, follow the 160°C isotherm up to 10000 kPa where h = 1650 kJ/kg. Following a line of constant enthalpy down to h = 1650 kJ/kg and 1000 kPa results a temperature of about 60°C, about a 100°C temperature drop. This is the Joule-Thompson effect and it can applied in a process to make light gases into liquids. All gases at room temperature experience a temperature drop (though it may be very small) on expansion at constant enthalpy except for He, H_2, and Ne.

Referring to Figure 9.c.1, we start with a gas at point 1 which is likely near ambient temperature and pressure. This gas is compressed to a high pressure at point 2, then cooled back to near T_1 at point 3. The gas then flows through a heat exchanger, then through a throttling valve where it expands back to a pressure sightly greater than P_1 so that the gas at point 4 can flow back toward point 6 where it joins with fresh feed at point 1 to go into the compressor. Since the gas at point 3 experiences a pressure drop at constant enthalpy, Joule-Thompson cooling occurs (assuming the gas is in the correct P-V-T region) and the gas at point 4 is cold. Flowing back through the heat exchanger, the cold gas from point 4 cools the gas headed to the throttling valve. Over time, the gas coming from the throttling valve gets so cold, it begins to form liquid which falls to the bottom of the accumulation tank where it can be withdrawn as liquid product. At steady state, a material balance around the equipment inside the dotted line yields:

$$\dot{m}_6 + \dot{m}_5 = \dot{m}_3 \qquad 9.4$$

And an enthalpy balance gives:

$$\dot{m}_6 h_6 + \dot{m}_5 h_5 = \dot{m}_3 h_3 \qquad 9.5$$

or, dividing through by $\dot{m}_3$:

$$\frac{\dot{m}_6}{\dot{m}_3} h_6 + \frac{\dot{m}_5}{\dot{m}_3} h_3 = h_3 \qquad 9.6$$

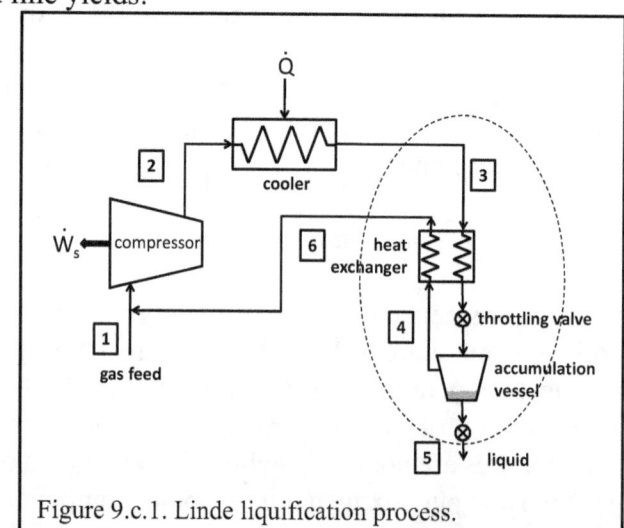

Figure 9.c.1. Linde liquification process.

With $z = \dot{m}_5/\dot{m}_3$ which is the fraction of stream 3 liquified and using equation 9.4:

$$(1-z)h_6 + zh_5 = h_3 \qquad 9.7$$

Or solving for z:

$$z = \frac{h_6 - h_3}{h_6 - h_5} \qquad 9.8$$

For a quick calculation using N_2, suppose we take saturated liquid at 1 bar at point 5, superheated vapor at 20°C and 1 bar at point 6, and superheated vapor at point at 40 bar and 20°C for point 3, from Appendix K.2, $h_5 = 77.07$ kJ/kg, $h_6 = 304.06$ kJ/kg, and $h_3 = 295.36$ kJ/kg, gives:

$$z = \frac{304.06 - 295.36}{304.06 - 77.07} = 0.0383 \qquad 9.9$$

Almost 4% of the N_2 entering the system can be converted to liquid under these conditions.

The Linde process could obviously be improved in terms of net work requirements by using an expander that produces work instead of the throttling valve, but a direct substitution would require an expander that operates on a two-phase mixture which is difficult. The Claude process is set up differently and allows some expansion work to be realized. Students may refer to the open literature for a description of the Claude process.

9.d   Chapter 9 Example Problems

9.1 A conventional vapor compression refrigeration cycle operates with R-134a refrigerant (see Figure 9.a.1). The evaporator operates at 200 kPa and the condenser at 1000 kPa. For a unit with a 1 ton refrigeration capacity, find the flowrate of R-134a, the work requirement for the compressor assuming it operates adiabatically and reversibly, the coefficient of performance, and the heat load on the condenser. Make reasonable assumptions for information that is missing.

Solution:

Referring to Figure 9.a.1 and the R-134a table in Appendix H, the following information is available or can be assumed:

Point 1:   Assume sat. liquid at $P_1 = 1000$ kPa, $h_1 = 255.50$ kJ/kg
Point 2: $P_2 = 200$ kPa, $h_2 = h_1 = 255.50$ kJ/kg
Point 3:   Assume sat. vapor at $P_3 = 200$ kPa, $h_3 = 392.62$ kJ/kg, $s_3 = 1.7334$ kJ/kg·K
Point 4: $P_4 = 1000$ kPa, $s_4 = s_3 = 1.7334$ kJ/kg·K, $h_4 = 426.18$ kJ/kg (via interpolation)

The refrigeration capacity is:

$$\dot{Q}_L = 1\,\text{ton} = \dot{m}(h_3 - h_2).$$

1 ton of refrigeration is also 3.517 kW = 3.517 kJ/s. Plugging in:

$$3.517\,\text{kJ/s} = \dot{m}(392.62 - 255.50)\,\text{kJ/kg}.$$

This gives $\dot{m} = \mathbf{0.0256\,kg/s}$.

The compressor work requirement is:

$$-\dot{W}_s = \dot{m}(h_4 - h_3) = 0.0256\,\text{kg/s}\,(426.18 - 392.62) = 0.859\,\text{kW} = 1.15\,\text{hp}.$$

The coefficient of performance is:

$$\omega = \frac{h_3 - h_2}{h_4 - h_3} = \frac{392.62 - 255.50}{426.18 - 392.62} = 4.09$$

A little over 4 units of energy can be transferred out of the refrigerated compartment for every unit of energy put into the compressor.

Finally:

$$\dot{Q}_H = \dot{m}(h_1 - h_4) = 0.0256\,\text{kg/s}\,(255.50 - 426.18) = -4.37\,\text{kW}$$

which is the heat load on the condenser. Heat is rejected to the outside of the compartment.

9.2 A cascaded vapor compression refrigeration cycle like that shown in Figure 9.a.3 needs to be designed to liquify krypton. The device needs to liquify 1 mole/s of krypton, and considering krypton's heat of liquification and that it needs to be cooled to below it's critical temperature before it can be liquified at all, we will need a low temperature cooling capacity of 10 kW. The critical temperature of krypton is about 209.48 K so we will design the low temperature evaporator, evap$_2$, to operate below that at 193 K or -80°C. Cond$_2$ will operate at -20°C, and it will reject heat to evap$_1$ operating at -30°C. Cond$_1$ rejects heat to 30°C, a little above the surroundings for the device. Use R-23 for the low temperature refrigerant and R-134a for refrigerant in the high temperature system. Find $\dot{m}_1$ and $\dot{m}_2$, the heat load on the evap$_1$/cond$_2$ heat exchanger, $\dot{Q}_H$, the heat load on cond$_1$, the work requirements for both compressors, $\dot{W}_{s1}$ and $\dot{W}_{s2}$, and the overall coefficient of performance defined as $|\dot{Q}_L|/|\dot{W}_{s1} + \dot{W}_{s2}|$, which is the heat transferred out of the low temperature compartment divided by the total work required.

Solution:

Points 1, 2, 3 and 4 are R-134a:

Point 1: Assume sat. liquid at $T_1 = 30°C$, $P_1 = 770.20$ kPa, $h_1 = 241.72$ kJ/kg
Point 2: liq/vap mix, $T_2 = -30°C$, $P_2 = 84.38$ kPa, $h_2 = h_1 = 241.72$ kJ/kg
Point 3: Assume sat. vapor at $P_3 = P_2 = 84.38$ kPa, $h_3 = 380.32$ kJ/kg, $s_3 = 1.7515$ kJ/kg·K
Point 4: $P_4 = P_1 = 770.20$ kPa, $s_4 = s_3 = 1.7515$ kJ/kg·K, $h_4 = 426.13$ kJ/kg (via double interpolation)

The key to doing the double interpolation is to first create a block (or a few rows in the block) for the needed pressure by interpolating between the two blocks of pressure above and below the needed pressure. Once that block corresponding to the desired pressure has been created, interpolation between rows of the new block proceeds like the usual single dimensional interpolation. For this specific example, we created the P = 770.30 kPa block by interpolating between the P = 600 kPa and P = 800 kPa blocks, then between the rows s = 1.7486 kJ/kg·K and s = 1.7806 kJ/kg·K (which were the result of the interpolation that created the block) for s = 1.7515 kJ/kg·K.

Points 5, 6, 7 and 8 are R-23:

Point 5: Assume sat. liquid at $T_5 = -20°C$, $P_5 = 1395.30$ kPa, $h_5 = 167.68$ kJ/kg
Point 6: sat. liq/vap mix, $T_6 = -80°C$, $P_6 = 113.70$ kPa, $h_6 = h_5 = 167.68$ kJ/kg
Point 7: Assume sat. vapor at $P_7 = P_6 = 113.70$ kPa, $h_7 = 324.81$ kJ/kg, $s_7 = 1.7543$ kJ/kg·K
Point 8: $P_8 = P_5 = 1395.30$ kPa, $s_8 = s_7 = 1.7543$ kJ/kg·K, $h_8 = 387.34$ kJ/kg (via double interpolation)

Now, we need to solve the remainder of the low temperature loop first because that loop has the refrigeration capacity specification of 10 kW. $\dot{Q}_L = 10\,kW = \dot{m}_2(h_7 - h_6)$ = $\dot{m}_2$ (324.81 - 167.68) kJ/kg. Solving, **$\dot{m}_2 = 0.06364$ kg/s.**

The heat load on cond$_2$ = 0.06364 kg/s · (h$_5$ - h$_8$) = 0.06364 · (167.68 - 387.34) = -13.98 kW. This is negative heat because it leaves the R-23 and is transferred to the R-134a in the high temperature loop. So, +13.98 kW is transferred to the R-134a in evap$_1$.

In the upper, high T loop with R-134a, $\dot{m}_1(h_3 - h_2) = 13.98$ kW. $\dot{m}_1$ (380.32 - 241.72) kJ/kg = 13.98 kW. **$\dot{m}_1 = 0.1009$ kg/s.** $\dot{Q}_H = \dot{m}_1(h_1 - h_4) = 0.1009$ kg/s (241.72 - 426.13) kJ/kg. This gives **$\dot{Q}_H = -18.61$ kW.**

$-\dot{W}_{s2} = \dot{m}_2(h_8 - h_7) = 0.06364$ kg/s (387.34 - 324.81) kJ/kg = 3.97 kW. **$\dot{W}_{s2} = -3.97$ kW.**

$-\dot{W}_{s1} = \dot{m}_1(h_4 - h_3) = 0.1009$ kg/s $(426.13 - 380.32)$ kJ/kg $= 4.62$ kW. $\dot{W}_{s2} = -4.62$ kW.
$\omega = 10$ kW$/(3.97 + 4.62) = 1.16$.

9.3 NASA plans to launch a probe to land on a celestial body where the ambient temperature is 250°F. An experiment onboard the lander is designed to study the effects of the atmosphere that exists on the celestial body on an earth microbe that can only withstand 120°F. Design a vapor-compression cycle (like Figure 9.a.1) that maintains the microbe's incubator at 120°F while rejecting heat to the celestial body's atmosphere at 250°F. Use R-11 (Appendix P) for the refrigerant. Assume the compressor has an isentropic efficiency of 0.80. For a unit with a 0.01 ton refrigeration capacity, find the flowrate of R-11, the work requirement for the compressor, the coefficient of performance, and the heat load on the condenser. Make reasonable assumptions for information that is missing, but you only have one shot at this experiment so it better be good!

Solution:

Referring to Figure 9.a.1 and the R-11 table in Appendix P, the following information is available or can be assumed:

Point 1: Assume sat. liquid at 270°F. We need about 20°F above the temperature of the atmosphere so that heat can transfer from the condenser to the environment at 250°F. $P_1 = 226.1$ psia, $h_1 = 68.19$ BTU/lbm (via interpolation).

Point 2: We will have a vapor/liquid equilibrium mixture, and now we need about 20°F lower than the incubator so that heat can transfer from the incubator to the evaporator at 120°F, so assume $T_2 = 100°F$. $P_2 = 23.581$ psia, $h_2 = h_1 = 68.19$ BTU/lbm.

Point 3: Assume sat. vapor at $T_2 = 100°F$. $P_3 = P_2 = 23.581$ psia, $h_3 = 104.53$ BTU/lbm, $s_3 = 0.19435$ BTU/lbm·°F.

Point 4': $P_{4'} = P_1 = 226.1$ psia, $s_{4'} = s_3 = 0.19435$ BTU/lbm·°F, $h_{4'} = 122.67$ BTU/lbm (via double interpolation).

Now, find the actual conditions - Point 4. $\Delta h' = -w'_s = h_4' - h_3 = (122.67 - 104.53)$ BTU/lbm $= 18.14$ BTU/lbm. $\Delta h = -w_s = 18.14/0.8 = 22.68$ BTU/lbm. $h_4 = 104.53 + 22.68 = 127.21$ BTU/lbm.

Point 4: $h_4 = 127.21$ BTU/lbm, $P_4 = P_1 = 226.1$ psia

The refrigeration capacity is $\dot{Q}_L = 0.01$ ton $= \dot{m}(h_3 - h_2)$. 2 BTU/min $= 0.01$ ton $= \dot{m}(104.53 - 68.19)$ BTU/lbm. $\dot{m} = 0.0550$ **lbm/min.**

The compressor work requirement is:

$-\dot{W}_s = \dot{m}(h_4 - h_3) = 0.0550$ lbm/min $(127.21 - 104.53)$ BTU/lbm $= 1.24$ BTU/min $=$

0.0292 hp.

The coefficient of performance is:

$$\omega = \frac{h_3 - h_2}{h_4 - h_3} = \frac{104.53 - 68.19}{127.21 - 104.53} = 1.60$$

$\dot{Q}_H = \dot{m}(h_1 - h_4) = 0.0550(68.19 - 127.21) = $ **−3.25 BTU / min** which is the heat load on the condenser.

9.4 A liquification process like in Figure 9.c.1 is used to liquify N_2. Fresh N_2 enters at 2 bar and 30°C and is compressed to 40 bar in the compressor which has an isentropic efficiency of 90%. The N_2 is cooled back to 30°C in the cooler before going into the heat exchanger. Unliquified N_2 at 30°C and 2 bar is recycled and saturated liquid at 2 bar is taken as product. Find the fraction liquified, the compressor work requirement, and the heat load on the cooler.

Solution:

Using Appendix K for N_2 properties, we find the following in reference to Figure 9.c.1:

Point 1: $T_1 = 30°C$, $P_1 = 2$ bar, so $h_1 = 314.26$ kJ/kg, $s_1 = 6.6501$ kJ/kg·K
Point 2': 40 bar, $s_{2'} = s_1 = 6.6501$ kJ/kg·K so $h_{2'} = 741.51$ kJ/kg

to find the actual conditions at point 2, $\Delta h_{1-2'} = 741.51 - 314.26 = 427.25$ kJ/kg. $\Delta h_{1-2} = 427.25/0.9 = $ **474.72 kJ/kg** $= -w_{s,compressor}$. $h_2 = 314.26 + 474.72 = $ **788.98 kJ/kg**.

Point 2: $h_2 = 788.98$ kJ/kg, $P_2 = 40$ bar
Point 3: $P_3 = 40$ bar, $T_3 = 30°C$, so $h_3 = 238.97$ kJ/kg
Point 4: sat. vapor at $P_4 = 2$ bar, so $h_4 = 81.52$ kJ/kg
Point 5: sat. liquid at $P_5 = 2$ bar, so $h_5 = -109.04$ kJ/kg
Point 6: $P_6 = 2$ bar, $T_6 = 30°C$, so $h_6 = 314.26$ kJ/kg

From equation 9.8, the fraction liquified is:

$$z = \frac{h_6 - h_3}{h_6 - h_5} = \frac{314.26 - 238.97}{314.26 - (-109.04)} = 0.178$$

The heat load on the cooler is:

$q = h_3 - h_2 = 238.97 - 788.98 = $ **−550.01 kJ/kg**

heat is transferred out of the system as expected.

Student Problems:

1. Using the results from Problem 9.1, mark points 1,2,3, and 4 on the P-H diagram for R-134a.
2. Repeat Problem 9.1 but assume the thermal efficiency of the compressor is 0.85.
3. For Problem 9.2, add a third low temperature stage using krypton as the refrigerant. You will have to get thermodynamic data for krypton out of the open literature. Make an estimate of the lowest reasonable temperature you can achieve with this three-stage system considering good engineering design limits.
4. Replace the single compressor of problem 9.4 referencing Figure 9.c.1 with a two-stage compressor system with interstage cooling. Find the savings in work requirements that would result.

Chapter 10 - Fugacity of Pure Fluids and Ideal Mixtures

This chapter of the book begins the process of developing the necessary background to deal with mixtures. The goal is to be able to understand multi-component phase equilibrium and other kinds of equilibrium processes applied to mixtures. On the microscopic scale, mixtures represent a new challenge that has not been addressed yet and that challenge relates to the interactions of molecules that are not alike. We start by considering how to characterize equilibrium processes for the general case.

10.a Characterization of Processes Moving Toward Equilibrium

Consider a frictionless piston/cylinder arrangement that contains an arbitrary number of phases and an arbitrary number of components. We require that the contents be at a uniform temperature and pressure initially and throughout a process, though the values corresponding to T and P of the system could change during the process. Initially, the contents of the device will not be at equilibrium and the whole device will be insulated with the piston locked in place. A few examples of systems that are not in equilibrium include undissolved NaCl (salt) crystals dropped into pure water, a pure liquid component 1 with pure vapor component 2 above the liquid that have not yet mixed, and two liquids that are segregated because they have not been fully mixed. Now, any changes which occur must go toward equilibrium. For example, the salt dropped in water would spontaneously begin to dissolve.

Consider now allowing this system to reversibly exchange heat and do work on the surroundings by removing the insulation and/or unlocking the piston. The Clausius inequality, equation 5.44, must apply for spontaneous changes toward equilibrium, and can be slightly rearranged to:

$$Tds^{tot} \geq dq \qquad \qquad 10.1$$

We use the superscript "tot" to indicate this change is for the sum of the all the phases and components. Even though the heat exchange between the system and surroundings has been defined to be reversible for the process we are dealing with, we still expect to see an increase in entropy of the universe because the driving forces pushing the system toward equilibrium are not differentially different and the "=" part of the inequality in equation 10.1 will not apply as we usually assume when we have reversible processes. For example, any salt crystal initially has a much higher concentration of salt than the pure water that surrounds it, and that macroscopic concentration difference drives the dissolution process. Because the concentration gradient is much greater than differentially different, the entropy of the universe will increase as the crystal dissolves.

Now, the First Law to this point in this book has always been applied to changes from an equilibrium system to another equilibrium state. If we reconsider using equation 10.1 instead of $Tds = dq_{rev}$ when we developed equation 6.1, we find instead:

$$du^{tot} \leq Tds^{tot} - Pdv^{tot} \qquad 10.2$$

This equation now applies for a system not at equilibrium that is moving toward equilibrium. Now, consider a few cases by applying equation 10.2:

1. Suppose T and V are constant while the process goes toward equilibrium. This could be accomplished by removing the insulation and thermostatting the device in a constant temperature bath while keeping the piston locked in place.

$$du^{tot} - Tds^{tot} \leq 0$$
$$d\left(u^{tot} - Ts^{tot}\right) \leq 0$$
$$da^{tot} \leq 0 \qquad 10.5$$

The Helmholtz Free Energy goes toward a minimum.

2. Suppose T and P are held constant

$$du^{tot} - Tds^{tot} + Pdv^{tot} \leq 0$$
$$d\left(u^{tot} + Pdv^{tot} - Ts^{tot}\right) \leq 0$$
$$dg^{tot} \leq 0 \qquad 10.6$$

The Gibbs Free Energy goes toward a minimum.

3. System is isolated so u and v are constant.

$$Tds^{tot} \geq 0$$
$$ds^{tot} \geq 0 \qquad 10.7$$

The entropy tends to a maximum.

The result that the Gibbs Free Energy tends to a minimum for a constant T and P process headed toward equilibrium is especially important for engineering considerations. Note that we can apply the result $dg^{tot} = 0$ as a criterion for equilibrium for any process taking place at constant T and P.

10.b Relationships Among Thermodynamic Properties for Multi-Component Systems

In this section, we revisit Chapter 6, section 6.a "Fundamental Relationships" with the intent to generalize the relationships for multi-component mixtures rather than pure components. We refer to systems of multi-component mixtures as *open systems* because these systems allow mass to move across the system boundaries. For closed systems, we wrote $U = f(S,V)$ but we now have the possibility of many components, so we write $U = f(S, V, n_1, n_2, n_3,..., n_m)$ for m components. Then, the total derivative is:

$$dU = \left(\frac{\partial U}{\partial S}\right)_{V,n_{j's}} dS + \left(\frac{\partial U}{\partial V}\right)_{S,n_{j's}} dV + \sum_{i=1}^{m} \left(\frac{\partial U}{\partial n_i}\right)_{S,V,n_{j,j\neq i}} dn_i \qquad 10.8$$

To clarify, let's look at how the n's work in the variables being held constant in the partial derivatives. In the first two terms of equation 10.8 we have "$n_{j's}$". This means moles of every species are held constant for the derivative or equivalently, the derivatives are for composition being held constant. Inside the sum in the third term, the derivative has "$n_{j,j\neq i}$". That tells us that for, say, the 4rd term in the sum, the derivative is taken with moles of all components *except* component 4 held constant. Moles of component 4 is with respect to what the derivative is taken.

In equation 10.8, when we compare with equation 6.2 (dU = TdS - PdV), we recognize the partial derivatives in the first two terms on the right-hand side as T and -P respectively. The derivative in the third term on the right-hand side is new to us. Let's see how we can choose a terminology for this derivative by considering our terminology for equation 6.2 which is for a pure component:

| Change in Energy Content of the System | = | Thermal Potential | x | Thermal Change | + | Mechanical Potential | x | Mechanical Change |

In light of this, it makes sense that we write the same idea for equation 10.8 as:

| ΔE of the System | = | Thermal Pot. | x | Thermal Change | + | Mech. Pot. | x | Mech. Change | + | Chemical Potential | x | Chemical Change |

The terminology above has "*chemical change*" as dn_i in equation 10.8. We will equate the symbol μ with the terminology *chemical potential*. We don't have to use just T and V to change the energy content of a system, we can also change the moles of any particular species. With this nomenclature, we write equation 10.8 as:

$$dU = TdS - PdV + \sum_{i=1}^{m} \mu_i dn_i \qquad 10.9$$

Similarly, for the other energy functions we get:

$$dH = TdS + VdP + \sum_{i=1}^{m} \mu_i dn_i \qquad 10.10$$

$$\boxed{dA = -SdT - PdV + \sum_{i=1}^{m} \mu_i dn_i} \qquad 10.11$$

$$\boxed{dG = -SdT + VdP + \sum_{i=1}^{m} \mu_i dn_i} \qquad 10.12$$

From equations 10.9 - 10.12, we see that:

$$\mu_i = \left.\frac{\partial H}{\partial n_i}\right)_{S,P,n_{j,j\neq i}} \qquad 10.13$$

$$\mu_i = \left.\frac{\partial A}{\partial n_i}\right)_{T,V,n_{j,j\neq i}} \qquad 10.14$$

$$\mu_i = \left.\frac{\partial G}{\partial n_i}\right)_{T,P,n_{j,j\neq i}} \qquad 10.15$$

Notice in equation 10.15 that the chemical potential is derived from the Gibbs Free Energy when *T and P are held constant*. This leads us to a definition of *partial molar properties* which are defined as properties derived from the partial of the property with respect to dn_i *at constant T and P*. Thus, by this definition, μ_i *is the partial molar Gibbs Free Energy*. In this chapter, we deal only briefly with partial molar properties, but they become extremely important later in this book. We give the partial molar properties a special nomenclature and that is a bar over the property. Thus:

$$\mu_i = \left.\frac{\partial G}{\partial n_i}\right)_{T,P,n_{j\neq i}} = \overline{G}_i \qquad 10.16$$

Note that we usually write molar properties as lower-case letters. However, partial molar properties, which are molar properties by their very nature, are written as capital letters in this book. The distinguishing feature is the over-bar that designates the property as a partial molar property. Other authors may use different nomenclature. There is no clear standard to choose, and we have chosen to use the capital letter because the partial molar property is most easily derived from the total property (as we will see), not from the molar property.

Recall now the Maxwell relations which were derived from the same equations written for pure components via the property of exact functions that if $z = f(x,y)$ with differential written $dz = Mdx + Ndy$:

$$\left.\frac{\partial M}{\partial y}\right)_x = \left.\frac{\partial N}{\partial x}\right)_y \qquad 5.6$$

We can pair any two sets of derivatives for multi-component mixtures via this equation. We now have derivatives from these equations involving every dn_i that did not appear in the Maxwell equations for pure components. Because there are so many to deal with, we will not list all the new Maxwell relations and we leave them to the interested student, but we will point out the important new Maxwell relations based upon partials with respect to moles:

$$\boxed{\left.\frac{\partial \mu_i}{\partial P}\right)_{T,n_{j's}} = \left.\frac{\partial V}{\partial n_i}\right)_{T,P,n_{j,j\neq i}} = \overline{V}_i} \qquad 10.17$$

$$\boxed{\left.\frac{\partial \mu_i}{\partial T}\right)_{P,n_{j's}} = -\left.\frac{\partial S}{\partial n_i}\right)_{T,P,n_{j,j\neq i}} = -\overline{S}_i} \qquad 10.18$$

Equations 10.17 and 10.18 tell us how the chemical potential depends upon T and P.

10.c Definition of Fugacity and Activity and the Phi-Gamma Formulation of VLE

The chemical potential represents a rigorous property that can used to compute properties of multi-component mixtures. However, there are two other functions we will now consider which are more convenient for calculations and more readily associated with properties that are directly observable. These two properties, fugacity and activity, derive from the chemical potential. Consider a pure component and let's evaluate:

$$d\overline{G}_i = d\mu_i = -\overline{S}_i dT + \overline{V}_i dP \qquad 10.19$$

Equation 10.19 is easily derived from equation 6.8 by taking the partial of each term in the equation with respect to n_i holding T and P constant to form partial molar properties for G, S, and V. Note that partial molar properties automatically have units per mole because of the nature of the derivative. When forming partial molar properties, take the derivative of the total property, not the molar property. Now, consider $\overline{V}_i$ as it relates to a pure component:

$$\overline{V}_i = \left.\frac{\partial V}{\partial n_i}\right)_{T,P,n_{j,j\neq i}}$$

Dissecting this, we see $\overline{V}_i$ is the rate at which volume increases as moles of i increases holding all other moles constant. In this case, there is only one component, so note that the volume increase is the pure component molar volume which we call v_i°. To get a better grasp of this concept, think about 10,000 moles of a pure component, then add a small amount of the same component and make the amount small enough so that it very closely represents a differential amount, say 1 mole to the 10,000 moles. dn = 1 mole in the denominator of the derivative, what is dV in the numerator? Clearly, that is the volume of one mole of the pure component which is v_i°. Now, divide equation 10.19 by dP and hold temperature constant to get:

$$\left(\frac{\partial \mu_i}{\partial P}\right)_T = v_i^\circ \qquad 10.20$$

Now, suppose the pure component we are dealing with is an ideal gas, then $v_i^\circ = RT/P$ so:

$$\left(\frac{\partial \mu_i}{\partial P}\right)_T = v_i^\circ = \frac{RT}{P} \qquad \text{(ideal gas) } 10.21$$

Integrate equation 10.21 at constant T to get:

$$\mu_i - \mu_i^\circ = RT\ln\left(\frac{P}{P_0}\right) \qquad \text{(ideal gas) } 10.22$$

where μ_i° is a reference chemical potential at pressure P_0. *We have now related the chemical potential of an ideal gas to pressure.* This relationship takes an abstract property, μ, and relates it to an understandable property, P. With this analysis for an ideal gas in mind. we define *fugacity* for a real fluid as:

$$\mu_i - \mu_i^\circ = RT\ln\left(\frac{f_i}{f_i^\circ}\right) \qquad 10.23$$

where f_i° is the standard state fugacity at a reference pressure which we generally take to be 1 bar. Students should always remember that the *fugacity of an ideal gas is it's pressure*. For a real gas, the fugacity differs from the pressure just like the volume of a real gas differs from the volume of an ideal gas. However, pressure and fugacity are very closely related for gases and just as the volume of a real gas approaches the volume of an ideal gas as pressure goes to zero, the fugacity of a real gas must approach the pressure as the pressure goes to zero and the real gas behavior approaches ideal gas behavior. *Fugacity has units of pressure*. The word "fugacity" means "escaping tendency". A component dissolved in a liquid that has a high fugacity has a tendency to go to the gas phase.

Using the relationship of the pressure of an ideal gas to the fugacity, we note for a pure gas:

$$\frac{f_i}{P} \to 1 \text{ as } P \to 0 \qquad 10.24$$

and for a mixture of ideal gases, we relate this result to the partial pressures of the components:

$$\frac{f_i}{y_i P} = \frac{f_i}{P_i} \to 1 \text{ as } P \to 0 \qquad 10.25$$

Now, let's go back to equation 10.23, and apply the equation to two phases, α and β, that are in equilibrium:

$$\mu_i^\alpha - \mu_i^{\circ\alpha} = RT\ln\left(\frac{f_i^\alpha}{f_i^{\circ\alpha}}\right) \qquad 10.26$$

$$\mu_i^\beta - \mu_i^{\circ\beta} = RT\ln\left(\frac{f_i^\beta}{f_i^{\circ\beta}}\right) \qquad 10.27$$

Consider the chemical potential of the ith species in the α and β phase. Recall that μ_i is the partial molar Gibbs Free Energy, and Gibbs Free Energy tends to a minimum when a system that is away from equilibrium moves toward equilibrium. Therefore, at equilibrium, the Gibbs Free Energy of each species must become the same in each phase. Otherwise, some of the species would move from the phase where it is at higher Gibbs Free Energy to the phase of lower Gibbs Free Energy until they are the same. We conclude that $\mu_i^\alpha = \mu_i^\beta$ at equilibrium. Equating these in equations 10.26 and 10.27 and combining, we have:

$$\mu_i^{\circ\alpha} + RT\ln\left(\frac{f_i^\alpha}{f_i^{\circ\alpha}}\right) = \mu_i^{\circ\beta} + RT\ln\left(\frac{f_i^\beta}{f_i^{\circ\beta}}\right) \qquad 10.28$$

if we take the standard states to be the same for both phases, $\mu_i^{\circ\alpha} = \mu_i^{\circ\beta}$ and $f_i^{\circ\alpha} = f_i^{\circ\beta}$ so we find $f_i^\alpha = f_i^\beta$. We can also generalize the standard states somewhat by taking them at the same temperature, but not necessarily the same pressure. To do this, we apply equation 10.23 to the two standard states. Then:

$$\mu_i^{\circ\alpha} - \mu_i^{\circ\beta} = RT\ln\left(\frac{f_i^{\circ\alpha}}{f_i^{\circ\beta}}\right) \qquad 10.29$$

and plugging back into equation 10.28, we again find $f_i^\alpha = f_i^\beta$. This is an important result and is a general result for all phase equilibrium processes. *The fugacity of each component in a mixture is the same in every phase at equilibrium.* This result derives through the chemical potential to the general result for all equilibrium processes that Gibbs Free Energy goes to a minimum.

Another direction for defining fugacity goes through this logic. For an ideal gas:

$$dG = RT\frac{dP}{P} = RT\, d\ln(P) \qquad 10.30$$

For a pure real gas, the *definition* of fugacity is:

$$dG_i = RT\, d\ln(f_i) \qquad 10.31$$

Here, the subscript i is not strictly required since there is only one species, but we generally keep this subscript to be able to identify more than one pure species. Now, for a real gas in a mixture:

$$d\overline{G}_i = RT\, d\ln(\overline{f}_i) \qquad 10.32$$

Next, we define the *fugacity coefficient*, φ_i according to:

$$\varphi_i = \frac{\overline{f}_i}{y_i P} \qquad 10.33$$

With this definition, $\varphi_i \to 1$ as $P \to 0$. Also, φ_i *for an ideal gas is 1*. We also define the *activity* as:

$$a_i = \frac{\overline{f}_i}{f_i^\circ} \qquad 10.34$$

We will more often apply the concept of activity to liquids than to gases. Note that at phase equilibrium, we have $\overline{f}_i^\alpha = \overline{f}_i^\beta$ but $a_i^\alpha \neq a_i^\beta$ unless $f_i^{\circ\alpha} = f_i^{\circ\beta}$ which is not the usual case. We will consider our choices for standard states later. Note that unlike fugacity, *activity is unitless*.

Consider the application of equation 10.34 to a multi-component liquid. If we choose f_i° to be the pure component liquid i (which we usually do), then clearly, the mole fraction of a species in the mixture is a first approximation to the activity of component i. With this in mind, we define the *activity coefficient* γ_i as:

$$\gamma_i = \frac{a_i}{x_i} = \frac{\overline{f}_i}{f_i^\circ x_i} \qquad 10.35$$

Let's see how all this goes together. Suppose we have a multi-component mixture at some fixed temperature and pressure that is allowed to reach vapor-liquid equilibrium (VLE). If we equate the fugacity in the vapor phase written in terms of the fugacity coefficient and partial pressure and equate the fugacity of the liquid through the activity coefficient, we get:

$$f_i^V = f_i^L \qquad 10.36$$

$$\boxed{\varphi_i y_i P = \gamma_i x_i f_i^\circ} \qquad 10.37$$

Equation 10.37 is known as the *phi-gamma formulation for VLE*. Note that in equation 10.37 we have chosen to write the vapor phase fugacity in the fugacity coefficient form while we have written the liquid phase fugacity in the activity coefficient form. The reason for this will become more apparent as we delve deeper into phase equilibrium, but we can say for now that the evaluation of fugacity coefficients for a component in a liquid mixture is difficult from equations of state for the component. In the development of equation 10.37, we note that the standard state fugacity for the vapor phase is the ideal gas at 1 bar, while for the liquid phase, the standard state fugacity is the pure liquid at the temperature and pressure of the system as the standard state. Since the standard states are different, the activity of a component in one phase is not the same as it's activity in the other phase. However as in equations 10.36 and 10.37, the fugacities are the same.

10.d Raoult's Law

Consider some simple cases. Suppose we have an ideal gas in the vapor phase, then $\varphi_{i,s} = 1$. We will define an ideal liquid phase as one in which $\gamma_{i's} = 1$, and this leaves us with:

$$y_i P = x_i f_i^\circ \qquad 10.38$$

Let's see if we can determine an approximate expression for f_i° which is the *fugacity of the pure component liquid at the temperature and pressure of the system*. For the pure component at the same temperature as the multi-component system, if the vapor above the pure component is an ideal gas, then $f_i^\circ = P_i^s$, the vapor pressure or saturation pressure of the pure component. This is true because the vapor and liquid are at equilibrium for the pure component so the fugacities are the same in each phase, and the fugacity of the vapor is the same as it's pressure for an ideal gas. All that we would need is to apply a pressure correction from the vapor pressure to the pressure of the multi-component mixture VLE. We will take up the exact calculation to this correction shortly, but for now we will assume that the pressure correction is small and we will use $f_i^\circ = P_i^s$. Then:

$$\boxed{y_i P = x_i P_i^s} \qquad \text{(ideal solutions)} \quad 10.39$$

Equation 10.39 is known as *Raoult's Law*. It is applicable for low pressures (ideal gas region) and for ideal liquid solutions. We should use equation 10.39 only sparingly or for illustrative purposes (which we do shortly) because liquid solutions are only near ideal for mixtures of very similar components under limited conditions.

Using Raoult's Law, we can get an overview of the kinds of important calculations we can do for VLE systems. Suppose we have a liquid of known composition or a gas of known composition. In either case, we often want to know what temperature or pressure represents the onset of VLE and at that point, we want to know the composition of the differentially small amount of the other phase that exists. These calculations breakdown into 4 categories:

1. Bubble point pressure - Calculate P and $y_{i's}$ given T and $x_{i's}$
2. Bubble point (temperature) - Calculate T and $y_{i's}$ given P and $x_{i's}$
3. Dew point pressure - Calculate P and $x_{i's}$ given T and $y_{i's}$
4. Dew point (temperature) - Calculate T and $x_{i's}$ given P and $y_{i's}$

Category 2 and 4 above are generally referred to simply as "bubble point" and "dew point" calculations, respectively, as this terminology by convention implies that the temperature is being sought.

Calculations get a bit messy for multi-component mixtures, so let focus on a binary (two component) system. Then, Raoult's Law gives:

$$x_1 P_1^s = y_1 P = P_1 \qquad 10.40$$

$$x_2 P_2^s = y_2 P = P_2 \qquad 10.41$$

Here we have included $y_i P = P_i$ which is also known as the partial pressure. Summing:

$$P = x_1 P_1^s + x_2 P_2^s \quad \left(\text{multi} - \text{component:} \quad P = \sum_i x_i P_i^S\right) \qquad 10.42$$

The bubble point pressure calculation (category 1 above) is the simplest calculation. This is the pressure at which the first (infinitesimally small) bubble of vapor would appear as the pressure above a subcooled liquid is slowly lowered while holding the temperature constant. Implied in the equations is the assumption that the bubble is in phase equilibrium with the liquid. Since we know T, we know P_1^s and P_2^s which can be determined by an Antoine Equation calculation (see Appendix E). The total pressure as a function of x_1 then comes from equation 10.42, noting that $x_2 = 1 - x_1$. We can see from equation 10.42 that P is linear in x_1, and this is reflected in the line "P vs x_1" on Figure 10.d.1. At $x_1 = 0$, we

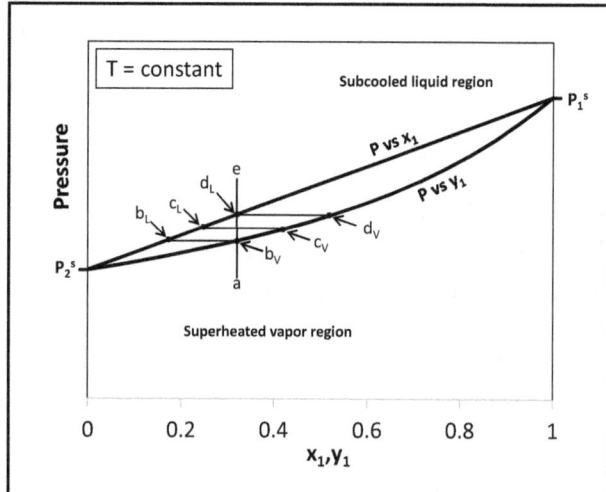

Figure 10.d.1. Bubble point pressure calculations using Raoult's Law.

have pure component 2, and P is the vapor pressure of component 2 at that point. For $x_1 = 1$ we have pure component 1, and the pressure is the vapor pressure of component 1 at that point. These are also reflected in Figure 10.d.1.

Equation 10.42 is also generally used to solve the bubble point problem (category 2 above). The bubble point is the temperature at which the first (infinitesimally small) bubble of vapor would appear when a subcooled liquid mixture is heated slowly while holding pressure constant. Knowing P and x's (the composition of the liquid), everything in equation 10.42 except for the saturation pressures are known. However, saturation pressures are known to be functions of temperature (Appendix E) but they are not simple functions of T. So, all we need to do is to iterate on temperature until equation 10.42 is solved. We can put this into a single equation as:

$$P = x_1 \exp\left(A_1 - \frac{B_1}{T+C_1}\right) + x_2 \exp\left(A_2 - \frac{B_2}{T+C_2}\right) \qquad 10.43$$

Antoine constants A_i's, B_i's, and C_i's come from Appendix E for appropriate components, x_1 and x_2 are known, and P is known. T is the only unknown in equation 10.43. Because equation 10.43 is not simple and solvable for T directly, we iterate on T (in K, since the Antoine equation in Appendix E requires K) until the desired pressure is reproduced. Once T is known, P_1^s and P_2^s are known and the vapor composition comes from equations 10.40 and 10.41. The curve generated by this function is shown in Figure 10.d.2 as "T vs x_1".

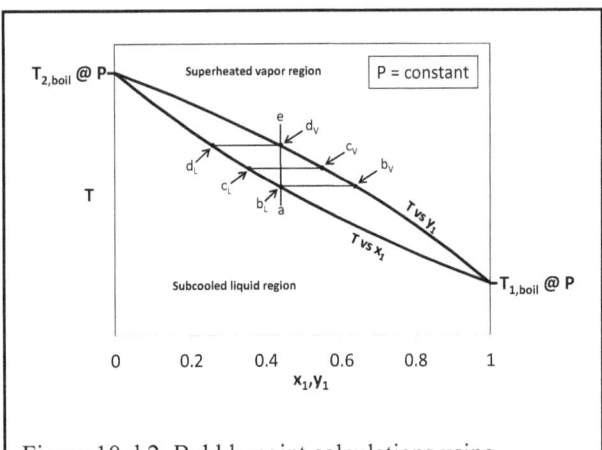

Figure 10.d.2. Bubble point calculations using Raoult's Law.

For dew point calculations (categories 3 and 4 above) we know y's instead of x's. Solving equation 10.40 and 10.41 for x's:

$$x_1 = \frac{y_1 P}{P_1^s} \quad \text{and} \quad x_2 = \frac{y_2 P}{P_2^s} \qquad 10.44$$

summing and using $x_1 + x_2 = 1$:

$$\frac{y_1 P}{P_1^s} + \frac{y_2 P}{P_2^s} = 1 \quad \left(\text{multi-component:} \quad P\sum \frac{y_i}{P_i^s} = 1\right) \qquad 10.45$$

If we know the temperature and we are looking for pressure (category 3 above), using P_1^s and P_2^s determined using the known T, we can solve equation 10.45 directly for pressure as that remains the only unknown. Varying y's from 0 to 1, this yields the curve "P vs y_1" in Figure 10.d.1.

The dew point calculation (category 4 above) is similar to the bubble point calculation. We can still use equation 10.45, but since T is unknown and P is known instead, we must iterate on temperature to get values for P_1^s and P_2^s to put into equation 10.45. We iterate on temperature until the left-hand side of equation 10.45 equals 1. Varying x's from 0 to 1 yields the curve "T vs y_1" in Figure 10.d.2.

Let's dissect Figures 10.d.1 and 10.d.2 beginning with 10.d.1. Imagine a binary solution beginning at point "a" in the superheated vapor region, and imagine the solution being contained in a frictionless piston/cylinder arrangement. All of Figure 10.d.1 is constructed for a single temperature, so the piston/cylinder arrangement is thermostated in a constant temperature bath. Force on the piston establishes the pressure of the system, and initially, the pressure is below the saturation pressure for the vapor mixture. Now, consider what happens as we begin increasing P, moving toward point "e". We follow the line a-e toward point b_V initially as the composition of the vapor doesn't change while P goes up, but eventually the pressure is high enough to reach b_V. At that point, an infinitesimally small droplet of liquid forms and b_V is known as the dew point. The composition of that liquid is given at b_L. Follow directly down and read x_1 that corresponds to b_L for the composition of that liquid. The segment b_V-b_L is known as a *tie-line*. Any horizontal line cutting through the P vs x_1 and P vs y_1 curves gives the composition of the liquid where it intersects the P vs x_1 curve in equilibrium the vapor where it intersects the P vs y_1 curve.

Going back to the line a-e, as we continue to increase the pressure headed toward d_L, macroscopic amounts of liquid have formed and along the way, the c_V-c_L tie-line represents the state of the system. Eventually, we come to point d_L where the last bubble of vapor is disappearing. This is the bubble point. The composition of the last bubble of vapor has the composition given by d_V. If we continue to increase the pressure as we head toward point "e", we are simply compressing a liquid headed into the subcooled liquid region.

Many similarities of Figure 10.d.1 with the constant P diagram in Figure 10.d.2 exist. Here, we will begin with a subcooled liquid at point "a" and we begin heating the mixture holding the pressure on the piston constant. When we get to b_L, the first bubble of vapor appears and that vapor has the composition given by b_V. As we continue to heat, we pass the c_L-c_V tie-line until we get to d_V. At that point, which is known as the dew point, the last droplet of liquid, which has composition given by d_L, is disappearing. As we continue to heat, we move into the superheated vapor region to point "e".

As an overview of these calculations for the binary system with Raoult's Law, note that when we know T and are looking for P, the calculation is direct and simple. When we know P and are looking for T, the process requires us to iterate to find T because the temperature functionality of the Antoine Equation is not simple and we need the saturation pressures of each component for the calculation.

Consider Raoult's Law as applied to systems with more than 2 components. These systems cannot be displayed easily on plots such as given in Figures 10.d.1 and 10.d.2 because they are multi-dimensional. However, the solution methods are generally the same as the binary case. When we know x's (the liquid composition), for determining P when T is known, equation 10.42 can be extended to multi-component mixtures:

$$P = \sum_i x_i P_i^s \qquad 10.46$$

When we know T and the set of x's, we know the set of P_i^s and P is readily determined. If we know P and x's, we have to iterate on T with an appropriate set of Antoine constants to solve equation 10.46.

If we know the set of y's and are doing dew point calculations, equation 10.45 can be written:

$$\sum_i \frac{y_i}{P_i^s} = \frac{1}{P} \qquad 10.47$$

Again, if we know T, a direct calculation of P is possible since we know the set of P_i^s. Otherwise, we must iterate on T to solve equation 10.47 if we know P and are looking for T.

We can also use *relative volatilities* with Raoult's law to sometimes simplify calculations. Relative volatility of component i with respect to n, denoted $\alpha_{i,n}$, is defined by:

$$\alpha_{i,n} = \frac{P_i^s}{P_n^s} \qquad 10.48$$

To use this quantity, start with:

$$y_i = \frac{x_i P_i^s}{\sum_j x_j P_j^s} \qquad 10.49$$

divide by P_n^s:

$$y_i = \frac{x_i \alpha_{i,n}}{\sum_j x_j \alpha_{j,n}} \qquad 10.50$$

The advantage of using relative volatility is that it is less temperature sensitive than the raw vapor pressures of the individual components making iterations in temperature more easily accomplished in many cases.

We also define *K values* as:

$$K_i = \frac{y_i}{x_i} \qquad 10.51$$

K values can be tabulated or depicted in various graphical forms. If Raoult's Law holds:

$$K_i = \frac{P_i^s}{P_i} \qquad 10.52$$

and values of K_i can be determined given the pressure and temperature (vapor pressures come from the Antoine equation). Then:

$$\alpha_{i,n} = \frac{K_i}{K_n} \qquad 10.53$$

$$\sum K_i x_i = 1 \qquad 10.54$$

$$\sum \frac{y_i}{K_i} = 1 \qquad 10.55$$

Consider an equilibrium flash process depicted in Figure 10.d.3. In this process, a relatively high temperature and pressure multi-component mixture flows through a "flash" valve to lower its pressure. The temperature of the mixture also drops as more vapor is formed in the lower pressure environment. The VLE mixture then flows to a flash drum where heavier components are taken off at the bottom while lighter components go the vapor overhead stream. We assume that VLE exists between the liquid and vapor streams leaving the device, Raoult's law holds for VLE, and the process is at steady state. An overall mole balance yields:

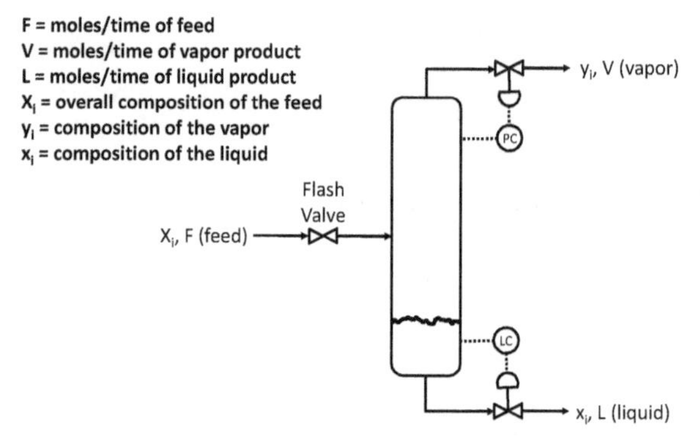
F = moles/time of feed
V = moles/time of vapor product
L = moles/time of liquid product
X_i = overall composition of the feed
y_i = composition of the vapor
x_i = composition of the liquid
y_i, V (vapor)
PC
Flash
Valve
X_i, F (feed)
LC
x_i, L (liquid)

Figure 10.d.3. Schematic of an equilibrium flash process

$$F = V + L \qquad 10.56$$

and an individual component balance yields:

$$x_i L + y_i V = X_i F = (L + V) X_i \qquad 10.57$$

dividing through by $x_i L$ and using equation 10.51:

$$1 + K_i\left(\frac{V}{L}\right) = \frac{X_i}{x_i}\left(1 + \frac{V}{L}\right) \qquad 10.58$$

$$x_i = \frac{X_i\left(1 + \frac{V}{L}\right)}{1 + K_i\left(\frac{V}{L}\right)} \qquad 10.59$$

similarly:

$$y_i = \frac{X_i\left(1 + \frac{L}{V}\right)}{1 + K_i\left(\frac{L}{V}\right)} \qquad 10.60$$

$\left(\frac{K_i V}{L}\right)$ is known as the stripping factor and $\left(\frac{L}{K_i V}\right)$ is known as the absorption factor. For any given component, these factors determine if components will concentrate more in the vapor or liquid products.

Even though Raoult's Law does not accurately represent actual VLE systems except in rare cases, the process of understanding the calculations and methods of solution using Raoult's Law can be extended to non-ideal systems. Raoult's Law is therefore very important to the student for understanding the future directions of this book. Remember that the goal is the phi-gamma formulation of VLE represented by equation 10.37. The solution to Raoult's Law for a system is often useful as a first guess for solving equation 10.37.

10.e Fugacities of Pure Components

10.e.i Fugacities of Pure Gases

Let's go back to equation 10.31 which was written for a pure component and defined fugacity:

$$dG_i = RT d\ln(f_i) \qquad 10.31$$

If we integrate this equation from P = 0 where we have an ideal gas to P we get:

$$G^R = G_i - G_i^{IG} = RT\ln\left(\frac{f_i}{P}\right) = RT\ln(\varphi_i) \qquad 10.61$$

We have substituted G^R into the equation because the definition of a residual property is the difference between the actual function and the ideal function which is what appears in equation 10.61, and we have used the definition of the fugacity coefficient as well.

Now if we use the residual enthalpy and entropy given in equations 6.31 and 6.32 to form $G^R = H^R - TS^R$, we have an elimination of two terms when combining, and using equation 10.61, we find:

$$\ln(\varphi_i) = \int_0^P (z-1)\frac{dP}{P} \qquad \text{(pure component, constant T)} \quad 10.62$$

Again, this is for the pure component i and there is a reminder that the integral is done at constant T. Just as a check, we should note that for an ideal gas, $z = 1$ and equation 10.62 results in $\varphi_i = 1$ as expected. Just as we can evaluate z using several different equations of state including the generalized virial coefficient of Pitzer and the Lee-Kesler tables, we can evaluate equation 10.62 to find values for φ for a pure component. Also, the Lee-Kesler tables, having evaluated equation 10.62 using the generalized compressibility factor correlation, include parameters φ^0 and φ^1 as functions of T_r and P_r. These can be combined to get the fugacity coefficient according to:

$$\varphi = \varphi^0 (\varphi^1)^\omega \qquad 10.63$$

The fugacity of the pure gas then comes from $f_i = \varphi_i P$. If we consider the Pitzer generalized virial coefficient given by equation 3.33 and substitute into equation 10.62, we can evaluate the integral to get:

$$\ln(\varphi_i) = \frac{P_r}{T_r}\left(B^0 + \omega B^1\right) \qquad \text{(pure component)} \quad 10.64$$

These generalized correlations along with equations 10.63 and 10.64 represent good ways to get fugacities for pure components and they are generally accurate enough for engineering calculations.

10.e.ii Fugacities of Pure Liquids and Solids (Condensed Phases)

Consider the fugacity of a pure liquid at T and P where P is greater than the saturation pressure, which is equivalent to a subcooled liquid. To evaluate this fugacity, we will start with the pure liquid in equilibrium with the pure vapor at T. Since the liquid and vapor are at equilibrium at saturation conditions, $f_i^V = f_i^L = \varphi_i^s P_i^s$. Now, all we need to do is apply the pressure correction going from $P_i^s \rightarrow P$ to the liquid. Going back to equation 10.20 applied to a pure component:

$$\left.\frac{\partial \mu_i}{\partial P}\right)_{T,n_j} = v_i^\circ \qquad 10.20$$

Previously, we used this equation to determine the fugacity of a pure ideal gas by substituting RT/P for the volume. In this case, we are dealing with a liquid so v_i° is the molar volume of the liquid and equation 10.20 tells us how to correct the chemical potential for pressure. Multiply equation 10.20 through by dP and integrate from $P_i^s \to P$ at constant T (since the partial derivative in equation 10.20 is at constant T):

$$\mu_i - \mu_i^s = RT \ln\left(\frac{f_i}{f_i^s}\right) = \int_{P_i^s}^{P} v_i^\circ dP \qquad 10.65$$

with $f_i^s = \varphi_i^s P_i^s$ we can rearrange this to:

$$f_i^c = \varphi_i^s P_i^s \exp\left(\int_{P_i^s}^{P} \frac{v_i^c}{RT} dP\right) \qquad 10.66$$

Here, we have used the superscript "c" to indicate a condensed phase, as equation 10.66 could also apply to a pure solid that exhibits a vapor pressure (solid CO_2, "dry ice" comes to mind). Do not lose sight of the fact that the "s" superscript is for saturation conditions. Generally, except at very high pressures, v_i^c can be considered constant with pressure and assuming that is the case we arrive at the final result:

$$\boxed{f_i^c = \varphi_i^s P_i^s \exp\left[\frac{v_i^c}{RT}(P - P_i^s)\right]} \qquad 10.67$$

The exponential term in equation 10.67 is called the Poynting correction factor. Depending upon the size of the condensed volume, for P not too far from P_i^s (say 10 bar or less) the Poynting correction factor can generally be ignored. For higher pressure differences, it can become very important, and when $P \gg P_i^s$ (~1000 bar difference) the Poynting correction factor is generally greater than one order of magnitude or more. A rule of thumb is that the Poynting correction factor should be evaluated for pressure differences above 10 bar.

Equation 10.67 says that for low pressures, the Poynting correction factor is negligible and the fugacity of a condensed phase is probably not far from it's saturation pressure at the same temperature ($\varphi_i^s \approx 1$), and in the case where the saturated vapor above the saturated liquid is an ideal gas, the pure condensed phase fugacity is equal to it's vapor pressure. Go back to section 10.d and it should now be clear why we used the substitution $f_i^\circ = P_i^s$ to get Raoult's Law from the phi-gamma formulation of VLE.

10.f Properties of Mixtures of Ideal Gases

Consider a mixture of ideal gases, and let's compute the partial molar volume. Now clearly, since individual molecules of a pure ideal gas do not have any energetic interactions, a mixture of ideal gases would also have neither interactions of like molecules, nor unlike molecules. Thus:

$$V^{IG} = n_T \frac{RT}{P} = \sum n_i \frac{RT}{P} \qquad 10.68$$

A common derivative we find often in calculations going forward is:

$$\left(\frac{\partial n_T}{\partial n_i}\right)_{T,P,n_{j,j\neq i}} = \frac{\partial(n_1 + n_2 + n_3 + ...)}{\partial n_i}\bigg)_{T,P,n_{j,j\neq i}} = 1 \qquad 10.69$$

For this derivative, every n in the sum is held constant except for the ith term, so we wind up with that term in the sum being the only one contributing to the derivative. Applying this derivative to equation 10.68:

$$\overline{V}_i^{IG} = \frac{\partial V^{IG}}{\partial n_i}\bigg)_{T,P,n_{j,j\neq i}} = \frac{RT}{P} \qquad 10.70$$

Consider the enthalpy of an ideal gas mixture. Since H^{IG} is independent of pressure and volume because of the lack of molecular interactions, and since in a mixture of ideal gases, there must be no interactions of either like or unlike molecules:

$$H^{IG} = \sum y_i H_i^{IG} = \sum n_i h_i^{IG} \qquad 10.71$$

Entropy is a function of pressure for an ideal gas, and:

$$dS_i^{IG} = -nR d\ln(P) \qquad \text{(constant T)} \quad 10.72$$

Consider individual ideal gases consisting of Σn_i moles of unmixed gases at T and P. Let these gases mix at constant T and P until we have a homogeneous mixture, and $P = \Sigma y_i P$. Each individual species has thus experienced a change from P to P_i or from P to $y_i P$. Let's integrate equation 10.72 for this case:

$$S_i^{IG}(T,P) - S_i^{IG}(T,P_i) = -nR\ln\left(\frac{P}{y_i P}\right) = nR\ln(y_i) \qquad 10.73$$

Then, for a mixture of ideal gases:

$$S^{IG} = \sum y_i S_i^{IG} - nR\sum y_i \ln(y_i) = \sum n_i s_i^{IG} - R\sum n_i \ln(y_i) \qquad 10.74$$

Note that in equation 10.71, h_i^{IG} is independent of pressure but s_i^{IG} in equation 10.74 is dependent on pressure. The final term in equation 10.74 is the entropy change upon mixing the ideal gases, not the entire pressure dependence of s^{IG}.

10.g Chapter 10 Example Problems

10.1 A gaseous mixture of $y_1 = 0.3$ mole fraction of cyclopentane/ $y_2 = 0.70$ mole fraction of cyclohexane is at 0.900 MPa. Assume Raoult's Law is valid for the following calculations.
a. What is the dew point of the mixture?
b. What are the fugacities of cyclohexane at the dew point in *both* the gas and (infinitesimally small droplet of) liquid phases - i.e., $f_2^V = ?$ and $f_2^L = ?$

c. What is the composition of the (infinitesimally small droplet of) liquid in equilibrium with the gas at its dew point?

Solution:

a. This is what is refereed to as a dew point calculation. Raoult's Law is written as:

$$x_1 P_1^s = y_1 P = P_1 \quad \text{and} \quad x_2 P_2^s = y_2 P = P_2$$

or solving for x_1 and x_2, then summing using $x_1 + x_2 = 1$ gives:

$$\frac{y_1 P}{P_1^s} + \frac{y_2 P}{P_2^s} = 1 \quad \text{plug in y's and P:} \quad \frac{0.3 \cdot 900 \text{kPa}}{P_1^s} + \frac{0.7 \cdot 900 \text{kPa}}{P_2^s} = 1 \qquad 10.75$$

We will use an Antoine equation for each component from Appendix E:

constants for $\ln(P^{sat}$ in kPa$) = A - B/[T(K) + C]$:

	A	B	C
cyclopentane (1)	14.3549	2887.91	-26.343
cyclohexane (2)	14.1996	3063.83	-34.723

When we plug the Antoine equation with constants for each component into equation 10.75, we are left with an equation in which only T is unknown. We will need to iterate on T until equation 10.75 is solved. Since this is the first calculation where we are doing this, we will show a couple of trials. Try T = 300 K. The Antoine equation calculations then yield:

	T (K)	P_i^s (kPa)
cyclopentane (1)	300	44.79
cyclohexane (2)	300	14.15

Plug these P_i^s's into equation 10.75 and we get the LHS = 50.538 whereas we want it to be 1. Clearly, these saturation pressures are too small so we want a higher T. Try T = 400 K:

	T (K)	P_i^s (kPa)
cyclopentane (1)	400	754.58
cyclohexane (2)	400	334.23

Now the LHS of equation 10.75 is 2.234. However, we need a still higher T. Continuing the iteration, we find:

	T (K)	P_i^s (kPa)
cyclopentane (1)	439.58	1581.96
cyclohexane (2)	439.58	758.9

and the LHS of equation 10.75 is 1.00. So, the dew point is at **T = 439.58 K**.

b. Raoult's Law is built upon $f_i^V = f_i^L$ - fugacities of each component are the same in each phase - so for cyclohexane $\mathbf{f_2^V = f_2^L}$. $\mathbf{f_2^V = y_2 P = 0.7 \cdot 900 kPa = 630\ kPa = f_2^L}$.

c. To find the composition of the (infinitesimally small) droplet of liquid:

$$x_1 P_1^s = y_1 P = P_1 \quad \text{and} \quad x_2 P_2^s = y_2 P = P_2$$

We now know everything in these equations except for x_1 an x_2. Solving we find $\mathbf{x_1 = 0.171}$ and $\mathbf{x_2 = 0.829}$.

10.2 A liquid mixture of $x_1 = 0.3$ mole fraction of CCl_4 (carbon tetrachloride)/ $x_2 = 0.70$ mole fraction of $CHCl_3$ (chloroform) is at 100 kPa. Assume Raoult's Law is valid for the following calculations.
a. What is the bubble point of the mixture?
b. What is the composition of the (infinitesimally small) bubble in equilibrium with the liquid at its bubble point?

Solution:

a. This is known as a bubble point calculation. For Raoult's Law:

$$P = y_1 P + y_2 P = x_1 P_1^s + x_2 P_2^s$$

Plugging in values that we know, and using Antoine's equation:

$$100\,\text{kPa} = 0.3\exp\left(14.287 - \frac{3090.47}{T-30.993}\right) + 0.7\exp\left(14.881 - \frac{3232.895}{T-20.011}\right)$$

Solving iteratively for T, we get **T = 338.48 K**.

b. With T = 338.48 K, P_1^s = 69.13 kPa. $y_1 P = x_1 P_1^s$ so y_1 = 0.3 · 69.13 kPa/100 kPa = **0.2074**. Similarly, y_2 = **0.7926**.

10.3 A mixture of 0.60 mole fraction acetone/ 0.40 mole fraction of acetonitrile is at 50°C. Assume Raoult's Law is valid for the following calculations:
a. If the mixture is a liquid, what is the bubble point pressure and what is the composition of the (infinitesimally small bubble of) vapor at the bubble point pressure?
b. If the mixture is a vapor, what is the dew point pressure and what is the composition of the (infinitesimally small droplet of) liquid at the dew point pressure?

Solution:

Since we are computing pressure in both instances in this problem, iteration will not be required.

a. x_1 = 0.60, x_2 = 0.40. At T = 50°C = 323.15 K, using the Antoine constants from Appendix E for acetone (1) and acetonitrile (2), P_1^s = 80.36 kPa and P_2^s = 33.44 kPa.

Then P_1 = 0.60 · 80.36 = 48.22 kPa and P_2 = 0.40 · 33.44 = 13.37 kPa. The bubble point pressure is $P_1 + P_2$ = **P = 61.59 kPa.**

With $y_1 P = x_1 P_1^s$ or y_1 · 61.59 = 0.60 · 80.36, **y_1 = 0.783** and similarly, **y_2 = 0.217** which is composition of the vapor phase.

b. y_1 = 0.60, y_2 = 0.40. At T = 50°C = 323.15 K, the vapor pressures are the same as in part "a", P_1^s = 80.36 kPa and P_2^s = 33.44 kPa. Plugging numbers into equation 10.45: is:

$$\frac{y_1 P}{P_1^s} + \frac{y_2 P}{P_2^s} = \frac{0.6 \cdot P}{80.36} + \frac{0.4 \cdot P}{33.44} = 1$$

Solving for P we find the dew point pressure, **P = 51.47 kPa**.

Then with $y_1 P = x_1 P_1^s$ or 0.6 · 51.47 = x_1 · 80.36 giving **x_1 = 0.348** and similarly, **x_1 = 0.616** which is the composition of the liquid phase.

10.4 A liquid mixture of x_1 = 0.25 mole fraction of n-pentane/ x_2 = 0.75 mole fraction of n-hexane is at 400 kPa. Assume Raoult's Law is valid for the following calculations:

a. What is the bubble point of the mixture?
b. What is the composition of the (infinitesimally small bubble of) vapor in equilibrium with the liquid at its bubble point?

Solution:

a. This problem is exactly like problem 10.2 which is a bubble point calculation. When we plug in Antoine constants for n-pentane and n-hexane instead of for the components in problem 10.2 and iterate on T until we get P = 400 kPa, we find **T = 381.02 K**. At this T, P_1^s = 706.88 kPa and P_2^s = 297.71 kPa. Students can check their calculations against these numbers.

b. y_1 = 0.25 · 706.88 kPa/400 kPa = 0.442. Similarly, y_2 = 0.75 · 297.71 kPa/400 kPa = 0.558

10.5 Draw the VLE diagram P vs. x_1 and P vs. y_1 for the system dichloromethane (1) /1-propanol (2) at 85°C using Raoult's Law.

Solution:

Using Antoine's equation with constants from Appendix E for the two components at 85°C (= 358.15 K) yields P_1^s = 399.58 kPa and P_2^s = 63.96 kPa. Repeatedly performing the dew point pressure and bubble point pressure calculation like in problem 10.4 for values of x_1 and y_1 ranging from 0 - 1 yields Figure 10.g.1.

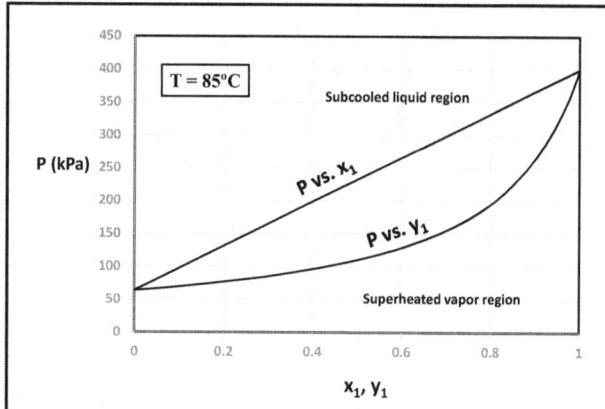

Figure 10.g.1. P vs. x_1, y_1 for the system dichloromethane(1)/1-propanol(2) at T = 85°C using Raoult's Law.

10.6 Draw the VLE diagram T vs. x_1 and T vs. y_1 for the system dichloromethane (1) /1-propanol (2) at 250 kPa using Raoult's Law.

Solution:

This requires considerably more effort than problem 10.5 because for each

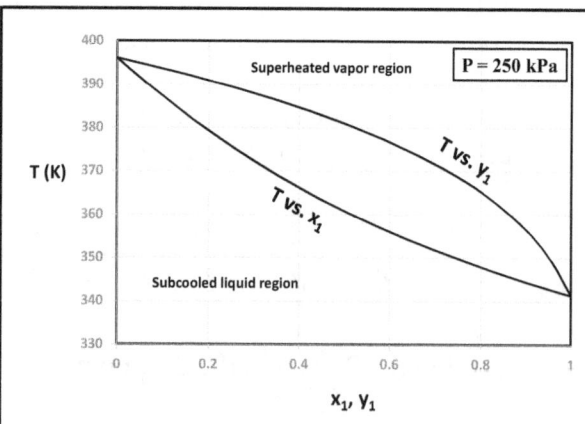

Figure 10.g.2. T vs. x_1, y_1 for the system dichloromethane(1)/1-propanol(2) at 250 kPa using Raoult's Law.

value of x_1 or y_1, we have to perform an iterative calculation to find T. Nonetheless, the calculations are not difficult using good computing tools.

10.7 Find the fugacity of pure n-hexane at 500°C and 20 atm. Assume the Poynting correction factor is 1.

Solution:

First, we need to decide if n-hexane is a gas or liquid at these conditions. Appendix C gives $T_c = 507.8$ K $= 234.7$°C and $P_c = 30.42$ bar for n-hexane, so it is a gas at 500°C and 20 atm.

Next, let's compute reduced properties.

$P_r = 20 \cdot 1.0133/30.42 = 0.6662$ and $T_r = (500 + 273.15)/507.8 = 1.522$.

These fall in the region (0.8 <Tr < 4.0 and Pr < 2.0) where the Pitzer generalized virial coefficient correlation is useful, so we will use that correlation to find φ using equation 10.64:

$$\ln(\varphi_i) = \frac{P_r}{T_r}\left(B^0 + \omega B^1\right)$$

From equations 3.36 and 3.37 which give the components of the virial coefficient in terms of reduced temperature:

$B^0 = -0.1325$ and $B^1 = 0.1097$

With ω = 0.307 for n-hexane from Appendix C:

$$\ln(\varphi) = \frac{0.6662}{1.522}\left(-0.1325 + 0.307 \cdot 0.1097\right) = -0.04326$$

φ = 0.9577

f = 0.9577 · 20 atm = 19.15 atm.

We could also have used the Lee-Kesler Tables or another EOS to evaluate this fugacity.

10.8 Calculate the fugacity of liquid chloroform (trichloromethane) at 100°C and 100 atm. The density of liquid chloroform may be taken to be constant at 1.4 g/cm³.

Solution:

For a liquid (condensed phase), we need to evaluate equation 10.67:

$$f_i^c = \varphi_i^s P_i^s \exp\left[\frac{v_i^c}{RT}(P - P_i^s)\right]$$

Using the constants for chloroform for the Antoine equation in Appendix E:

$$P_{CHCl_3}^s = \exp\left(14.8810 - \frac{3232.9}{373.15K - 20.011}\right) = 306.81 \text{kPa} = 3.028 \text{atm}$$

We need to evaluate $\varphi_{CHCl_3}^s$ at 100°C = 373.15 K and 3.028 atm = 3.0681 bar (*not at 100 atm!*).

We will use the Lee-Kesler table (Appendix D) to illustrate it's use for determining fugacities. For chloroform, T_c = 536.1 K, P_c = 56.00 bar, and ω = 0.238.

T_r = 373.15/536.1 = 0.696 and P_r = 3.0681/56.00 = 0.0548

Looking in the Lee-Kesler tables at T_r = 0.696 and P_r = 0.0548

φ^0 = 0.947 and φ^1 = 0.946

Plugging into equation 10.63:

$$\varphi_{CHCl_3}^s = 0.947 \cdot 0.946^{0.238} = 0.935.$$

The last term we need to evaluate in equation 10.67 is the molar volume of the liquid:

$$v_{CHCl_3}^c = \frac{1 \text{cm}^3}{1.4 \text{g}} \times \frac{119.369 \text{g}}{\text{mole}} = 85.2 \frac{\text{cm}^3}{\text{mole}}$$

Plugging everything we know into equation 10.67 and making sure we have the units correct:

$$f_{CHCl_3} = 0.935 \cdot 3.028 \text{atm} \cdot \exp\left[\frac{85.2 \frac{\text{cm}^3}{\text{mole}}}{82.06 \frac{\text{cm}^3 \cdot \text{atm}}{\text{mole} \cdot K} \cdot 373.15K}(100 - 3.028) \text{atm}\right]$$

This gives $\mathbf{f_{CHCl_3} = 3.71}$ **atm.**

The Poynting correction factor (the exponential term in the equation above) is 1.31, so ignoring it would result in a 30% error. 100 atm is too high a pressure to ignore the Poynting correction. Note, however, the fugacity of $CHCl_3$ is not too far from it's vapor pressure of 3.028 atm. Raoult's Law assumes the vapor pressure is the fugacity of the pure liquid.

10.9 **Find the fugacity of pure liquid benzene at 50 bar and 73°C.**

Solution:

For benzene, P^s at 73°C = 346.15 K is 0.7917 bar using the Antoine equation and constants in Appendix E.

For φ^s, we want to evaluate at saturation conditions.

T_r = 346.15/562.0 = 0.616. P_r = 0.7917/49.06 = 0.0161

This region is where the Pitzer generalized virial coefficient is useful so we will use that EOS.

B^0 = -0.8334 and B^1 = -1.1780. Then:

$$\ln(\varphi^s) = \frac{P_r}{T_r}(B^0 + \omega B^1) = \frac{0.0161}{0.616}(-0.8334 + 0.217 \cdot (-1.1780)) = -0.0285$$

φ^s = 0.972.

We need a molar volume of liquid benzene to evaluate the Poynting correction (PC) and we will get that from the open literature. We find v^c = 94.2 cm³/mole. Another alternative would be to use a correlation such as the Racket equation (section 3.f) or the COSTALD equation. Then:

$$PC = \exp\left[\frac{94.2 \frac{cm^3}{mole}}{83.14 \frac{cm^3 \cdot bar}{mole \cdot K} \cdot 346.15K}(50 - 0.7917)bar\right]$$

This yields PC = 1.175.

f = 0.972·0.7917·1.175 = 0.904 bar.

10.10 **Find the fugacity of pure methanol at a) 50°C and 200 bar, and b) 350°C and 15 bar.**

Solution:

We need to consider the phase of the methanol for each of these conditions. Let's compute the vapor pressure at each temperature. With Antoine constants for methanol from Appendix E, $P^s_{50°C}$ = 0.5556 bar so for part a) with P = 200 bar, methanol is a liquid. We will evaluate the fugacity using equation 10.67 for a pure liquid. For part b), $P^s_{350°C}$ = 336.6 bar, but we are actually above the critical temperature of methanol, so methanol is a vapor. We will use an EOS for a pure vapor to solve part b).

a. For equation 10.67, we need to evaluate φ^s at $P^s_{50°C} = 0.5556$ bar and 323.15 K. We again have a choice but choose Pitzer generalized virial coefficient for the EOS.

$T_r = 0.923$, $P_r = 0.00676$, $B^0 = -0.396$, $B^1 = -0.106$

$\ln(\varphi^s) = -0.00332$, $\varphi^s = 0.997$

φ^s being so close to 1 indicates we could have chosen to treat the vapor as an ideal gas, and the low pressure suggests that.

To evaluate the Poynting correction factor (PC), we need v^c (volume of a condensed phase). Rather than going to the open literature for a value as we have done previously, let's use the Rackett equation (equation 3.38):

$$v^{sat} = v_c z_c^{(1-T_r)^{2/7}}$$

The critical volume, $v_c = 113.9$ cm³/mole is listed in Appendix C along with z_c, the critical compressibility factor, $z_c = 0.219$. Plugging in, we get $v^{sat} (= v^c) = 36.31$ cm³/mole. For PC:

$$PC = \exp\left[\frac{36.31 \frac{cm^3}{mole}}{83.14 \frac{cm^3 \cdot bar}{mole \cdot K} \cdot 323.15K}(200 - 0.5556)bar\right] = 1.309$$

then f = 0.997 · 0.5556 bar · 1.309 = **0.7251 bar**.

b. For this phase, we need φ at 350°C and 15 bar.

$T_r = 1.214$, $P_r = 0.1825$, $B^0 = -0.227$, $B^1 = 0.0628$, $\ln(\varphi) = -0.0287$, $\varphi = 0.977$

f = 0.997 · 15 bar = 14.58 bar.

10.11 The compressibility of a pure gas may be represented by:

$$z = \frac{Pv}{RT} = 1 + bP + cP^2$$

where a, b, and c are functions of T only. For CH_4 gas at -10°C:

b = -5.52 x 10^{-4} bar^{-1}
c = 4.83 x 10^{-6} bar^{-2}

What is the fugacity of at CH_4 at 200 bar and -10°C using this equation?

Solution:

We will use the compressibility factor correlation given in the problem as the EOS. We need to evaluate equation 10.62:

$$\ln(\varphi_i) = \int_0^P (z-1)\frac{dP}{P}$$

Plug in the expression for z and integrate at constant T (so b and c are constants):

$$\ln(\varphi) = \int_0^P (bP + cP^2)\frac{dP}{P} = \int_0^P (b + cP)dP = bP + \frac{c}{2}P^2$$

$$\ln(\varphi) = -5.52 \times 10^{-4} \text{ bar}^{-1} \cdot 200 \text{bar} + \frac{4.83 \times 10^{-6}}{2} \text{bar}^{-2} \cdot 200^2 \text{bar}^2 = -0.0138$$

φ = exp(-0.0138) = 0.9863. **f = 0.9863 · 200 bar = 197.3 bar.**

10.12 Diethylether (1), ethanol (2), and acetaldehyde (3) are in a gaseous mixture with y_1 = 0.2, y_2 = 0.3 and y_3 = 0.5 at 150 kPa. Find the dew point and the composition of the liquid in equilibrium with the gas. Assume Raoult's Law holds.

Solution:

Gather Antoine constants from Appendix E:

	diethylether (1)	ethanol (2)	acetaldehyde (3)
A	14.3944	16.0810	14.7934
B	2700.14	3338.53	2848.98
C	-31.842	-59.953	-14.948

For the dew point calculation, we want to use the multi-component version of equation 10.45:

$$\frac{y_1 P}{P_1^s} + \frac{y_2 P}{P_2^s} + \frac{y_3 P}{P_3^s} = \frac{0.2 \cdot 150}{P_1^s} + \frac{0.3 \cdot 150}{P_2^s} + \frac{0.5 \cdot 150}{P_3^s} = 1$$

For each P_i^s, substitute the Antoine equation, then only T is unknown. Iterate on T until the LHS is equal to 1 and find **T = 339.75 K.** The result of the calculation yields:

T (K)	P_1^s (kPa)	P_2^s (kPa)	P_3^s (kPa)	$1/\sum y_i/P_i^s$ (kPa)
339.75	277.3394	63.3823	412.4204	150.0

To find the composition of the (infinitesimally small droplet of) liquid:

$$x_i = \frac{y_i P}{P_i^s}$$

resulting in:

x_1	x_2	x_3
0.1082	0.7100	0.1819

Student Problems:

1. Compute the fugacity of pure water at:

   a. 100°C and 600 bar
   b. 300°C and 20 bar

State your assumptions in performing the calculations and use good judgement in your assumptions.

2. The compressibility of a gas is given by:

$$z = A + BP + CP^2 + DP^3$$

where A, B, C and D are constants at 0°C for N_2:

$A = 1.00$, $B = -5.314 \times 10^{-4}$ atm^{-1}, $C = 4.276 \times 10^{-6}$ atm^{-2} and $D = -3.292 \times 10^{-9}$ atm^{-3}

Find the fugacity of N_2 at 0°C and 300 atm using this equation.

3. Find an expression for the fugacity of a pure gas that obeys the Van der Waals EOS. Answer:

$$f = P \exp\left[\left(\frac{b}{v-b} - \frac{2a}{RTv} - \ln\left[z\left(1-\frac{b}{v}\right)\right]\right)\right]$$

4. Repeat problem 10.12, but find the bubble point of the liquid mixture of the same composition as the gas in problem 10.12.

5. In problem 10.7, the problem statement said to assume the Poynting correction factor is 1. Estimate the error that assumption caused.

Chapter 11 - Fugacity of Mixtures of Real Gases

This chapter of the book is focused on fugacity of mixtures of gases. If we think back to the phi-gamma formulation of VLE (equation 10.37), we note that this represents one side of the equation. In Chapter 12, we will take up activity coefficients for liquid mixtures which represents the other side of the equation.

11.a Partial Molar Gibbs Free Energy of Mixtures

In previous chapters, we showed how to get fugacities of pure components and we related ideal gas partial pressure to fugacities for mixtures of ideal gases. We would now like to consider how to determine fugacities of individual components in real gas mixtures. To do this, we need to find G for mixtures of real gases. In order to form G for a mixture of gases in terms of P, V and T, we need to recall and partly rewrite H and S from chapter 6. For H, recall equation 6.25:

$$H^R = \int_0^P \left[V - T\left(\frac{\partial V}{\partial T}\right)_P \right] dP \quad \text{or} \quad h^R = \int_0^P \left[v - T\left(\frac{\partial v}{\partial T}\right)_P \right] dP \qquad \text{(constant T) 6.25}$$

Using the definition of residual properties:

$$H - H^{IG} = H - \sum n_i h_i^{IG} = \int_0^P \left[V - T\left(\frac{\partial V}{\partial T}\right)_{P,n_T} \right] dP \qquad 11.1$$

We have added n_T to what is being held constant in the partial derivative to indicate it is taken at constant composition since we now have a mixture to deal with. We can also write this equation as:

$$H_{T,P} = \int_0^P \left[V - T\left(\frac{\partial V}{\partial T}\right)_{P,n_T} \right] dP + \sum n_i h_i^\circ \qquad 11.2$$

We have summed the enthalpies of the individual components according to equation 10.71 to give the enthalpy of the ideal gas reference state. $H_{T,P}$ is the enthalpy of the mixture of gases at any temperature and pressure and we have used h_i° as a reference enthalpy for the individual components which is at the same temperature as $H_{P,T}$. Clearly, the reference pressure for h_i° doesn't matter since the enthalpy of an ideal gas is independent of pressure. For S, go back to equation 6.26:

$$S^R = -\int_0^P \left[\left(\frac{\partial V}{\partial T}\right)_P - \frac{nR}{P} \right] dP \quad \text{or} \quad s^R = -\int_0^P \left[\left(\frac{\partial v}{\partial T}\right)_P - \frac{R}{P} \right] dP \qquad \text{(constant T) 6.26}$$

and rewrite in a similar fashion to H using equation 10.74:

$$S - S^{IG} = -\int_0^P \left[\left(\frac{\partial V}{\partial T}\right)_P - \frac{n_T R}{P}\right] dP - R \sum n_i \ln(y_i) \qquad 11.3$$

However, unlike enthalpy, the entropy of an ideal gas is a function of pressure, so we write:

$$S_{T,P} = -\int_0^P \left[\left(\frac{\partial V}{\partial T}\right)_P - \frac{n_T R}{P}\right] dP - R \sum n_i \ln(y_i) + \sum n_i s_i^\circ - R \sum n_i \ln(P) \qquad 11.4$$

In equation 11.4, s_i° is the ideal gas reference state which is at the same T as $S_{T,P}$ but at 1 bar pressure in the ideal gas state. We can combine terms in equation 11.4 to give the final result:

$$S_{T,P} = -\int_0^P \left[\left(\frac{\partial V}{\partial T}\right)_P - \frac{n_T R}{P}\right] dP - R \sum n_i \ln(y_i P) + \sum n_i s_i^\circ \qquad 11.5$$

Let's now form G = H - TS from equations 11.2 and 11.5, combining some of the terms under the integral:

$$G_{T,P} = \int_0^P \left(V - \frac{n_T RT}{P}\right) dP + RT \sum n_i \ln(y_i P) + \sum n_i \left(h_i^\circ - T s_i^\circ\right) \qquad 11.6$$

or:

$$G_{T,P} = \int_0^P \left(V - \frac{n_T RT}{P}\right) dP + RT \sum n_i \ln(y_i P) + \sum n_i g_i^\circ \qquad 11.7$$

Finding the partial molar Gibbs Free Energy will give us the chemical potential:

$$\mu_i = \overline{G}_i = \left(\frac{\partial G}{\partial n_i}\right)_{T,P,n_{j,j\neq i}} = \int_0^P \left(\overline{V}_i - \frac{RT}{P}\right) dP + RT \ln(y_i P) + g_i^\circ \qquad 11.8$$

With $g_i^\circ = \mu_i^\circ$:

$$\mu_i - \mu_i^\circ = RT \ln\left(\frac{f_i}{f_i^\circ}\right) = \int_0^P \left(\overline{V}_i - \frac{RT}{P}\right) dP + RT \ln(y_i P) \qquad 11.9$$

We choose the standard state fugacity to be the ideal gas state at 1 bar pressure, then $f_i^\circ = 1$ bar and rearranging:

$$RT\ln\left(\frac{f_i}{y_i P}\right) = \int_0^P \left(\overline{V}_i - \frac{RT}{P}\right) dP \qquad 11.10$$

Which gives the final result for the fugacity coefficient:

$$\boxed{RT\ln(\varphi_i) = \int_0^P \left(\overline{V}_i - \frac{RT}{P}\right) dP} \qquad \text{(constant T)} \quad 11.11$$

Notice that this reduces to equation 10.62 for a pure component since $\overline{V}_i = v_i^\circ = zRT/P$ for a pure component.

11.b Fugacity Coefficients for Simple Cases

Let's evaluate φ_i using equation 11.11 for two simple cases:

1. The mixture obeys Dalton's Law:

$$P = \sum y_i P = \sum P_i \quad \text{or} \quad V = \sum n_i \frac{RT}{P} \qquad 11.12$$

From equation 11.12, we get $\overline{V}_i = RT/P$ and plugging into equation 11.11, we get $\varphi_i = 1$ as we expected.

2. The solution obeys Amagat's Law:

$$V = \sum n_i v_i^\circ \qquad 11.13$$

This is equivalent to a solution which is formed with no volume change on mixing the pure components. From equation 11.13, $\overline{V}_i = v_i^\circ$ and we have:

$$RT\ln(\varphi_i) = RT\ln\left(\frac{f_i}{y_i P}\right) = \int_0^P \left(v_i^\circ - \frac{RT}{P}\right) dP \qquad 11.14$$

But for a pure component:

$$RT\ln\left(\frac{f_i^{Pure}}{P}\right) = \int_0^P \left(v_i^\circ - \frac{RT}{P}\right)dP \qquad 11.15$$

Here, we have noted the pure component fugacity to clearly distinguish it from the fugacity of a component in a mixture. Equating equation 11.14 with 11.15, we see:

$$f_i = y_i f_i^{Pure} \qquad \text{(approximate)} \quad 11.16$$

Equation 11.16 is known as the *Lewis Fugacity Rule*. This rule approximates the fugacity of a component in a mixture as the pure component fugacity multiplied by it's mole fraction in the mixture. Thus, according to this rule, we could get the pure component fugacity by using, for example, equations 10.62, 10.63, or 10.64, then use equation 11.16 to get the fugacity of that component in a mixture. The inaccuracy built into this law is that it assumes that interactions of unlike molecules are the same as interactions of like molecules. This is not generally true, and is sometimes a poor assumption.

11.c Evaluation of Fugacity Coefficients in Mixtures Using EOS

Consider evaluation of f_i for a mixture that obeys VDW's EOS:

$$\left(P + \frac{a}{v^2}\right)(v-b) = RT \qquad 3.8$$

First, if we are going to apply equation 11.11, we will need to write V = f(P, T) to find the partial molar volume, then we can substitute and integrate equation 11.11. However, we cannot solve VDW's equation explicitly for volume, so let's approximate VDW's equation. If we solve equation 3.8 for P then write it out as a series by doing long division, we find:

$$P = \frac{RT}{v}\left(1 + \frac{b}{v} + \frac{b^2}{v^2} + \frac{b^3}{v^3} + \ldots\right) - \frac{a}{v^2} \qquad 11.17$$

Truncating at the second term and rearranging:

$$\frac{Pv}{RT} = 1 + \left(b - \frac{a}{RT}\right)\frac{1}{v} \qquad 11.18$$

and this approximate form may be solved for V explicitly:

$$V = n_T \frac{RT}{P} + n_T b - n_T \frac{a}{RT} \qquad 11.19$$

Using this truncated form of VDW as an approximate representation to get the fugacity coefficient, first form the partial molar volume:

$$\overline{V}_i = \frac{\partial V}{\partial n_i}\bigg)_{T,P,n_{j,j\neq i}} = \frac{RT}{P} + n_T \frac{\partial b}{\partial n_i}\bigg)_{T,P,n_{j,j\neq i}} + b - \frac{n_T}{RT}\frac{\partial a}{\partial n_i}\bigg)_{T,P,n_{j,j\neq i}} - \frac{a}{RT} \qquad 11.20$$

This is going to be very messy for a multi-component mixture, so let's work with a binary mixture:

$$\overline{V}_1 = \frac{\partial V}{\partial n_1}\bigg)_{T,P,n_2} = \frac{RT}{P} + b - \frac{a}{RT} + n_T \frac{\partial b}{\partial n_1}\bigg)_{T,P,n_2} - \frac{n_T}{RT}\frac{\partial a}{\partial n_1}\bigg)_{T,P,n_2} \qquad 11.21$$

Clearly, in equation 11.21, a and b are VDW constants for the *mixture* of gases. Whatever these values are, when used in equation 11.19, the equation should at least approximately represent the P-V-T behavior of the *mixture*. Where will we get those values? Closely related to this question is what we will do with the partial derivatives of a and b in equation 11.21.

Suppose the mixture is made leaner and leaner in component 2 so $y_2 \to 0$, then we expect a $\to a_1$ and similarly $b \to b_1$. Also, for $x_1 \to 0$, $a \to a_2$ and $b \to b_2$. For b, this represents the volume occupied by the molecules in the mixture. A reasonable way to take this average as a function of composition is:

$$b = y_1 b_1 + y_2 b_2 \qquad 11.22$$

Equation 11.22 is known as a *mixing rule*. It tells us how we wish to obtain the constants for the mixture from the pure component constants. Note that indeed equation 11.22 gives us $b = b_1$ for $y_2 = 0$ and $b = b_2$ for $y_1 = 0$. Also, its clear that we could get b_1 and b_2 for the pure components from the critical properties just as we did when we were using VDW equation for a pure gas - see equation 3.15.

The mixing rule for a's should be different from b's because a represents interactions of two molecules. Let's write down a plausible mixing rule that incorporates the relative number of times like and unlike molecules encounter each other:

$$a = \sum_i \sum_j y_i y_j a_{ij} = y_1^2 a_{11} + 2 y_1 y_2 a_{12} + y_2^2 a_{22} \qquad 11.23$$

The basic form of equation 11.23 agrees with the exact P-V-T representation of mixtures of gases that results from the virial equation applied to mixtures. In equation 11.23, obviously, $a_{11} = a_1$ and $a_{22} = a_2$ because a_1 represents the 1-1 interaction and a_2 represents the 2-2 interaction. The coefficients $y_1^2, y_1 y_2$, and y_2^2 represent the relative number of encounters of 1-1, 1-2, and 2-2 molecules assuming similar sizes for the molecules. It is clear that a_{12} is an unlike pair interaction parameter, but where will we get a value for it? There is not an easy answer to this question, and the best answers will go to advanced statistical thermodynamic calculations based upon molecular simulations. Such calculations are evolving, but let's note that such results would be greatly advanced beyond the likely accuracy of this approximate equation we are using anyway. Kwak and Mansoori [1] have discussed some options, but we will use the VDW's rule:

$$a_{12} = \sqrt{a_1 a_2} \qquad 11.24$$

which has little theoretical foundation. Equation 11.24 sometimes has an empirical correction applied so that:

$$a_{12} = (1 - k_{12})\sqrt{a_1 a_2} \qquad 11.25$$

The parameter k_{12} has been determined for some binary systems, but to complete our analysis, we take $k_{12} = 0$. The final mixing rule for a is:

$$a = y_1^2 a_1 + 2 y_1 y_2 \sqrt{a_1 a_2} + y_2^2 a_2 \qquad 11.26$$

The pure component values for a's can be found, and given the composition of the mixture, we can find a value of a for the mixture. Now that we have the mixing rules giving the constants as a function of composition, the meaning of the partial derivatives of a and b in equation 11.21 is clear. All that is left is formal mathematical operations.

We can better do the partial derivatives by writing a and b in terms of moles instead of mole fractions. Thus:

$$a = \frac{n_1^2}{n_T^2} a_1 + 2 \frac{n_1 n_2}{n_T^2} \sqrt{a_1 a_2} + \frac{n_2^2}{n_T^2} a_2 \qquad 11.27$$

$$b = \frac{n_1}{n_T} b_1 + \frac{n_2}{n_T} b_2 \qquad 11.28$$

From here, we take the partial derivatives of a and b. The algebra is a bit messy, but well within the capability of an engineering student and is left for interested students to do:

$$\left(\frac{\partial a}{\partial n_1}\right)_{T,P,n_2} = \frac{1}{n_T}\left(2 y_1 a_1 + 2 y_2 \sqrt{a_1 a_2} - 2a\right) \qquad 11.29$$

$$\left(\frac{\partial b}{\partial n_1}\right)_{T,P,n_2} = \frac{b_1 - b}{n_T} \qquad 11.30$$

Plugging into equation 11.21 and simplifying:

$$\overline{V}_1 = \frac{RT}{P} + b_1 - \frac{1}{RT}\left[\left(2y_1 a_1 + 2y_2\sqrt{a_1 a_2}\right) - a\right] \qquad 11.31$$

This goes into equation 11.11 for the fugacity coefficient. Upon simplification, we get:

$$RT\ln(\varphi_1) = \int_0^P \left\{ b_1 - \frac{1}{RT}\left[\left(2y_1 a_1 + 2y_2\sqrt{a_1 a_2}\right) - a\right]\right\} dP \qquad 11.32$$

Integrate and rearrange to get the final result:

$$\varphi_1 = \frac{f_1}{y_1 P} = \exp\left[\left(b_1 - \frac{a_1}{RT}\right)\frac{P}{RT}\right] \exp\left[\frac{\left(a_1^{1/2} - a_2^{1/2}\right)}{(RT)^2} y_2^2 P\right] \qquad 11.33$$

Finding the pure component fugacity, take $y_1 = 1$ and $y_2 = 0$ in equation 11.33:

$$f_1^{Pure} = P\exp\left[\left(b_1 - \frac{a_1}{RT}\right)\frac{P}{RT}\right] \qquad 11.34$$

Substituting equation 11.32 into 11.31, we find:

$$f_1 = y_1 f_1^{Pure} \exp\left[\frac{\left(a_1^{1/2} - a_2^{1/2}\right)}{(RT)^2} y_2^2 P\right] \qquad 11.35$$

Notice that the exponential term in equation 11.35 can be viewed as a correction factor to the Lewis Fugacity Rule, $f_i = y_i f_i^{Pure}$. Even this simple case that we have solved shows the shortcomings of the Lewis Fugacity Rule. The exponential term in equation 11.35 clearly relates to the interaction of component 1 with 2 which has been identified as a physical reality that the Lewis Fugacity Rule does not address. However, also note that the exponential term in equation 11.35 is near 1 for cases where a_1 is near a_2.

What we learn from this analysis is how the formalities are done. Like other calculations we have done with VDW's equation, the overall result is not accurate enough to be that useful to us other than the instructive nature of how to use an EOS for this analysis. As we noted in the beginning of this section, equation 11.11 is difficult to evaluate when we are unable to solve V = f(P,T) as is the case with VDW's equation and the obvious question is how to use pressure explicit cubic equations of state such as VDW, Peng-Robinson, and Redlich-Kwong without truncating the equations. To do this, we can apply a mathematical transformation to equation 11.11 to get the volume integral equivalent. The result is:

$$RT\ln(\varphi_i) = \int_V^\infty \left[\left(\frac{\partial P}{\partial n_i}\right)_{T,V,n_{j,j\neq i}} - \frac{RT}{P}\right]dV - RT\ln(z) \qquad 11.36$$

Application of this equation using the untruncated VDW's equation results in:

$$\ln(\varphi_i) = \ln\left(\frac{v}{v-b}\right) + \frac{b_i}{v-b} - \frac{2\sqrt{a_i}\sum_{j=1}^m y_j\sqrt{a_j}}{vRT} - \ln(z) \qquad 11.37$$

Consider how to use equation 11.37. Suppose we know the temperature, pressure and composition of a mixture of gases, and we want to find the fugacity coefficient for one component in the mixture. To proceed, we find a_i's and b_i's for the individual components, then apply equations 11.22 and 11.26 to find the constants for the mixture. We next have to solve the full VDW's equation for v and for z using a and b for the mixture. Since VDW's equation is cubic, we will generally be looking for the largest of three roots which corresponds to the vapor root for v and z. (See section 14.c for an exact determination of when to use the upper real root as a vapor root.) Once we have these, we can plug into equation 11.37 to find φ_i.

Equations similar to equation 11.37 for the Peng-Robinson, Redlich-Kwong and other EOS are reported in the open literature; e.g., Walas [2]. Many of these more advanced equations are quite useful for engineering calculations.

11.d Fugacity Coefficients from Generalized Correlations

In chapter 8, we considered the calculation of fugacity of pure components using the Lee-Kesler Tables and the Pitzer Generalized Viral Coefficient correlation. If we wish to use these correlations for the calculation of fugacities of components in a mixture, there are two basic routes we can use. The first is to compute the pure component fugacity according to the methods described in chapter 10 then apply the Lewis Fugacity rule ($f_i = y_i f_i^{Pure}$). This method will generally result in a rough value that may be within the desired accuracy for engineering calculations.

The second method relies upon using the Pitzer Generalized Virial Coefficient with pseudo properties of the mixture to determine the molecular interaction parameters. Let's go back and rewrite equations 3.34 to 3.37 but we will now include an "ij" subscript for appropriate parameters. With this nomenclature, whenever the equations call for the "ii" component (like molecule interaction), we use the pure component properties for species i. Whenever the equations call for the "ij" (unlike molecule interaction) component, we will use pseudo properties defined below. With this nomenclature, equations 3.34 to 3.37 become:

$$B_{r,ij} = \frac{B_{ij}P_{c,ij}}{RT_{c,ij}} \qquad 11.38$$

$$B_{r,ij} = B_{ij}^0 + \omega_{ij}B_{ij}^1 \qquad 11.39$$

$$B_{ij}^0 = 0.083 - \frac{0.422}{T_{r,ij}^{1.6}} \qquad 11.40$$

$$B_{ij}^0 = 0.139 - \frac{0.172}{T_{r,ij}^{4.2}} \qquad 11.41$$

Instead of deriving the i-j cross parameters from the i-i pure component parameters like we did for VDW's equation, we instead derive pseudo critical properties for the i-j pair, then use the pseudo critical properties in equations 11.38 to 11.41. The pseudo critical properties are defined by:

$$\omega_{ij} = \frac{\omega_i + \omega_j}{2} \qquad 11.42$$

$$T_{c,ij} = \left(T_{c,i}T_{c,j}\right)^{1/2}\left(1-k_{ij}\right) \qquad 11.43$$

$$z_{c,ij} = \frac{z_{c,i} + z_{c,j}}{2} \qquad 11.44$$

$$v_{c,ij} = \left(\frac{v_{c,i}^{1/3} + v_{c,j}^{1/3}}{2}\right)^3 \qquad 11.45$$

$$P_{c,ij} = \frac{z_{c,ij}RT_{c,ij}}{v_{c,ij}} \qquad 11.46$$

k_{ij} in equation 11.43 is an empirical constant that has been determined for some binary pairs, but it is not generally available. In the absence of data, use $k_{ij} = 0$. Also, be aware that pseudo properties are not intended to be considered the actual properties of the mixture.

Once all the i-j component pair parameters have been determined using the pseudo critical properties and the pure component parameters have been determined, we use:

$$\ln(\varphi_i) = \frac{P}{RT}\left[B_{ii} + \tfrac{1}{2}\sum_j\sum_k y_j y_k \left(2\delta_{ji} - \delta_{jk}\right)\right] \qquad 11.47$$

where:

$$\delta_{ji} = \delta_{ij} = B_{ij} - B_{ii} - B_{jj} \quad \text{and} \quad \delta_{ii} = 0 \qquad 11.48$$

For the case of a binary mixture, equation 11.47 reduces to:

$$\ln(\varphi_1) = \frac{P}{RT}\left(B_{11} + y_2^2 \delta_{12}\right) \qquad 11.49$$

$$\ln(\varphi_2) = \frac{P}{RT}\left(B_{22} + y_1^2 \delta_{12}\right) \qquad 11.50$$

11.e Fugacity Coefficients in the Phi-Gamma Formulation of VLE

Now that we have various options for computing fugacity coefficients for mixtures of real gases, consider the application to the phi-gamma formulation of VLE (equation 10.37). Apply the same approximations to equation 10.37 to arrive at Raoult's Law, and write $f_i^\circ = P_i^s$ but do not assume an ideal gas vapor phase. We arrive at:

$$\varphi_i y_i P = x_i P_i^s \qquad 11.51$$

If we try to solve this equation for a binary solution in a similar fashion to the solution for Raoult's Law, we find some difficulties. For instance, if we know P and x's (the liquid phase composition) and we want to do a bubble point calculation, we follow the same procedure as we did in section 8.d of chapter 8, but we arrive at the modified form of equation 10.42:

$$P = \frac{x_1 P_1^s}{\varphi_1} + \frac{x_2 P_2^s}{\varphi_2} \qquad 11.52$$

When dealing with Raoult's Law calculations, we plugged in Antoine equations for the saturation pressures and iterated on temperature until equation 10.42 gave the actual pressure. However, we cannot do this in equation 11.52 until we have values for the fugacity coefficients, and we cannot get the fugacity coefficients unless we know the composition of the vapor phase which we cannot get until we solve equation 11.52. Generally, the simplest way around this problem is to begin by assuming ideal gases so that φ's are 1, then solving equation 11.52 just as we did the Raoult's Law case, continuing the calculations until we get the vapor phase composition. Now we have an approximate vapor phase composition to get a first trial for φ's, then we use those φ's in equation

11.52 and find the bubble point temperature for the second iteration. We continue to loop, finding T, finding the vapor composition, then computing new values for φ's until the loop closes. This loop generally closes very quickly.

In reviewing the difficulties which might arise in other types of calculations, we see the same difficulty for bubble pressure calculations as we have considered above, but dew point temperature and dew point pressure calculations are somewhat easier because in such cases, we know the composition of the vapor phase. However, we are missing either the temperature or the pressure initially, and these are needed for fugacity coefficient calculations. We again have to assume either a temperature or a pressure to initiate the calculation, or we can do a Raoult's Law calculation first to establish an initial estimate of the missing variable to get the iterative process started.

11.f Chapter 11 Example Problems

11.1 Compute the fugacity coefficients and fugacities for n-heptane (1)/ hydrogen (2) for a 0.2/0.8 mole fraction mixture at 500°C and 10 atm using a) the ideal EOS, b) the Lewis Fugacity Rule with the Pitzer generalized virial coefficient EOS and c) the Pitzer generalized virial coefficient EOS for mixtures.

Solution:

a. For the ideal gas, the fugacity coefficients are 1. $\varphi_{heptane} = \varphi_{hydrogen} = 1$. $f_{heptane} = 0.2 \cdot 10$ bar = 2 bar and $f_{hydrogen} = 0.8 \cdot 10$ bar = 8 bar.

b. To find pure component fugacity coefficients using the Pitzer generalized virial coefficient, we gather critical properties then start with equations 3.34 - 3.37 (the pure component versions of equations 11.38 - 11.41) to find parameters.

	n-heptane (1)	hydrogen (2)
T_c (K)	541.2	33.1
T_r	1.43	23.33
P_c	27.74	12.96
P_r	0.3605	0.7716
B^0	-0.1555	0.0803
B^1	0.1005	0.1390

Along with the critical properties, we use these parameters in equation 10.64:

$$\ln(\varphi_i) = \frac{P_r}{T_r}\left(B^0 + \omega B^1\right)$$

to get $\varphi_{heptane} = 0.9702$ and $\varphi_{hydrogen} = 1.0017$. Using the Lewis Fugacity Rule, $f_{heptane} = 0.9702 \cdot 0.2 \cdot 10$ bar = 1.940 bar and $f_{hydrogen} = 1.0017 \cdot 0.8 \cdot 10$ bar = 8.013 bar.

c. We now need to evaluate equations 11.38 - 11.41 for the 1-2 (i-j) interaction parameters. To do this, we need the pseudo critical properties from equations 11.42 - 11.46.

$$\omega_{ij} = \frac{\omega_i + \omega_j}{2} = \frac{0.353 + (-0.217)}{2} = 0.068$$

We do not have a value for k_{ij} in equation 11.43, so we take it to be zero:

$$T_{c,ij} = \left(T_{c,i} T_{c,j}\right)^{\frac{1}{2}} \left(1 - k_{ij}\right) = (541.2 \cdot 33.1)^{\frac{1}{2}} = 133.9 \, K$$

$$z_{c,ij} = \frac{z_{c,i} + z_{c,j}}{2} = \frac{0.275 + 0.304}{2} = 0.289$$

$$v_{c,ij} = \left(\frac{v_{c,i}^{\frac{1}{3}} + v_{c,j}^{\frac{1}{3}}}{2}\right)^3 = \left(\frac{445.6^{\frac{1}{3}} + 64.6^{\frac{1}{3}}}{2}\right)^3 = 197.7 \, cm^3/mole$$

$$P_{c,ij} = \frac{z_{c,ij} RT_{c,ij}}{v_{c,ij}} = \frac{0.289 \cdot 83.14 \frac{cm^3 \cdot bar}{mole \cdot K} \cdot 133.9 \, K}{197.7 \frac{cm^3}{mole}} = 16.3 \, bar$$

using these pseudo critical properties, we can compute the parameters we need for equations 11.49 and 11.50:

i-j	1-1	2-2	1-2
T_r	1.43	23.33	5.77
B^0	-0.1555	0.0803	0.0575
B^1	0.1005	0.1390	0.1389
B_r	-0.1200	0.0500	0.0669
B (cm³/mole)	-194.66	10.64	45.68
δ (cm³/mole)	0	0	275.39

$$\ln(\varphi_1) = \frac{P}{RT}\left(B_{11} + y_2^2 \delta_{12}\right) = \frac{10 \, bar}{83.14 \frac{cm^3 \cdot bar}{mole \cdot K} \cdot 773.15 \, K}\left(-194.66 + 0.8^2 \cdot 275.39\right)\frac{cm^3}{mole} = -0.0029$$

$$\ln(\varphi_2) = \frac{P}{RT}\left(B_{22} + y_1^2 \delta_{12}\right) = \frac{10 \, bar}{83.14 \frac{cm^3 \cdot bar}{mole \cdot K} \cdot 773.15 \, K}\left(10.64 + 0.2^2 \cdot 275.39\right)\frac{cm^3}{mole} = 0.0034$$

This gives $\varphi_{heptane} = 0.9971$ and $\varphi_{hydrogen} = 1.0034$. $f_{heptane} = 0.9971 \cdot 0.2 \cdot 10 \, bar = 1.994 \, bar$ and $f_{hydrogen} = 1.0034 \cdot 0.8 \cdot 10 \, bar = 8.027 \, bar$.

In this problem, we see that the components behave nearly as ideal gases resulting in all the fugacity coefficients near 1.0 regardless of the method used to compute them.

11.2 A ternary mixture consists of methane (1)/ ethane (2)/ propane (3) with mole fractions 0.17/0.35/0.48, respectively at 40°C and 20 bar. Find the fugacity coefficients and fugacity for each component using the generalized viral coefficient of Pitzer EOS.

Solution:

This problem is similar to problem 11.1, but we now have a ternary mixture. This results in three interaction parameters 1-2, 1-3, and 2-3 in addition to the pure component parameters. Calculations proceed in a similar fashion to the calculations in problem 11.1.

		methane	ethane	propane			
T(°C)	40						
T(K)	313.15						
P (bar)	20						
component #		1	2	3			
y_i		0.17	0.35	0.48			
T_c (K)		190.56	305.32	369.89			
P_c (bar)		45.99	48.72	42.51			
ω		0.015	0.103	0.157			
z_c		0.286	0.280	0.277			
v_c (cm³/mole)		98.7	145.9	200.1			
Equations 11.42 - 11.46:							
i-j interaction		1-1	2-2	3-3	1-2	1-3	2-3
$T_{c,ij}$ (K)		190.6	305.3	369.9	241.2	265.5	336.1
ω_{ij}		0.015	0.103	0.157	0.059	0.086	0.130
$z_{c,ij}$		0.286	0.280	0.277	0.283	0.281	0.278
$v_{c,ij}$ (cm³/mole)		98.7	145.9	200.1	120.7	143.5	171.6
$P_{c,ij}$ (bar)		45.99	48.72	42.51	47.04	43.31	45.32
Equations 11.38 - 11.41:							
$T_{r,ij}$		1.6433	1.0256	0.8466	1.2982	1.1795	0.9318
B^0_{ij}		-0.1076	-0.3222	-0.4678	-0.1949	-0.2410	-0.3895
B^1_{ij}		0.1176	-0.0157	-0.2072	0.0815	0.0530	-0.0924
$B_{r,ij}$		-0.1058	-0.3239	-0.5004	-0.1901	-0.2365	-0.4015
B_{ij} (cm³/mole)		-36.5	-168.7	-362.0	-81.1	-120.5	-247.5
Equations 11.49 and 11.50:							
δ_{ij}		0	0	0	43.08	157.41	35.71
$\ln(\varphi_i)$		0.0252	-0.1273	-0.2643			
φ_i		**1.0255**	**0.8805**	**0.7677**			
f_i (bar)		**3.487**	**6.163**	**7.370**			

11.3 Use the Peng-Robinson Equation of State to determine a) the fugacity (in bar) of pure chloroform (trichloromethane) and pure cyclohexane at 225°C and 10 bar total pressure, b) the fugacity (in bar) of pure liquid chloroform and pure liquid cyclohexane at 225°C and 40 bar total pressure, and c) the fugacities (in bar) of chloroform and cyclohexane in a 40/60 mole % mixture when T = 225°C and P = 10 bar. Compare these fugacities to the fugacities given by the Lewis Fugacity Rule. The molar volume of pure liquid chloroform is 85.63 cm³/mole and of cyclohexane is 108.16 cm³/mole.

The Peng-Robinson (P-R) Equation of State:

$$P = \frac{RT}{v-b} - \frac{a}{v(v+b)+b(v-b)}$$

$$\alpha^{0.5} = 1 + \kappa(1 - T_r^{0.5}) \quad \kappa = 0.37464 + 1.54226\omega - 0.26992\omega^2$$

$$a = \frac{0.45724 R^2 T_c^2 \alpha}{P_c} \quad b = \frac{0.07780 RT_c}{P_c}$$

also, for shorthand for the P-R equation, use:

$$A = \frac{aP}{R^2T^2} \quad B = \frac{bP}{RT}$$

Mixing rules:

$$a = \sum_i \sum_j y_i y_j a_{ij} \quad b = \sum_i y_i b_i \quad a_{ij} = (1-\delta_{ij}) a_i^{0.5} a_j^{0.5}$$

where δ_{ij} is an empirical constant for the chloroform/cyclohexane mixture:

$$\delta_{ii} = 0 \quad \delta_{ij} = 0.01$$

Polynomial form:

$$z^3 - (1-B)z^2 + (A - 3B^2 - 2B)z - (AB - B^2 - B^3) = 0$$

Fugacity coefficient for the first component in a mixture of j components:

$$\ln\left(\frac{f_1}{y_1 P}\right) = \ln(\phi_1) =$$

$$\frac{b_1}{b}(z-1) - \ln(z-B) - \frac{A}{2\sqrt{2}B}\left(\frac{2\sum_j y_j a_{1j}}{a} - \frac{b_1}{b}\right) \ln\left(\frac{z + 2.414B}{z - 0.414B}\right)$$

Note that z in the above equation is the compressibility factor for the mixture. Fugacity coefficient for a pure component:

$$\ln(f/P) = \ln(\phi) = (z-1) - \ln(z-B) - \left(\frac{A}{2\sqrt{2}B}\right) \ln\left(\frac{z+2.414B}{z-0.414B}\right)$$

Solution:

At 225°C = 498.15 K, many parameters that apply to all parts of this problem can be calculated or obtained from the Appendix:

	Chloroform	Cyclohexane
MW (g/mole)	119.369	84.162
T_c (K)	536.1	553.6
P_c (bar)	56.00	40.75
z_c	0.302	0.273
ω	0.238	0.215
A	14.8810	14.1996

B	3232.90	3063.83
C	-20.011	-34.723
T_r	0.9292	0.8998
α	1.0621	1.1074
κ	0.6866	0.7602
a (bar-lt^2/mole^2)	20.7261	28.5738
b (lt/mole)	0.07262	0.09121
P_i^s (bar)	33.59	19.75

These results are used repeatedly below.

a. Find the fugacity (in bar) of pure Chloroform and pure Cyclohexane at 225 °C and 10 bar total pressure. Find the parameters for the polynomial form of the compressibility factor, and solve for the largest real root:

	Chloroform	Cyclohexane	
T (K)	498.15		
P (bar)	10		
T_r	0.9292	0.8998	
P_r	0.1786	0.2454	
A	0.0996	0.1486	
B	0.0150	0.0212	
a1	1	1	Coefficients of z
a2	-0.9850	-0.9788	polynomial for
a3	0.06901	0.10485	Peng-Robinson
a4	-0.001262	-0.002694	equation
z	0.9108	0.8606	
φ (pure)	0.9160	0.8742	
f (bar) (pure)	**9.160**	**8.742**	

b. The fugacity (in bar) of pure liquid Chloroform and pure liquid Cyclohexane at 225°C and 40 bar total pressure. We need to solve equation 10.67 which includes φ_i^s at 225°C and $P = P_i^s$ (not 40 bar!) and the Poynting correction factor.

	Chloroform	Cyclohexane	
$P (= P_i^s)$ (bar)	33.59	19.75	
P_r	0.5999	0.4847	
A	0.33454	0.29356	
B	0.05023	0.04190	
a1	1	1	Coefficients for z
a2	-0.949771	-0.958096	polynomial for
a3	0.226516	0.204480	Peng-Robinson
a4	-0.014154	-0.010472	equation

z	0.6244	0.6811
φ_i^s (pure)	0.7273	0.7571
v^c (cm³/mole)	85.63	108.16
Poynting correction	1.013	1.054
f (bar) (pure)	**24.76**	**15.77**

c.   The fugacities (in bar) of Chloroform and Cyclohexane in a 40/60 mole % mixture when T = 225°C and P = 10 bar.

	Chloroform	Cyclohexane	
P_r	0.1786	0.2454	
y_i	0.4	0.6	
a^{mix}	21.8281		
b^{mix}	0.07749		
A^{mix}	0.12725		
B^{mix}	0.01871		
a1	1		Coefficients for z
a2	-0.981289		polynomial for
a3	0.088783		Peng-Robinson
a4	-0.002024		equation
z (for the mixture)	0.8834		
φ_i	0.9187	0.8757	
f_i	**3.675**	**5.254**	

The Lewis Fugacity Rule uses the pure component fugacities which we computed in part a) of this problem: $f_{CHCl_3}^{Pure}$ = 9.160 bar and $f_{C_6H_{12}}^{Pure}$ = 8.742 bar. By the Lewis Fugacity Rule: $\mathbf{f_{CHCl_3} = 0.4 \cdot 9.160 = 3.664}$ **bar** and $\mathbf{f_{C_6H_{12}} = 0.6 \cdot 8.742 = 5.245}$ **bar.** The Lewis Fugacity Rule result is close to the result derived from the P-R equation applied to the mixture in this example.

11.4   **For the EOS:**

$$P(v-b) = RT \quad \text{or} \quad v = \frac{RT}{P} + b \quad \text{where } b = \frac{RT_c}{P_c}$$

Find a) an expression for the fugacity of a pure gas, b) an expression for the fugacity coefficient of component 1 in a binary gas mixture using the mixing rule:

$$b = \left(y_1 b_1^{1/3} + y_2 b_2^{1/3}\right)^3$$

and c) the fugacity of ethane in a 30 mole %/ 70 mole % mixture of ethane/propane at 40 bar and 400 K using the result from part b).

Solution:

a. $$RT\ln(\varphi) = \int_0^P \left(v_i^\circ - \frac{RT}{P}\right)dP = \int_0^P \left(\frac{RT}{P} + b - \frac{RT}{P}\right)dP = bP$$

$$\varphi^{Pure} = \exp\left(\frac{bP}{RT}\right)$$

b. write out b in terms of total moles:

$$n_T b = \left(n_1 b_1^{1/3} + n_2 b_2^{1/3}\right)^3 \left(\frac{1}{n_T}\right)^2$$

then:

$$\left(\frac{\partial(n_T b)}{\partial n_1}\right)_{T,P,n_2} = \left(n_1 b_1^{1/3} + n_2 b_2^{1/3}\right)^3 \frac{\partial\left(\frac{1}{n_T}\right)^2}{\partial n_1} + \left(\frac{1}{n_T}\right)^2 \frac{\partial}{\partial n_1}\left(n_1 b_1^{1/3} + n_2 b_2^{1/3}\right)^3$$

$$\left(\frac{\partial(n_T b)}{\partial n_1}\right)_{T,P,n_2} = \left(n_1 b_1^{1/3} + n_2 b_2^{1/3}\right)^3 \left(\frac{-2}{n_T^3}\right) + \left(\frac{1}{n_T^2}\right)\left(n_1 b_1^{1/3} + n_2 b_2^{1/3}\right)^2 3 b_1^{1/3}$$

$$\left(\frac{\partial(n_T b)}{\partial n_1}\right)_{T,P,n_2} = 3 b_1^{1/3} b^{2/3} - 2b$$

with $V = n_T RT + n_T b$:

$$\overline{V}_1 = \left(\frac{\partial V}{\partial n_1}\right)_{T,P,n_2} = \frac{RT}{P} + 3 b_1^{1/3} b^{2/3} - 2b$$

$$RT\ln(\varphi_1) = \int_0^P \left(\overline{V}_i - \frac{RT}{P}\right)dP = \int_0^P \left(3 b_1^{1/3} b^{2/3} - 2b\right)dP = \left(3 b_1^{1/3} b^{2/3} - 2b\right)P$$

c. calculations proceed as follows:

	Ethane	Propane
T (K)	400	
P (bar)	40	
y_i	0.3	0.7
T_c (K)	305.3	369.9
P_c (bar)	48.72	42.51
b_i (cm³/mole)	65.128	90.428
b (cm³/mole)	82.254	
$RT\ln(\varphi_1)$ (cm³-bar/mole)	2551.2	
$\ln(\varphi_1)$	0.07671	
φ_1	1.080	
f_1 (bar) $= \varphi_1 y_1 P$	12.96	

11.5 Calculate the fugacities of the following at 290 K and 300 bar for a) pure O_2, b) pure

N_2, and c) air. Use the Lewis Fugacity Rule with the Lee-Kesler table for the EOS.

Solution:

a. For O_2, $T_c = 154.6$ K, $P_c = 50.46$ bar so $T_r = 1.88$ and $P_r = 5.95$. Interpolating in the Lee-Kesler table (double interpolation) we find $\varphi^0 = 0.9572$ and $\varphi^1 = 1.509$. By equation 10.63, **$\varphi_{O2} = 0.967$. $f_{O2} = 290.19$ bar**.

For N_2, $T_c = 126.2$ K, $P_c = 33.96$ bar so $T_r = 2.30$ and $P_r = 8.83$. Interpolating in the Lee-Kesler table (double interpolation) we find $\varphi^0 = 1.017$ and $\varphi^1 = 1.632$. By equation 10.63, **$\varphi_{N2} = 1.038$. $f_{N2} = 311.34$ bar**.

Use the Lewis Fugacity rule for air. **$f_{air} = 0.21 \cdot 290.19 + 0.79 \cdot 311.34 = 306.9$ bar**.

11.6 For a 0.40/0.60 mixture of chloroform(1)/ethanol(2), find the bubble point at 10 bar. You may assume that the liquid phase is ideal (γ's = 1) but consider the vapor phase to not be an ideal gas. Use the Pitzer generalized virial coefficient for the EOS for the vapor phase.

Solution:

The simplified version of the phi-gamma formulation that we want to solve is equation 11.52. Equation 11.52 is like Raoult's Law, but it includes φ_i's. We immediately run into a problem. How can we compute φ_i's for the vapor phase when we don't know the composition of the vapor phase? So to solve this problem, we begin by taking $\varphi_i = 1$. Now we have a Raoult's Law problem, and once hat is solved, we have an approximate gas phase composition to get values of φ_i's.

Gather data we need:

	T_c K	P_c bar	ω	v_c cm^3/mole	z_c	\multicolumn{3}{c}{Constants for Antoine}		
						A	B	C
1	536.1	56.00	0.238	240.7	0.302	14.881	3232.90	-20.011
2	515.4	63.00	0.635	214.5	0.315	16.081	3338.53	-59.953

$$P = 10\,\text{bar} = \frac{x_1 P_1^s}{\varphi_1} + \frac{x_2 P_2^s}{\varphi_2} = 0.4 \cdot P_1^s + 0.6 \cdot P_2^s \quad \text{with } \varphi\text{'s} = 1 \text{ first iteration}$$

Now we need T to find the saturation pressures with Antoine's equation, and this is just like the bubble point calculations we did in chapter 10. Loop on T until we get P = 10 bar. This results in:

T = 424.4 K, $P_1^s = 9.796$ bar, $P_2^s = 10.136$ bar, $y_1 = 0.392$, $y_2 = 0.608$

Next, we compute φ_1 and φ_2 for T = 424.4 K, P = 10 bar, $y_1 = 0.392$, $y_2 = 0.608$. We need the pseudo critical properties from equations 11.42 - 11.46 for the interaction parameter.

	$T_{c1,2}$ (K)	$P_{c1,2}$ (bar)	$\omega_{1,2}$	$v_{c1,2}$ (cm^3/mole)	$z_{c1,2}$
1-2 interaction	525.6	59.38	0.436	227.3	0.309

	T_r	B^0	B^1	P_r
1 (chloroform)	0.792	-0.5302	-0.3197	0.179
2 (ethanol)	0.824	-0.4928	-0.2498	0.159
1-2 interaction	0.807	-0.5112	-0.2833	0.168
virial coefficients:	1-1	2-2	1-2	
B_{ij} (cm³/mole)	-482.47	-443.09	-467.22	
δ_{12}	-8.89			
φ_1	0.8714			
φ_2	0.8817			

Now we can plug these in and do a new loop:

$$10\,\text{bar} = \frac{0.4 \cdot P_1^s}{0.8714} + \frac{0.6 \cdot P_2^s}{0.8817}$$

Solving for T we get T = 418.8 K, y_1 = 0.402, and y_2 = 0.598. T has changed about 6 K, and y's have changed about 0.01. Recalculating with the new T and yi's, we have φ_1 = 0.8659 and φ_2 = 0.8756. We probably need one more iteration:

$$10\,\text{bar} = \frac{0.4 \cdot P_1^s}{0.8659} + \frac{0.6 \cdot P_2^s}{0.8756}$$

Now, **T = 418.6 K, y_1 = 0.402, and y_2 = 0.598.** T has changed 0.2 K and the y's have not changed to the third decimal place. Therefore, we take these values as the answer.

Student Problems:

1. Using the Peng-Robinson EOS, find the fugacities and fugacity coefficients for a 0.40/0.60 gaseous mixture of benzene/toluene at 250°C = 523.15 K and 15 bar. For this mixture, take k_{ij} = 0.01. (k_{ij} is the parameter which modifies the a_{ij} interaction parameter.)

Answer: φ_1 = 0.8625 and φ_2 = 0.8051.

2. Beginning with the truncated form of Van der Waals equation (equation 11.18):

$$\frac{Pv}{RT} = 1 + \left(b - \frac{a}{RT}\right)\frac{1}{v}$$

using the mixing rules for a binary solution:

$$a = y_1^2 a_1 + 2 y_1 y_2 \sqrt{a_1 a_2} + y_2^2 a_2$$

$$b = \left(y_1 b_1^{1/3} + y_2 b_2^{1/3}\right)^3$$

find an expression for the fugacity coefficient of component 1 in the mixture.

3. A ternary solution consists of 1-propanol (1)/2-propanol (2)/1-butanol with mole fractions 0.40/0.28/0.32, respectively at 230°C and 15 bar. Find the fugacity coefficients and fugacity for each component using the generalized virial coefficient of Pitzer EOS.

Answer: $\varphi_1 = 0.8723$, $\varphi_2 = 0.8890$ and $\varphi_3 = 0.8289$.

11.g References

1. T. Y. Kwak and G. A. Mansoori, <u>Chemical Engineering Science 41</u>, 1303 (1986).
2. Walas, S. M.; "Phase Equilibrium in Chemical Engineering", Butterworth Publishers, Boston (1985).

Chapter 12 - Activity and Fugacity of Liquid Mixtures

This chapter of the book is focused on activity and fugacity of liquid mixtures. This is where we tackle the right-hand side of the phi-gamma formulation for VLE (equation 10.37).

12.a More on Partial Molar Properties

We begin this section by introducing *Euler's Theorem*:

$$\boxed{m = \sum x_i \overline{M}_i} \qquad 12.1$$

Euler's theorem can be applied to any property such as V, H, U, S, or G. So to understand this equation a little better, consider applying it to volume. If we take 100 cm³ of water, and 100 cm³ of methanol and mix the two, we will not get exactly 200 cm³ of solution. Exactly what volume do we get? The answer can be found using equation 12.1 and in order to apply it, we need the partial molar volumes at the desired composition of the mixture, not the molar volumes to get the exact result. A solution that mixes with no property change would give:

$$m = \sum x_i m_i^\circ \qquad 12.2$$

Under this rule, if we mix 100 cm³ of component 1 and 100 cm³ of component 2, we get exactly 200 cm³ of the mixture. Similarly, the enthalpy of the mixture would be a mole fraction weighted average of the enthalpies of the pure components. In order for this to be true, there would have to be no heat of mixing. Note that equation 12.2 cannot apply to entropy because the mixing process itself increases the entropy of the system.

Referring to Figure 12..a.1, we have drawn a hypothetical molar property m as a function of x_1 for a binary mixture. As drawn in this figure, m is not linear in x_1, and equation 12.2 is not the appropriate function to model this molar property. Euler's Theorem gives us the appropriate function. We can find the partial molar property by taking a derivative of m with x_1 at a given point, deriving the tangent to the curve at that point, and extrapolating the tangent at that point to the axes. This is depicted in Figure 12.a.1 as a spot at approximately $x_1 = 0.7$ where the tangent intersects and has been extrapolated. From the intersection of the tangent with the $x_1 = 1$ axis, we get $\overline{M}_1$ and from the intersection with the x_1

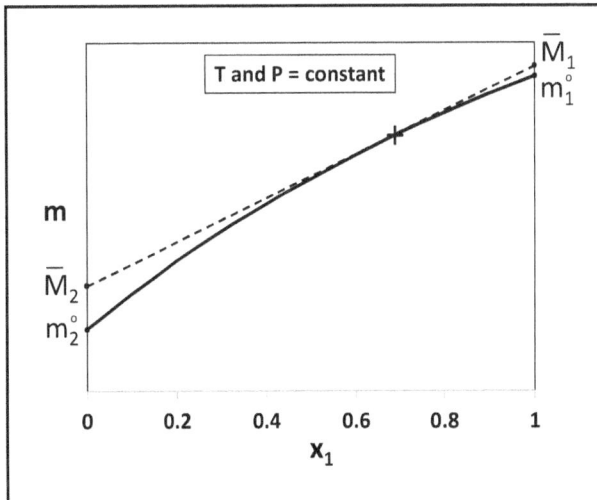

Figure 12.a.1. Graphical depiction of Euler's Theorem for a binary mixture.

$= 0$ ($x_2 = 1$) axis, we get $\overline{M}_2$. We immediately see that $\overline{M}_i = m_i^\circ$ for the pure components (at $x_1 = 1$ and $x_2 = 1$) as we have noted earlier. Also note that if we drew more tangents at other values of x_1, $\overline{M}_1$ and $\overline{M}_2$ would take on different values as a function of x_1. These could be combined using Euler's Theorem (equation 12.1) to generate the m vs. x_1 curve. However, let's look into these relationships more closely.

We can write a total derivative of a (total) property of a mixture M as:

$$dM = \left(\frac{\partial M}{\partial T}\right)_{P,n_{j's}} dT + \left(\frac{\partial M}{\partial P}\right)_{T,n_{j's}} dP + \sum \overline{M}_i dn_i \qquad 12.3$$

But, taking the derivative of 12.1 we get an alternative for dM:

$$dM = \sum \overline{M}_i dn_i + \sum n_i d\overline{M}_i \qquad 12.4$$

Equating equation 12.3 and 12.4, we get:

$$\boxed{\left(\frac{\partial M}{\partial T}\right)_{P,n_{j's}} dT + \left(\frac{\partial M}{\partial P}\right)_{T,n_{j's}} dP - \sum n_i d\overline{M}_i = 0} \qquad 12.5$$

Equation 12.5 is known as the *Gibbs-Duhem Equation*. If we apply equation 12.5 to a constant T and P process, it reduces to:

$$\sum n_i d\overline{M}_i = 0 \qquad 12.6$$

Dividing equation 12.6 by total moles gives an equivalent form at constant T and P:

$$\sum x_i d\overline{M}_i = 0 \qquad 12.7$$

Equation 12.7 is especially useful for relating partial molar properties. Writing Euler's Theorem for a binary mixture:

$$m = x_1 \overline{M}_1 + x_2 \overline{M}_2 \qquad 12.8$$

Taking the derivative with respect to x_1, and using $dx_2 = -dx_1$:

$$\frac{dm}{dx_1} = x_1 \frac{d\overline{M}_1}{dx_1} + \overline{M}_1 + x_2 \frac{d\overline{M}_2}{dx_1} - \overline{M}_2 \qquad 12.9$$

Using equation 12.7 with 12.9 yields:

$$\frac{dm}{dx_1} = \overline{M}_1 - \overline{M}_2 \qquad 12.10$$

Combining equations 12.8 and 12.10 yields:

$$\boxed{\overline{M}_1 = m + x_2 \frac{dm}{dx_1} \qquad 12.11 \qquad \overline{M}_2 = m - x_1 \frac{dm}{dx_1} \qquad 12.12}$$

Equations 12.11 and 12.12 are the proper equations to apply for a binary mixture when determining the partial molar properties via derivatives with respect to x_i (mole fraction) rather than with respect to n_i (moles). For multi-component mixtures, the equation is:

$$\overline{M}_i = \left.\frac{\partial m}{\partial n_i}\right)_{T,P,n_{j,j\neq i}} = m - \sum_{k \neq i} x_k \left.\frac{\partial m}{\partial x_k}\right)_{T,P,x_i} \qquad 12.13$$

So, it is often easier to derive the property as a function of n_i taking the derivative with respect to n_i rather than writing the property as a function of x_i taking the derivative with respect to x_i then applying equation 12.13 to get the partial molar function. However, both methods give the same result when applied correctly.

When we take equation 12.7 for a binary mixture, divide through by dx_1, and use $dx_1 = -dx_2$, we find:

$$x_1 \left.\frac{\partial \overline{M}_1}{\partial x_1}\right)_{T,P} = x_2 \left.\frac{\partial \overline{M}_2}{\partial x_2}\right)_{T,P} \qquad 12.14$$

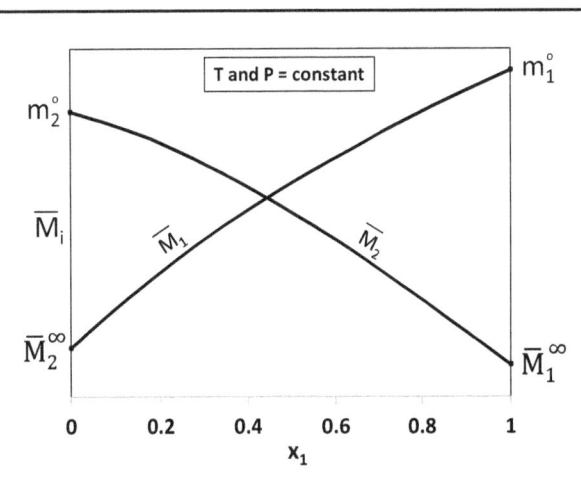

Figure 12.a.2. Partial molar properties for a binary mixture.

Equation 12.14 (which is derived from the Gibbs-Duhem Equation) shows that the derivative of one partial molar property is restricted to a close relationship with the derivative of the other partial molar property. Referring to Figure 12.a.2 where we have a hypothetical pair of partial molar properties for a binary mixture, we note that equation 12.14 tells us that the derivatives of the $\overline{M}_1$ and $\overline{M}_2$ curves are restricted so that features in one curve, such a maximum, minimum, or rapid change must be reflected in the derivative of the other curve. $\overline{M}_1$ and $\overline{M}_2$ in Figure 12.a.2 are both smooth and both reflect the qualitative nature of the other curve given by equation 12.14. Note also that Figure 12.a.2 depicts the *infinite dilution partial molar property*, $\overline{M}_i^\infty$:

$$\overline{M}_i^\infty = \lim_{x_i \to 0} \overline{M}_i \qquad 12.15$$

12.b Excess Properties

We define Δm, the *deviation property* as:

$$\Delta m = m - \sum x_i m_i^\circ = \sum x_i \overline{M}_i - \sum x_i m_i^\circ = \sum x_i \left(\overline{M}_i - m_i^\circ \right) \qquad 12.16$$

If we use:

$$\overline{M}_i - m_i^\circ = \Delta \overline{M}_i \qquad 12.17$$

then:

$$\Delta m = \sum x_i \Delta \overline{M}_i \qquad 12.18$$

$\Delta \overline{M}_i$ represents the change in the property when one mole of a pure component becomes a component in solution. Equation 12.18 defines the mixing changes such as free energy or enthalpy of mixing for a solution.

Let's apply this to chemical potential. Going back to equation 10.23:

$$\mu_i - \mu_i^\circ = RT \ln \left(\frac{f_i}{f_i^\circ} \right) \qquad 10.23$$

With $\mu_i = \overline{G}_i$ and $a_i = f_i / f_i^\circ$:

$$\overline{G}_i - g_i^\circ = RT \ln a_i \qquad 12.19$$

summing to make a solution:

$$\sum x_i \left(\overline{G}_i - g_i^\circ \right) = \sum x_i RT \ln a_i \qquad 12.20$$

or using our definition of deviation property:

$$\frac{\Delta g}{RT} = \sum x_i \ln a_i \qquad (12.21)$$

Similarly, we can show:

$$\frac{P\Delta v}{RT} = \sum \left[x_i \left(\frac{\partial \ln a_i}{\partial \ln P} \right)_{T,x} \right] \qquad (12.22)$$

$$\frac{\Delta h}{RT} = -\sum \left[x_i \left(\frac{\partial \ln a_i}{\partial \ln T} \right)_{P,x} \right] \qquad (12.23)$$

$$\frac{\Delta s}{R} = -\sum x_i \ln a_i - \sum \left[x_i \left(\frac{\partial \ln a_i}{\partial \ln T} \right)_{P,x} \right] \qquad (12.24)$$

for an ideal solution, we have $\gamma_i = 1$, and:

$$f_i^{ID} = x_i f_i^\circ \qquad (12.25)$$

$$a_i^{ID} = x_i \qquad (12.26)$$

Applying this definition to equations 12.21 to 12.24:

$$\frac{\Delta g^{ID}}{RT} = \sum x_i \ln x_i \qquad (12.27)$$

$$\Delta v^{ID} = 0 \qquad (12.28)$$

$$\Delta h^{ID} = 0 \qquad (12.29)$$

$$\frac{\Delta s^{ID}}{R} = -x_i \ln x_i \qquad (12.30)$$

So, an ideal solution is not only characterized by $\gamma_i = 1$, equations 12.27 to 12.30 are also characteristics of the ideal solution. This leads us to the definition of an *excess property*:

excess property = actual property - ideal property

$$m^E = m - m^{ID} \qquad 12.31$$

With this definition, $g^E = v^E = h^E = v^E = 0$ for an ideal solution. To emphasize for an ideal solution, even though Δh and Δv are zero, neither Δg nor Δs are zero, but *all excess properties are zero*.

Excess properties have the same relationships as the total property:

$$h^E = u^E + Pv^E \qquad 12.32$$

$$g^E = h^E - Ts^E \qquad \text{etc...} \qquad 12.33$$

$$dg^E = v^E dP - s^E dT \qquad \text{etc...} \qquad 12.34$$

$$\left.\frac{\partial g^E}{\partial P}\right)_{T,x} = v^E \qquad \text{etc...} \qquad 12.35$$

Excess functions can be positive or negative. A *positive deviation solution* is defined as g^E is positive, and a *negative deviation solution* is defined as g^E is negative.

12.c Partial Molar Excess Properties

We have already defined a general partial molar property, B as:

$$\overline{B}_i = \left.\frac{\partial B}{\partial n_i}\right)_{T,P,n_{j,j\neq i}} \qquad 12.36$$

The partial molar excess property derives simply from the definition above:

$$\overline{B}_i^E = \left.\frac{\partial B^E}{\partial n_i}\right)_{T,P,n_{j,j\neq i}} \qquad 12.37$$

Applying Euler's Theorem:

$$B^E = \sum n_i \overline{B}_i^E \qquad 12.38$$

$$b^E = \frac{B^E}{n} = \frac{\sum n_i \overline{B}_i^E}{n} = \sum x_i \overline{B}_i^E \qquad 12.39$$

Equation 12.21 gives us the Gibbs Free Energy change of mixing, and if we substitute $a_i = \gamma_i x_i$ into that equation, we get:

$$\Delta g = RT \sum x_i \ln x_i + RT \sum x_i \ln \gamma_i \qquad 12.40$$

Equation 12.27 gives the free energy change of mixing for an ideal solution ($\gamma_i = 1$) so combining equations 12.40 and 12.27 to form the excess free energy:

$$g^E = \Delta g - \Delta g^{ID} = RT \sum x_i \ln \gamma_i \qquad 12.41$$

$$\boxed{G^E = ng^E = nRT \sum n_i \ln \gamma_i} \qquad 12.42$$

Now let's form the partial molar excess Gibbs Free Energy by taking the derivative of equation 12.42 with respect to n_i at constant T, P, and $n_{j, j \neq i}$. In the sum on the right hand side of equation 12.42, all of the n's are constant except for the ith term. The ith term will therefore give two terms upon differentiation whereas the other terms will give only a single term. The result is:

$$\overline{G}_i^E = \left(\frac{\partial G^E}{\partial n_i}\right)_{T,P,n_{j,j \neq i}} = RT \ln \gamma_i + \sum n_j \left(\frac{\partial \ln \gamma_j}{\partial n_j}\right)_{T,P,n_{k, k \neq j}} \qquad 12.43$$

Now, $\ln \gamma_i$ is a property, so the final term in equation 12.43 is a molar weighted sum of partial molar properties which sums to zero when the Gibbs-Duhem Equation (equation 12.6) is applied. This gives the result:

$$\boxed{\overline{G}_i^E = RT \ln \gamma_i} \qquad 12.44$$

Equation 12.44 is very useful to us because if we can form an excess free energy function as a function of moles of the components that make up the mixture, we can use equation 12.44 to extract the activity coefficients from that function. We can then use these activity coefficients in the phi-gamma formulation of VLE.

12.d   Regular Solutions

Consider a binary solution and consider what an excess free energy function might look like. If we use $g^E = 0$, we get an ideal solution and equation 12.44 tells us $\gamma_1 = 1$ and $\gamma_2 = 1$. This is a trivial result. Now at $x_1 = 1$ and $x_2 = 1$, we have the two pure components, and since pure components can be thought of as ideal single component solutions, $g^E\big)_{x_1=1} = g^E\big)_{x_2=1} = 0$. So, the excess free energy function must start at zero at $x_1 = 0$ and end at zero at $x_1 = 1$. What simple function could satisfy these endpoint requirements? In Figure 12.d.1, we have plotted a simple parabolic function:

$$g^E = A x_1 x_2 \qquad 12.45$$

where A is a constant. Note that this function meets the endpoint requirements that $g^E = 0$ at $x_1 = 1$ and at $x_2 = 0$ where the pure components exist. This function has another property which is important and that is that the excess Gibbs Free Energy is symmetrical about $x_1 = 0.5$. Let's extract the partial molar free energy and then the activity coefficients for this function.

$$G^E = n g^E = n A \frac{n_1}{n} \frac{n_2}{n} = A \frac{n_1 n_2}{n} \qquad 12.46$$

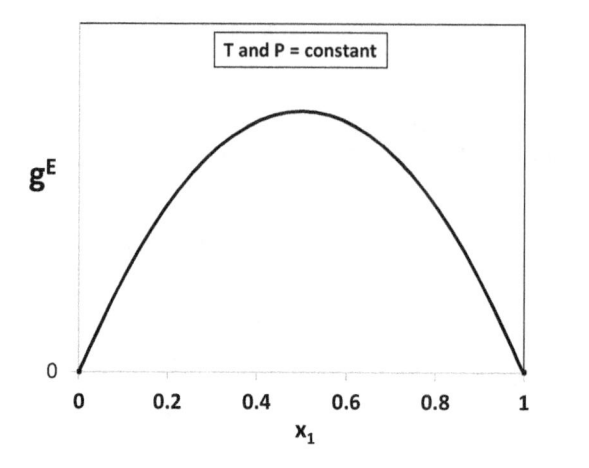

Figure 12.d.1. Excess Gibbs Free Energy function for a regular solution.

$$\overline{G}_1^E = \frac{\partial G^E}{\partial n_1} \bigg)_{T,P,n_{j,j\neq i}} = A \frac{n_2}{n} - A \frac{n_1 n_2}{n} \qquad 12.47$$

$$\overline{G}_1^E = A \frac{n_2}{n} \left(1 - \frac{n_2}{n} \right) = A x_2^2 \qquad 12.48$$

similarly:

$$\overline{G}_2^E = A x_1^2 \qquad 12.49$$

Using equation 12.44:

$$\boxed{\ln \gamma_1 = \frac{A}{RT} x_2^2 \qquad \ln \gamma_2 = \frac{A}{RT} x_1^2} \qquad 12.50$$

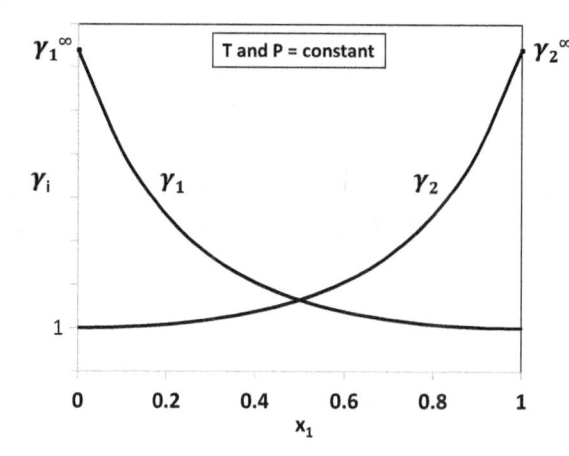

Figure 12.d.2. Regular solution activity coefficients for a binary mixture.

These activity coefficients are plotted in Figure 12.d.2. Note that the two curves are symmetrical about $x_1 = 0.5$ where the two curves cross and that the infinite dilution activity coefficients, γ_1^∞ and γ_2^∞, are the same value given by:

$$\gamma_1^\infty = \gamma_2^\infty = \exp\left(\frac{A}{RT}\right) \qquad 12.51$$

The excess free energy function and corresponding activity coefficients given in equation 12.50 are often called a *regular solution*. With the regular solution model, we can derive other properties that result from the model. Starting with:

$$dg^E = v^E dP - s^E dT + \sum \mu_i^E dn_i \qquad 12.52$$

we see that:

$$s^E = -\left(\frac{\partial g^E}{\partial T}\right)_{P,x} = -x_1 x_2 \left(\frac{\partial A}{\partial T}\right)_{P,x} \qquad 12.53$$

when we take the derivative of equation 12.45. Similarly:

$$h^E = g^E + Ts^E = \left[A - T\left(\frac{\partial A}{\partial T}\right)_{P,x}\right] x_1 x_2 \qquad 12.54$$

$$v^E = \left(\frac{\partial g^E}{\partial P}\right)_{T,x} = x_1 x_2 \left(\frac{\partial A}{\partial P}\right)_{T,x} \qquad 12.55$$

The question that arises is whether or not A is a pure constant, or if it is dependent on T or P or both. We know that activity coefficients are mild functions of temperature. Some temperature functionality is evident in equation 12.50 for the activity coefficients but by experience, we know A is also a function of T. Nonetheless, temperature functionality can often be neglected over small temperature ranges. Activity coefficients are very weak functions of pressure just as liquid volumes change very little with pressure. Pressure effects on activity coefficients can be safely neglected in virtually all situations. If we neglect both T and P effects on the constant A, we see:

$$s^E = 0 \qquad 12.56$$

$$v^E = 0 \qquad 12.57$$

$$h^E = A x_1 x_2 \qquad 12.58$$

A regular solution where A is treated as a pure constant has an enthalpy change on mixing, but no volume change on mixing, and the entropy change is identical to that of an ideal solution.

12.e Modified Raoult's Law Solutions

Consider the phi-gamma formulation of VLE, then let's simplify it assuming that we have ideal gases above the liquid so φ's are 1, and the standard state fugacity is P_i^s (Poynting corrections factor is 1 and φ_i^s is 1). We are left with:

$$y_i P = \gamma_i x_i P_i^s \qquad 12.59$$

Equation 12.59 is sometimes referred to as the modified Raoult's Law. Suppose we know A, the single constant in the regular solution theory, leading to values for the activity coefficients as functions of liquid phase composition. Let's look at what equation 12.59 predicts in terms of VLE using the regular solution model for the liquid in a binary solution:

$$P = y_1 P + y_2 P = \gamma_1 x_1 P_1^s + \gamma_2 x_2 P_2^s \qquad 12.60$$

Substituting equation 12.50 in equation 12.60:

$$P = y_1 P + y_2 P = x_1 P_1^s \exp\left(\frac{A x_2^2}{RT}\right) + x_2 P_2^s \exp\left(\frac{A x_1^2}{RT}\right) \qquad 12.61$$

Consider a bubble point pressure calculation using equation 12.61. If we substitute an Antoine Equation for the saturation pressures, knowing T and the composition of the liquid, we can directly calculate P. A typical P vs. x_1 diagram is given in Figure 12.e.1. The dotted lines in the figure represent the Raoult's Law solution or the solution for γ's = 1.

Equation 12.61 can also be solved for T knowing P and the composition of the liquid. Here, upon substituting Antoine equations for saturation pressure and putting in values for P and x's, the only unknown is temperature. Since the temperature functionality is complex, an iterative approach to finding the temperature is needed.

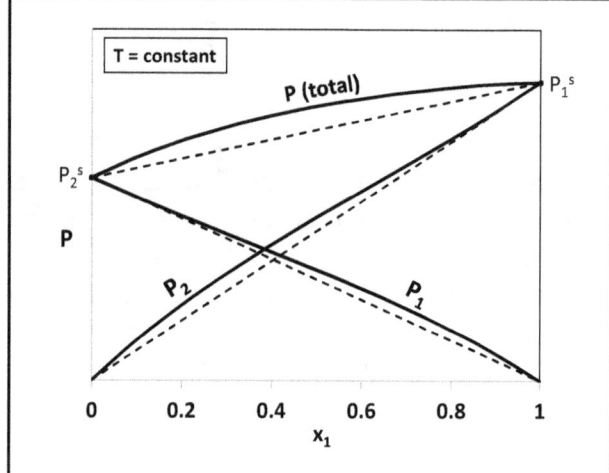

Figure 12.e.1. Pressure vs x_1 VLE diagram for a regular solution.

Dew point calculations with the modified Raoult's Law mirror the solution procedures for Raoult's Law. However, in this case, we need to have the liquid composition to calculate activity coefficients, and for dew point calculations, we have the vapor composition instead. Generally, to solve this case, we do a Raoult's Law calculation first to get a seed for an iterative process. Once we have a trial solution from Raoult's Law, we can compute all the parameters for equation 12.61, then we can test the calculated pressure (dew point pressure calculation) or calculated temperature (dew point calculation) to see if the results are correct, then adjust and iterate further as necessary.

12.f Higher Order Excess Gibbs Free Energy Functions

Consider the situation where the regular solution model does not adequately represent the physical situation, which is generally the case. Rarely do we find infinite dilution coefficients for both components in a binary solution to be the same, and regular solution theory limits us to that condition.

The Redlich-Kister expansion allows fitting virtually any excess free energy function:

$$\frac{g^E}{RT} = x_1 x_2 \left[A + B(x_1 - x_2) + C(x_1 - x_2)^2 + D(x_1 - x_2)^3 + \ldots \right] \qquad 12.62$$

The algebraic arrangement of the expansion with the product $x_1 x_2$ outside the series guarantees that the function results in $g^E = 0$ at $x_2 = 0$ and $x_1 = 0$ as required for the pure components. Figure 12.f.1 shows the first few terms of the expansion. The first term is the regular solution term and is symmetric about $x_1 = x_2 = 0.5$. The second term skews the regular solution term either left (as shown in Figure 12.f.1) or right depending upon whether B is positive or negative. The third term flattens or sharpens the curve, while the fourth term skews either right or left. As we see, even terms flatten or sharpen while odd terms skew either right or left. Any viable g^E function could be fitted with enough terms.

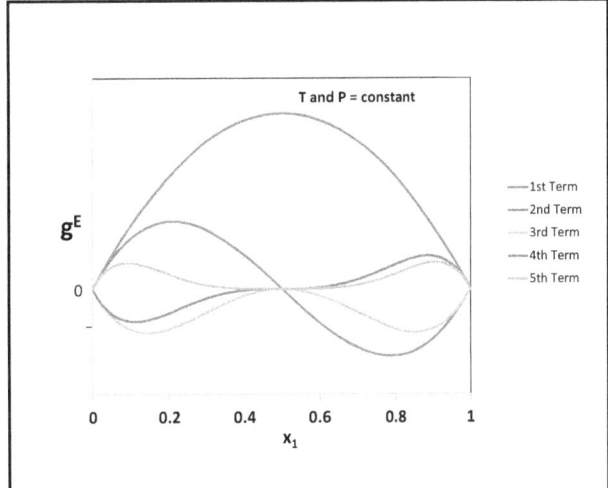

Figure 12.f.1. Terms in the Redlich-Kister expansion for excess Gibbs Free Energy.

Extracting the partial molar Gibbs Free Energy, and using equation 12.44, we get the activity coefficients for a binary solution derived from the Redlich-Kister expansion:

$$\ln \gamma_1 = a^{(1)} x_2^2 + b^{(1)} x_2^3 + c^{(1)} x_2^4 + \ldots \qquad 12.63$$

$$\ln \gamma_2 = a^{(2)} x_1^2 + b^{(2)} x_1^3 + c^{(2)} x_1^4 + \ldots \qquad 12.64$$

$$a^{(1)} = A + 3B + 5C + 7D + \ldots \qquad 12.65$$

$$a^{(2)} = A - 3B + 5C - 7D + \ldots \qquad 12.66$$

etc.

The Redlich-Kister expansion is most widely used in truncated forms. Obviously, if $b^{(1)} = c^{(1)} \ldots =$

$b^{(2)} = c^{(2)}... = 0$, we get the regular solution model which is also known as the Margules 2-suffix equation. For $c^{(1)} = d^{(1)}... = c^{(2)} = d^{(2)}... = 0$, we get the activity coefficients for the *Margules 3-suffix equation*:

$$\boxed{\ln \gamma_1 = x_2^2\left[A + 2(B-A)x_1\right]} \quad \boxed{\ln \gamma_2 = x_1^2\left[B + 2(A-B)x_2\right]} \qquad 12.67$$

Note that for the Margules 3-suffix equation:

$$\gamma_1^\infty = \exp(A) \qquad \gamma_2^\infty = \exp(B) \qquad 12.68$$

The infinite dilution activity coefficients are different, so a wider range of solutions can be modeled with this equation.

The Redlich-Kister expansion and Margules 3-suffix equation have little theoretical underpinning, but the Redlich-Kister expansion can model virtually any excess free energy function with enough terms in the equation and the Margules 3-suffix equation can often represent binary solution behavior very well with just two constants. Multi-component forms of both equations exist in the open literature.

12.g Excess Gibbs Free Energy Functions Derived from Molecular Arguments

As an example of how molecular arguments might be used to derive activity coefficient correlations from excess free energy functions, we consider a binary solution and assume that $v^E = s^E = 0$. Since:

$$g^E = u^E + Pv^E - Ts^E \qquad 12.69$$

these assumptions lead us to $g^E = u^E$. Let's consider how we might compute u^E. Follow this sequence which takes place at constant T:

| x_1 moles of pure liquid 1 / x_2 moles of pure liquid 2 | Δu_I vaporize $\rightarrow$ | x_1 moles of pure ideal gas 1 / x_2 moles of pure ideal gas 2 | Δu_{II} mix $\rightarrow$ | 1 mole of ideal gas mixture | Δu_{III} condense $\rightarrow$ | 1 mole of condensed liquid mixture |

Because the overall process converts two pure liquids to the same liquids in a solution, $\Delta u = u^E = \Delta u_I + \Delta u_{II} + \Delta u_{III} = g^E$. For step I:

$$\left(\frac{\partial u}{\partial v}\right)_T = T\left(\frac{\partial P}{\partial T}\right)_v - P \qquad 12.70$$

which we can get from equation 6.2 also using the Maxwell relations in a fashion similar to that used in chapter 6 to determine the effect of pressure on enthalpy. Using Van der Waals equation for the EOS, the student can verify that:

$$\left.\frac{\partial u}{\partial v}\right)_T = \frac{a}{v^2} \qquad 12.71$$

then:

$$\Delta u_I = \int_{v^{Liq}}^{\infty} \frac{a}{v^2} dv = x_1 \left(-\frac{a_1}{v}\right)\Big|_{v_1^{Liq}}^{\infty} + x_2 \left(-\frac{a_2}{v}\right)\Big|_{v_2^{Liq}}^{\infty} \qquad 12.72$$

To approximate the liquid volumes, we will use $v^{Liq} = b$, the excluded volume for Van der Waals equation. This gives:

$$\Delta u_I = \frac{x_1 a_1}{v_1^{Liq}} + \frac{x_2 a_2}{v_2^{Liq}} \approx \frac{x_1 a_1}{b_1} + \frac{x_2 a_2}{b_2} \qquad 12.73$$

Ideal gases mix with no internal energy change so $\Delta u_{II} = 0$. Step III is very similar to the opposite of step I, but now we are dealing with the mixture, not the individual components. Instead of using a_1, a_2, b_1 and b_2, we want to use:

$$a = x_1^2 a_1 + 2 x_1 x_2 \sqrt{a_1 a_2} + x_2^2 a_2 \quad \text{and} \quad b = x_1 b_1 + x_2 b_2 \qquad 12.74$$

which, with the appropriate modifications to equation 12.73, gives:

$$\Delta u_{III} = -\frac{x_1^2 a_1 + 2 x_1 x_2 \sqrt{a_1 a_2} + x_2^2 a_2}{x_1 b_1 + x_1 b_2} \qquad 12.75$$

Adding up the contributions, $\Delta u = u^E = \Delta u_I + \Delta u_{II} + \Delta u_{III} = g^E$, and simplifying we get the result:

$$g^E = \frac{x_1 x_2 b_1 b_2}{x_1 b_1 + x_2 b_2} \left(\frac{\sqrt{a_1}}{b_1} - \frac{\sqrt{a_2}}{b_2}\right)^2 \qquad 12.76$$

Extracting the partial molar Gibbs Free Energy using equation 12.76, applying equation 12.44 to get the activity coefficients, and simplifying with:

$$A = \frac{b_1}{RT}\left(\frac{\sqrt{a_1}}{b_1} - \frac{\sqrt{a_2}}{b_2}\right)^2 \quad \text{and} \quad B = \frac{b_2}{RT}\left(\frac{\sqrt{a_1}}{b_1} - \frac{\sqrt{a_2}}{b_2}\right)^2 \qquad 12.77$$

we get the *Van Laar equation*:

$$\ln\gamma_1 = \frac{A}{\left(1 + \frac{A\,x_1}{B\,x_2}\right)^2} \qquad \ln\gamma_2 = \frac{B}{\left(1 + \frac{B\,x_2}{A\,x_1}\right)^2} \qquad 12.78$$

Though in principle, constants A and B can be computed from Van der Waals constants which can be obtained from critical properties of the pure components, the result is generally not accurate enough for even simple engineering calculations, so A and B are treated as adjustable parameters.

Wohl's expansion [1] was derived from considerations of local compositions. Instead of mole fractions, Wohl's expansion uses volume fractions:

$$z_i = \frac{x_i q_i}{\sum x_i q_i} \qquad 12.79$$

where z_i's are volume fractions and q_i's are effective volumes of the molecules. For a binary solution, Wohl's expansion can be written:

$$\frac{g^E}{RT(x_1 q_1 + x_2 q_2)} = 2a_{12}z_1 z_2 + 3a_{112}z_1^2 z_2 + 3a_{1222}z_1 z_2^2 + \\ 4a_{1112}z_1^3 z_2 + 4a_{1222}z_1 z_2^3 + \ldots \qquad 12.80$$

Other equations can be derived from Wohl's expansion. Truncating at the second term, the Van Laar equation is obtained with Van Laar parameters $A = 2q_1 a_{12}$ and $B = 2q_2 a_{12}$. For $q_1 = q_2$ (similar molecular sizes), the Margules 3-suffix equation may be extracted after truncating the series.

Wilson' equation has been derived from the assumption that the local concentration in the vicinity of a molecule differs from the bulk composition. Walas [2] gives a fairly detailed development of this equation. Wilson's equation for the activity coefficients for a binary solution may be written:

$$g^E = -RT\left[x_1 \ln(x_1 + \Lambda_{12} x_2) + x_2 \ln(x_2 + \Lambda_{21} x_1)\right] \qquad 12.81$$

$$\ln\gamma_1 = -\ln(x_1 + \Lambda_{12}x_2) + x_2\left(\frac{\Lambda_{12}}{x_1 + \Lambda_{12}x_2} - \frac{\Lambda_{21}}{x_2 + \Lambda_{21}x_1}\right) \qquad 12.82$$

$$\ln\gamma_2 = -\ln(x_2 + \Lambda_{21}x_1) + x_1\left(\frac{\Lambda_{12}}{x_1 + \Lambda_{12}x_2} - \frac{\Lambda_{21}}{x_2 + \Lambda_{21}x_1}\right) \qquad 12.83$$

The parameters are:

$$\Lambda_{12} = \frac{v_2^\circ}{v_1^\circ}\exp\left(-\frac{\lambda_{12}}{RT}\right) \text{ and } \Lambda_{12} = \frac{v_1^\circ}{v_2^\circ}\exp\left(-\frac{\lambda_{21}}{RT}\right) \qquad 12.84$$

λ_{12} and λ_{21} are adjustable constants. Note that built into Wilson's equation is temperature functionality for the activity coefficients. Wilson's equation is therefore useful over wide temperature ranges and it often does an excellent job correlating activity coefficients.

Other excess Gibbs Free Energy functions with corresponding activity coefficient correlations that have found widespread utility are the NTRL equation [3], and UNIQUAC [4]. UNIQUAC not only includes a term for the excess enthalpy in the excess free energy, it also has a modification for the excess entropy. UNIQUAC also can be coupled with UNIFAC [5] which is a group contribution method for estimating binary solution parameters for UNIQUAC based upon the molecular structure of the components of the mixture. Thus, UNIQUAC/UNIFAC can often be applied where no VLE data is available. Students are referred to the open literature for further information.

12.h   Determination of Parameters for Activity Coefficient Correlations

In most circumstances, we need some VLE data to determine constants in activity coefficient models. We will look at several models beginning with the method of Carlson and Colburn for binary mixtures which requires only boiling point data for liquid mixtures of known composition. Let's assume low pressure so that fugacity coefficients are 1 and $f_i^\circ = P_i^s$. Then, the modified Raoult's law model holds including equation 12.60. Solving equation 12.60 for γ_1:

$$\gamma_1 = \frac{P - \gamma_2 x_2 P_2^s}{x_1 P_1^s} \qquad 12.85$$

By measuring the boiling points of liquid mixtures of known composition under constant pressure, we know T, P and x's, so we know everything in equation 12.85 except for γ_1 and γ_2. However, we do know $\gamma_2 = 1$ as $x_1 \to 0$. Also, as $x_1 \to 0, \gamma_1 \to \gamma_1^\infty$. This is mirrored when we consider $x_2 \to 0$. In light of these notes, let's define functions:

$$\gamma_1^{APP} = \frac{P - x_2 P_2^s}{x_1 P_1^s} \quad \text{and} \quad \gamma_2^{APP} = \frac{P - x_1 P_1^s}{x_2 P_2^s} \qquad 12.86$$

Comparing equation 12.86 with 12.85, since the pure component activity coefficients are 1, these equations become correct when extrapolated to infinite dilution. The strategy then is to measure the boiling points of several solutions over the range $0 < x_1 < 1$. We then plot the γ_1^{APP} and γ_2^{APP} functions, extrapolating each function to infinite dilution to obtain γ_1^∞ and γ_2^∞. We can then use A $= \ln \gamma_1^\infty$ and B $= \ln \gamma_2^\infty$ in either the Margules 3-suffix equation (see equation 12.68) or the Van Laar equation (equation 12.68 also holds for Van Laar) to predict activity coefficients at any value of x_1. If we further assume that activity coefficients are independent of T for small temperature ranges, we can use the values of A and B for other temperatures and pressures. We can also use the modified Raoult's Law to find the vapor phase compositions. Example data for the method of Carlson and Colburn (ethanol/benzene mixtures) is below. Vapor pressures have been determined based upon the boiling temperature using an Antoine equation.

x_1	BP (°C)	P_1^s (mmHg)	P_2^s (mmHg)	γ_1^{APP}	γ_2^{APP}
0.00	79.7	802.2	750.0		1.000
0.04	75.2	671.5	652.0	4.619	1.155
0.11	70.8	560.7	565.7	3.998	1.367
0.28	68.3	504.9	520.8	2.653	1.623
0.43	67.8	494.3	512.2	2.155	1.841
0.61	68.3	504.9	520.8	1.776	2.176
0.80	70.1	544.6	552.8	1.468	2.843
0.89	72.4	599.1	595.9	1.284	3.308
0.94	74.4	650.1	635.6	1.165	3.643
1.00	78.0	749.9	711.2	1.000	

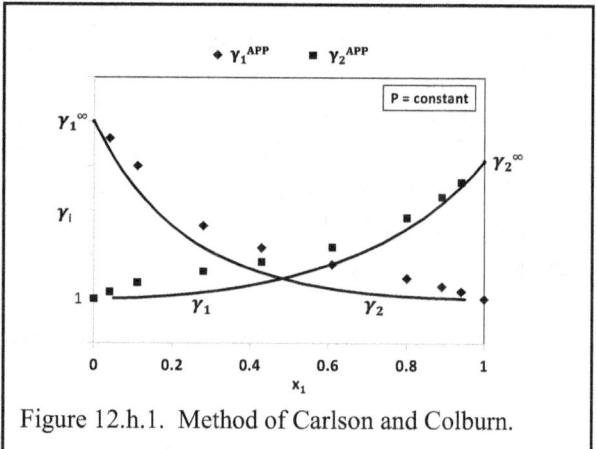

Figure 12.h.1. Method of Carlson and Colburn.

Using this data, the strategy is implemented in Figure 12.h.1. Diamonds and squares represent the data points in the table. These must be extrapolated to $x_1 = 1$ and $x_1 = 0$ to get the infinite dilution activity coefficients, which have then been used in the Van Laar equation to get activity coefficients over the whole range $0 \le x_1 \le 1$. The advantage of using this strategy is that it requires only boiling point data. The liquid compositions can be determined by weight or volumetric procedures used to create the solutions from pure components, and vapor compositions are not required.

More often when gathering data to determine constants for an activity coefficient correlation, we choose to gather full VLE data. Such a measurement includes determination of x's, y's, T, and P. Knowing full sets of these data, we can compute the activity coefficients directly from the phi-gamma formulation of VLE:

$$\gamma_i = \frac{\varphi_i y_i P}{x_i f_i^\circ} \approx \frac{y_i P}{x_i P_i^s} \qquad 12.87$$

Often, for low pressures, we can assume $\varphi_i = 1$, but since we know the composition of the gas phase and T and P, φ_i can be computed directly. Also at low pressure, we can very often take $f_i^\circ = P_i^s$ but this can also be computed exactly knowing T and P if necessary. For the remainder of this section, we will assume low pressures and take $\varphi_i = 1$ and $f_i^\circ = P_i^s$. We will also work only with binary solutions.

If we have one VLE measurement that gives γ_1 and γ_2 (from equation 12.87), we have the minimum amount of data required to get A and B for Margules 3-suffix and Van Laar equations. If we solve Margules 3-suffix for A and B from the activity coefficients:

$$A = \frac{x_2 - x_1}{x_2^2}\ln\gamma_1 + 2\frac{\ln\gamma_2}{x_1} \quad \text{and} \quad B = \frac{x_1 - x_2}{x_1^2}\ln\gamma_2 + 2\frac{\ln\gamma_1}{x_2} \qquad 12.88$$

and for Van Laar:

$$A = \ln\gamma_1\left(1 + \frac{x_2 \ln\gamma_2}{x_1 \ln\gamma_1}\right)^2 \quad \text{and} \quad B = \ln\gamma_2\left(1 + \frac{x_1 \ln\gamma_1}{x_2 \ln\gamma_2}\right)^2 \qquad 12.89$$

Using these values of A and B, we can get the activity coefficients for the entire range $0 \leq x_1 \leq 1$ and solve the phi-gamma formulation for VLE (or Modified Raoult's Law) for the compositions of both phases. The VLE data that was used in equation 12.88 or 12.89 would be echoed exactly using the phi-gamma formulation of VLE. However, the further away from the original data point that the calculation is done, the higher the possible level of error. With only a single data point, we also cannot observe whether the Margules 3-suffix, the Van Laar, or other correlations might be a better correlation for the entire composition range.

Let's rewrite equation 12.42 for a binary mixture:

$$g^E = RT(x_1 \ln\gamma_1 + x_2 \ln\gamma_2) \qquad 12.90$$

Then rearrange by dividing through by RTx_1x_2:

$$\frac{g^E}{RTx_1x_2} = \frac{x_1 \ln\gamma_1 + x_2 \ln\gamma_2}{x_1 x_2} \qquad 12.91$$

We can determine values for this function as a function of x_1 given full sets of VLE data from which we can extract γ_1 and γ_2. Now consider the complexity of the data we might get using this equation.

If:

$$\frac{g^E}{RTx_1x_2} \text{ is constant with } x_1 \qquad 12.92$$

Then, we have a regular solution. Even if there is some scatter in the data (as we always have with experimental measurements), individual values of $\frac{g^E}{RTx_1x_2}$ can be averaged to get a value of A for the regular solution. Now consider:

$$\frac{g^E}{RTx_1x_2} \text{ is linear in } x_1 \qquad 12.93$$

then:

$$\frac{g^E}{RTx_1x_2} = \alpha + \beta x_1 \qquad 12.94$$

or:

$$\frac{g^E}{RT} = x_1x_2(\alpha + \beta x_1) = x_1x_2[A + (B - A)x_1] \qquad 12.95$$

Equation 12.95 is the excess free energy form of the Margules 3-suffix equation. Therefore, if we encounter $\frac{g^E}{RTx_1x_2}$ is linear in x_1, we can use linear regression to get the slope, which is (B - A) and the intercept A for the Margules 3-suffix equation. For:

$$\frac{g^E}{RTx_1x_2} = \frac{1}{\alpha + \beta x_1} \qquad 12.96$$

we find that equation 12.96 can be rearranged to yield the Van Laar equation. On a plot of $\frac{g^E}{RTx_1x_2}$ vs. x_1, we would find that the plot is non-linear. It is possible to set up a linearized form of equation 12.96 by taking the reciprocal, then constants A and B could be extracted by linear regression. However, the Van Laar equation can also be put in the linear form:

$$\sqrt{\ln\gamma_1} = \sqrt{A} - \sqrt{\frac{A}{B}}\sqrt{\ln\gamma_2} \qquad 12.97$$

so that using linear regression on $\sqrt{\ln\gamma_1}$ vs. $\sqrt{\ln\gamma_2}$ gives A = intercept² and A/B = -slope². This is generally the preferred regression method for getting constants for the Van Laar equation.

However, wrapping up this section, be sure to take away that the complexity of the $\frac{g^E}{RTx_1x_2}$ vs. x_1 function gives us an indication of what activity coefficient correlation we should be using to model the data.

12.i Azeotropes

Azeotropic behavior in mixtures is relatively common. It occurs when a mixture that is boiled has the same liquid composition as the vapor composition. Because of this behavior, azeotropes are also known as constant boiling mixtures. Once a mixture reaches it's azeotropic point, it will boil at constant composition and temperature until it is completely vaporized.

There are two types of azeotropes which are minimum boiling and maximum boiling. If we consider holding pressure constant, minimum boiling azeotropes always have positive deviation excess Gibbs Free Energies while maximum boiling azeotropes are excess Gibbs Free Energy negative deviation solutions. Minimum boiling azeotropes are the most common. We will only consider binary solutions in considering azeotropic behavior in a mathematical sense. In order for an azeotrope to exist mathematically, it must have a boiling point either higher than both pure components (maximum boiling when holding pressure constant) or below both boiling points of the pure components (minimum boiling when holding pressure constant).

Figure 12.i.1 compares a negative deviation azeotrope with a positive deviation azeotrope. Note that the azeotropic point is at the extrema in both cases. The x_1 vs y_1 graph shows the azeotropic point clearly where the equilibrium curve crosses the $x_1 = y_1$ reference line. Note that the graphs in Figure 12.i.1 are for T = constant. P is minimum at the azeotrope for the negative deviation mixture and is at the maximum at the corresponding point in the positive deviation mixture. If the corresponding P = constant diagrams were included, T is minimum at the azeotrope for the positive deviation mixture and is at the maximum at the corresponding point in the negative deviation mixture.

We can take advantage of the conditions at the azeotrope for determination of activity coefficients. Go back to the modified Raoult's Law:

$$y_i P = \gamma_i x_i P_i^s \qquad 12.59$$

and apply this to an azeotrope. For the binary azeotrope, $x_1 = y_1$ and $x_2 = y_2$. Solving for the activity coefficient then gives:

$$\gamma_1^{az} = \frac{P}{P_1^s} \quad \text{and} \quad \gamma_2^{az} = \frac{P}{P_2^s} \qquad 12.98$$

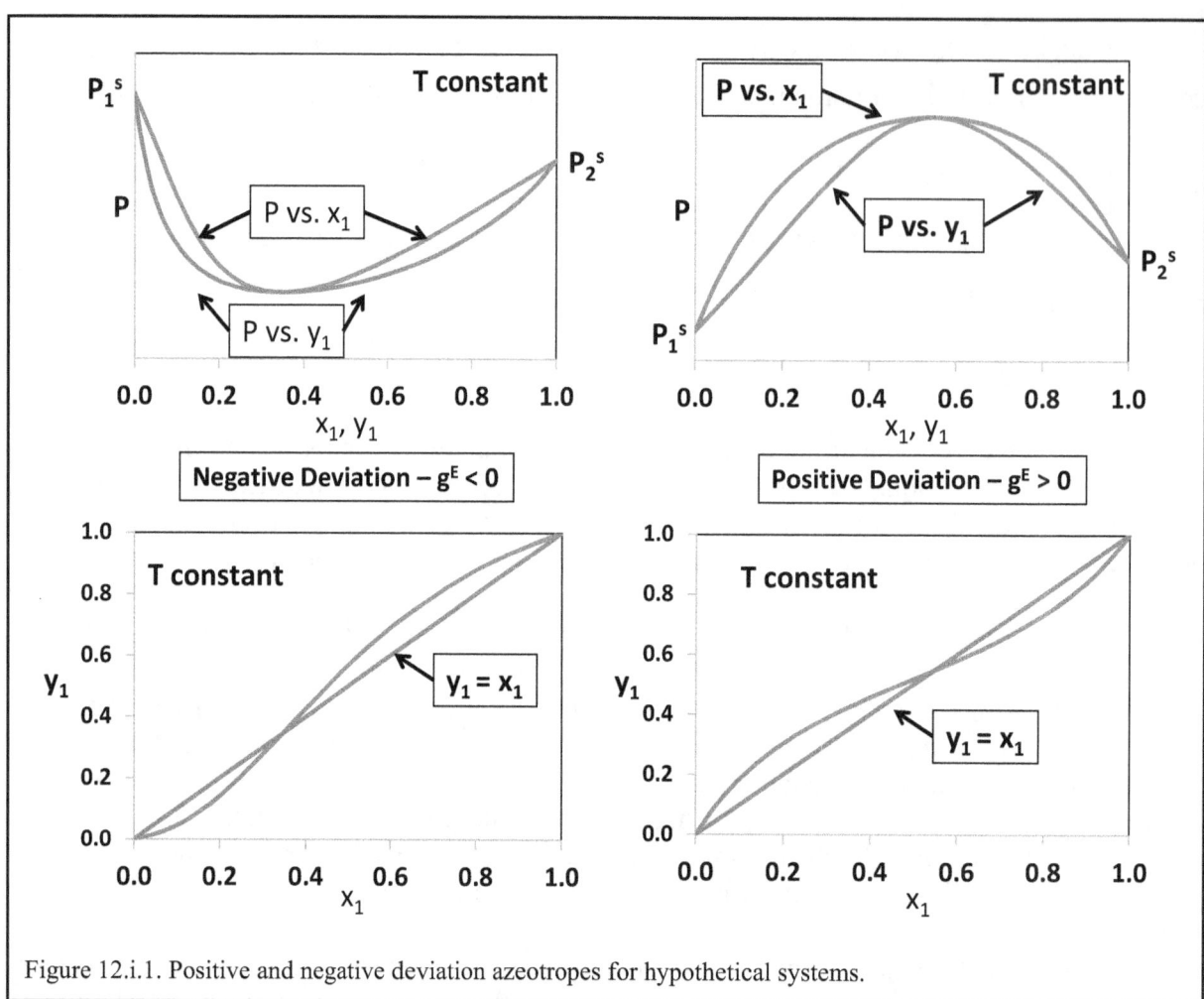

Figure 12.i.1. Positive and negative deviation azeotropes for hypothetical systems.

Knowing T and P at an azeotropic point, equation 12.98 can be used to extract the activity coefficients. Assuming we also know the composition of the azeotrope, we can apply either equation 12.88 or equation 12.89 to determine A and B for the Margules 3-suffix or Van Laar equations. With these values of A and B, the activity coefficient correlations would give back the azeotropic conditions exactly. Equation 12.59 can then be used with the activity coefficient correlation to generate the full equilibrium diagrams for $0 \leq x_1 \leq 1$. We would expect for the diagram to hold well in the vicinity of the azeotropic point, but the further we move away, the less reliable the calculation would be. However, hundreds of azeotropes are known and azeotropic compositions with T and P can be found in the open literature. This analysis can get a ballpark model for these systems with appropriate consideration to the dangers of extrapolation.

12.j   Thermodynamic Consistency

12.j.i   The Differential Test

The Gibbs-Duhem equation can be used to relate properties of one component to another, and this relationship can be used to check the consistency of data and/or equations. In a binary solution with the partial molar property being $\ln(\gamma_i)$, equation 12.7 yields:

$$x_1 \left.\frac{\partial \ln \gamma_1}{\partial x_1}\right)_{T,P} = x_2 \left.\frac{\partial \ln \gamma_2}{\partial x_2}\right)_{T,P} \qquad 12.99$$

This relates the derivatives of the activity coefficients. In observing Figure 12.j.1 where we have hypothetical data points marked with diamonds, we see that equation 12.99 requires a close relationship between the slopes of the two curves on the figure. Any extrema in one curve would have to also have an extrema in the other curve, and rapid changes in one curve must be reflected in the other curve. Testing activity coefficient data over the whole range of composition may be accomplished by differentiating the data. However, differentiation of experimental data is often very difficult especially when there is noise in the data, so it may be difficult to assess whether or not equation 12.99 is satisfied or if any differences that might be observed between the left-hand and right-hand sides of equation 12.99 is simply noise.

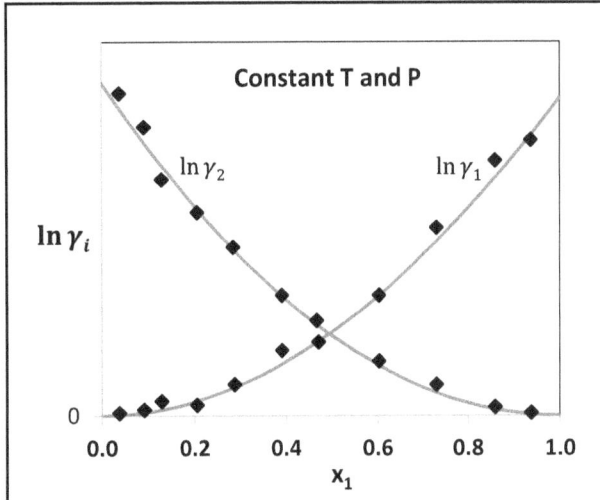

Figure 12.j.1. Plot for consideration of the differential test applied to activity coefficients.

12.j.ii The Integral Test

If we write out the excess Gibbs Free Energy for a binary solution using activity coefficients:

$$\frac{g^E}{RT} = x_1 \ln \gamma_1 + x_2 \ln \gamma_2 \qquad 12.100$$

take the partial by x_1 holding T and P constant:

$$\left.\frac{\partial \frac{g^E}{RT}}{\partial x_1}\right)_{T,P} = x_1 \left.\frac{\partial \ln \gamma_1}{\partial x_1}\right)_{T,P} + \ln \gamma_1 + x_2 \left.\frac{\partial \ln \gamma_2}{\partial x_1}\right)_{T,P} + \ln \gamma_2 \qquad 12.101$$

But in light of equation 12.99, the first and third terms on the right-hand side sum to zero, so:

$$\left.\frac{\partial \frac{g^E}{RT}}{\partial x_1}\right)_{T,P} = \ln\gamma_1 - \ln\gamma_2 \qquad 12.102$$

Integrating:

$$\int_{x_1=0}^{x_1=1} d\left(\frac{g^E}{RT}\right) = \frac{g^E}{RT}\bigg|_{x_1=1} - \frac{g^E}{RT}\bigg|_{x_1=0} = \int_{x_1=0}^{x_1=1} \ln\frac{\gamma_1}{\gamma_2} dx_1 \qquad \text{(constant T and P)} \quad 12.103$$

since g^E starts at 0 at $x_1 = 0$ and ends at 0 at $x_1 = 1$, the left-hand side of the integral is zero and we are left with:

$$\int_0^1 \ln\frac{\gamma_1}{\gamma_2} dx_1 = 0 \qquad \text{(constant T and P)} \quad 12.104$$

Therefore, a set of activity coefficients derived from VLE measurements must satisfy equation 12.104 or the data are not consistent with the Gibbs-Duhem equation. The accuracy and utility of such data should be considered carefully. Consider how we apply equation 12.104. Figure 12.j.2 gives a set of hypothetical VLE data from which activity coefficients have been calculated, then $\ln\frac{\gamma_1}{\gamma_2}$ computed and plotted vs x_1 as solid diamonds. In order for the data to be consistent, $|A_1|$ must equal $|A_2|$ so the sum $A_1 + A_2 = 0$. This is the integral test of experimental VLE data. In determining if this test is satisfied, note that we should compare the result $A_1 + A_2$ (which, do to some inevitable experimental noise, will not be exactly zero) with $|A_1|$ and $|A_2|$, not to a absolute value of zero.

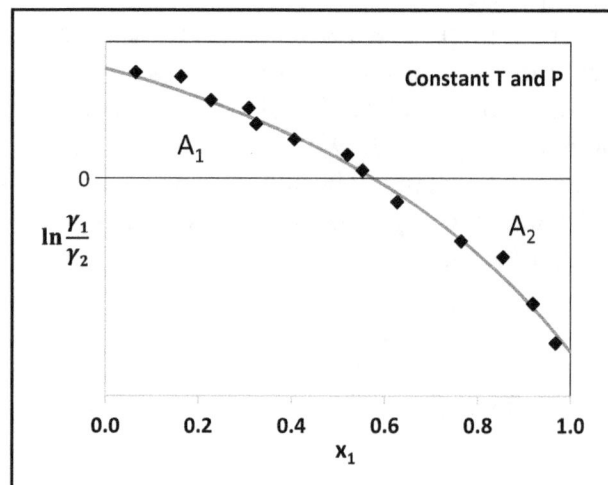

Figure 12.j.2. The integral test of experimental VLE data.

The astute student has already noted that we cannot have binary VLE data over a range of composition at both constant T and P. We can do experiments for one or the other constant, not both. However, if the experiments are done at constant T allowing P to change, very little error is interjected as activity coefficients are virtually independent of pressure. If the experiments are done at constant P but not constant T, we have to rely on a small enough temperature range so that activity coefficients are essentially constant over that range, or we cannot expect the result to be reliable.

It is also possible to re-derive equation 12.104 for both the constant T and constant P case. The results are:

$$\int_0^1 \ln\frac{\gamma_1}{\gamma_2} dx_1 = \int_{T|_{x_1=0}}^{T|_{x_1=1}} \frac{h^E}{RT^2} dT \qquad \text{(constant P)} \quad 12.105$$

$$\int_0^1 \ln\frac{\gamma_1}{\gamma_2} dx_1 = -\int_{P|_{x_1=0}}^{P|_{x_1=1}} \frac{v^E}{RT} dP \qquad \text{(constant T)} \quad 12.106$$

To put the integrals on the right-hand side of both equations 12.105 and 12.106 into some perspective, note that a regular solution has $v^E = 0$ but $h^E \neq 0$. When considering the importance of the right-hand side of equation 12.105 (or 12.106), we need to compare it to $|A_1|$ and $|A_2|$, not to zero.

When comparing the differential test to the integral test, the integral test has inherent smoothing of experimental data built into the process. However, the differential test is the more stringent test. This observation has been proven because there are equations that satisfy the integral test but not the differential test. In most applications, because of the difficulty in differentiation of experimental data, the integral test is applied.

12.k   More on the Gibbs-Duhem Equation

If we write equation 12.7 for a binary solution using $\ln(\gamma_i)$ as the partial molar property:

$$x_1 d\ln\gamma_1 + x_2 d\ln\gamma_2 = 0 \qquad 12.107$$

For low P, $\varphi_i = 1$ and $f_i^\circ = P_i^s$ which results in the modified Raoult's Law equation (section 12.e), we can substitute for γ_i:

$$x_1 d\ln\left(\frac{y_1 P}{x_1 P_1^s}\right) + x_2 d\ln\left(\frac{y_2 P}{x_2 P_2^s}\right) = 0 \qquad 12.108$$

Since P_i^s is constant at constant T, and since $x_1 d\ln x_1 + x_2 d\ln x_2 = 0$:

$$x_1 \frac{d\ln y_1 P}{dx_1} = x_2 \frac{d\ln y_2 P}{dx_2} \quad \text{or} \quad x_1 \frac{d\ln P_1}{dx_1} = x_2 \frac{d\ln P_2}{dx_2} \qquad 12.109$$

Equation 12.109 relates (and restricts) the slopes of the partial pressure vapor curves in binary VLE.

Consider *Henry's Law*:

$$f_i = k_i x_i \tag{12.110}$$

Henry's Law is often applied near infinite dilution, and it says that fugacity is proportional to mole fraction near infinite dilution. The Henry's Law constant, k, is considered to be constant. Henry's Law can also be applied to activity or other properties as well. If Henry's Law holds for the fugacity for component 2 in a binary mixture at VLE, what does the Gibbs-Duhem equation tell us about component 1? Let's assume ideal gases in the gas phase so $\varphi_i = 1$. Since we have VLE, $f_i^V = f_i^L$, and writing this for the Henry's Law component (component 2):

$$y_2 P = k_2 x_2 \tag{12.111}$$

$$\ln y_2 P = \ln k_2 + \ln x_2 \tag{12.112}$$

differentiating equation 12.112 by x_1 and noting that k_2 is constant:

$$\frac{d \ln y_2 P}{dx_1} = \frac{d \ln x_2}{dx_1} = \frac{1}{x_2} \frac{dx_2}{dx_1} = -\frac{1}{x_2} \tag{12.113}$$

substituting the result from equation 12.113 into equation 12.109, we get:

$$x_1 \frac{d \ln y_1 P}{dx_1} - 1 = 0 \tag{12.114}$$

Note that by using equation 12.114, we have transferred focus from component 2 to component 1. Rearranging and integrating equation 12.114:

$$\ln y_1 P = \ln x_1 + \ln C \tag{12.115}$$

where lnC is a constant of integration. Now, for component 2 being near infinite dilution, $x_1 \to 1$, $y_1 \to 1$, and $P \to P_1^s$. Therefore, C in equation 12.115 must be P_1^s. Plugging in to equation 12.115 and exponentiating:

$$y_1 P = x_1 P_1^s \tag{12.116}$$

which is Raoult's Law. So, if Henry's Law holds near infinite dilution for one component in a binary mixture at VLE, the other component must satisfy Raoult's Law in that same region. Look back at Figure 12.e.1 which is a modified Raoult's Law equilibrium curve for a regular solution. Even though we did not assume activity coefficients were one in that figure, we find that the actual P_1 and P_2 curves become identical to the Raoult's Law solution which are the dotted lines in Figure 12.e.1 near infinite dilution.

Similar arguments can also be used to test equations for compliance with the Gibbs-Duhem equation. Let's then ask if the regular solution satisfies the integral and differential tests. For the integral test:

$$RT\ln\gamma_1 = Ax_2^2 \text{ and } RT\ln\gamma_2 = Ax_1^2 \qquad 12.117$$

yielding:

$$\ln\frac{\gamma_1}{\gamma_2} = A\left(x_2^2 - x_1^2\right) \qquad 12.118$$

plugging into the integral test and integrating:

$$\int_0^1 \ln\frac{\gamma_1}{\gamma_2}dx_1 = \int_0^1 A\left(x_2^2 - x_1^2\right)dx_1 \qquad 12.119$$

$$\int_0^1 \ln\frac{\gamma_1}{\gamma_2}dx_1 = A\int_0^1 (1-2x_1)dx_1 \qquad 12.120$$

$$\int_0^1 \ln\frac{\gamma_1}{\gamma_2}dx_1 = A\left(x_1\Big|_0^1 - x_1^2\Big|_0^1\right) = 0 \qquad 12.121$$

Therefore, it does satisfy the integral test. Consider the differential test. Plugging into equation 12.99:

$$x_1\frac{d(Ax_2^2)}{dx_1} \stackrel{?}{=} x_2\frac{d(Ax_1^2)}{dx_2} \qquad 12.122$$

$$x_1(2Ax_2)\frac{dx_2}{dx_1} \stackrel{?}{=} x_2(2Ax_1)\frac{dx_1}{dx_2} \qquad 12.123$$

We see that equation 12.123 is satisfied, and it does satisfy the differential test. We could continue and test Margules 3-suffix equation, Van Laar, Wilson's, etc. and we would find that these also satisfy the differential and integral tests. In fact, some of the rather awkward ways that constants appear in these equations is because they would not satisfy the tests if the constants are not properly arranged.

12.1 Chapter 12 Example Problems

12.1 In the acetone(1)/methanol(2) system at 55°C and 96.365 kPa, vapor-liquid equilibrium exists with $x_1 = 0.4480$, $y_1 = 0.5512$. a) Find g^E under these conditions, b) find γ_1 and γ_2 for a liquid mixture with $x_1 = 0.40$ at $T = 55°C$ assuming A and B in the Margules equation are independent of pressure and c) find T, y_1, and y_2 in equilibrium with the liquid of composition $x_1 = 0.40$ at $P = 100$ kPa assuming A and B are independent of T.

Solution:

We note that we are at a low pressure, and we will assume ideal gases so fugacity coefficients are 1 and we take $f_i^\circ = P_i^s$ which are the general assumptions we make for all example problems.

a. $\quad \gamma_1 = \dfrac{y_1 P}{x_1 P_1^s} = \dfrac{0.5512 \cdot 96.365 \text{ kPa}}{0.4480 \cdot 95.75 \text{ kPa}} = 1.2383 \qquad$ similarly, $\gamma_2 = 1.1404$

$\dfrac{g^E}{RT} = \sum x_i \ln \gamma_i = 0.4480 \ln(1.2383) + 0.5520 \ln(1.1404) \qquad g^E = 459.01$ J/mole

b. We can find values of A and B for Margules 3-suffix equation by using the single measurement of the VLE data from part a) and equation 12.88:

$A = \dfrac{x_2 - x_1}{x_2^2} \ln \gamma_1 + 2 \dfrac{\ln \gamma_2}{x_1} = \dfrac{0.5520 - 0.4480}{0.5520^2} \ln(1.2383) + 2 \dfrac{\ln(1.1404)}{0.4480} = 0.6593$

$B = \dfrac{x_1 - x_2}{x_1^2} \ln \gamma_2 + 2 \dfrac{\ln \gamma_1}{x_2} = \dfrac{0.4480 - 0.5520}{0.4480^2} \ln(1.1404) + 2 \dfrac{\ln(1.2383)}{0.5520} = 0.7062$

Now, plug these values of the parameters into the Margules 3-suffix equation and evaluate it at $x_1 = 0.40$ to get activity coefficients. Using Margules 3-suffix equation (equation 12.67):
$\ln \gamma_1 = x_2^2 [A + 2(B-A)x_1] = 0.60^2 [0.6593 + 2(0.7062 - 0.6593)0.40] = 0.25087$
$\ln \gamma_2 = x_1^2 [B - 2(A-B)x_2] = 0.40^2 [0.7062 - 2(0.6593 - 0.7062)0.60] = 0.10399$
$\gamma_1 = 1.2851$ and $\gamma_2 = 1.1096$

c. $\quad P = 100 \text{ kPa} = \gamma_1 x_1 P_1^s + \gamma_2 x_2 P_2^s = 1.2851 \cdot 0.40 \cdot P_1^s + 1.1096 \cdot 0.60 \cdot P_2^s$
This is a bubble point calculation, so iterate on T until P = 100 kPa. This results in $T = 329.52$ K. $P_1^s = 72.72$ and $P_2^s = 56.37$

$y_1 = \dfrac{\gamma_1 x_1 P_1^s}{P} = \dfrac{1.2851 \cdot 0.40 \cdot 72.72 \text{ kPa}}{100 \text{ kPa}} = 0.5158$

$y_2 = \dfrac{\gamma_2 x_2 P_2^s}{P} = \dfrac{1.1096 \cdot 0.60 \cdot 56.37 \text{ kPa}}{100 \text{ kPa}} = 0.4842$

12.2 a) Using the data in the table below for the binary system Acetone (1), Cyclohexane (2) at 308.15 K, compute γ_1, γ_2 and the functions g^E/RT and g^E/RTx_1x_2 and enter the numbers table. Assume that φ_i's are 1.0.

P (kPa)	x_1	y_1	γ_1	γ_2	g^E/RT	g^E/RTx_1x_2
19.625	0.000	0.000	----		----	----
37.877	0.098	0.482				
45.476	0.198	0.601				
49.489	0.387	0.665				
50.969	0.598	0.709				
50.196	0.895	0.841				
45.863	1.000	1.000		----	----	----

b) Plot the values of g^E/RT and g^E/RTx_1x_2 vs. x_1
c) Which activity coefficient correlation: regular, Margules 3-suffix, or Van Laar, do you think would model this set of data best? Briefly explain why you chose the one you did.
d) Find values of A and B for the model you choose.

Solution:

Note that for $x_1 = 0$, $P = P_2^s = 19.625$ kPa and for $x_1 = 1$, $P = P_1^s = 45.863$ kPa. We do not need an Antoine equation calculation for this problem and in fact, if the Antoine equation gave slightly different values for P_1^s and P_2^s and we used those values, we would find inconsistencies in the data in that γ_1 and γ_2 would not compute to be equal to 1 for the respective pure components. Activity coefficients can be calculated like the following example for the $P = 37.625$ row and the remainder also computed and added to the table:

$$\gamma_1 = \frac{x_1 P_1^s}{y_1 P} = \frac{0.098 \cdot 45.863}{0.482 \cdot 37.877} = 4.062 \qquad \gamma_2 = \frac{x_2 P_2^s}{y_2 P} = \frac{(1-0.098) \cdot 19.625}{(1-0.482) \cdot 37.877} = 1.108$$

P (kPa)	x_1	y_1	γ_1	γ_2	g^E/RT	g^E/RTx_1x_2
19.625	0.000	0.000	----	1.000	----	----
37.877	0.098	0.482	4.062	1.108	0.2302	2.6040
45.476	0.198	0.601	3.010	1.153	0.3322	2.0922
49.489	0.387	0.665	1.854	1.378	0.4356	1.8360
50.969	0.598	0.709	1.318	1.880	0.4187	1.7418
50.196	0.895	0.841	1.028	3.873	0.1673	1.7800
45.863	1.000	1.000	1.000	----	----	----

The g^E/RT and g^E/RTx_1x_2 column come from:

$$\frac{g^E}{RT} = x_1 \ln\gamma_1 + x_2 \ln\gamma_2 \qquad \frac{g^E}{RTx_1x_2} = \frac{x_1\ln\gamma_1 + x_2\ln\gamma_2}{x_1x_2}$$

b. See Figure 12.1.1

c. Because $\frac{g^E}{RTx_xx_2}$ is not constant with x_1 (which would be indicative of a regular solution) and is not linear in x_1 (which would be indicative of the Margules 3-suffix equation), a better choice would be Van Laar equation.

d. For Van Laar, the best way to find the find A and B is to plot $\sqrt{\ln\gamma_1}$ vs. $\sqrt{\ln\gamma_2}$ and use the slopes and intercept to compute A = intercept^2 and A/B = -slope^2. A plot is given in Figure 12.1.2.

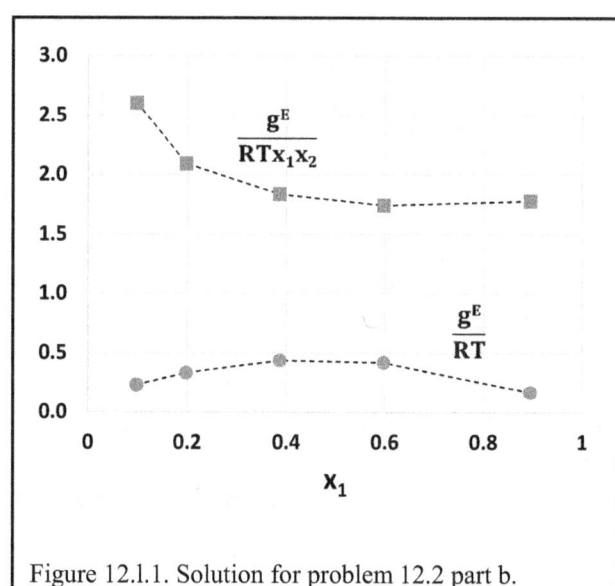

Figure 12.1.1. Solution for problem 12.2 part b.

From the plot, intercept = 1.5025 and slope = -1.1794. **A = 2.2576 and B = 1.6229.**

12.3 Using the Van Laar equation, given that an azeotrope exists at atmospheric pressure (101.3 kPa) with chloroform(1)/n-hexane(2) at $x_1 = 0.72$ and T = 60.0°C, a) find g^E in J/mole at the azeotrope, b) find γ_1 and γ_2 at $x_1 = 0.35$ at T = 60.0°C and c) find P, y_1, and y_2 in equilibrium with the liquid of composition $x_1 = 0.35$ at T=60.0°C.

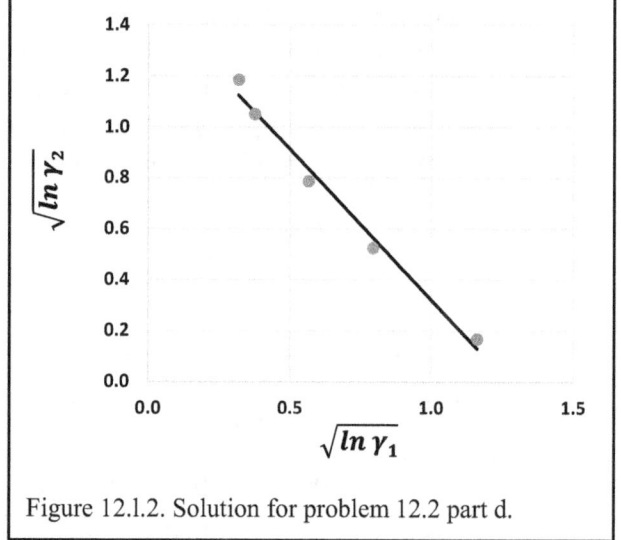

Figure 12.1.2. Solution for problem 12.2 part d.

Solution:

a. At 60.0°C = 333.15 K using Antoine constants from Appendix E, P_1^s = 95.28 kPa and P_2^s = 75.06 kPa. At the azeotrope where $x_1 = y_1 = 0.72$, γ_1 = 101.3 kPa/95.28 kPa = 1.0632 and γ_2 = 1.3495. This gives g^E/RT = 0.72·ln(1.0632) + 0.28·ln(1.3495) = 0.1281. **g^E = 354.70 J/mole.**

b. Using equation 12.89 with the activity coefficients at the azeotrope:

$$A = \ln 1.0632\left(1+\frac{0.28\ln 1.3495}{0.72\ln 1.0632}\right)^2 = 0.5162 \quad \text{and} \quad B = \ln 1.3495\left(1+\frac{0.72\ln 1.0632}{0.28\ln 1.3495}\right)^2 = 0.6978$$

Now use the Van Laar constants in the Van Laar equation at $x_1 = 0.35/x_2 = 0.65$. $\gamma_1 = \mathbf{1.3021}$ and $\gamma_2 = \mathbf{1.0583}$.

c. $P = \gamma_1 x_1 P_1^s + \gamma_2 x_2 P_2^s = \mathbf{95.06}$ kPa. $y_1 = \gamma_1 x_1 P_1^s/P = \mathbf{0.4568}$ and $y_2 = \gamma_2 x_2 P_2^s/P = \mathbf{0.5432}$.

12.4 Experimental VLE data for chloroform (1)/ethanol (2) system at 35°C are as follows:

P [kPa]	x_1	y_1
13.703	0.0000	0.0000
13.982	0.0062	0.0254
14.840	0.0241	0.0991
15.147	0.0297	0.1210
16.471	0.0542	0.2154
16.775	0.0594	0.2343
19.766	0.1109	0.3885
23.678	0.1730	0.5304
27.422	0.2361	0.6207
30.006	0.2873	0.6747
30.563	0.3014	0.6870
31.531	0.3227	0.7009
33.783	0.3845	0.7370
34.035	0.3922	0.7412
35.684	0.4384	0.7646
36.536	0.4827	0.7797
38.923	0.6185	0.8181
39.587	0.6783	0.8327
40.403	0.7746	0.8554
40.715	0.8265	0.8698
40.679	0.8423	0.8752
40.830	0.8483	0.8783
40.803	0.9315	0.9161
40.646	0.9560	0.9363
40.553	0.9586	0.9385
40.489	0.9600	0.9403
40.518	0.9616	0.9414
39.345	1.0000	1.0000

Using this data:
a. Calculate values for $\ln \gamma_1$ and $\ln \gamma_2$ and $g^E/x_1 x_2 RT$ and plot these vs x_1

b. Compute values of A and B for the Margules 3-suffix equation from $\ln \gamma_1^\infty$ and $\ln \gamma_2^\infty$ (extrapolate $\ln \gamma_1$ and $\ln \gamma_2$ to $x_1 = 0$ and $x_2 = 0$ to get these), and plot $\ln \gamma_1$ and $\ln \gamma_2$ and g^E/x_1x_2RT from the Margules 3-suffix equation using these values of A and B all on the same graph.

c. Plot x_1 vs y_1 from the experimental data and add the diagonal (a line for $x_1 = y_1$) and estimate the azeotrope composition.

d. Perform an integral test on the experimental data. Plot $\ln(\gamma_1/\gamma_2)$ vs. x_1 using the experimental data and make some quantitative assessment of whether or not the data is consistent.

Solution:

a. Columns for γ_1, γ_2, and g^E/x_1x_2RT come from:

$$\gamma_1 = \frac{y_1 P}{x_1 P_1^s} \quad \text{and} \quad \gamma_2 = \frac{y_2 P}{x_2 P_2^s} \quad \text{and} \quad \frac{g^E}{x_1 x_2 RT} = \frac{x_1 \ln \gamma_1 + x_2 \ln \gamma_2}{x_1 x_2}$$

P (kPa)	x_1	y_1	γ_1	γ_2	$\ln \gamma_1$	$\ln \gamma_2$	g^E/x_1x_2RT
13.703	0.0000	0.0000	-	1.000	-	0.0000	-
13.982	0.0062	0.0254	1.456	1.001	0.3756	0.0006	0.4823
14.840	0.0241	0.0991	1.551	1.000	0.4389	-0.0003	0.4392
15.147	0.0297	0.1210	1.568	1.001	0.4501	0.0014	0.5099
16.471	0.0542	0.2154	1.664	0.997	0.5090	-0.0029	0.4853
16.775	0.0594	0.2343	1.682	0.997	0.5198	-0.0035	0.4945
19.766	0.1109	0.3885	1.760	0.992	0.5653	-0.0079	0.5641
23.678	0.1730	0.5304	1.845	0.981	0.6125	-0.0190	0.6309
27.422	0.2361	0.6207	1.832	0.994	0.6056	-0.0064	0.7657
30.006	0.2873	0.6747	1.791	0.999	0.5828	-0.0005	0.8158
30.563	0.3014	0.6870	1.771	0.999	0.5713	-0.0007	0.8155
31.531	0.3227	0.7009	1.741	1.016	0.5542	0.0160	0.8680
33.783	0.3845	0.7370	1.646	1.053	0.4982	0.0521	0.9449
34.035	0.3922	0.7412	1.635	1.058	0.4915	0.0560	0.9514
35.684	0.4384	0.7646	1.582	1.092	0.4586	0.0876	1.0163
36.536	0.4827	0.7797	1.500	1.135	0.4054	0.1271	1.0470
38.923	0.6185	0.8181	1.309	1.354	0.2689	0.3033	1.1953
39.587	0.6783	0.8327	1.235	1.502	0.2112	0.4071	1.2567
40.403	0.7746	0.8554	1.134	1.892	0.1258	0.6374	1.3808
40.715	0.8265	0.8698	1.089	2.230	0.0853	0.8019	1.4618
40.679	0.8423	0.8752	1.074	2.349	0.0717	0.8541	1.4684
40.830	0.8483	0.8783	1.074	2.390	0.0718	0.8715	1.5006
40.803	0.9315	0.9161	1.020	3.647	0.0197	1.2939	1.6769
40.646	0.9560	0.9363	1.012	4.294	0.0117	1.4573	1.7905
40.553	0.9586	0.9385	1.009	4.396	0.0090	1.4808	1.7633

			γ	γ		γ	γ
40.489	0.9600	0.9403	1.008	4.410	0.0079	1.4839	1.7439
40.518	0.9616	0.9414	1.008	4.512	0.0081	1.5068	1.7791
39.345	1.0000	1.0000	1.000	-	0.0000	-	-

These are plotted in Figure 12.1.3.

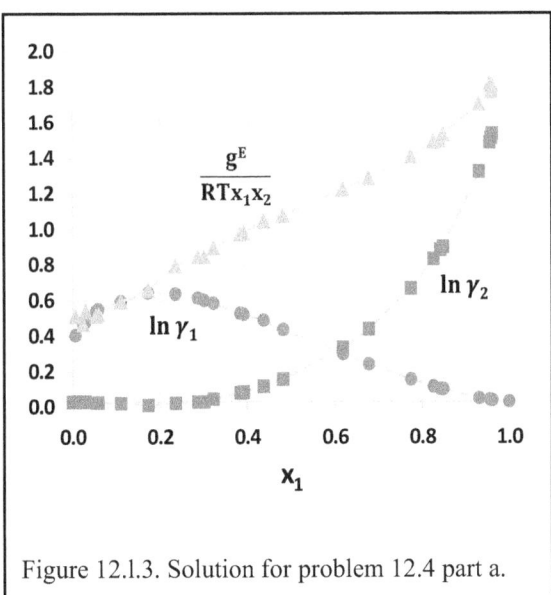

Figure 12.1.3. Solution for problem 12.4 part a.

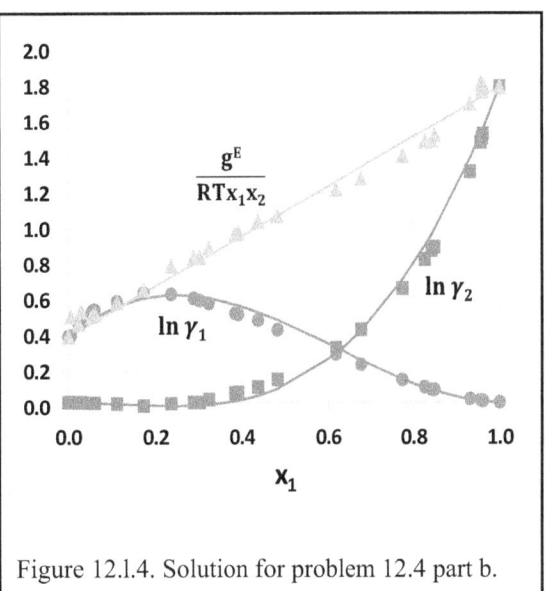

Figure 12.1.4. Solution for problem 12.4 part b.

b. Extrapolating $\ln \gamma_1$ and $\ln \gamma_2$ to the respective axis gives **A = $\ln \gamma_1^\infty$ = 1.77 and B = $\ln \gamma_2^\infty$ = 0.37**. Using these values in the Margules 3-suffix equation, theoretical calculation of $\ln \gamma_1$, $\ln \gamma_2$, and $g^E/x_1 x_2 RT$ can be made. These functions are plotted against the experimental values already determined in part a. Results are shown in Figure 12.1.4. We see that there is reasonable agreement between the theoretical values computed form the Margules 3-suffix equation and the experimental values.

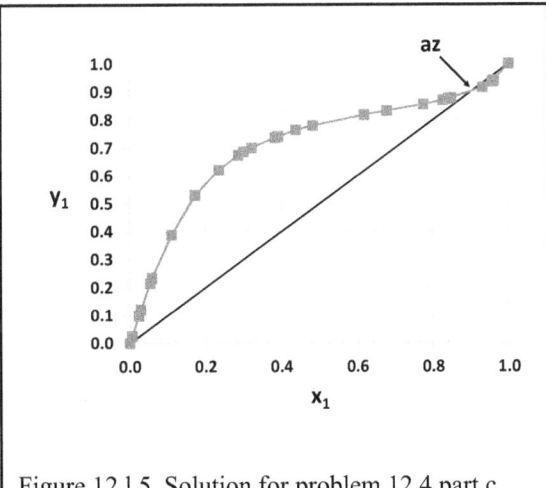

Figure 12.1.5. Solution for problem 12.4 part c.

c. A plot of the experimental data with the $x_1 = y_1$ line is given in Figure 12.1.5. The azeotrope is close to $x_1 = y_1 = 0.9$.

d. A plot of $\ln(\gamma_1/\gamma_2)$ vs. x_1 is given in Figure 12.1.6. To perform the integral test, we want to determine if equation 12.121 is satisfied, so, we need to evaluate whether or not $A_1 = -A_2$ within experimental error in Figure 12.1.6. Using a trapezoidal rule to perform the integration, we find $A_1 = 0.265$ and $A_2 = -0.280$ giving $A_1 + A_2 = 0.015$. This represents an error of approximately 3% of the total area and is small enough to be considered experimental error. Equation 12.121 is satisfied to within experimental error.

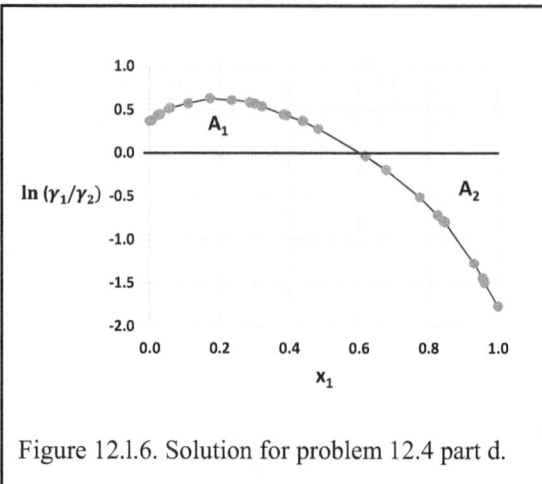

Figure 12.1.6. Solution for problem 12.4 part d.

Student Problems:

1. Repeat problem 12.1 but use the Van Laar equation instead of the Margules 3-suffix equation. Compare answers.

2. Using the results of problem 12.2, plot experimental values of γ_1 and γ_2 vs. x_1 then use A and B that were determined in the Van Laar equation to create γ_1 and γ_2 theoretical functions vs. x_1 and compare to the experimental results.

3. Using the data from problem 12.4, evaluate constants for the Margules 3-suffix equation via a linear regression of g^E/RTx_1x_2 and application of equation 12.95. Use the results of this calculation to produce a graph like Figure 12.1.4 comparing the theoretical Margules 3-suffix equation to the experimental data points. Make a visual assessment of whether Van Laar of Margules 3-suffix better represents the experimental data by comparing your graph with Figure 12.1.4.

4. Using the data from problem 12.2, perform an integral test by plotting $\ln \gamma_1/\gamma_2$ vs. x_1 and determining the area like in problem 12.4 part d. Comment on difficulties you experience, especially the number of data points you have available to do the calculations.

12.m   References

1. Wohl, K.; Chem. Eng. Prog. 49, 218 (1953).
2. Walas, S. M.; "Phase Equilibrium in Chemical Engineering", Butterworth Publishers, Boston (1985), pg. 192.
3. Renon H., and Prausnitz J. M., "Local Compositions in Thermodynamic Excess Functions for Liquid Mixtures", AIChE J. 14, 135 (1968).
4. Abrams, D. S., and Prausnitz, J. M. "Statistical thermodynamics of liquid mixtures: A new expression for the excess Gibbs energy of partly or completely miscible systems", AIChE J. 21, 116 (1975).
5. Fredenslund, A., Jones, R. L., and Prausnitz, J. M., "Group-contribution estimation of activity coefficients in nonideal liquid mixtures", AIChE J. 21, 1086 (1975).

Chapter 13 - Chemical Reaction Equilibrium

Chemical reactions are an extremely important aspect of chemical engineering practice because almost every useful product for industrial or consumer use has been transformed one way or another through a chemical reaction, and often multiple chemical reactions. The design of industrial scale reactors is heavily dependent upon chemical engineering practice as other engineering disciplines do not study the subject area.

In thermodynamics, we consider only the equilibrium states and limitations on reactions imposed by thermodynamics, not the rates at which reactions occur. We would like to think that any reaction we write down is possible, but that is not at all the case. Even with limited thermodynamic study, students should be able to recognize that reactions are subject to the same laws of thermodynamics as any other process, and hopefully, students are already thinking about entropies and free energies associated with reactions. Engineers first encountered limits on the progress of industrial chemical reactions in the development of the Haber process which synthesizes ammonia from nitrogen and hydrogen. This is still an important process and successful operation of reactors for this process balances high rates of reaction at high temperatures against high favorability of the reaction thermodynamics at low temperatures. Thermodynamic principles are therefore extremely important in engineering considerations.

13.a The Reaction Coordinate and Extent of Reaction

Let's begin by recalling generalized stoichiometry represented by equation 4.8:

$$0 = \sum v_i A_i \qquad 4.8$$

Let's generalize this stoichiometry for multiple reactions:

$$0 = \sum_j \sum_i v_{i,j} A_i \qquad 13.1$$

where j is the number of reactions and i represents components. In order for equation 13.1 to give unique results, the set of reactions must be independent which means no linear combinations of equations results in one of the other equations. For example, the set of reactions:

1. $CO + \tfrac{1}{2} O_2 \rightarrow CO_2$
2. $CO_2 \rightarrow C + O_2$
3. $C + \tfrac{1}{2} O_2 \rightarrow CO$

is not an independent set because adding reactions 2 and 3 gives the reverse of reaction 1, but we could pick any two of the three equations, which would be independent, to represent the reaction system. Suppose we chose reactions 1 and 2. Then:

$$\nu_{1,CO}=-1 \quad \nu_{1,O_2}=-\tfrac{1}{2} \quad \nu_{1,CO_2}=1 \quad \nu_{1,C}=0$$

$$\nu_{2,CO}=0 \quad \nu_{2,O_2}=1 \quad \nu_{2,CO_2}=-1 \quad \nu_{2,C}=1$$

Each reaction has a unique stoichiometry and therefore has a unique enthalpy of reaction which we have already developed in chapter 4:

$$\Delta h^o_{r,T} = \Delta h^o_{r,298} + \int_{298}^{T} \sum_{all} \nu_i C^{IG}_{p,i} dT = \sum \nu_i \Delta h^o_{f,i,298} + \int_{298}^{T} \sum \nu_i C^{IG}_{p,i} dT \qquad 4.16$$

Consider any single reaction. We can use the reaction stoichiometry to relate the moles of each species as:

$$\frac{\Delta n_i}{\Delta n_j} = \frac{\nu_i}{\nu_j} \qquad 13.2$$

Putting in differential form and rearranging:

$$\frac{dn_i}{\nu_i} = \frac{dn_j}{\nu_j} = \frac{dn_k}{\nu_k}\ldots \qquad 13.3$$

This leads to the definition of the reaction coordinate or *extent of reaction*:

$$d\zeta = \frac{dn_i}{\nu_i} \qquad 13.4$$

To complete the definition of ζ, the extent of reaction, we need to choose when $\zeta = 0$ which generally corresponds to the initial reaction mixture. To emphasize this concept, note that there is one ζ for any one reaction but we can compute a value for the extent of reaction from any of the components using equation 13.4. Extent of reaction has units of moles just like any particular species.

If we have multiple reactions, then we have multiple extents of reaction, one extent for each reaction. Individual components of the reaction mixture may participate in more than one reaction, so we need to sum up possible participation in all the reactions. Then:

$$dn_i = \sum_{j=1}^{r} \nu_{i,j} d\zeta_j \qquad 13.5$$

where j represents each reaction and there are a total of r reactions. In equation 13.5, we see that any particular species i can be produced or used up in any given reaction as is required by the set of reaction stoichiometries.

Consider a single reaction example. For the reaction:

$$H_2S + 2H_2O \rightarrow 3H_2 + SO_2$$

with initial moles $n_{H_2S} = 1$, $n_{H_2O} = 3$, $n_{H_2} = 0$, and $n_{SO_2} = 0$ at $\zeta = 0$, let's see how we can find mole fractions as a function of extent. From equation 13.4, we have:

$$\frac{dn_{H_2S}}{-1} = \frac{dn_{H_2O}}{-2} = \frac{dn_{H_2}}{3} = \frac{dn_{SO_2}}{1} = d\zeta \qquad 13.6$$

And we can find moles of any species by solving:

$$\int_1^{n_{H2S}} (-1) dn_{H_2S} = \int_0^{\zeta} d\zeta \qquad 13.7$$

Solving equation 13.7 for H_2S and the same equations for the other species, we get:

$$n_{H_2S} = 1-\zeta, \; n_{H_2O} = 3-2\zeta, \; n_{H_2} = 3\zeta, \; n_{SO_2} = \zeta \qquad 13.8$$

Adding up moles of each species, we get $n_T = 4 + \zeta$. Therefore, we find:

$$y_{H_2S} = \frac{1-\zeta}{4+\zeta}, \; y_{H_2O} = \frac{3-2\zeta}{4+\zeta}, \; y_{H_2} = \frac{3\zeta}{4+\zeta}, \; y_{SO_2} = \frac{\zeta}{4+\zeta} \qquad 13.9$$

13.b Gibbs Free Energy of Reaction

We have already developed the enthalpy of reaction in chapter 4, and we now need to consider the entropy of reaction and Gibbs Free Energy change of reaction. For entropy, using what we know for the entropy change for an ideal gas and comparing to the enthalpy of reaction, we see that for any single reaction:

$$\Delta s_{r,T}^{\circ} = \Delta s_{r,298}^{\circ} + \int_{298}^{T} \sum_{all} \nu_i C_{p,i}^{IG} \frac{dT}{T} = \sum \nu_i \Delta s_{f,i,298}^{\circ} + \int_{298}^{T} \sum \nu_i C_{p,i}^{IG} \frac{dT}{T} \qquad 13.10$$

In equation 13.10, $\Delta s_{r,298}^{\circ}$ is the entropy change when a stoichiometric mixture of reactants reacts completely to give stoichiometric products at 298 K, $\Delta s_{r,T}^{\circ}$ is the entropy change when a stoichiometric mixture of reactants reacts completely to give stoichiometric products at an arbitrary temperature T, and $\Delta s^\circ_{f,298}$ is the entropy change of formation at 298 K for the components of the reaction. If we had entropies of formation just like the enthalpies of formation, we could compute the standard entropy of reaction. However, the usual protocol is to tabulate free energies of formation along with enthalpies of formation as is the case in Appendix F.

Using the entropy and enthalpy of reaction, we can form the free energy of reaction using $\Delta g = \Delta h - T\Delta s$:

$$\Delta g^\circ_{r,T} = \Delta h^\circ_{r,298} - 298 \cdot \Delta s^\circ_{r,298} + \int_{298}^{T} \sum_{all} v_i C^{IG}_{p,i} dT - T\int_{298}^{T} \sum_{all} v_i C^{IG}_{p,i} \frac{dT}{T}$$

$$= \Delta g^\circ_{r,298} + \int_{298}^{T} \sum_{all} v_i C^{IG}_{p,i} dT - T\int_{298}^{T} \sum_{all} v_i C^{IG}_{p,i} \frac{dT}{T} \qquad 13.11$$

With equation 13.11, given enthalpies and free energies of reaction and the ideal gas heat capacities for each component in a reaction, we can compute the standard free energy of reaction at any temperature. This free energy change is a very important number that we will use as the driving force toward evaluating when $\Delta g = 0$, but we need to recognize that the standard free energy of reaction is the free energy change when stoichiometric reactants react *completely* to form stoichiometric products. What do we do when the reaction is not complete? What is the equilibrium conversion we expect?

13.c The Reaction Equilibrium Constant

As the student has likely surmised by now, a reaction which proceeds at constant T and P will continue until the free energy of the reaction mixture reaches a minimum. At first glance, one might make a mistake and say that if the standard free energy change of reaction is negative, the reaction will proceed to completion because the free energy is heading down in going from reactants to products and that if the standard free energy change of reaction is positive, the reaction will not go at all. This ignores two important aspects. First, the standard free energy change does not have built into it the free energy of mixing. It is simply the difference in free energy of unmixed reactants and products. Secondly, the standard free energy change is for gases at 1 bar pressure. How do we account for other pressures?

Let's go back to some fundamentals starting with the chemical potential from chapter 8:

$$\boxed{dG = -SdT + VdP + \sum_{i=1}^{m} \mu_i dn_i} \qquad 8.12$$

for constant T and P:

$$dG = \sum \mu_i dn_i \qquad \text{(constant T and P)} \qquad 13.12$$

Now in the context of reactions, we can use $v_i d\zeta$ for each of the species since they are related through reaction stoichiometry:

$$dG = \sum v_i \mu_i d\zeta \qquad \text{(constant T and P)} \quad 13.13$$

This free energy change is now constrained by the reaction stoichiometry. Since we want to find a minimum in free energy as a function of extent of reaction, we have:

$$\left.\frac{\partial G}{\partial \zeta}\right)_{T,P} = \sum v_i \mu_i = 0 \qquad 13.14$$

Via the definition of activity:

$$\mu_i = \mu_i^\circ + RT \ln a_i \qquad 13.15$$

We now choose the standard state to be the pure component at some arbitrary temperature and 1 bar pressure in the ideal gas state. Substituting equation 13.15 into 13.14:

$$\sum v_i \left(\mu_i^\circ + RT \ln a_i\right) = \sum v_i \mu_i^\circ + RT \sum v_i \ln a_i = 0 \qquad 13.16$$

or:

$$\sum v_i \mu_i^\circ = -RT \sum v_i \ln a_i = -RT \ln \prod a_i^{v_i} \qquad 13.17$$

Consider $\sum v_i \mu_i^\circ$. This is the difference between the chemical potential of the stoichiometric products and the stoichiometric reactants. If we chose this to be at 298 K, we have already chosen the pressure to be 1 bar, so this is the same as the standard state Gibbs Free Energy of reaction. If we further define K_a as:

$$\boxed{K_a = \prod a_i^{v_i}} \qquad 13.18$$

then:

$$\boxed{\Delta g_r^\circ = -RT \ln K_a} \qquad 13.19$$

Let's dissect this result some. First, the "a" subscript in K_a stands for "activity" because it is written in terms of activity. This is the most fundamental way to write the reaction equilibrium constant. Also, the equilibrium constant is not really a constant as it depends upon the temperature, but this is the customary name. We see that Δg_r° is only a function of T by examining equation 13.11. However, from fundamental chemistry classes, students should know that pressure affects reaction equilibrium as well as temperature. The effect of pressure is built into the activities of the individual species. Finally to link this up with our knowledge of chemistry, let's write a simple chemical reaction and plug into equation 13.18:

$$H_2S + 2H_2O \rightarrow 3H_2 + SO_2$$

$$K_a = \frac{a_{H_2}^3 \, a_{SO_2}^1}{a_{H_2S}^1 \, a_{H_2O}^2} \qquad 13.20$$

Because we defined stoichiometric coefficients as positive for products and negative for reactants, the activity terms for the products appear in the numerator of equation 13.20 and the reactants appear in the denominator. If we wrote the chemical reaction in reverse, we would swap numerator and denominator which would result in the reciprocal equilibrium constant compared to that shown in equation 13.20. If we multiply the stoichiometric equation through by a constant such as 2, we would get the square of the constant. Therefore, it is extremely important that we write down the balanced chemical reaction for which we are doing the computations and we must be sure the equation is properly balanced.

13.d Gaseous Reaction Mixtures

Suppose we have a gaseous mixture of components undergoing a chemical reaction. Using:

$$a_i = \frac{f_i}{f_i^\circ} \qquad 8.34$$

we can substitute into equation 13.18, and find:

$$K_a = \frac{\prod f_i^{v_i}}{\prod f_i^{\circ v_i}} = \frac{K_f}{K_{f^\circ}} \qquad 13.21$$

K_f is the equilibrium constant written in terms of fugacity and so forth. Recall that we already decided to use 1 bar for the standard state pressure, so f_i°'s are all 1 bar, resulting in $K_{f^\circ} = 1$ but the units are $bar^{\sum v_i}$ and we need to put the fugacities in K_f in bar for units to work properly. With $K_{f^\circ} = 1$, we see:

$$K_a = K_f \qquad 13.22$$

For a reacting mixture composed of ideal gases, $f_i = y_i P$ and:

$$K_a = K_f = \prod (y_i P)^{\nu_i} = K_y P^{\sum \nu_i} \qquad 13.23$$

In equation 13.23, P must be in bar assuming K_{f° was constructed using bar. If the gases are not ideal:

$$K_a = \prod (\varphi_i y_i P)^{\nu_i} = K_\varphi K_y P^{\sum \nu_i} \qquad 13.24$$

Note that in equations 13.23 and 13.24, K_y can be written as a function of ζ (the extent of reaction). Knowing a value of K_a then allows us to compute ζ at equilibrium.

13.e The Van't Hoff Equation

We have (at least) two important equations for G:

$$\boxed{G = H - TS} \qquad 6.6$$

$$\boxed{dG = -SdT + VdP} \qquad 6.8$$

We can combine these beginning by differentiating g/RT:

$$d\left(\frac{g}{RT}\right) = \frac{dg}{RT} - \frac{g dT}{RT^2} \qquad 13.25$$

then substituting equations 6.6 and 6.8 for g (= G/n) and dg (= dG/n), respectively:

$$d\left(\frac{g}{RT}\right) = \frac{-sdT + vdp}{RT} - \frac{h - Ts}{RT^2} dT = \frac{vdP}{RT} - \frac{h}{RT^2} dT \qquad 13.26$$

If we apply equation 13.26 to the standard Gibbs Free Energy change for reaction, we note that the standard state reaction is carried out at constant P = 1 bar, so dP = 0 and:

$$d\left(\frac{\Delta g_r^\circ}{RT}\right) = -\frac{\Delta h_r^\circ}{RT^2} dT \qquad 13.27$$

or using equation 13.19:

$$\boxed{\frac{d \ln K_a}{dT} = \frac{\Delta h_r^\circ}{RT^2}} \qquad 13.28$$

Equation 13.28 is known as the *van't Hoff equation* and it relates the reaction equilibrium constant to the standard enthalpy of reaction.

For a small temperature range, or for situations where Δh_r° is approximately constant with T, we can integrate equation 13.28 to yield:

$$\ln \frac{K_{a,1}}{K_{a,2}} = -\frac{\Delta h_r^\circ}{R}\left(\frac{1}{T_1} - \frac{1}{T_2}\right) \qquad \text{(approximate)} \quad 13.29$$

According to equation 13.29, a plot of ln K_a vs. 1/T is linear with slope $\frac{-\Delta h_r^\circ}{R}$. Knowing one value of K_a at T and the standard heat of reaction, equation 13.29 allows us to compute K_a's at different T's.

So, we have several options for computing K_a for a reaction at an arbitrary T. We can use equation 13.11 with values of $\Delta g_{r,298}^\circ$ and $\Delta h_{r,298}^\circ$ from Appendix F (or from the open literature) at T = 298 K to evaluate $\Delta s_{r,298}^\circ$. Then, with heat capacity data from Appendix B (or the open literature), we can evaluate the remaining unknown parameters, integrating equation 13.11 to yield Δg_r° as a function of temperature. If we have a measurement of the equilibrium constant or a single value of free energy of reaction, we can use that to calculate a single value of K_a using equation 13.19 then with the heat of reaction, apply equation 13.29. Sometimes we may only have a little information so we need to recognize limitations when using method such as just described.

13.f Liquid and Heterogeneous Reactions

The example we have done assumed a gas phase reaction, and section 13.c developed K_a in terms of mole fractions in the gas phase for application primarily to gaseous systems. Consider what we should do with a liquid phase reaction. First, we need to consider standard states for the condensed phases. When we choose a pure solid or liquid as the standard state, generally for low to moderate pressures, we use $a_i = 1$. This is because $a_i = x_i \gamma_i$ and for a pure components, $x_i = 1$ and $\gamma_i = 1$. However, at high pressures, we may need to make a Poynting type correction for the standard state, resulting in $a_i \neq 1$. Sandler [1] has a listing of standard states and the corresponding activities which result for virtually all situations at high pressure and the limiting value of activities for low to moderate pressures. We will assume low to moderate pressures so $a_i = 1$ for pure components for our discourse.

Assuming low to moderate pressures, if we go back to equation 13.18, we see that K_a was originally written in terms of activity, and we generally handle liquid phase calculations with activity coefficients and liquid phase mole fractions:

$$K_a = \prod a_i^{\nu_i} = \prod (\gamma_i x_i)^{\nu_i} = K_\gamma K_x \qquad 13.30$$

We need a trial composition to compute γ's so we generally take $K_\gamma = 1$ to get K_x for the first trial, then compute γ's and K_γ then recompute K_x and loop until the loop closes. For ideal liquid solutions, $K_\gamma = 1$ and $K_a = K_x$.

Consider a heterogeneous system, for example consisting of a liquid which condenses out of a reacting gaseous reaction. To solve this system exactly, we would need to consider small amounts of the gas phase components dissolved in the liquid phase and small amounts of the product liquid remaining in the gas phase. However, suppose we have the following situation. No gas is dissolved in the liquid phase and no liquid is vaporized to the gas phase. Then, in equation 13.18, the activity of the liquid falls out ($a_i = 1$ for a pure liquid), and we are left with:

$$K_a = \prod_{\text{gases}} a_i^{\nu_i} \qquad 13.31$$

The calculation of K_a follows the same procedure as the homogenous gas phase reaction through free energy of reaction, and $\varphi_i y_i P$ substituted for the activities of the gas phase components, then the system solves just like a homogenous system. However, there will not be a mole fraction to solve for in the gas phase for the liquid component. Our assumptions have that component in the liquid phase with $x = 1$.

Suppose we have a solid-gas reaction. For example, if we heat a metal carbonate (such as $CaCO_3$) to break it down to CO_2 and the metal oxide (such as CaO, lime) in the solid phase, the CO_2 gas will leave the system as a pure component and its fugacity will be $\varphi_{CO_2} P$. Now, we have to think about the solid phase to figure out the activities of $CaCO_3$ and CaO. If $CaCO_3$ and CaO completely mix in the solid phase to form a solid solution, then we would have to write $a_i = \gamma_i z_i$ (z_i is the mole fraction of species in a solid) for each of the components. However, it would be unusual for the phases to mix. The mixture would need to be at a high enough temperature such that molecules of each species can jump out of their positions in the crystal lattice and move to other empty lattice sites in order to mix. Such behavior has characteristics of a liquid, and usually we do not have this situation unless we are near the melting points of the solids. The unreacted $CaCO_3$ likely remains as a pure solid and CaO likely forms small islands of another phase that is pure CaO. In this case, since the solids are unmixed and individually pure phases, their activities can be taken as 1. The result is $K_a = \varphi_{CO_2} P$. The pressure that we get from this equation is the pressure of CO_2 in the gas phase that is in equilibrium with the solid phase.

13.g Multiple Reactions

Equation 13.5 gives the relationship of moles of each species to the extents of reaction in a multi-reaction equilibrium system:

$$dn_i = \sum_{j=1}^{r} \nu_{i,j} d\zeta_j \qquad 13.5$$

In this equation, recall there are r reactions, and each reaction has an extent of reaction associated with it, ζ_j. One values of K_a is associated with each reaction in the set and the value of K_a can be computed via the free energy of each individual reaction. Call this set $K_{a,j}$, $1 \leq j \leq r$ and recall that all reactions must be independent for the equations to work properly. Let's consider how to set up a multi-reaction system to solve for the composition of the equilibrium mixture.

Each species will have an associated initial number of moles which we call $n_{i,0}$. We might also have a set of inert species (not participating in any reaction) that we can call n_I. When we integrate equation 13.5 for each individual species, we get;

$$n_i = n_{i,0} + \sum_{j=1}^{r} \nu_{i,j} \zeta_j \qquad 13.32$$

If there are m components, then adding up all the reacting species and adding in all the inert species, we get the total number of moles, n_T, as a function of the extents of all reactions:

$$n_T = \sum_{i=1}^{m} n_{i,0} + \sum_{i=1}^{m} \sum_{j=1}^{r} \nu_{i,j} \zeta_j + \sum_{\text{all inerts}} n_I \qquad 13.33$$

With equations 13.32 and 13.33, we can get the mole fraction of any species:

$$y_i = \frac{n_i}{n_T} = \frac{n_{i,0} + \sum_{j=1}^{r} \nu_{i,j} \zeta_j}{\sum_{i=1}^{m} n_{i,0} + \sum_{i=1}^{m} \sum_{j=1}^{r} \nu_{i,j} \zeta_j + \sum_{\text{all inerts}} n_I} \qquad 13.34$$

The mole fractions given by equation 13.34 can be plugged into each $K_{y,j}$ (remembering that there are j reactions) to be part of:

$$K_{a,j} = K_{\phi,j} K_{y,j} P^{\sum_{i=1}^{m} \nu_{i,j}} \qquad 13.35$$

Here, as with single reactions, we have used $K_{f^\circ,j} = 1$ when we use 1 bar for our standard states and P must then be in bar. $K_{\phi,j}$ is based upon the fugacity coefficients of all the species in the jth reaction, but we would compute the individual fugacity coefficients based upon the composition of the entire reacting mixture, not just the species participating in the jth reaction. Many times, we can make the approximation that $K_{\phi,j} = 1$ if we are dealing with mixtures of gases near the ideal gas region..

When we write out all the equations substituting the mole fractions from equation 13.34 into $K_{y,j}$ in equation 13.35, we will have r equations, r values for $K_{a,j}$ based upon the free energies of each reaction, and all the equations together will have r extents of reaction (ζ_j's). This gives a set of r equations with r unknowns (ζ_j's). This algebraic set of equations will be independent if all the chemical reactions are independent. Therefore, we should be able to solve simultaneously for the ζ_j's, and use equation 13.34 to get the mole fractions of each species at equilibrium.

This discourse is necessarily full of abstract equations. A simplification of the process dealing with two simultaneous reactions is given in the example problems below which should help make the abstract more tenable for the confused.

Let's also consider reaction pathways for both single and multiple reactions. When we compute an equilibrium composition using methods outlined here, this composition corresponds to the minimum free energy *of that set of components*. All we can say about the reaction pathway is that every step of the way, the free energy must go down. It is quite possible that when we have multiple reactions taking place, one reaction may be the dominant reaction initially. We could stop the reaction before other reactions become important and get a reaction mixture that does not look like the equilibrium mixture at all. There is no thermodynamic problem with such an occurrence as long as the free energy is continuously going down along the dominant reaction pathway.

An example of this is the partial oxidation of a hydrocarbon. Ethylene plus oxygen to give ethylene oxide is a specific example and is industrially important. Byproducts of this reaction are CO_2 and H_2O and are undesired. If we set up reaction sets giving all the observed products which would include ethylene oxide, CO_2 and H_2O and computed the mole fractions of each species at reaction equilibrium, we would find the products were virtually all CO_2 and H_2O, no ethylene oxide. Yet ethylene oxide is lower free energy than ethylene + oxygen, so there is no violation of thermodynamic principles if we stop the reaction prior to the reaction continuing to CO_2 and H_2O.

Think of a ball rolling down the inside of a cone that has it's tip pointed down, headed toward the tip of the cone at the bottom. The free energy of the ball will be at a minimum when it rests at the tip of the cone. We could let the ball roll straight down the side by setting it on the edge and simply letting go, or we could give it an initial push to the left or right, so that it goes around the cone in circles while headed down toward the tip of the cone. Either pathway is guaranteed to have the ball with at least some downward component while headed toward the tip at any instant. As long as there is some downward component, the free energy is falling, and the ball will eventually rest at the bottom. This is also true for reactions. Any reaction that leads toward the free energy going down is possible.

13.h Chapter 13 Example Problems

13.1 Consider the hydration of ethylene to give ethanol at 400K and 1 bar. First, write a balanced equation:

$$C_2H_4 \text{ (g)} + H_2O \text{ (g)} \rightarrow C_2H_5OH \text{ (g)}$$

Compute the composition of the equilibrium mixture at 400K and 1 bar assuming the we start with 1 mole of ethylene, 3 moles of water, and no ethanol.

Solution:

Let's begin by evaluating equation 13.11, finding values from Appendix B for heat capacity equation constants and Appendix F for enthalpy and free energy of formation:

	$\Delta h^\circ_{f,298}$ J/mole	$\Delta g^\circ_{f,298}$ J/mole	A	B	C	D	E
C_2H_4	52467	68421	1.9993	0.0139160	-5.3561×10^{-6}	7.3142×10^{-10}	-42295.0
H_2O	-241800	-228400	3.1590	0.0020385	-2.1753×10^{-7}	-1.8960×10^{-11}	25105.3
C_2H_5OH	-234800	-168300	3.9711	0.0194736	-7.2137×10^{-6}	9.3590×10^{-10}	-108102.8

Using these numbers we get:

$$\Delta h^\circ_{r,298} = \sum v_i \Delta h^\circ_{f,298} = 1(-234800) - 1(52467) - 1(-241800) = -45467 \text{ J/mole}$$
$$\Delta g^\circ_{r,298} = \sum v_i \Delta g^\circ_{f,298} = 1(-168300) - 1(68421) - 1(-228400) = -8321 \text{ J/mole}$$

plugging these into equation 13.11 at 298.15K, we find $\Delta s^\circ_{r,298} = -124.6 \dfrac{J}{mole \cdot K}$. For the heat capacity equation we get:

$$\sum v_i A_i = -1.1872 \qquad \sum v_i B_i = 0.0035191 \qquad \sum v_i C_i = -1.6501 \times 10^{-6}$$
$$\sum v_i D_i = 2.2344 \times 10^{-10} \qquad \sum v_i E_i = -90913.1$$

Plugging these into equation 13.11, and evaluating the integrals:

$$\Delta h^\circ_{r,400} = \Delta h^\circ_{r,298} + \int_{298}^{400} \sum v_i C^{IG}_{p,i} dT = -45467 \frac{J}{mole} - 772.81 \frac{J}{mole} = -46239.8 \frac{J}{mole}$$

$$\Delta s^\circ_{r,400} = \Delta s^\circ_{r,298} + \int_{298}^{400} \sum v_i C^{IG}_{p,i} \frac{dT}{T} = -124.6 \frac{J}{mole \cdot K} - 2.27 \frac{J}{mole \cdot K}$$
$$= -126.9 \frac{J}{mole \cdot K}$$

with these numbers, equation 13.11 yields:

$$\Delta g^\circ_{r,400} = -46239.8 \frac{J}{mole} - 400K \left(-126.9 \frac{J}{mole \cdot K} \right) = 4504.2 \frac{J}{mole}$$

equation 13.19 is used to calculate the equilibrium constant:

$$4504.2 \frac{J}{mole} = -8.314 \frac{J}{mole \cdot K} \cdot 400K \cdot \ln K_a$$

$K_a = 0.2581$

Next, let's set up a mole table so we can compute y's from the extent of reaction:

	Start	Finish
C_2H_4	1	$1 - \zeta$
H_2O	3	$3 - \zeta$
C_2H_5OH	0	ζ
	total:	$4 - \zeta$

computing y's:

$$y_{C_2H_4} = \frac{1-\zeta}{4-\zeta} \quad y_{H_2O} = \frac{3-\zeta}{4-\zeta} \quad y_{C_2H_5OH} = \frac{\zeta}{4-\zeta}.$$

Let's assume ideal gases so $K_\varphi = 1$ and since standard state fugacities are all 1 bar, $K_{f^\circ} = 1$ bar^{-1}. Then:

$$K_a = 0.2581 = \frac{K_\varphi K_y}{K_{f^\circ}} P^{\Sigma v_i} = \frac{\zeta(4-\zeta)}{(1-\zeta)(3-\zeta)(1\,bar)^{-1}}(1\,bar)^{-1} \qquad 13.36$$

solving for ζ and plugging back to get mole fractions:

$\zeta = 0.1606, \quad y_{C_2H_4} = 0.2186, \quad y_{H_2O} = 0.7395, \quad y_{C_2H_5OH} = 0.0418$

With this example, an important observation can be made. Le Chatelier's principle is evident given the pressure term. A reaction which creates more molecules than it uses up will have $\Sigma v_i > 0$ and high pressure would tend to shift the reaction to favor reactants and vice versa, $\Sigma v_i < 0$ corresponds to a reaction that creates fewer molecules than it uses up and high pressure favors products. Since the example reaction has $\Sigma v_i < 0$, Le Chatelier's principle tells us high pressure favors products. Suppose we ran this reaction at 10 bar instead of 1 bar. Then the numerator in equation 13.36 must be 10 times larger than the denominator to get the same K_a. This requires a larger value of ζ. Students should try this calculation and confirm this reasoning.

Also, if the reaction conditions were such that an assumption of ideal gases was not reasonable, at this point we have a trial composition that could be used in the evaluation of φ's, then K_φ could be computed, and a new composition computed from equation 13.36. We can continue this loop until it closes.

13.2 The dehydrogenation of 1-butene to butadiene is generally carried out in steam which helps keep the catalyst clean. The reaction is:

1-butene ⇆ butadiene + H_2 or

A ⇌ B + H_2

Steam does not participate in the reaction. Given that 1 mole of A (1-butene) is fed for every 5 moles of steam (water) to the reactor at 500°C.
a. **Find the equilibrium constant K_a for the reaction at 500°C.**
b. **Find the maximum pressure at which the reactor can operate giving a conversion of 1-butene of at least 30% at reaction equilibrium.**

Solution:
a. getting data from Appendix F:

Data	1-butene	H_2	butadiene
ν (stoich. coeff.)	-1	1	1
$\Delta h°_{f,298}$ (J/mole)	-100	0.0	110200
$\Delta g°_{f,298}$ (J/mole)	71300	0.0	150700
$\Delta s°_{f,298}$ (J/mole-K)	-239.6	0.0	-135.9
Constants for $C_p/R = A + B·T + C·T^2 + D·T^3 + E/T^2$			
A	4.2919	3.5395	-0.2922
B	0.0289537	-0.0002114	0.042418
C	-1.1168×10^{-5}	4.0418×10^{-7}	-2.8287×10^{-5}
D	1.5150×10^{-9}	-7.7486×10^{-11}	7.1085×10^{-9}
E	-138915.7	-4341.0	-34048.4

For the reaction:

$\Delta h°_{r,298}$ (J/mole)   110300

$\Delta g°_{r,298}$ (J/mole)   79400

$\Delta s°_{r,298}$ (J/mole-K)   103.6

so for the heat capacity equation:

$\sum \nu_i A_i = \Delta A$ -1.0446

$\sum \nu_i B_i = \Delta B$ 0.0132529

$\sum \nu_i C_i = \Delta C$ 1.6715×10^{-5}

$\sum v_i D_i = \Delta D$ 5.5160×10^{-9}

$\sum v_i E_i = \Delta D$ 100526.3

Setting up the heat capacity integrals:

$$R\int_{298}^{500+273}\left(\Delta A + \Delta B \cdot T^2 + \Delta C \cdot T^3 + \Delta D \cdot T^3 + \Delta E/T^2\right)dT = 9456.7 \text{ J/mole}$$

$$R\int_{298}^{500+273}\frac{\left(\Delta A + \Delta B \cdot T^2 + \Delta C \cdot T^3 + \Delta D \cdot T^3 + \Delta E/T^2\right)}{T}dT = 19.363 \frac{\text{J}}{\text{mole} \cdot \text{K}}$$

$\Delta s°_{r,773} = 103.6 + 19.363 = 123.0$ J/mole·K

$\Delta h°_{r,733} = 110300 + 9456.7 = 119755$ J/mole

$\Delta g°_{r,733} = 119755 - 773 \cdot 123.0 = 24674$ J/mole

$$K_a = \exp\left(\frac{-\Delta g°_{R,773}}{RT}\right) = \exp\left(\frac{-24674}{8.314 * 773}\right) = \mathbf{0.02151}$$

b.   set up a mole table using the inert steam as 5 moles:

	start	finish
n_A	1	$1-\zeta$
n_B	0	ζ
n_{H2}	0	ζ
n_{H20}	5	5
	total:	$6+\zeta$

$$0.02151 = \left(\frac{1-\zeta}{6+\zeta}\right)^{-1}\left(\frac{\zeta}{6+\zeta}\right)^{1}\left(\frac{\zeta}{6+\zeta}\right)^{1} P^{1+1-1}$$

with $\zeta = 0.3$ (given) solve for P. **P = 1.054 bar**.

13.3   For the dehydrogenation of n-hexane to 1-hexene followed by dehydrocyclization of 1-hexene to benzene at 480°C and 25 bar:

rxn 1: $C_6H_{14} \rightarrow C_6H_{12} + H_2$

rxn 2: $C_6H_{12} \rightarrow C_6H_6 + 3H_2$

Calculate the equilibrium composition given that the starting components are 1 mole of n-hexane and 1 mole of hydrogen and no other species present. Assume that the components are ideal gases, but explain the process you would use for the case where the gases are not ideal.

Solution:

Given the temperature of the reaction, values for $K_{a,1}$ and $K_{a,2}$ can be calculated via the standard free energy of the reaction. This is accomplished by evaluating the terms in equation 13.11 using heat capacity data from Appendix B with energies of formation from Appendix F resulting in the free energy of reaction at 480°C = 753 K. Below are intermediate numbers for rxn 1 and 2:

	rxn 1	rxn 2
$\Sigma v_i A_i$	4.98989	10.2516
$\Sigma v_i B_i$	-0.0092495	-0.0170597
$\Sigma v_i C_i$	4.96718x10^{-6}	9.5395x10^{-6}
$\Sigma v_i D_i$	-1.15649x10^{-9}	-1.9013x10^{-9}
$\Sigma v_i E_i$	-49239.6	-125471.0
$\Delta h^0_{r,298}$ (J/mole)	122400	127600
$\Delta g^0_{r,298}$ (J/mole)	87780	42220
$\Delta s^0_{r,298}$ (J/mole·K)	116.12	286.37
$\Delta h^0_{r,753}$ (J/mole)	126816	139697
$\Delta g^0_{r,753}$ (J/mole)	31752	-95830
$\Delta s^0_{r,753}$ (J/mole·K)	126.22	312.72
$K_{a,i,753}$	0.0062767	4430832

Now, let's write $K_{a,1}$ and $K_{a,2}$ in terms of mole fractions of each component in the reactions as in equation 13.35:

$$K_{a,1} = \frac{\varphi_{C_6H_{12}} \varphi_{H_2}}{\varphi_{C_6H_{14}}} \frac{y_{C_6H_{12}} y_{H_2}}{y_{C_6H_{14}}} P \quad \text{and} \quad K_{a,2} = \frac{\varphi_{C_6H_6} \varphi_{H_2}^3}{\varphi_{C_6H_{12}}} \frac{y_{C_6H_6} y_{H_2}^3}{y_{C_6H_{12}}} P^3 \qquad 13.37$$

Since we are dealing with ideal gases, all φ's will be equal to one in equation 13.37.

Equation 13.32 can be used to evaluate a mole table where we have r = 2 and there are 4 species so m = 4.

	Start	Finish	Plug In Values
C_6H_{14}	$n_{0,C_6H_{14}} = 1$	$n_{0,C_6H_{14}} - \zeta_1$	$1 - \zeta_1$
C_6H_{12}	$n_{0,C_6H_{12}} = 0$	$n_{0,C_6H_{12}} + \zeta_1 - \zeta_2$	$\zeta_1 - \zeta_2$
H_2	$n_{0,H_2} = 1$	$n_{0,H_2} + \zeta_1 + 3\zeta_2$	$1 + \zeta_1 + 3\zeta_2$
C_6H_6	$n_{0,C_6H_6} = 0$	$n_{0,C_6H_6} + \zeta_2$	ζ_2
	total:	$\Sigma n_{0,i} + \zeta_1 + 3\zeta_2$	$2 + \zeta_1 + 3\zeta_2$

Get the algebraic expressions for mole fractions using equation 13.34:

$$y_{C_6H_{14}} = \frac{1-\zeta_1}{2+\zeta_1+3\zeta_2} \qquad y_{C_6H_{12}} = \frac{\zeta_1-\zeta_2}{2+\zeta_1+3\zeta_2}$$

$$y_{H_2} = \frac{1+\zeta_1+3\zeta_2}{2+\zeta_1+3\zeta_2} \qquad y_{C_6H_6} = \frac{\zeta_2}{2+\zeta_1+3\zeta_2} \qquad 13.38$$

Plug these into equation 13.37, use $K_{\varphi,1} = K_{\varphi,2} = 1$ and simplify:

$$\frac{K_{a,1}}{P} = \frac{0.0062767}{25\,\text{bar}} = 2.5107 \times 10^{-4} = \frac{(\zeta_1-\zeta_2)(1+\zeta_1+3\zeta_2)}{(2+\zeta_1+3\zeta_2)(1-\zeta_1)} \qquad 13.39$$

$$\frac{K_{a,2}}{P^3} = \frac{4430832}{25^3\,\text{bar}^3} = 283.573 = \frac{\zeta_2(1+\zeta_1+3\zeta_2)^3}{(2+\zeta_1+3\zeta_2)^3(\zeta_1-\zeta_2)} \qquad 13.40$$

We use pressure in bar because that is the standard state pressure and in the same units as K_{f°. We now know everything in equations 13.39 and 13.40 except ζ_1 and ζ_2 and we have two equations and two unknowns so we can solve:

$\zeta_1 = 0.260115 \qquad \zeta_2 = 0.259838$

Plugging ζ_1 and ζ_2 back into equation 13.38 yields the composition of the equilibrium mixture:

$y_{C_6H_{14}} = 0.2434129 \qquad y_{C_6H_{12}} = 9.113 \times 10^{-5}$

$y_{H_2} = 0.6710125 \qquad y_{C_6H_6} = 0.0854835$

We have carried more digits than usual because the mole fraction of 1-hexene is small and based upon the difference between ζ_1 and ζ_2, so back-checking numbers to confirm solution of the equations requires a lot of significant digits. This same fact also makes the equations somewhat difficult to solve.

If we cannot assume ideal gases and we need to compute values for $K_{\varphi,1}$ and $K_{\varphi,2}$, we can use this first calculation as a trial composition, calculate values of φ and plug φ's into equation 13.36, then loop until the solution closes.

13.4 Ammonia synthesis is an extremely important process for the production of fertilizers for agricultural use. This process occurs at relatively high pressure over a potassia promoted iron oxide catalyst. For a particular reaction, a 30/70 mole % mixture of N_2/H_2, respectively, enters a reactor. The reaction:

$N_2 + 3H_2 \rightarrow 2NH_3$

proceeds readily on the catalyst so that products exit the reactor at 300°C and 25 bar essentially at equilibrium. For this reaction, a) calculate K_a for the reaction at 300°C, b) set up a mole balance to give mole fractions (y's) of each species in terms of extent of reaction (ζ), and c) compute the mole fractions of each species at equilibrium at 300°C and 25 bar assuming the gas phase is a mixture of ideal gases.

Solution:

Gathering data from Appendix F and using the heats and free energies of formation to compute heat and free energies of reaction:

	N_2	H_2	NH_3
ν_i	-1	-3	2
$\Delta h°_{f,298}$ (J/mole)	0.0	0.0	-45600.0
$\Delta g°_{f,298}$ (J/mole)	0.0	0.0	-16300.0
$\Delta s°_{f,298}$ (J/mole-K)	0.0	0.0	-98.3
$\Delta h°_{R,298}$ (J/mole)		-91200.0	
$\Delta g°_{R,298}$ (J/mole)		-32600.0	
$\Delta s°_{R,298}$ (J/mole-K)		-196.55	

For $C_p/R = A + B \cdot T + C \cdot T^2 + D \cdot T^3 + E/T^2$:

	N_2	H_2	NH_3
A	2.819	3.540	2.406
B	0.001553	-0.000211	0.005944
C	-5.278E-07	4.042E-07	-1.765E-06
D	6.384E-11	-7.749E-11	1.906E-10

	E	22375.7	-4341.0	23272.6
	ΔA		-8.6242	
	ΔB		0.0109682	
	ΔC		-4.215E-06	
	ΔD		5.497E-10	
	ΔE		37192.5	

Careful attention needs to paid to the setup of mole fractions:

$$y_{N_2} = \frac{30-\zeta}{100-2\zeta} \qquad y_{H_2} = \frac{70-3\zeta}{100-2\zeta} \qquad y_{NH_3} = \frac{2\zeta}{100-2\zeta}$$

and to the setup of the equilibrium equation:

$$K_a = \frac{y_{NH_3}^2}{y_{N_2} y_{H_2}^3} P^{\sum v_i} = \frac{y_{NH_3}^2}{y_{N_2} y_{H_2}^3} P^{-2}$$

$\int C_p \, dT$	-10070.7
$\int C_p/T \, dT$	-24.5
$\Delta h^\circ_{R,573K}$ (J/mole)	-101270.7
$\Delta g^\circ_{R,573K}$ (J/mole)	25401.80
$\Delta s^\circ_{R,573K}$ (J/mole-K)	-221.0
K_a	0.00484

$$K_a = 0.00484 = \frac{y_{NH_3}^2}{y_{N_2} y_{H_2}^3} P^{-2} = \frac{4\zeta^2 (100-2\zeta)^2}{(30-\zeta)(70-3\zeta)^3} 25^{-2}$$

solving:

ζ	10.97
y_{N2}	0.2438
y_{H2}	0.4751
y_{NH3}	0.2812

13.5 It is proposed to clean the oxide layer off of tungsten wire by heating it in hydrogen gas at 1 bar of pressure at 800 K. The reaction is:
$WO_2 (s) + 2 H_2 (g) \rightarrow W(s) + 2 H_2O (g)$

You can use the following free energies for your calculations:
$WO_2 (s)$: $\quad \Delta g^\circ_f$ (J/mole) = -550,640 + 153.14 T $\qquad$ (T in Kelvin)
$H_2O (g)$: $\quad \Delta g^\circ_f$ (J/mole) = -246,449 + 54.81 T $\qquad$ (T in Kelvin)

based upon the elements in their standard states at 1 bar pressure. a) Can this process work?

b) What is the maximum water vapor partial pressure that can exist in the hydrogen for it to work?

Solution:

Here we have a solid-fluid reaction. We are going to assume that elemental W and WO_2 are separate pure solid phases (not mixed). This results in the activities of each solid species being 1. Therefore, writing the equilibrium equation in terms of activity, then substituting fugacities for the gas phase:

$$K_a = \frac{a_W a^2_{H_2O}}{a_{WO_2} a^2_{H_2}} = \frac{y^2_{H_2O} P^2}{y^2_{H_2} P^2} = \frac{y^2_{H_2O}}{y^2_{H_2}}$$

The free energies of elemental W and H_2 are zero because they are elements in their standard states. Missing the heat capacities of the tungsten species as a function of T, we will take $\Delta C_p = 2C_{p,H2O} + C_{p,W} - 2C_{p,H2} + C_{p,WO2} = 0$ as an estimate which should be reasonable. So:

$$\Delta g^0_{R,800K} = 2g^0_{f,H_2O,800K} - g^0_{f,WO_2,800K} = 22926 \frac{J}{mole}$$

This results in $K_a = 0.03184 = y_{H2O}^2/y_{H2}^2$. $y_{H2O}/y_{H2} = 0.1784$ and with P = 1 bar, $y_{H2} = 0.8486$ bar and $y_{H2O} = 0.1514$ bar. All that needs to be done is to flow enough pure H_2 to keep the water mole fraction below 15% for the thermodynamics to be feasible, and this is not difficult to accomplish.

13.6 Suppose we have a series of reactions that add to an overall reaction:

$$
\begin{array}{ll}
A + B \rightarrow C + D & K_{a1} \\
C + D \rightarrow E + F & K_{a2} \\
E + F \rightarrow G + H & K_{a3} \\
G + H \rightarrow I + J & K_{a4} \\
\hline
A + B \rightarrow I + J & K_{a5}
\end{array}
$$

a) How is K_{a5} related to the other K_a's? b) How many independent reactions are there?

Solution:

a) If you write out $K_a = \sum \varphi_i y_i^{\nu_i} P$ for each reaction, you will find that is the same as $K_{a1} \cdot K_{a2} \cdot K_{a3} \cdot K_{a4} = K_{a5}$.

b) There are 4 independent reactions. Obviously, the addition of all 4 reaction together gives equation 5 so that one is not independent. Also, the products of each reaction are uniquely formed by only one of the four reactions so those are each independent because no other reaction or linear combination of reactions can form those products.

Student Problems:

1. Suppose in problem 13.4 we had written the reaction as ½ N_2 + 3/2 H_2 → NH_3 instead of the way it is written (multiplied through by 2). How does it affect the heats and free energies of reaction? How does it affect the mole balance? How does it affect the value of K_a? How does it affect the computed equilibrium composition?

2. At 300°C, compute K_a for:

C_2H_4 (ethylene) + 3 O_2 → 2 CO_2 + 2 H_2O

and:

C_2H_4 (ethylene) + ½ O_2 → C_2H_4O (ethylene oxide)

Is it possible to make ethylene oxide from the partial oxidation of ethylene using oxygen?

3. Find the equilibrium composition for the isomerization of 1-butene, cis-2-butene, and trans-2-butene at 50°C and 1 bar. Each isomer can react to form each of the other two isomers. Show that it doesn't matter with which isomer you start because the equilibrium composition will be the same.

4. Find the maximum pressure which allows at least 30% conversion of cyclohexane for:

cyclohexane → benzene + 3 H_2

at 500°C starting with a 1/1 molar mixture of cyclohexane and hydrogen.

5. Find the equilibrium composition for:

CH_4 + Cl_2 → CH_3Cl + HCl

at 300°C and 5 bar starting with a stoichiometric mixture of reactants.

13.i References

1. Sandler, S. I.; Chemical and Engineering Thermodynamics, 5th ed., Wiley and Sons (2017).

Chapter 14 - Other Topics in Phase Equilibrium

14.a Liquid-Liquid Equilibrium

Many binary systems exhibit liquid-liquid equilibrium behavior which is equivalent to limited solubility of one component in another. In Figure 14.a.1, a typical binary liquid-liquid equilibrium system is depicted. Many regions are shown on the figure. At low temperatures, we see an α-β equilibrium region where two liquid phases coexist. At low values of x_1, an α region exists which is a single liquid phase rich in component 2. Similarly, at high values of x_1, a β region exists which is a single liquid phase rich in component 1. In the α-β equilibrium region, a tie line at any temperature connects the composition of the α phase in equilibrium with the composition of the β phase. At higher temperatures, the α and β

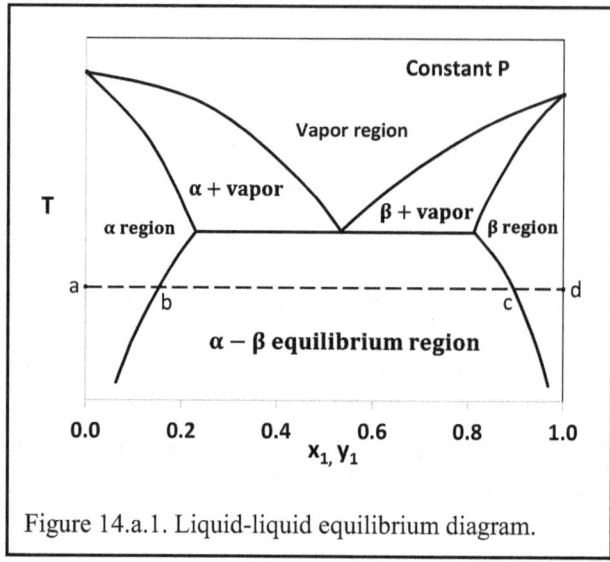

Figure 14.a.1. Liquid-liquid equilibrium diagram.

liquids coexist with vapor. There is a single point where two liquid phases and a vapor phase coexist. The single point is in accordance with the phase rule with phases = 3, components = 2, and P is fixed, leaving no more degrees of freedom and that fixes T. In all the regions, compositions of liquid phases and the vapor phases that are in equilibrium can be found using tie lines.

Now, consider pure liquid component 2 at point a, and let's follow a constant temperature (horizontal) line toward point d as we add component 1 to component 2. A single α liquid phase rich in component 2 exists until we get to point b. The composition at point b represents the solubility of component 1 in component 2 at that particular temperature. Adding more component 1 does not change the composition of the α liquid because a second phase corresponding to the composition of the β liquid at point c appears and grows. As we continue to add component 1, the fraction of the α liquid gets smaller while the fraction of the β liquid grows, and we eventually have enough component 1 to completely dissolve the α liquid when we reach point c. If we were to start with pure component 1 at point d and start adding component 2, we get to point c at the solubility limit of component 2 in component 1. α and β liquids at points b and c, respectively, are two liquid phases in equilibrium with each other.

Suppose we start in the α-β equilibrium region at low temperature. Let's use $x_1 = 0.4$ as an example and follow a vertical line at constant x_1 by heating the system. As we heat, we will continue to have two liquids until we reach the point at which the α + vapor region is shown. At this point, continued heating increases the amount of vapor relative to the liquid, and this continues until we get to the vapor region where no liquid at all exists.

Now, let's consider what's happening with the free energy of a system in the α-β equilibrium region of Figure 14.a.1. Three functions are shown in Figure 14.a.2 for g^E, g^{MIX} and g^{ID} (the free energy of the ideal solution). The g^{MIX} function is the sum of g^E and g^{ID}. Note that the g^{ID} function for the ideal solution is smooth with a smooth second derivative with respect to x_1 that is always positive (students can prove this themselves). All of the g^{ID} function is negative, so if we have an ideal solution, starting with two pure components and mixing them will always result in the free energy going down. Ideal solutions (binary and multi-component) have components which are completely miscible. Now, the particular non-ideal solution depicted in Figure 14.a.2 has a completely positive excess free energy which is what all positive deviation solutions (which are very common) exhibit. All solutions obeying the regular, van Laar and Margules 3-suffix equations with positive constants (A and B) are positive deviation solutions. So when g^E and g^{ID} are summed for this solution, we see a feature in the g^{MIX} function where $0.025 < x1 < 0.660$ (approximately). A dashed line tangent to g^{MIX} at $x_1 = 0.025$ and $x_1 = 0.660$ is shown and let's consider this line in relation to the g^{MIX} function.

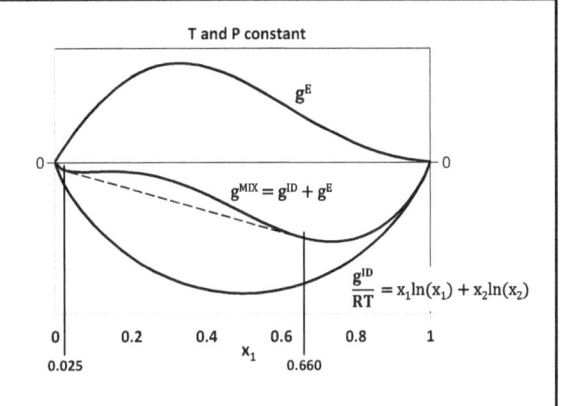

Figure 14.a.2. g^{ID}, g^E, and g^{MIX} for a hypothetical binary system.

Suppose we have a solution that is near the composition $x_1 = 0.3$. Which free energy is lower, the function representing the g^{MIX} function or the dotted line? Clearly, if a solution followed the g^{MIX} function at $x_1 = 0.3$, the free energy could be lower if it switched to the dotted line, and that would be favorable since the free energy would go down. What does the dotted line represent? It represents a linear combination of $g^{MIX}\big|_{x_1=0.025}$ and $g^{MIX}\big|_{x_1=0.660}$ in x_1 and these two solutions are two separate liquid phases. So, if we started with pure component 2 and started adding component 1, when we reached $x_1 = 0.025$ and continued, the solution would split into two phases, the α phase of Figure 14.a.1 and the β phase in the same figure with α composition $x_1 = 0.025$ and β composition $x_1 = 0.660$ (Figure 14.a.1 and 14.a.2 do not match in x_1 but show the same trends). This is the same as following the a → b segment in Figure 14.a.1 then heading into the α-β equilibrium region. As we follow along the dotted line in Figure 14.a.2, it is the same as following along a tie line in the α-β equilibrium region of Figure 14.a.1. The system follows the dotted line in Figure 14,a.2 instead of the g^{MIX} function in this region because the free energy is lower along the dotted line. Once we reach $x_1 = 0.660$, the solution again becomes a single liquid solution corresponding to the β region in Figure 14.a.1.

Let's look at the underlying theory. We will have phase stability when:

$$\frac{d^2\left(\frac{g^{MIX}}{RT}\right)}{dx_1^2} > 0 \qquad 14.1$$

Otherwise, a region (however large or small) such as $0.025 < x1 < 0.660$ in Figure 14.a.2 will exist.

So let's derive the second derivative for a binary mixture:

$$\frac{g^{MIX}}{RT} = x_1 \ln x_1 + x_2 \ln x_2 + \frac{g^E}{RT} \qquad 14.2$$

$$\frac{g^{MIX}}{RT} = x_1 \ln x_1 + x_2 \ln x_2 + x_1 \ln \gamma_1 + x_2 \ln \gamma_2 \qquad 14.3$$

$$\frac{d\left(\frac{g^{MIX}}{RT}\right)}{dx_1} = x_1 \frac{d\ln x_1}{dx_1} + \ln x_1 + x_2 \frac{d\ln x_2}{dx_1} - \ln x_2 + $$
$$x_1 \frac{d\ln \gamma_1}{dx_1} + \ln \gamma_1 + x_2 \frac{d\ln \gamma_2}{dx_1} - \ln \gamma_2 \qquad 14.4$$

Using the Gibbs-Duhem equation, we can eliminate four terms from equation 14.4, then:

$$\frac{d\left(\frac{g^{MIX}}{RT}\right)}{dx_1} = \ln x_1 - \ln x_2 + \ln \gamma_1 - \ln \gamma_2 = \ln(\gamma_1 x_1) - \ln(\gamma_2 x_2) \qquad 14.5$$

$$\frac{d^2\left(\frac{g^{MIX}}{RT}\right)}{dx_1^2} = \frac{1}{x_1} + \frac{1}{x_2} + \frac{d\ln \gamma_1}{dx_1} - \frac{d\ln \gamma_2}{dx_1} \qquad 14.6$$

Therefore, we will have phase stability (a single liquid phase) when:

$$\frac{1}{x_1} + \frac{1}{x_2} > \frac{d\ln \gamma_2}{dx_1} - \frac{d\ln \gamma_1}{dx_1} \qquad 14.7$$

Equation 14.7 has many possible outcomes. The left-hand side is always positive regardless of the value of x_1. It has a minimum value of 4 at $x_1 = \frac{1}{2}$. Let's consider various solution models to see if they can exhibit phase instability which results in two liquid phases.

First, consider Raoult's Law which assumes an ideal liquid phase with γ's = 1. Since $\ln(1) = 0$, the right-hand side of Equation 14.7 is zero for the ideal solution. Therefore, since the left-hand side is always positive, any ideal solution and any solution which obeys Raoult's Law cannot show phase instability for any composition range. *Raoult's Law is only useful for completely miscible systems.*

Consider a regular solution. Recalling Equation 12.50 for the activity coefficients given by regular solution theory:

$$\boxed{\ln \gamma_1 = \frac{A}{RT} x_2^2 \quad \ln \gamma_2 = \frac{A}{RT} x_1^2} \qquad 10.50$$

Let's compute the right-hand side of equation 14.7:

$$\frac{d \ln \gamma_1}{dx_1} = \frac{A}{RT}(-2x_2) \quad \text{and} \quad \frac{d \ln \gamma_2}{dx_1} = \frac{A}{RT}(2x_1) \qquad 14.8$$

Plugging into equation 14.7 for stability:

$$\frac{1}{x_1} + \frac{1}{x_2} > 2 \frac{A}{RT} \qquad 14.9$$

Since the minimum value of the left-hand side is 4, equation 14.9 will turn false in the vicinity of the minimum at $x_1 = \frac{1}{2}$ when:

$$2 \frac{A}{RT} > 4 \quad \text{or} \quad \frac{A}{RT} > 2 \qquad 14.10$$

So, a regular solution can exhibit liquid-liquid equilibrium behavior if A is large enough. Looking at Figure 14.a.3, we see how g^{MIX}/RT for a regular solution changes for different values of A/RT. Looking at the function for A/RT = 2 which is the onset of two phase behavior, the curve is exactly flat (first and second derivatives = 0) at $x_1 = \frac{1}{2}$. Functions for A/RT < 2 show behavior equivalent to completely miscible components. For A/RT = 2.5 we expect and see a feature which is symmetric about $x_1 = \frac{1}{2}$ indicating that two liquid phases coexist at the two minima (where the minima in this case equate with $d^2(g^{MIX}/RT)/dx^2 = 0$) in the g^{MIX}/RT curve. For A/RT = 3, the minima move closer to the respective axes.

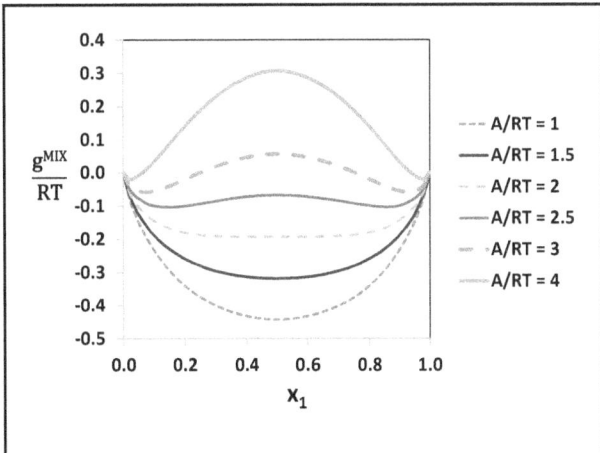

Figure 14.a.3. Regular solution free energies of mixing for several values of A/RT.

We can compute the composition of the minima as a function of A/RT and the result is given in Figure 14.a.4. We see that the function lies in the region A/RT > 2.0 which is the threshold for the onset of liquid-liquid equilibrium (equation 14.10). At $A/RT = 2.0^+$, $x_1 = 0.5$ and the entire curve is symmetric about $x_1 = 0.5$. Moving up from $A/RT = 2.0$, we see two compositions corresponding to any one value. For example, at $A/RT = 2.3$ (approximately), we see $x_1^\alpha = 0.2$ and $x_1^\beta = 0.8$ at the same value of A/RT. These are the compositions of the two liquid phases in equilibrium with each other predicted by regular solution theory for $A/RT = 2.3$ and these values are symmetric about $x_1 = 0.5$. Therefore, the composition of the two phases in equilibrium is $(x_1^\alpha, x_2^\alpha) = (0.2, 0.8) = (x_2^\beta, x_1^\beta)$. This is what we would term *symmetric liquid-liquid phases*. If we know the composition of the α phase, we need only equate x_1^α with x_2^β to

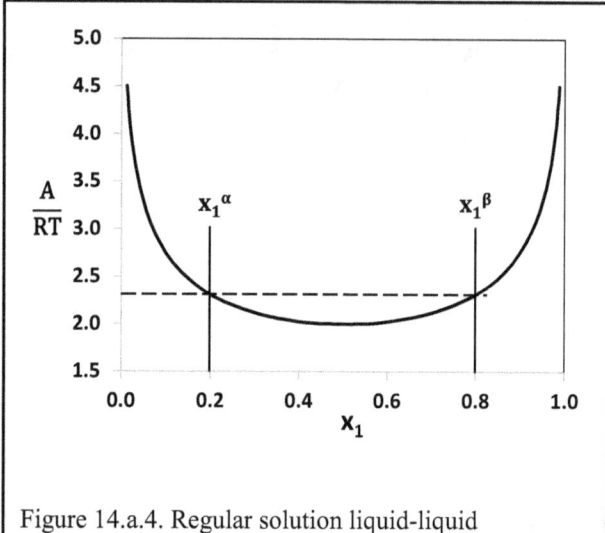

Figure 14.a.4. Regular solution liquid-liquid equilibrium phase composition.

find the composition of the β phase that is equilibrium with the α phase. This is not typical for liquid-liquid equilibrium systems. For example, in the water (1)/n-butanol (2) system at 20°C, $(x_1^\alpha, x_2^\alpha) = (0.4940, 0.5060)$ and $(x_2^\beta, x_1^\beta) = (0.0182, 0.9818)$. The regular solution model cannot give a result with non-symmetric liquid-liquid phases such as the water/n-butanol system.

The van Laar and Margules 3-suffix models can handle non-symmetric liquid-liquid equilibrium systems when the parameters $A/RT = \ln \gamma_1^\infty$ and $B/RT = \ln \gamma_2^\infty$ are different. This gives these models much greater applicability to liquid-liquid equilibrium systems. Diagrams for α-β phase compositions as functions of A and B can be found in Walas [1] for van Laar, Margules 3-suffix, and a number of other activity coefficient models.

Let's go back to equation 14.5 and transform over to fugacity. We can substitute for activity coefficients:

$$\gamma_i = \frac{f_i}{x_i f_i^\circ} \qquad 14.11$$

getting:

$$\frac{d\left(\frac{g^{MIX}}{RT}\right)}{dx_1} = \ln\frac{f_1}{f_1^\circ} - \ln\frac{f_2}{f_2^\circ} = \ln f_1 - \ln f_1^\circ - \ln f_2 + \ln f_2^\circ \qquad 14.12$$

Since standard states are constants, forming the second derivative drops them out so:

$$\frac{d^2\left(\frac{g^{MIX}}{RT}\right)}{dx_1^2} = \frac{d\ln f_1}{dx_1} - \frac{d\ln f_2}{dx_1} \qquad 14.13$$

To eliminate one of the two fugacities in equation 14.13, we use the Gibbs-Duhem equation:

$$\frac{d\ln f_2}{dx_1} = -\frac{x_1}{x_2}\frac{d\ln f_1}{dx_1} \qquad 14.14$$

Plugging equation 14.14 into 14.13 and simplifying:

$$\frac{d^2\left(\frac{g^{MIX}}{RT}\right)}{dx_1^2} = \frac{1}{x_2}\frac{d\ln f_1}{dx_1} = \frac{1}{x_1}\frac{d\ln f_2}{dx_2} \qquad 14.15$$

Since $1/x_1$ and $1/x_2$ are both always positive, for phase stability:

$$\frac{d\ln f_1}{dx_1} > 0 \quad \text{and} \quad \frac{d\ln f_2}{dx_2} > 0 \qquad 14.16$$

Therefore, f_1 must continuously increase with x_1 and f_2 must continuously increase with x_2 for phase stability. If we apply the criterion of stability to a liquid phase in equilibrium with an ideal gas phase, we can substitute y_iP for f_i, and equation 14.13 yields:

$$\frac{d^2\left(\frac{g^{MIX}}{RT}\right)}{dx_1^2} = \frac{d\ln y_1 P}{dx_1} - \frac{d\ln y_2 P}{dx_1} \qquad 14.17$$

For constant P, this can be rearranged to:

$$\frac{d^2\left(\frac{g^{MIX}}{RT}\right)}{dx_1^2} = \frac{1}{y_1 y_2}\frac{dy_1}{dx_1} \qquad 14.18$$

Since $1/y_1y_2$ is always positive, equation 14.18 tells us that for phase stability, y_1 must continuously increase with x_1. Finally, if we go back to equation 14.17 and don't take a constant P system, we can show that:

$$\frac{1}{P(y_1-x_1)}\frac{dP}{dx_1} = \frac{1}{y_1 y_2}\frac{dy_1}{dx_1} > 0 \text{ for phase stability} \qquad 14.19$$

So dP/dx_1 and $(y_1 - x_1)$ must have the same sign for stability. Note that at an azeotropic point, $x_1 = y_1$ and $dP/dx_1 = 0$ (pressure is either at a maximum or minimum for an azeotrope). Azeotropic behavior might then be characterized as being on the cusp of liquid-liquid equilibrium behavior. Figure 14.a.5 compares $y_1 = f(x_1)$ against the function $y_1 = x_1$ for three cases regarding the question of stability of the liquid phase.

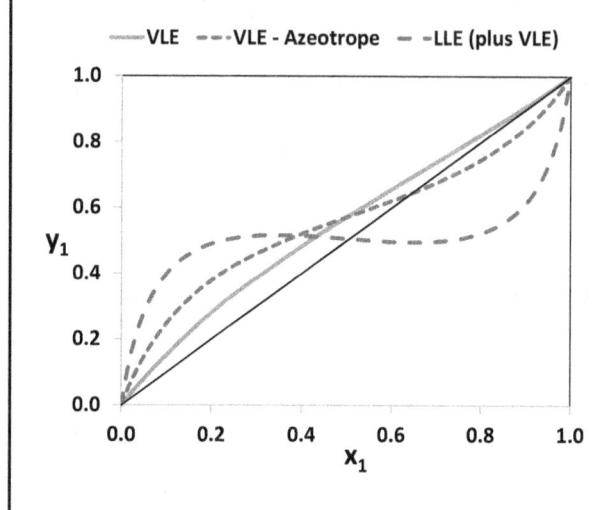

Figure 14.a.5. x1-y1 diagrams for different phase stability scenarios.

The question may arise that if we know the composition of each phase in a liquid-liquid equilibrium system, how do we find regular solution, van Laar, Margules, or other activity coefficient model parameters? This can be approached beginning with noting that for phases which are in equilibrium:

$$f_i^\alpha = f_i^\beta \qquad 14.20$$

$$\gamma_i^\alpha x_i^\alpha f_i^{\circ \alpha} = \gamma_i^\beta x_i^\beta f_i^{\circ \beta} \qquad 14.21$$

Since for any component, the standard state fugacities are the same for both liquid phases:

$$\gamma_i^\alpha x_i^\alpha = \gamma_i^\beta x_i^\beta \qquad 14.22$$

If we have a binary solution, this problem becomes finding activity coefficient parameters such as A and B given x_1^α and x_1^β. Suppose we are interested in finding A (or A/RT) for a regular solution:

$$\ln \gamma_1 = A x_2^2 \text{ and } \ln \gamma_2 = A x_1^2 \qquad 14.23$$

Plugging into equation 14.22, taking the log and applying to component 1 and 2, we have:

$$A\left(x_2^\alpha\right)^2 + \ln x_1^\alpha = A\left(x_2^\beta\right)^2 + \ln x_1^\beta \qquad 14.24$$

$$A\left(x_1^\alpha\right)^2 + \ln x_2^\alpha = A\left(x_1^\beta\right)^2 + \ln x_2^\beta \qquad 14.25$$

Thus, we have two equations but only unknown (A). Solving both equations:

$$A = \frac{\ln\left(\frac{x_1^\beta}{x_1^\alpha}\right)}{\left(x_2^\alpha\right)^2 - \left(x_2^\beta\right)^2} = \frac{\ln\left(\frac{x_2^\beta}{x_2^\alpha}\right)}{\left(x_1^\alpha\right)^2 - \left(x_1^\beta\right)^2} \qquad 14.26$$

Both ways to calculate A in equation 14.26 should yield exactly the same result, but the student may verify that this will be true only for symmetric liquid phases; i.e., $x_1^\alpha = x_2^\beta$. We have already seen the reason for this requirement - regular solution theory can only give symmetric solutions in liquid-liquid equilibrium.

Suppose we know parameters for an activity coefficient model such as van Laar or Margules 3-suffix equations. Assuming it is a LLE system, how do we find the compositions of the two liquid phases? The first method we will consider is hand determination by dual slopes method. Knowing the parameters for the activity coefficients, we begin by forming g^{MIX}/RT from equation 14.3 or other suitable equation as a function of x_1. A plot is then prepared that would have features such as that in Figure 14.a.6. We then draw the dual tangents by hand - a ruler might suffice - as indicated in the dashed line in Figure 14.a.6. This dual tangents line is, as the name suggests, tangent to the g^{MIX}/RT function at two points

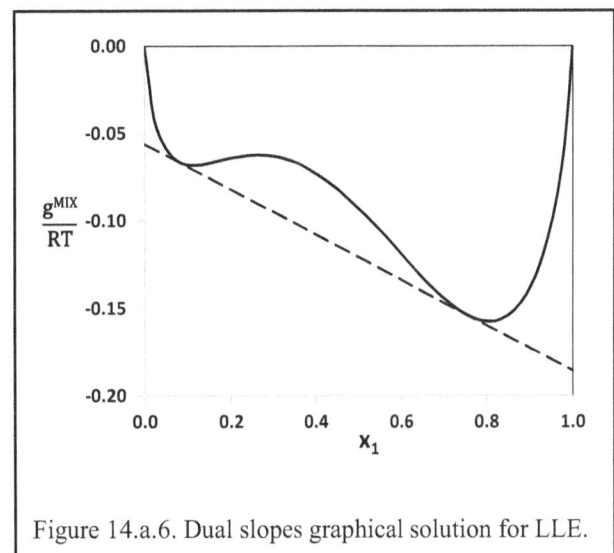

Figure 14.a.6. Dual slopes graphical solution for LLE.

such that the g^{MIX}/RT curve lies above the points. Note that we do not choose the two minima in the g^{MIX}/RT function as it should be evident from Figure 14.a.6 that the two minima are not representative of the dual tangents line and the intersection with g^{MIX}/RT. The two points of intersection of the dual tangents line with the g^{MIX}/RT function represent the two compositions of the liquid phases in equilibrium with each other. The accuracy of this method then becomes the accuracy of our ability to read the graph.

If we want to consider how to accomplish this solution process by algebraic means, we can picture the dual tangents method as a helpful aid in setting up the algebraic equations. One way to set it up is to consider four unknowns, x_1^α, $\left(\frac{g^{MIX}}{RT}\right)^\alpha$, x_1^β and $\left(\frac{g^{MIX}}{RT}\right)^\beta$ which are the two sets of x-y coordinates for the dual tangent points. We then need four equations. One possible set is as follows. From equation 14.3:

1. $$\left(\frac{g^{MIX}}{RT}\right)^{\alpha} = x_1^{\alpha} \ln\left(\gamma_1^{\alpha} x_1^{\alpha}\right) + x_2^{\alpha} \ln\left(\gamma_2^{\alpha} x_2^{\alpha}\right) \qquad 14.27$$

2. $$\left(\frac{g^{MIX}}{RT}\right)^{\beta} = x_1^{\beta} \ln\left(\gamma_1^{\beta} x_1^{\beta}\right) + x_2^{\beta} \ln\left(\gamma_2^{\beta} x_2^{\beta}\right) \qquad 14.28$$

γ's are evaluated using the activity coefficient model and these equations require the two points to be on the g^{MIX}/RT curve. Equation 14.5 has the first derivative of g^{MIX}/RT with respect to x_1, but given the dual tangents line, the derivatives are exactly the same at the dual tangents points, so we can compute both derivatives as:

$$\frac{d\left(\frac{g^{MIX}}{RT}\right)}{dx_1} = \frac{\left(\frac{g^{MIX}}{RT}\right)^{\alpha} - \left(\frac{g^{MIX}}{RT}\right)^{\beta}}{x_1^{\alpha} - x_1^{\beta}} \qquad 14.29$$

which is the slope of the dual tangent line. With equation 14.5 and 14.29:

3. $$\frac{\left(\frac{g^{MIX}}{RT}\right)^{\alpha} - \left(\frac{g^{MIX}}{RT}\right)^{\beta}}{x_1^{\alpha} - x_1^{\beta}} = \ln\left(\gamma_1^{\alpha} x_1^{\alpha}\right) - \ln\left(\gamma_2^{\alpha} x_2^{\alpha}\right) \qquad 14.30$$

4. $$\frac{\left(\frac{g^{MIX}}{RT}\right)^{\alpha} - \left(\frac{g^{MIX}}{RT}\right)^{\beta}}{x_1^{\alpha} - x_1^{\beta}} = \ln\left(\gamma_1^{\beta} x_1^{\beta}\right) - \ln\left(\gamma_2^{\beta} x_2^{\beta}\right) \qquad 14.31$$

The g^{MIX}/RT function in Figure 14.a.6 has been generated with the van Laar activity coefficient model with A = 3 and B = 2. With these parameters, these four equations can be solved simultaneously to give $x_1^{\alpha} = 0.0786$, $x_1^{\beta} = 0.7560$, $\left(\frac{g^{MIX}}{RT}\right)^{\alpha} = -0.0663$, and $\left(\frac{g^{MIX}}{RT}\right)^{\beta} = -0.1541$.

The other problem is to find parameters, such as A and B, for models such as van Laar or Margules 3-suffix equations given the composition of two liquid phases in LLE. The graphs in Walas [1] discussed earlier may also be read for values A ad B for the Margules 3-suffix, van Laar, and other equations given the compositions of the two liquid phases in equilibrium with each other. However, reading graphs often does not give sufficient accuracy, so consider algebraic solutions for the same problem. The equations are algebraically messy, but for the Margules 3-suffix equation:

$$\frac{A}{B} = \frac{2a_2 \ln\left(\frac{x_2^\alpha}{x_2^\beta}\right) + b_1 \ln\left(\frac{x_1^\alpha}{x_1^\beta}\right)}{2a_1 \ln\left(\frac{x_1^\alpha}{x_1^\beta}\right) + b_2 \ln\left(\frac{x_2^\alpha}{x_2^\beta}\right)} \qquad 14.32$$

where:

$$a_i = \left(x_i^\beta\right)^2 - \left(x_i^\alpha\right)^2 - \left(x_i^\beta\right)^3 + \left(x_i^\alpha\right)^3 \qquad 14.33$$

$$b_i = \left(x_i^\beta\right)^2 - \left(x_i^\alpha\right)^2 - 2\left(x_i^\beta\right)^3 + 2\left(x_i^\alpha\right)^3 \qquad 14.34$$

Once the value of the A/B ratio has been found, it can be substituted into:

$$\ln\left(\frac{\gamma_1^\beta}{\gamma_1^\alpha}\right) = \ln\left(\frac{x_1^\alpha}{x_1^\beta}\right) = A\left[\left(x_2^\beta\right)^2\left(1 + 2\left(\frac{B}{A} - 1\right)x_1^\beta\right) - \left(x_2^\alpha\right)^2\left(1 + 2\left(\frac{B}{A} - 1\right)x_1^\alpha\right)\right] \qquad 14.35$$

In equation 14.35, the only unknown is A then B = A/(B/A).

For the van Laar equation:

$$\frac{A}{B} = \frac{\left(\frac{x_1^\beta}{x_2^\beta} + \frac{x_1^\alpha}{x_2^\alpha}\right)\left(\frac{\ln\left(x_1^\alpha/x_1^\beta\right)}{\ln\left(x_2^\beta/x_2^\alpha\right)}\right) - 2}{\left(\frac{x_1^\beta}{x_2^\beta} + \frac{x_1^\alpha}{x_2^\alpha}\right) - \frac{2x_1^\alpha x_1^\beta \ln\left(x_1^\alpha/x_1^\beta\right)}{x_2^\alpha x_2^\beta \ln\left(x_2^\beta/x_2^\alpha\right)}} \qquad 14.36$$

$$A = \frac{\ln\left(\frac{x_1^\alpha}{x_1^\beta}\right)}{\frac{1}{\left(1 + \frac{Ax_1^\beta}{Bx_2^\beta}\right)^2} - \frac{1}{\left(1 + \frac{Ax_1^\alpha}{Bx_2^\alpha}\right)^2}} \qquad 14.37$$

then B = A/(A/B).

14.b Osmotic Pressure

Suppose we have a semipermeable membrane that allows the transmission of only one of two components in a binary mixture, as depicted in Figure 14.b.1. Component 2 might be water and component 1 a soluble salt such as NaCl. If both chambers of the device are at the same pressure, then component 2 will tend to move through the membrane to dilute the solution in the other chamber. This thinking follows naturally from Le Chatelier's Principle. Since increasing pressure increases the fugacity of a liquid, there is some pressure P' > P that

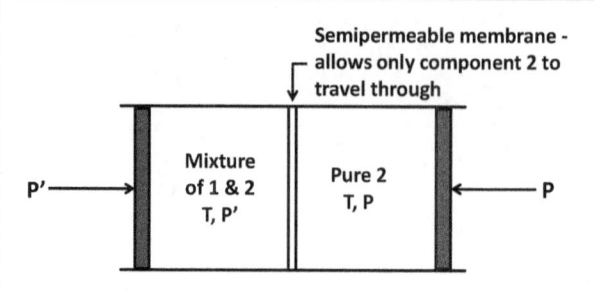

Figure 14.b.1. Semipermeable membrane schematic for consideration of osmotic pressure.

will bring the system to equilibrium across the membrane such that the net flow of component 2 across the membrane is zero. When equilibrium is established, the extra force given by P' exactly balances the tendency of component 2 to move into the chamber where the mixture exists. (P' - P) = π is known as the *osmotic pressure* and the passage of pure component 2 into the mixture when (P' - P) < π is known as *osmosis*. If (P' - P) > π, then pure component 2 tends to pass out of the mixture through the membrane creating more pure component 2. This process is known as *reverse osmosis*.

Clearly, the fugacities of component 2 will be the same on both sides of the membrane at equilibrium:

$$f_2(x_2, P') = f_2(x_2 = 1, P) \qquad 14.38$$

Writing each fugacity on each side of the membrane as a liquid fugacity, being sure to include the Poynting correction factor:

$$\gamma_2 x_2 f_2^\circ \Big|_{\text{left side}} = f_2^\circ \Big|_{\text{right side}} \qquad 14.39$$

$$\gamma_2 x_2 \varphi_2^S P_2^S \exp\left(\int_{P_2^S}^{P'} \frac{v_2^L}{RT} dP\right) = \varphi_2^S P_2^S \exp\left(\int_{P_2^S}^{P} \frac{v_2^L}{RT} dP\right) \qquad 14.40$$

Several terms divide out, then rearranging and taking the log:

$$\ln(\gamma_2 x_2) = \int_{P_2^S}^{P} \frac{v_2^L}{RT} dP - \int_{P_2^S}^{P'} \frac{v_2^L}{RT} dP = \int_{P'}^{P} \frac{v_2^L}{RT} dP \qquad 14.41$$

Assuming the volume of the liquid is constant:

$$-\ln(\gamma_2 x_2) = \frac{v_2^L}{RT}(P' - P) = \frac{v_2^L \pi}{RT} \qquad 14.42$$

or:

$$\pi = -\frac{RT}{v_2^L}\ln(\gamma_2 x_2)$$ 14.43

Equation 14.43 can be simplified with some approximations. First, for a dilute solution of component 1, component 2 is almost pure and $\gamma_2 \to 1$ for that case. Also, $\ln(x_2) = \ln(1-x_1)$, and for x_1 small, $\ln(1-x_1) \to -x_1$. These two approximations result in:

$$\pi = \frac{x_1 RT}{v_2^L}$$ (approximate) 14.44

If the solute is an electrolyte that disassociates into m ions upon dissolution, we need to take that into account since the number of moles of species in solution is greater that the number of moles of pure solute in its undissolved state, so:

$$\pi = \frac{m x_1 RT}{v_2^L}$$ (approximate) 14.45

For seawater, depending upon the salinity, the osmotic pressure is approximately 27 bar. Thus, pure water could be obtained from seawater with an appropriate membrane and pressure exceeding about 27 bar. This application has already been accomplished, but application on an extremely large scale that may be necessary for something like farming operations or water needs for a large city is not often economically feasible.

14.c Vapor Pressure from an EOS

In section 6.d.i, we mentioned that when using a cubic equation of state to get compressibility factors, the cubic equations often give three roots for z (or V) at a single P and T. We noted that in the case where we have three roots, the largest root is the root to use for the vapor compressibility factor. We would like now to expand on this discussion so that we can analyze and get the exact answer to what the roots of the cubic equations of state mean.

In order to accomplish this purpose, let's put some specific numbers to Van der Waals equation. We will use critical properties for methylchloride which are $T_c = 416.2$ K and $P_c = 6714$ kPa $= 67.14$ bar resulting in a =

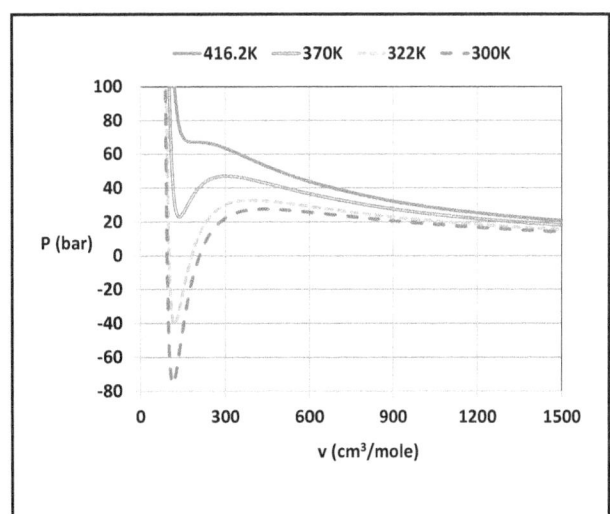

Figure 14.c.1. Van der Waals equation for methylchloride.

7523616 cm⁶-bar/mole² and b = 64.42 cm³/mole. Figure 14.c.1 shows four isotherms given by Van der Waals equation using these values of a and b. Note the critical isotherm (T = 416.2 K) has an inflection point at P = 67.14 bar as expected for the critical point. The other three isotherms, all at temperatures below the critical point exhibit the possibility of three roots for particular pressure ranges. The 322K and 300K isotherms exhibit some negative pressure behavior which is clearly impossible, but that's how the equation works and that is an essential feature of cubic equations of state.

Now, let's inspect the 322K isotherm more closely. Consider the volumes which are roots to the equation for T = 322K and P = 20 bar. Wherever the 322K isotherm crosses the 20 bar horizontal line will correspond to a solution, and clearly by inspection, there are three possibilities (roots). The smallest root is at 97.3 cm³/mole, the intermediate at 232.1 cm³/mole, and the upper root is at 1073.6 cm³/mole. Roughly, the smallest root represents the volume of the liquid as predicted by Van Der Waals equation while the upper root represents roughly the vapor volume root. The intermediate root has no physical significance. The reason that we used the term "roughly" for categorization of the three roots is that the vapor and liquid roots do not actually correspond to both the liquid and vapor values *unless the system is at vapor-liquid equilibrium*, and we need to investigate exactly what the mathematical implication is when the system is at VLE.

For VLE, $dg = vdP - sdT = 0$. When we follow an isotherm, we are holding T constant, so $dT = 0$ and $vdP = 0$ is a condition for equilibrium. Let's integrate assuming we are using an EOS from v_L to v_V and set equal to zero. Since Van der Waals and other cubic equations are generally explicit in pressure, it is easier to write $d(Pv) = vdP + Pdv$, or $vdP = 0 = d(Pv) - Pdv$. In evaluating the integral, note when the equations are satisfied and VLE exists, we are integrating from the liquid side of the saturation dome to the vapor side.

$$\int_{Pv_L}^{Pv_V} d(Pv) - \int_{v_L}^{v_V} P\,dv = Pv\Big|_{Pv_L}^{Pv_V} - \int_{v_L}^{v_V} P\,dv = 0 \qquad 14.46$$

$$P^s(v_V - v_L) - \int_{v_V}^{v_L} P\,dv = 0 \qquad 14.47$$

When equation 14.47 is satisfied, the system is at VLE so the corresponding pressure is the saturation pressure as we have used in the equation. What equation 14.47 tells us is that we need to find a pressure such that the negative area of the isotherm's lobe of the cubic equation below that pressure is exactly the same absolute area as the positive area of the isotherm's lobe above that pressure for VLE to exist.

At this point, we could plug in any EOS, but we are using Van Der Waals for illustrative purposes, so let's substitute that into equation 14.47 for P in the integral and integrate:

$$P^s(v_V - v_L) - \int_{v_L}^{v_V} \left(\frac{RT}{v-b} - \frac{a}{v^2} \right) dv = 0 \qquad 14.48$$

$$P^s(v_V - v_L) - RT\ln\left(\frac{v_V - b}{v_L - b}\right) - a\left(\frac{1}{v_V} - \frac{1}{v_L}\right) = 0 \qquad 14.49$$

Recalling that we are working at a single T (corresponding to an isotherm), the saturation pressure corresponds to the P which satisfies equation 14.49. When VLE exists, then the largest root to Van der Waals equation at P^s corresponds to the vapor root, the lowest root corresponds to the liquid volume, and the intermediate root has no physical significance. When equation 14.49 is not satisfied at a particular P, if the result (left-hand side) is below zero, the largest root is a vapor root (the other two have no significance) and if the result is above zero, the lowest root is a liquid root (the other two have no significance). For the example we are currently working on, equation 14.49 is satisfied with $P^s = 22.0$ bar at T = 322K and the liquid volume (root) is 97.0 cm³/mole while the vapor volume is 937.7 cm³/mole. These are at VLE according to Van der Waals equation. For P < 22.0 bar at T = 322K, the largest root is the vapor root and for P > 22.0 bar, the smallest root is a liquid volume root. For methylchloride, the actual vapor pressure at T = 322 K is 10.92 bar so Van der Waals equation is off by a factor of more than 2. As we have mentioned previously, Van der Waals equation is particularly bad for predicting liquid volumes which likely contributes heavily to the poor quality of this calculation. The take-away is not to use Van der Waals equation except for instructional purposes (as we have noted previously).

Other possibilities according to cubic EOS are as follows. If the system is above the critical T, there is only a single positive real root to the EOS and the root is a gas or supercritical fluid. If the system is below the critical T and there is only one positive root which is real, the root can be either a liquid volume (see Figure 14.c.1 with T = 370K and P > about 50 bar) or a vapor volume (see Figure 14.c.1 with T = 370K and P ≈ 20 bar).

14.d Chapter 14 Example Problems

14.1 A binary solution obeys the Margules 3- suffix equation with A = 4 and B = 2. a) Find the compositions of the two phases (call them α and β) which coexist at equilibrium when 3 moles of component 1 are mixed with 7 moles of component 2, and b) find the relative amounts of each phase (i.e., moles of phase α and moles of phase β).

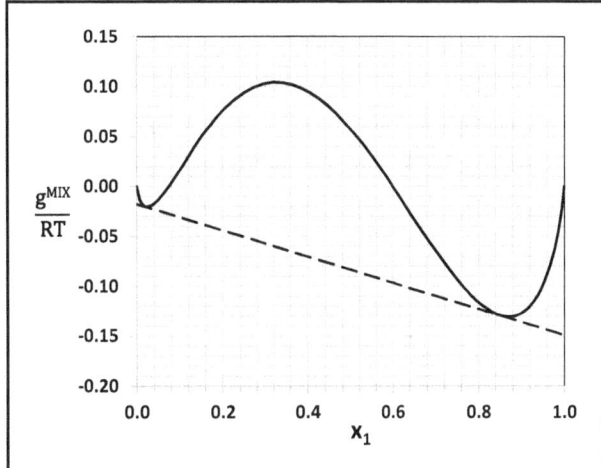

Figure 14.d.1. Graphical depiction of LLE with Margules 3-suffix for problem 14.1.

Solution:

a. For the Margules 3-suffix equation:

$$\frac{g^E}{RT} = x_1 x_2 [Ax_1 + Bx_2]$$

$$\frac{g^{MIX}}{RT} = x_1 x_2 [Ax_1 + Bx_2] + x_1 \ln(x_1) + x_2 \ln(x_2)$$

The g^{mix} function is plotted in Figure 14.d.1 for A = 4, B = 2. Visual inspection with the line for dual slopes suggests $x_1^\alpha \approx 0.02$ and $x_1^\beta \approx 0.85$. Algebraic refinement with simultaneously solving equations 14.27 - 14.31 yields $x_1^\alpha = \mathbf{0.01995}$ and $x_1^\beta = \mathbf{0.84984}$. The corresponding values for the free energies of mixing of these solutions in equilibrium with each other are:

$$\left(\frac{g^{MIX}}{RT}\right)^\alpha = -0.02042 \quad \text{and} \quad \left(\frac{g^{MIX}}{RT}\right)^\beta = -0.12944$$

b. The overall mole fraction for component 1 is 0.30 which lies between $x_1^\alpha = 0.01995$ and $x_1^\beta = 0.84984$, so the mixture will split into two phases with total moles of the alpha phase n^α and total moles of the beta phase n^β. For $n_1 = 3$, $n_2 = 7$, a material balance yields:

$$3 = 0.01995 n^\alpha + 0.84984 n^\beta \quad \text{and} \quad 7 = (1 - 0.01995) n^\alpha + (1 - 0.84984) n^\beta$$

These can be solved simultaneously to yield $n^\alpha = \mathbf{6.6255}$ and $n^\beta = \mathbf{3.3745}$.

14.2 Find van Laar equation constants A and B which correspond to the LLE water (1)/n-butanol (2) system at 20°C with $(x_1^\alpha, x_2^\alpha) = (0.4940, 0.5060)$ and $(x_2^\beta, x_1^\beta) = (0.0182, 0.9818)$.

Solution:

Using the compositions of the two LLE solutions, equation 14.36 yields A/B = 0.2818 and using this value of A/B, equation 14.37 yields **A = 1.124**. Then, **B = 3.988**. A plot of the g^{MIX} function for these values of A and B in the van Laar equation along with the dual slopes solution is given in Figure 14.d.2.

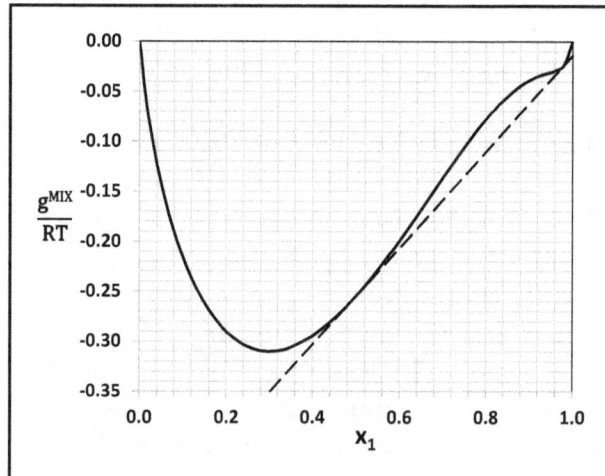

Figure 14.d.2. Graphical depiction of LLE with van Laar for problem 14.2.

14.3 Using Van der Waals equation for the EOS, find the vapor pressure of n-butane at 0°C. Compare the result with the vapor pressure given by the Antoine equation and data from Appendix E.

Solution:

For n-butane, $T_c = 425.12$ K, $P_c = 37.96$ bar. From these critical properties, we get a = 1.3884×10^7 cm^6-bar/mole^2 and b = 116.39 cm^3/mole for Van der Waals equation constants. We need to set up an iterative process guessing P^s then evaluating equation 14.49 to see if it is solved. We need v_V (the largest volume root of Van der Waals equation) and v_L (the smallest volume root of Van der Waals equation) for solution of equation 14.49 based upon whatever pressure we guess. These are obtained by solving Van der Waals equation as a cubic equation in volume, which yields 3 roots under the conditions of interest in this problem. Calculations yield:

$P^{s\text{-guess}}$ (bar)	v_L (cm³/mole)	v_V (cm³/mole)	LHS equation 14.49
4.000	156.0	5132.9	3745.4
4.500	155.7	4494.3	1422.6
4.600	155.9	4383.0	994.4
4.700	155.7	4276.3	576.9
4.800	155.2	4173.9	170.9
4.842	155.8	4132.2	2.2
4.843	155.7	4131.2	-1.8
4.900	155.7	4075.7	-226.8
5.000	155.9	3981.3	-613.9

Thus, **P^s = 4.843 bar** as given by Van der Waals equation. Using the Antoine constants for n-butane from Appendix E and evaluating at 0°C, we find **P^s = 1.013 bar**. This is very close to 1 atm and the normal boiling point of n-butane is known to be near 0°C, which agrees with the Antoine equation calculation. *Van der Waals equation gives a very poor result.*

14.4 Seawater is often reported to have, on average, 3.1–3.8% NaCl. For T = 300 K, compute the osmotic pressure range corresponding to this range of salt concentration. Also, convert the results from bar to height of liquid equivalent.

Solution:

We need to evaluate equation 14.45. NaCl dissociates into two ions (Na⁺ and Cl⁻) so m = 2. MW_{NaCl} = 58.44 g/mole, so in 100 g of solution, n_{NaCl} = 0.053 - 0.065 moles and n_{H2O} = 5.377 - 5.339 moles. x_{NaCl} = 0.0098 - 0.0120. We will take v_2^L = 1 cm³/g or 18 cm³/mole. This gives **π = 27.2 - 33.2 bar**. In feet of water, this range is equivalent to **π = 910 - 1110 feet of water**. For comparison purposes, Niagra Falls is a little shy of 200 feet.

Student Problems:

1. Use the Peng-Robinson equation to find the expression equivalent to equation 14.49 (which is for Van der Waals equation).

2. Using the results from student problem 1, solve problem 14.3 for the vapor pressure of n-butane at 0°C. Compare the result with Van der Waals equation and with the correct value.

3. Find constants for Margules 3-suffix equation for the water/n-butanol system of problem

14.2. Generate the g^{MIX} function and the dual tangents line using the results.

14.e References

1. Walas, S. M.; "Phase Equilibrium in Chemical Engineering", Butterworth Publishers, Boston (1985), pgs. 362 - 365.

Contents of the Appendix

1. Contents of the Appendix 293
   a. Appendix A: Conversion Factors and Values of R. 295
   b. Appendix B: Heat Capacities of Pure Substances 296
   c. Appendix C: Critical Constants 298
   d. Appendix D: Lee-Kesler Tables 301
      i. Table D.1 - z^0 301
      ii. Table D.2 - z^1 301
      iii. Table D.3 - h^{R0}/RT_c 302
      iv. Table D.4 - h^{R1}/RT_c 302
      v. Table D.5 - s^{R0}/R 303
      vi. Table D.6 - s^{R1}/R 303
      vii. Table D.7 - φ^0 304
      viii. Table D.8 - φ^1 304
   e. Appendix E: Antoine constants for vapor pressure of selected components . . . 305
   f. Appendix F: Enthalpies and Free Energies of Formation. 307

Properties of Selected Pure Components - Pure component data derives from the National Institute of Standards and Technology (NIST) digital databases REFPROP and Steam 3.0 and has been used by permission.

   g. Appendix G: Steam. 309
      i. G.1 - Saturated Steam Temperature Increment. 309
      ii. G.2 - Saturated Steam Pressure Increment 311
      iii. G.3 - Superheated Steam. 313
      iv. G.4 - Compressed Liquid Water 324
      v. P-H Diagram for Steam 326
      vi. Mollier Diagram for Steam. 327
      vii. T-S Diagram for Steam. 328
   h. Appendix H: Properties of R-134a 329
      i. H.1 - Saturated R-134a 329
      ii. H.2 - Superheated R-134a. 331
      iii. H.3 - P-H Diagram for R-134a 334
   i. Appendix I: Properties of R-23 335
      i. I.1 - Saturated R-23. 335
      ii. I.2 - Superheated R-23 337
      iii. I.3 - P-H Diagram for R-23. 344
   j. Appendix J: Properties of Ammonia. 345
      i. J.1 - Saturated Ammonia. 345
      ii. J.2 - Superheated Ammonia 347
      iii. J.3 - P-H Diagram for Ammonia. 359
   k. Appendix K: Properties of Nitrogen 360
      i. K.1 - Saturated Nitrogen. 360
      ii. K.2 - Superheated Nitrogen 362
   l. Appendix L: Properties of Oxygen. 376
      i. L.1 - Saturated Oxygen. 376
      ii. L.2 - Superheated Oxygen 377

- m.   Appendix M: Properties of Carbon Dioxide........................380
  - i.   M.1 - Saturated Carbon Dioxide.............................380
  - ii.   M.2 - Superheated Carbon Dioxide 381
- n.   Appendix N: Properties of Methane...............................387
  - i.   N.1 - Saturated Methane....................................387
  - ii.   N.2 - Superheated Methane 388
- o.   Appendix O: Properties of Hydrogen..............................391
  - i.   O.1 - Saturated Hydrogen...................................391
  - ii.   O.2 - Superheated Hydrogen 392
- p.   Appendix P: Properties of R-11 395
  - i.   P.1 - Saturated R-11 395
  - ii.   P.2 - Superheated R-11....................................397

Appendix A - Conversion Factors and Values of R

Conversion Factors

Energy:

$1\ J = 1\ N\text{-}m = 1\ kg\text{-}m^2/s^2$
$1\ J = 0.737562\ ft\text{-}lbf$
$1\ cal = 4.184\ J$
$1\ BTU = 1055.06\ J = 778.169\ ft\text{-}lbf = 252.164\ cal$

Mass:

$1\ kg = 2.2046226\ lbm$

Moles:

$1\ mole\ (a.k.a.\ gram\ mole) = 0.001\ kgmole = 0.002205\ lbmole$

Gravitational Constant:

$g = 9.80665\ m/s^2 = 32.174\ ft/s^2$

Mass to Force Conversion:

$g_c = 1\ kg\text{-}m/N\text{-}s^2 = 32.174\ ft\text{-}lbm/lbf\text{-}s^2$

Power:

$1\ W = 1\ J/s = 0.737562\ ft\text{-}lbf/s$
$1\ kW = 1000\ W = 1.341022\ HP$
$1\ HP = 550\ ft\text{-}lbf/s = 0.707260\ BTU/s$
$1\ ton\ of\ refrigeration = 12000\ BTU/h = 3.51685\ kW$

Pressure:

$1\ Pa = 1\ N/m^2 = 0.001\ kPa$
$1\ bar = 100\ kPa = 100000\ Pa$
$1\ atm = 101.325\ kPa = 1.01325\ bar = 760\ torr$
$1\ atm = 14.696\ lbf/in^2 = 33.8995\ ft\ of\ H_2O = 760\ mmHg$

Temperature:

when applied to changes in temperature: $1\ K = 1°C = 1.8°R = 1.8°F$
$T\ (°C) = T\ (K) - 273.15 = [T\ (°F) - 32]/1.8$
$T\ (K) = T\ (°R)/1.8$
$T\ (°F) = T\ (°R) - 459.67$

Volume:

$1\ m^3 = 1000\ L = 1 \times 10^6\ cm^3 = 35.3147\ ft^3$ $1\ gal = 3.785\ L = 231.0\ in^3 = 0.13368\ ft^3$

Values of R (universal gas constant)

$R = 8.314463\ J/mol\text{-}K = 1.986\ cal/mole\text{-}K = 83.14463\ cm^3\text{-}bar/mole\text{-}K = 0.08314463\ L\text{-}bar/mole\text{-}K$
$R = 0.082057\ L\text{-}atm/mole\text{-}K = 82.057\ cm^3\text{-}atm/mole\text{-}K = 1.986\ BTU/lbmole\text{-}°R = 1.986\ cal/mole\text{-}K$
$R = 0.73024\ ft^3\text{-}atm/lbmole\text{-}°R = 10.7317\ psi\text{-}ft^3/lbmole\text{-}°R$

Appendix B: Heat Capacities of Pure Substances

Heat Capacities of pure substances in the ideal gas state at 1 bar

Paraffins

Constants for $C_p/R = A + BT + CT^2 + DT^3 + ET^{-2}$ where T is in K

Name	Formula	MW (g/mole)	$C_{p,298}^{IG}/R$	A	B	C	D	E	T_{max} (K)
methane	CH_4	16.0430	4.2776	0.1591	0.0124190	-4.4710E-06	5.7324E-10	70944.1	1500
ethane	C_2H_6	30.0700	6.3167	-1.1426	0.0266530	-1.3532E-05	2.7138E-09	57215.5	1500
propane	C_3H_8	44.0970	8.8743	-1.7622	0.0399990	-2.2109E-05	4.8353E-09	48712.4	1500
n-butane	C_4H_{10}	58.1240	11.8796	-2.2226	0.0529100	-3.0160E-05	6.7624E-09	73678.5	1500
i-butane	C_4H_{10}	58.1240	11.6654	-2.3266	0.0539180	-3.1182E-05	7.0512E-09	44566.7	1500
n-pentane	C_5H_{12}	72.1510	14.4703	-4.0098	0.0686360	-3.9976E-05	9.1053E-09	118101.0	1500
isopentane	C_5H_{12}	72.1510	14.3178	-1.7887	0.0618970	-3.2426E-05	6.7245E-09	31661.5	1500
neopentane	C_5H_{12}	72.1510	14.5275	-0.0020	0.0568490	-2.5753E-05	4.7545E-09	-22821.8	1500
n-hexane	C_6H_{14}	86.1780	17.1895	-4.9602	0.0834510	-5.0879E-05	1.2123E-08	130697.6	1500
isohexane	C_6H_{14}	86.1780	17.1417	-2.6234	0.0776120	-4.4778E-05	1.0331E-08	29489.9	1500
3-methylpentane	C_6H_{14}	86.1780	16.8738	-3.0920	0.0777780	-4.4343E-05	1.0052E-08	40147.1	1500
2,2-dimethylbutane	C_6H_{14}	86.1780	17.0430	-2.8544	0.0762440	-4.0251E-05	8.6869E-09	45600.0	1500
2,3-dimethylbutane	C_6H_{14}	86.1780	16.7828	-1.3157	0.0723340	-3.8062E-05	8.0768E-09	-26536.1	1500
n-heptane	C_7H_{16}	100.2050	19.9291	-5.5468	0.0968510	-5.9858E-05	1.4353E-08	136927.7	1500
n-octane	C_8H_{18}	114.2320	22.6521	-6.4104	0.1109280	-6.9279E-05	1.6589E-08	151838.4	1500
n-nonane	C_9H_{20}	128.2590	25.3810	-7.6405	0.1264610	-8.0672E-05	1.9674E-08	174843.6	1500
n-decane	$C_{10}H_{22}$	142.2860	28.1282	-8.1610	0.1395490	-8.9109E-05	2.1608E-08	180561.6	1500

Heat Capacities of pure substances in the ideal gas state at 1 bar

Olefins and unsaturates

Constants for $C_p/R = A + BT + CT^2 + DT^3 + ET^{-2}$ where T is in K

Name	Formula	MW (g/mole)	$C_{p,298}/R$	A	B	C	D	E	T_{max} (K)
ethylene	C_2H_4	28.054	5.2158	1.9993	0.0139160	-5.3561E-06	7.3142E-10	-42295.0	3000
propylene	C_3H_6	42.081	7.8111	3.0834	0.0212870	-8.0856E-06	1.0804E-09	-82571.8	3000
1-butene	C_4H_8	56.108	10.4091	4.2919	0.0289537	-1.1168E-05	1.5150E-09	-138915.7	3000
cis-2-butene	C_4H_8	56.108	9.7549	2.9208	0.0304852	-1.1778E-05	1.5967E-09	-111148.7	3000
trans-2-butene	C_4H_8	56.108	10.6262	3.6988	0.0295760	-1.1375E-05	1.5359E-09	-81805.4	3000
isobutene	C_4H_8	56.108	10.6793	4.3509	0.0285700	-1.0880E-05	1.4572E-09	-112111.8	3000
1,3 butadiene	C_4H_6	54.092	9.6456	-0.2922	0.0424180	-2.8287E-05	7.1085E-09	-34048.4	1500
1-pentene	C_5H_{10}	70.135	13.0797	-2.5547	0.0597880	-3.6034E-05	8.3908E-09	70175.4	1500
1-hexene	C_6H_{12}	84.162	15.8171	-3.5098	0.0744129	-4.6316E-05	1.1044E-08	85799.0	1500
1-heptene	C_7H_{14}	98.189	18.5591	-4.5451	0.0892035	-5.6680E-05	1.3699E-08	105211.9	1500
1-octene	C_8H_{16}	112.216	21.3090	-5.6862	0.1044691	-6.7590E-05	1.6548E-08	126003.7	1500
styrene	C_8H_8	104.152	14.5463	-3.0182	0.0731977	-4.7358E-05	1.1573E-08	-31673.8	1500
acetylene	C_2H_2	26.038	5.3162	4.8828	0.0045947	-1.3971E-06	1.5837E-10	-72579.8	3000
methylacetylene	C_3H_4	40.065	7.3508	4.4877	0.0139661	-5.1716E-06	6.7871E-10	-76376.6	3000

Heat Capacities of pure substances in the ideal gas state at 1 bar

Aromatics and Cyclics

Constants for $C_p/R = A + BT + CT^2 + DT^3 + ET^{-2}$ where T is in K

Name	Formula	MW (g/mole)	$C_{p,298}/R$	A	B	C	D	E	T_{max} (K)
cyclopentane	C_5H_{10}	70.1350	10.1863	1.5010	0.0435666	-1.7560E-05	2.4701E-09	-249667.4	3000
cyclohexane	C_6H_{12}	84.1620	13.0261	1.9157	0.0560126	-2.3505E-05	3.4135E-09	-319201.0	3000
cyclohexene	C_6H_{10}	82.1460	12.2595	-8.0541	0.0789617	-4.9917E-05	1.2097E-08	78920.3	1500
benzene	C_6H_6	78.1140	9.9840	-3.8767	0.0579874	-3.7989E-05	9.3752E-09	-26649.0	1500
toluene	C_7H_8	92.1410	12.5405	-5.3021	0.0715495	-4.6126E-05	1.1304E-08	27630.2	1500
o-xylene	C_8H_{10}	106.1680	16.0058	0.1390	0.0638361	-3.2571E-05	6.1234E-09	-38483.2	2000
m-xylene	C_8H_{10}	106.1680	15.2447	-0.9635	0.0666150	-3.4901E-05	6.7306E-09	-64797.1	2000
p-xylene	C_8H_{10}	106.1680	15.2959	-0.6626	0.0666265	-3.5150E-05	6.8249E-09	-85556.3	2000
ethylbenzene	C_8H_{10}	106.1680	15.4003	-5.9771	0.0850993	-5.4684E-05	1.3335E-08	45576.2	1500
phenol	C_6H_6O	94.1130	12.5074	-0.1706	0.0557927	-3.6517E-05	8.9975E-09	-84354.5	1500

Heat Capacities of pure substances in the ideal gas state at 1 bar

Oxygen Containing Organics

Constants for $C_p/R = A + BT + CT^2 + DT^3 + ET^{-2}$ where T is in K

Name	Formula	MW (g/mole)	$C_{p,298}/R$	A	B	C	D	E	T_{max} (K)
methanol	CH_3OH	32.042	5.3285	1.7097	0.0132885	-4.9380E-06	6.4830E-10	6989.0	3000
ethanol	C_2H_5OH	46.069	7.9446	3.9711	0.0194736	-7.2137E-06	9.3590E-10	-108102.8	3000
ethylene oxide	C_2H_4O	44.053	5.7890	-1.9316	0.0297508	-1.8298E-05	4.3034E-09	32259.2	1500
ethylene glycol	$C_2H_6O_2$	62.068	9.9421	-2.8957	0.0501300	-3.8732E-05	1.2091E-08	90144.2	1000
1-propanol	C_3H_7OH	60.096	10.4212	4.4719	0.0282252	-1.0956E-05	1.5004E-09	-136178.7	3000
2-propanol	C_3H_7OH	60.096	10.8928	6.4816	0.0251536	-9.9623E-06	1.3940E-09	-199095.8	3000
1-butanol	C_4H_9OH	74.123	13.1823	6.3169	0.0353103	-1.3888E-05	1.9317E-09	-220365.9	3000
2-butanol	C_4H_9OH	74.123	13.5796	0.3195	0.0517328	-2.6192E-05	4.2875E-09	4503.0	1000
acetone	C_3H_6O	58.080	8.9768	-0.6184	0.0353177	-1.9378E-05	4.1232E-09	60312.6	1500
diethylether	$C_4H_{10}O$	74.123	14.3549	-0.8695	0.0539234	-3.0436E-05	6.7814E-09	148717.4	1500
methylethylketone	C_4H_8O	72.107	12.4287	0.6152	0.0435420	-2.3214E-05	4.8017E-09	68253.4	1500

Appendix B: Heat Capacities of Pure Substances

Name	Formula	MW	$C_{p,298}/R$	A	B	C	D	E	T_{max} (K)
formaldehyde	CH_2O	30.026	4.2552	1.6416	0.0086906	-3.4203E-06	4.7114E-10	27915.0	3000
acetaldehyde	C_2H_4O	44.053	6.7044	3.1590	0.0153423	-5.7720E-06	7.5583E-10	-47636.5	3000
acetic acid	$C_2H_4O_2$	60.052	7.6456	-0.5365	0.0315083	-1.8070E-05	3.8260E-09	26026.1	1500

Heat Capacities of pure substances in the ideal gas state at 1 bar
Miscellaneous Hydrocarbons

Constants for $C_p/R = A+BT+CT^2+DT^3+ET^{-2}$ where T is in K

Name	Formula	MW (g/mole)	$C_{p,298}/R$	A	B	C	D	E	T_{max} (K)
methylamine	CH_5N	31.0580	6.0237	-0.1189	0.0218424	-1.1640E-05	2.4854E-09	53257.3	1500
ethylamine	C_2H_7N	45.0850	8.6339	-0.2250	0.0338552	-1.9737E-05	4.6527E-09	35214.7	1500
acetonitrile	C_2H_3N	41.0530	6.2962	1.5395	0.0174381	-8.9436E-06	1.7813E-09	27142.4	1500
methylchloride	CH_3Cl	50.4850	4.9075	0.6603	0.0155139	-8.5838E-06	1.8806E-09	29772.0	1500
dichloromethane	CH_2Cl_2	84.9270	6.1438	2.5556	0.0152478	-9.6806E-06	2.3457E-09	-14184.1	1500
trichloromethane	$CHCl_3$	119.3690	7.8902	5.6256	0.0123374	-8.5990E-06	2.2018E-09	-62919.1	1500
carbon tetrachloride	CCl_4	153.8110	10.0687	9.6327	0.0074099	-5.9466E-06	1.6419E-09	-114508.6	1500
fluoromethane	CH_3F	34.0330	4.5377	1.2176	0.0121408	-4.7861E-06	6.6351E-10	9616.9	3000
difluoromethane	CH_2F_2	52.0230	5.2081	2.8914	0.0110589	-4.5622E-06	6.5400E-10	-52654.2	3000
R-134a	$C_2H_2F_4$	102.032	10.5606	9.8305	0.0134132	-5.7471E-06	8.4806E-10	-247177.4	3000

Heat Capacities of pure substances in the ideal gas state at 1 bar
Inorganics

Constants for $C_p/R = A+BT+CT^2+DT^3+ET^{-2}$ where T is in K

Name	Formula	MW (g/mole)	$C_{p,298}/R$	A	B	C	D	E	T_{max} (K)
water	H_2O	18.015	4.0294	3.1590	0.0020385	-2.1753E-07	-1.8960E-11	25105.3	3000
hydrogen	H_2	2.016	3.4615	3.5395	-0.0002114	4.0418E-07	-7.7486E-11	-4341.0	3000
oxygen	O_2	31.999	3.5364	3.0643	0.0016886	-6.8712E-07	1.0656E-10	2393.3	3000
nitrogen	N_2	28.014	3.4881	2.8185	0.0015534	-5.2779E-07	6.3840E-11	22375.7	3000
carbon dioxide	CO_2	44.009	4.4861	3.8803	0.0041571	-1.7388E-06	2.5319E-10	-43186.8	3000
carbon monoxide	CO	28.014	3.4914	2.7963	0.0017182	-6.3511E-07	8.2914E-11	21075.2	3000
helium	He	4.003	2.5010	2.5010	0.0000000	0.0000E+00	0.0000E+00	0.0	1000
argon	Ar	39.948	2.5010	2.5010	0.0000000	0.0000E+00	0.0000E+00	0.0	1000
ammonia	NH_3	17.031	4.2884	2.4064	0.0059437	-1.7653E-06	1.9055E-10	23272.6	3000
chlorine	Cl_2	70.900	4.0932	4.3731	0.0002017	-6.3715E-08	1.4975E-11	-29763.2	3000
hydrogen chloride	HCl	36.458	3.4921	2.9828	0.0009845	-1.7793E-07	4.6662E-12	20574.1	3000
hydrogen sulfide	H_2S	34.086	4.0858	2.5235	0.0042500	-1.3643E-06	1.6492E-10	36627.4	3000
hydrogen cyanide	HCN	27.026	4.3125	3.7880	0.0032348	-1.1776E-06	2.2742E-10	-30338.5	1000
carbon disulfide	CS_2	76.131	5.4730	4.3711	0.0060020	-4.5555E-06	1.2715E-09	-28117.6	1000
sulfur dioxide	SO_2	64.058	4.8494	4.5030	0.0034217	-1.4742E-06	2.2439E-10	-48774.3	3000
sulfur trioxide	SO_3	80.063	6.1740	5.9958	0.0053703	-2.3915E-06	3.6521E-10	-108452.0	3000
nitric oxide	NO	30.006	3.5813	2.8162	0.0018986	-7.7194E-07	1.0981E-10	23530.2	3000
nitrogen dioxide	NO_2	46.006	4.5440	3.6801	0.0045434	-1.9306E-06	3.0453E-10	-29081.2	3000
nitrous oxide	N_2O	44.013	4.6645	4.1155	0.0039480	-1.6379E-06	2.3672E-10	-43445.0	3000
dinitrogen tetroxide	N_2O_4	92.010	9.8079	9.6905	0.0071967	-3.2012E-06	4.7522E-10	-156127.9	3000

Heat Capacity of saturated liquids in the range 273 - 373 K

Constants for $C_p/R = A+BT+CT^2$ where T is in K

Name	Formula	MW (g/mole)	$C_{p,298}/R$	A	B	C
n-hexane	C_6H_{14}	86.18	23.337	18.1927	-0.0085075	8.6400E-05
n-heptane	C_7H_{16}	100.21	27.006	22.4545	-0.0141958	9.8820E-05
n-octane	C_8H_{18}	114.23	30.603	24.1284	-0.0054755	9.1203E-05
n-nonane	C_9H_{20}	128.26	34.073	22.3417	0.0216528	5.9352E-05
n-decane	$C_{10}H_{22}$	142.29	37.541	21.4111	0.0431033	3.6878E-05
cyclohexane	C_6H_{12}	84.16	18.298	-27.0800	0.2279190	-2.5397E-04
benzene	C_6H_6	78.11	16.319	18.2324	-0.0356793	9.8146E-05
toluene	C_7H_8	92.14	18.860	11.9758	0.0108150	4.1173E-05
o-xylene	C_8H_{10}	106.17	22.593	15.0814	0.0119965	4.4265E-05
m-xylene	C_8H_{10}	106.17	21.857	14.9961	0.0058433	5.7581E-05
p-xylene	C_8H_{10}	106.17	21.896	13.8602	0.0125979	4.8144E-05
ethylbenzene	C_8H_{10}	106.17	22.346	16.4873	-0.0002427	6.6724E-05
methanol	CH_3OH	32.04	9.763	16.1192	-0.0645983	1.4516E-04
ethanol	C_2H_5OH	46.04	14.280	-13.6755	0.1200035	-8.8011E-05
R-134a	$C_2H_2F_4$	102.03	17.107	89.3039	-0.5268781	9.5498E-04
water	H_2O	18.02	9.072	11.4000	-0.0146868	2.3070E-05

Appendix C: Critical Constants

P_c, T_c, V_c, and T_n (normal boiling point) derived from NIST data. Hydrogen and helium acentric factors (ω) have been found in open sources but other values of ω have been computed and better values may exist in the open literature.

Paraffins

Name	Formula	MW (g/mole)	T_c (K)	P_c (kPa)	ω	V_c (cm^3/mole)	z_c	T_n (K)
methane	CH_4	16.0430	190.6	4599	0.015	98.7	0.286	111.7
ethane	C_2H_6	30.0700	305.3	4872	0.103	145.9	0.280	184.6
propane	C_3H_8	44.0970	369.9	4251	0.157	200.1	0.277	231.0
n-butane	C_4H_{10}	58.1240	425.1	3796	0.206	255.0	0.274	272.7
i-butane	C_4H_{10}	58.1240	407.8	3629	0.189	257.8	0.276	261.4
n-pentane	C_5H_{12}	72.1510	469.7	3369	0.258	306.9	0.265	309.2
isopentane	C_5H_{12}	72.1510	460.4	3380	0.234	305.9	0.270	301.0
neopentane	C_5H_{12}	72.1510	433.7	3196	0.200	305.9	0.271	282.7
n-hexane	C_6H_{14}	86.1780	507.8	3042	0.307	386.8	0.279	341.9
isohexane	C_6H_{14}	86.1780	497.7	3043	0.287	368.4	0.271	333.4
3-methylpentane	C_6H_{14}	86.1780	504.6	3123	0.278	368.7	0.274	336.4
2,2-dimethylbutane	C_6H_{14}	86.1780	489.1	3102	0.238	365.0	0.278	322.9
2,3-dimethylbutane	C_6H_{14}	86.1780	500.2	3133	0.254	360.2	0.271	331.1
n-heptane	C_7H_{16}	100.2050	541.2	2774	0.353	445.6	0.275	371.5
n-octane	C_8H_{18}	114.2320	569.6	2507	0.401	501.9	0.266	398.8
n-nonane	C_9H_{20}	128.2590	594.6	2280	0.451	552.6	0.255	423.9
n-decane	$C_{10}H_{22}$	142.2860	617.7	2101	0.496	610.7	0.250	447.3

Olefins and unsaturates

Name	Formula	MW (g/mole)	T_c (K)	P_c (kPa)	ω	V_c (cm^3/mole)	z_c	T_n (K)
ethylene	C_2H_4	28.0540	282.4	5042	0.091	131.0	0.281	169.4
propylene	C_3H_6	42.0810	364.2	4555	0.151	183.3	0.276	225.5
1-butene	C_4H_8	56.1080	419.3	4010	0.198	235.9	0.271	266.8
cis-2-butene	C_4H_8	56.1080	435.8	4236	0.209	235.7	0.276	276.9
trans-2-butene	C_4H_8	56.1080	428.6	4019	0.215	237.4	0.268	274.0
isobutene	C_4H_8	56.1080	418.1	4020	0.199	239.9	0.277	266.2
1,3 butadiene	C_4H_6	54.0920	425.1	4270	0.194	222.6	0.269	268.6
1-pentene	C_5H_{10}	70.1350	464.7	3547	0.241	301.0	0.276	303.1
1-hexene	C_6H_{12}	84.1620	504.1	3210	0.292	347.8	0.266	336.6
1-heptene	C_7H_{14}	98.1890	537.5	2852	0.338	402.4	0.257	366.8
1-octene	C_8H_{16}	112.2160	566.6	2676	0.404	463.7	0.263	394.4
styrene	C_8H_8	104.1520	635.2	3880	0.309	356.7	0.262	418.4
acetylene	C_2H_2	26.0380	308.4	6206	0.199	117.8	0.285	189.2
methylacetylene	C_3H_4	40.0650	402.7	5658	0.209	161.2	0.272	248.0

Appendix C: Critical Constants

Aromatics and Cyclics

Name	Formula	MW (g/mole)	T_c (K)	P_c (kPa)	ω	V_c (cm³/mole)	z_c	T_n (K)
cyclopentane	C_5H_{10}	70.1350	511.7	4510	0.199	261.9	0.278	322.4
cyclohexane	C_6H_{12}	84.1620	553.6	4075	0.215	308.4	0.273	353.9
cyclohexene	C_6H_{10}	82.1460	560.5	4420	0.223	286.4	0.272	356.1
benzene	C_6H_6	78.1140	562.0	4906	0.217	256.4	0.269	353.2
toluene	C_7H_8	92.1410	591.7	4126	0.273	315.7	0.265	383.8
o-xylene	C_8H_{10}	106.1680	630.4	3746	0.318	370.7	0.265	417.5
m-xylene	C_8H_{10}	106.1680	616.9	3540	0.334	403.5	0.279	412.2
p-xylene	C_8H_{10}	106.1680	616.2	3524	0.330	377.3	0.260	411.5
ethylbenzene	C_8H_{10}	106.1680	617.1	3616	0.310	372.9	0.263	409.3
phenol	C_6H_6O	94.1130	694.3	5540	0.409	277.6	0.266	455.0

Oxygen Containing Organics

Name	Formula	MW (g/mole)	T_c (K)	P_c (kPa)	ω	V_c (cm³/mole)	z_c	T_n (K)
methanol	CH_3OH	32.0420	513.4	8220	0.565	113.9	0.219	337.6
ethanol	C_2H_5OH	46.0690	515.4	6300	0.635	214.5	0.315	351.4
ethylene oxide	C_2H_4O	44.0530	469.0	7220	0.200	137.7	0.255	283.6
ethylene glycol	$C_2H_6O_2$	62.0680	719.3	8200	0.526	180.4	0.247	470.6
1-propanol	C_3H_7OH	60.0960	536.7	5120	0.609	219.9	0.252	370.2
2-propanol	C3H7OH	60.0960	508.3	4750	0.657	223.7	0.251	355.4
1-butanol	C_4H_9OH	74.1230	561.9	4420	0.597	279.3	0.264	390.8
2-butanol	C_4H_9OH	74.1230	533.2	4240	0.613	268.2	0.256	372.6
acetone	C_3H_6O	58.0800	508.1	4690	0.313	212.8	0.236	329.2
diethylether	$C_4H_{10}O$	74.1230	466.8	3649	0.287	279.5	0.263	307.6
methylethylketone	C_4H_8O	72.1070	536.5	4172	0.327	274.2	0.256	352.7
formaldehyde	CH_2O	30.0260	416.5	6830	0.211	98.8	0.195	253.4
acetaldehyde	C_2H_4O	44.0530	462.1	4800	0.267	154.0	0.192	293.6
acetic acid	$C_2H_4O_2$	60.0520	593.0	5786	0.467	171.2	0.201	391.0

Miscellaneous Hydrocarbons

Name	Formula	MW (g/mole)	T_c (K)	P_c (kPa)	ω	V_c (cm³/mole)	z_c	T_n (K)
methylamine	CH_5N	31.0580	430.6	7610	0.297	138.5	0.294	266.7
ethylamine	C_2H_7N	45.0850	456.5	5640	0.283	184.8	0.275	289.8
acetonitrile	C_2H_3N	41.0530	545.4	4866	0.350	166.2	0.178	354.7
methylchloride	CH_3Cl	50.4850	416.2	6714	0.158	136.4	0.265	249.1
dichloromethane	CH_2Cl_2	84.9270	508.0	6360	0.232	182.6	0.275	313.0
trichloromethane	$CHCl_3$	119.3690	536.1	5600	0.238	240.7	0.302	334.3
carbon tetrachloride	CCl_4	153.8110	556.5	4580	0.200	278.8	0.276	349.8
fluoromethane	CH_3F	34.0330	317.3	5906	0.207	107.6	0.241	194.8
difluoromethane	CH_2F_2	52.0230	351.3	5783	0.283	122.7	0.243	221.5
R-134a	$C_2H_2F_4$	102.0320	374.2	4059	0.333	190.7	0.249	247.1

Appendix C: Critical Constants

Inorganics

Name	Formula	MW (g/mole)	T_c (K)	P_c (kPa)	ω	V_c (cm³/mole)	z_c	T_n (K)
water	H_2O	18.0150	647.1	22064	0.351	56.0	0.230	373.2
hydrogen	H_2	2.0160	33.1	1296	-0.217	64.6	0.304	20.4
oxygen	O_2	31.9990	154.6	5046	0.026	75.4	0.296	90.2
nitrogen	N_2	28.0140	126.2	3396	0.041	89.4	0.290	77.4
carbon dioxide	CO_2	44.0090	304.1	7377	0.272	94.2	0.275	194.7*
carbon monoxide	CO	28.0140	132.9	3498	0.053	92.2	0.292	81.6
helium	He	4.0030	5.2	228	-0.377	55.2	0.291	4.2
argon	Ar	39.9480	150.7	4863	0.001	74.6	0.290	87.3
ammonia	NH_3	17.0310	405.5	11359	0.261	75.8	0.255	239.8
chlorine	Cl_2	70.9000	416.9	7980	0.091	130.1	0.300	239.2
hydrogen chloride	HCl	36.4580	324.7	8260	0.129	86.4	0.264	188.3
hydrogen sulfide	H_2S	34.0860	373.1	8999	0.104	98.2	0.285	212.9
hydrogen cyanide	HCN	27.0260	456.7	4970	0.385	138.6	0.181	298.8
carbon disulfide	CS_2	76.1310	552.2	8430	0.157	168.1	0.309	319.4
sulfur dioxide	SO_2	64.0580	430.6	7880	0.261	122.3	0.269	263.1
sulfur trioxide	SO_3	80.0630	491.5	8491	0.434	126.5	0.263	318.0
nitric oxide	NO	30.0060	177.0	6430	0.678	70.6	0.308	121.4
nitrogen dioxide	NO_2	46.0060	431.0	10128	0.838	82.5	0.233	295.1
nitrous oxide	N_2O	44.0130	309.5	7245	0.166	97.2	0.274	184.7
dinitrogen tetroxide	N_2O_4	92.0100	431.3	10230	0.856	166.8	0.476	294.1

*sublimes

Appendix D - Lee-Kesler Tables (adapted from AIChE Journal 21, 515 - 522, 1975) used by permission

Table D.1: z^0   liquid values

T_r	P_r														
	0.0100	0.0500	0.1000	0.2000	0.4000	0.6000	0.8000	1.0000	1.2000	1.5000	2.0000	3.0000	5.0000	7.0000	10.0000
0.30	0.0029	0.0145	0.0290	0.0579	0.1158	0.1737	0.2315	0.2892	0.3479	0.4335	0.5775	0.8648	1.4366	2.0048	2.8507
0.35	0.0026	0.0130	0.0261	0.0522	0.1043	0.1564	0.2084	0.2604	0.3123	0.3901	0.5195	0.7775	1.2902	1.7987	2.5539
0.40	0.0024	0.0119	0.0239	0.0477	0.0953	0.1429	0.1904	0.2379	0.2853	0.3563	0.4744	0.7095	1.1758	1.6373	2.3211
0.45	0.0022	0.0110	0.0221	0.0442	0.0882	0.1322	0.1762	0.2200	0.2638	0.3294	0.4384	0.6551	1.0841	1.5077	2.1338
0.50	0.0021	0.0103	0.0207	0.0413	0.0825	0.1236	0.1647	0.2056	0.2465	0.3077	0.4092	0.6110	1.0094	1.4017	1.9801
0.55	0.9804	0.0098	0.0195	0.0390	0.0778	0.1166	0.1553	0.1939	0.2323	0.2899	0.3853	0.5747	0.9475	1.3137	1.8520
0.60	0.9849	0.0093	0.0186	0.0371	0.0741	0.1109	0.1476	0.1842	0.2207	0.2753	0.3657	0.5446	0.8959	1.2398	1.7440
0.65	0.9881	0.9377	0.0178	0.0356	0.0710	0.1063	0.1415	0.1765	0.2113	0.2634	0.3495	0.5197	0.8526	1.1773	1.6519
0.70	0.9904	0.9504	0.8958	0.0344	0.0687	0.1027	0.1366	0.1703	0.2038	0.2538	0.3364	0.4991	0.8161	1.1341	1.5729
0.75	0.9922	0.9598	0.9165	0.0336	0.0670	0.1001	0.1330	0.1656	0.1981	0.2464	0.3260	0.4823	0.7854	1.0787	1.5047
0.80	0.9935	0.9669	0.9319	0.8539	0.0661	0.0985	0.1307	0.1626	0.1942	0.2411	0.3182	0.4690	0.7598	1.0400	1.4456
0.85	0.9946	0.9725	0.9436	0.8810	0.0661	0.0983	0.1301	0.1614	0.1924	0.2382	0.3132	0.4591	0.7388	1.0071	1.3943
0.90	0.9954	0.9768	0.9528	0.9015	0.7800	0.1006	0.1321	0.1630	0.1935	0.2383	0.3114	0.4527	0.7220	0.9793	1.3496
0.93	0.9959	0.9790	0.9573	0.9115	0.8059	0.6635	0.1359	0.1664	0.1963	0.2405	0.3122	0.4507	0.7138	0.9648	1.3257
0.95	0.9961	0.9803	0.9600	0.9174	0.8206	0.6967	0.1410	0.1705	0.1998	0.2432	0.3138	0.4501	0.7092	0.9561	1.3108
0.97	0.9963	0.9815	0.9625	0.9227	0.8338	0.7240	0.5580	0.1779	0.2055	0.2474	0.3164	0.4504	0.7052	0.9480	1.2968
0.98	0.9965	0.9821	0.9637	0.9253	0.8398	0.7360	0.5887	0.1844	0.2097	0.2503	0.3182	0.4508	0.7035	0.9442	1.2901
0.99	0.9966	0.9826	0.9648	0.9277	0.8455	0.7471	0.6138	0.1959	0.2154	0.2538	0.3204	0.4514	0.7018	0.9406	1.2835
1.00	0.9967	0.9832	0.9659	0.9300	0.8509	0.7574	0.6355	0.2901	0.2237	0.2583	0.3229	0.4522	0.7004	0.9372	1.2772
1.01	0.9968	0.9837	0.9669	0.9322	0.8561	0.7671	0.6542	0.4648	0.2370	0.2640	0.3260	0.4533	0.6991	0.9339	1.2710
1.02	0.9969	0.9842	0.9679	0.9343	0.8610	0.7761	0.6710	0.5146	0.2629	0.2715	0.3297	0.4547	0.6980	0.9307	1.2650
1.05	0.9971	0.9855	0.9707	0.9401	0.8743	0.8002	0.7130	0.6026	0.4437	0.3131	0.3452	0.4604	0.6956	0.9222	1.2481
1.10	0.9975	0.9874	0.9747	0.9485	0.8930	0.8323	0.7649	0.6880	0.5984	0.4580	0.3953	0.4770	0.6950	0.9110	1.2232
1.15	0.9978	0.9891	0.9780	0.9554	0.9081	0.8576	0.8032	0.7443	0.6803	0.5798	0.4760	0.5042	0.6987	0.9033	1.2021
1.20	0.9981	0.9904	0.9808	0.9611	0.9205	0.8779	0.8330	0.7858	0.7363	0.6605	0.5605	0.5425	0.7069	0.8990	1.1844
1.30	0.9985	0.9926	0.9852	0.9702	0.9396	0.9083	0.8764	0.8438	0.8111	0.7624	0.6908	0.6344	0.7358	0.8998	1.1580
1.40	0.9988	0.9942	0.9884	0.9768	0.9534	0.9298	0.9062	0.8827	0.8595	0.8256	0.7753	0.7202	0.7761	0.9112	1.1419
1.50	0.9991	0.9954	0.9909	0.9818	0.9636	0.9456	0.9278	0.9103	0.8933	0.8689	0.8328	0.7887	0.8200	0.9297	1.1339
1.60	0.9993	0.9964	0.9928	0.9856	0.9714	0.9575	0.9439	0.9308	0.9180	0.9000	0.8738	0.8410	0.8617	0.9518	1.1320
1.70	0.9994	0.9971	0.9943	0.9886	0.9775	0.9667	0.9563	0.9463	0.9367	0.9234	0.9043	0.8809	0.8984	0.9745	1.1343
1.80	0.9995	0.9977	0.9955	0.9910	0.9823	0.9739	0.9659	0.9583	0.9511	0.9413	0.9275	0.9118	0.9297	0.9961	1.1391
1.90	0.9996	0.9982	0.9964	0.9929	0.9861	0.9796	0.9735	0.9678	0.9624	0.9552	0.9456	0.9359	0.9557	1.0157	1.1452
2.00	0.9997	0.9986	0.9972	0.9944	0.9892	0.9842	0.9796	0.9754	0.9715	0.9664	0.9599	0.9550	0.9772	1.0328	1.1516
2.20	0.9998	0.9992	0.9983	0.9967	0.9937	0.9910	0.9886	0.9865	0.9847	0.9826	0.9806	0.9827	1.0094	1.0600	1.1635
2.40	0.9999	0.9996	0.9991	0.9983	0.9969	0.9957	0.9948	0.9941	0.9936	0.9935	0.9945	1.0011	1.0313	1.0793	1.1728
2.60	1.0000	0.9998	0.9997	0.9994	0.9991	0.9990	0.9990	0.9993	0.9998	1.0010	1.0040	1.0137	1.0463	1.0926	1.1792
2.80	1.0000	1.0000	1.0001	1.0002	1.0007	1.0013	1.0021	1.0031	1.0042	1.0063	1.0106	1.0223	1.0565	1.1016	1.1830
3.00	1.0000	1.0002	1.0004	1.0008	1.0018	1.0030	1.0043	1.0057	1.0074	1.0101	1.0153	1.0284	1.0635	1.1075	1.1848
3.50	1.0001	1.0004	1.0008	1.0017	1.0035	1.0055	1.0075	1.0097	1.0120	1.0156	1.0221	1.0368	1.0723	1.1138	1.1834
4.00	1.0001	1.0005	1.0010	1.0021	1.0043	1.0066	1.0090	1.0115	1.0140	1.0179	1.0249	1.0401	1.0747	1.1136	1.1773

Table D.2: z^1   liquid values

T_r	P_r														
	0.0100	0.0500	0.1000	0.2000	0.4000	0.6000	0.8000	1.0000	1.2000	1.5000	2.0000	3.0000	5.0000	7.0000	10.0000
0.30	-0.0008	-0.0040	-0.0081	-0.0161	-0.0323	-0.0484	-0.0645	-0.0806	-0.0966	-0.1207	-0.1608	-0.2407	-0.3996	-0.5572	-0.7915
0.35	-0.0009	-0.0046	-0.0093	-0.0185	-0.0370	-0.0554	-0.0738	-0.0921	-0.1105	-0.1379	-0.1834	-0.2738	-0.4523	-0.6279	-0.8863
0.40	-0.0010	-0.0048	-0.0095	-0.0190	-0.0380	-0.0570	-0.0758	-0.0946	-0.1134	-0.1414	-0.1879	-0.2799	-0.4603	-0.6365	-0.8936
0.45	-0.0009	-0.0047	-0.0094	-0.0187	-0.0374	-0.0560	-0.0745	-0.0929	-0.1113	-0.1387	-0.1840	-0.2734	-0.4475	-0.6162	-0.8608
0.50	-0.0009	-0.0045	-0.0090	-0.0181	-0.0360	-0.0539	-0.0716	-0.0893	-0.1069	-0.1330	-0.1762	-0.2611	-0.4253	-0.5831	-0.8099
0.55	-0.0314	-0.0043	-0.0086	-0.0172	-0.0343	-0.0513	-0.0682	-0.0849	-0.1015	-0.1263	-0.1669	-0.2465	-0.3991	-0.5446	-0.7521
0.60	-0.0205	-0.0041	-0.0082	-0.0164	-0.0326	-0.0487	-0.0646	-0.0803	-0.0960	-0.1192	-0.1572	-0.2312	-0.3718	-0.5047	-0.6928
0.65	-0.0137	-0.0772	-0.0078	-0.0156	-0.0309	-0.0461	-0.0611	-0.0759	-0.0906	-0.1122	-0.1476	-0.2160	-0.3447	-0.4653	-0.6346
0.70	-0.0093	-0.0507	-0.1161	-0.0148	-0.0294	-0.0438	-0.0579	-0.0718	-0.0855	-0.1057	-0.1385	-0.2013	-0.3184	-0.4270	-0.5785
0.75	-0.0064	-0.0339	-0.0744	-0.0143	-0.0282	-0.0417	-0.0550	-0.0681	-0.0808	-0.0996	-0.1298	-0.1872	-0.2929	-0.3901	-0.5250
0.80	-0.0044	-0.0228	-0.0487	-0.1160	-0.0272	-0.0401	-0.0526	-0.0648	-0.0767	-0.0940	-0.1217	-0.1736	-0.2682	-0.3545	-0.4740
0.85	-0.0029	-0.0152	-0.0319	-0.0715	-0.0268	-0.0391	-0.0509	-0.0622	-0.0731	-0.0888	-0.1138	-0.1602	-0.2439	-0.3201	-0.4254
0.90	-0.0019	-0.0099	-0.0205	-0.0442	-0.1118	-0.0396	-0.0503	-0.0604	-0.0701	-0.0840	-0.1059	-0.1463	-0.2195	-0.2862	-0.3788
0.93	-0.0015	-0.0075	-0.0154	-0.0326	-0.0763	-0.1662	-0.0514	-0.0602	-0.0687	-0.0810	-0.1007	-0.1374	-0.2045	-0.2661	-0.3516
0.95	-0.0012	-0.0062	-0.0126	-0.0262	-0.0589	-0.1110	-0.0540	-0.0607	-0.0678	-0.0788	-0.0967	-0.1310	-0.1943	-0.2526	-0.3339
0.97	-0.0010	-0.0050	-0.0101	-0.0208	-0.0450	-0.0770	-0.1647	-0.0623	-0.0669	-0.0759	-0.0921	-0.1240	-0.1837	-0.2391	-0.3163
0.98	-0.0009	-0.0044	-0.0090	-0.0184	-0.0390	-0.0641	-0.1100	-0.0641	-0.0661	-0.0740	-0.0893	-0.1202	-0.1783	-0.2322	-0.3075
0.99	-0.0008	-0.0039	-0.0079	-0.0161	-0.0335	-0.0531	-0.0796	-0.0680	-0.0646	-0.0715	-0.0861	-0.1162	-0.1728	-0.2254	-0.2989
1.00	-0.0007	-0.0034	-0.0069	-0.0140	-0.0285	-0.0435	-0.0588	-0.0879	-0.0609	-0.0678	-0.0824	-0.1118	-0.1672	-0.2185	-0.2902
1.01	-0.0006	-0.0030	-0.0060	-0.0120	-0.0240	-0.0351	-0.0429	-0.0223	-0.0473	-0.0621	-0.0778	-0.1072	-0.1615	-0.2116	-0.2816
1.02	-0.0005	-0.0026	-0.0051	-0.0102	-0.0198	-0.0277	-0.0303	-0.0062	-0.0227	-0.0524	-0.0722	-0.1021	-0.1556	-0.2047	-0.2731
1.05	-0.0003	-0.0015	-0.0029	-0.0054	-0.0092	-0.0097	-0.0032	0.0220	0.1059	0.0451	-0.0432	-0.0838	-0.1370	-0.1835	-0.2476
1.10	0.0000	0.0000	0.0001	0.0007	0.0038	0.0106	0.0236	0.0476	0.0897	0.1630	0.0698	-0.0373	-0.1021	-0.1469	-0.2056
1.15	0.0002	0.0011	0.0023	0.0052	0.0127	0.0237	0.0396	0.0625	0.0943	0.1548	0.1667	0.0332	-0.0611	-0.1084	-0.1642
1.20	0.0004	0.0019	0.0039	0.0084	0.0190	0.0326	0.0499	0.0719	0.0991	0.1477	0.1990	0.1095	-0.0141	-0.0678	-0.1231
1.30	0.0006	0.0030	0.0061	0.0125	0.0267	0.0429	0.0612	0.0819	0.1048	0.1420	0.1991	0.2079	0.0875	0.0176	-0.0423
1.40	0.0007	0.0036	0.0072	0.0147	0.0306	0.0477	0.0661	0.0857	0.1063	0.1383	0.1894	0.2397	0.1737	0.1008	0.0350
1.50	0.0008	0.0039	0.0078	0.0158	0.0323	0.0497	0.0677	0.0864	0.1055	0.1345	0.1806	0.2433	0.2309	0.1717	0.1058
1.60	0.0008	0.0040	0.0080	0.0162	0.0330	0.0501	0.0677	0.0855	0.1035	0.1303	0.1729	0.2381	0.2631	0.2255	0.1673
1.70	0.0008	0.0040	0.0081	0.0163	0.0329	0.0497	0.0667	0.0838	0.1008	0.1259	0.1658	0.2305	0.2788	0.2628	0.2179
1.80	0.0008	0.0040	0.0081	0.0162	0.0325	0.0488	0.0652	0.0816	0.0978	0.1216	0.1593	0.2224	0.2846	0.2871	0.2576
1.90	0.0008	0.0040	0.0079	0.0159	0.0318	0.0477	0.0635	0.0792	0.0947	0.1173	0.1532	0.2144	0.2848	0.3017	0.2876
2.00	0.0008	0.0039	0.0078	0.0155	0.0310	0.0464	0.0617	0.0767	0.0916	0.1133	0.1476	0.2069	0.2819	0.3097	0.3096
2.20	0.0007	0.0037	0.0074	0.0147	0.0293	0.0437	0.0579	0.0719	0.0857	0.1057	0.1374	0.1932	0.2720	0.3135	0.3355
2.40	0.0007	0.0035	0.0070	0.0139	0.0276	0.0411	0.0544	0.0675	0.0803	0.0989	0.1285	0.1812	0.2602	0.3089	0.3459
2.60	0.0007	0.0033	0.0066	0.0131	0.0260	0.0387	0.0512	0.0634	0.0754	0.0929	0.1207	0.1706	0.2484	0.3009	0.3475
2.80	0.0006	0.0031	0.0062	0.0124	0.0245	0.0365	0.0483	0.0598	0.0711	0.0876	0.1138	0.1613	0.2372	0.2915	0.3443
3.00	0.0006	0.0029	0.0059	0.0117	0.0232	0.0345	0.0456	0.0565	0.0672	0.0828	0.1076	0.1529	0.2268	0.2817	0.3385
3.50	0.0005	0.0026	0.0052	0.0103	0.0204	0.0303	0.0401	0.0497	0.0591	0.0728	0.0949	0.1356	0.2042	0.2584	0.3194
4.00	0.0005	0.0023	0.0046	0.0091	0.0182	0.0270	0.0357	0.0443	0.0527	0.0651	0.0849	0.1219	0.1857	0.2378	0.2994

Appendix D - Lee-Kesler Tables (adapted from AIChE Journal 21, 515 - 522, 1975)

Table D.3: h^{R0}/RT_c liquid values

T_r	P_r = 0.0100	0.0500	0.1000	0.2000	0.4000	0.6000	0.8000	1.0000	1.2000	1.5000	2.0000	3.0000	5.0000	7.0000	10.0000
0.30	-6.045	-6.043	-6.040	-6.034	-6.022	-6.011	-5.999	-5.987	-5.975	-5.957	-5.927	-5.868	-5.748	-5.628	-5.446
0.35	-5.906	-5.904	-5.901	-5.895	-5.882	-5.870	-5.858	-5.845	-5.833	-5.814	-5.783	-5.721	-5.595	-5.469	-5.278
0.40	-5.763	-5.761	-5.757	-5.751	-5.738	-5.726	-5.713	-5.700	-5.687	-5.668	-5.636	-5.572	-5.442	-5.311	-5.113
0.45	-5.615	-5.612	-5.609	-5.603	-5.590	-5.577	-5.564	-5.551	-5.538	-5.519	-5.486	-5.421	-5.288	-5.154	-4.950
0.50	-5.465	-5.463	-5.459	-5.453	-5.440	-5.427	-5.414	-5.401	-5.388	-5.369	-5.336	-5.279	-5.135	-4.999	-4.791
0.55	-0.032	-5.312	-5.309	-5.303	-5.290	-5.278	-5.265	-5.252	-5.239	-5.220	-5.187	-5.121	-4.986	-4.849	-4.638
0.60	-0.027	-5.162	-5.159	-5.153	-5.141	-5.129	-5.116	-5.104	-5.091	-5.073	-5.041	-4.976	-4.842	-4.704	-4.492
0.65	-0.023	-0.118	-5.008	-5.002	-4.991	-4.980	-4.968	-4.956	-4.949	-4.927	-4.896	-4.833	-4.702	-4.565	-4.353
0.70	-0.020	-0.101	-0.213	-4.848	-4.838	-4.828	-4.818	-4.808	-4.797	-4.781	-4.752	-4.693	-4.566	-4.432	-4.221
0.75	-0.017	-0.088	-0.183	-4.687	-4.679	-4.672	-4.664	-4.655	-4.646	-4.632	-4.607	-4.554	-4.434	-4.303	-4.095
0.80	-0.015	-0.078	-0.160	-0.345	-4.507	-4.504	-4.499	-4.494	-4.488	-4.478	-4.459	-4.413	-4.303	-4.178	-3.974
0.85	-0.014	-0.069	-0.141	-0.300	-4.309	-4.313	-4.316	-4.316	-4.316	-4.312	-4.302	-4.269	-4.173	-4.056	-3.857
0.90	-0.012	-0.062	-0.126	-0.264	-0.596	-4.074	-4.094	-4.108	-4.118	-4.127	-4.132	-4.119	-4.043	-3.935	-3.744
0.93	-0.011	-0.058	-0.118	-0.246	-0.545	-0.960	-3.920	-3.953	-3.976	-4.000	-4.020	-4.024	-3.963	-3.863	-3.678
0.95	-0.011	-0.056	-0.113	-0.235	-0.516	-0.885	-3.763	-3.825	-3.865	-3.904	-3.940	-3.958	-3.910	-3.815	-3.634
0.97	-0.011	-0.054	-0.109	-0.225	-0.490	-0.824	-1.356	-3.658	-3.732	-3.796	-3.853	-3.890	-3.856	-3.767	-3.591
0.98	-0.010	-0.053	-0.107	-0.221	-0.478	-0.797	-1.273	-3.544	-3.652	-3.736	-3.806	-3.854	-3.829	-3.743	-3.569
0.99	-0.010	-0.052	-0.105	-0.216	-0.466	-0.773	-1.206	-3.376	-3.558	-3.670	-3.758	-3.818	-3.801	-3.719	-3.548
1.00	-0.010	-0.051	-0.103	-0.212	-0.455	-0.750	-1.151	-2.584	-3.441	-3.598	-3.706	-3.782	-3.774	-3.695	-3.526
1.01	-0.010	-0.050	-0.101	-0.208	-0.445	-0.721	-1.102	-1.796	-3.283	-3.516	-3.652	-3.744	-3.746	-3.671	-3.505
1.02	-0.010	-0.049	-0.099	-0.203	-0.434	-0.708	-1.060	-1.627	-3.039	-3.422	-3.595	-3.705	-3.718	-3.647	-3.484
1.05	-0.009	-0.046	-0.094	-0.192	-0.407	-0.654	-0.955	-1.359	-2.034	-3.030	-3.398	-3.583	-3.632	-3.575	-3.420
1.10	-0.008	-0.042	-0.086	-0.175	-0.367	-0.581	-0.827	-1.120	-1.487	-2.203	-2.965	-3.353	-3.484	-3.453	-3.315
1.15	-0.008	-0.039	-0.079	-0.160	-0.334	-0.523	-0.732	-0.968	-1.239	-1.719	-2.479	-3.091	-3.329	-3.329	-3.211
1.20	-0.007	-0.036	-0.073	-0.148	-0.305	-0.474	-0.657	-0.857	-1.076	-1.443	-2.079	-2.801	-3.166	-3.202	-3.107
1.30	-0.006	-0.031	-0.063	-0.127	-0.259	-0.399	-0.545	-0.698	-0.860	-1.116	-1.560	-2.274	-2.825	-2.942	-2.899
1.40	-0.005	-0.027	-0.055	-0.110	-0.224	-0.341	-0.463	-0.588	-0.716	-0.915	-1.253	-1.857	-2.486	-2.679	-2.692
1.50	-0.005	-0.024	-0.048	-0.097	-0.196	-0.297	-0.400	-0.505	-0.611	-0.774	-1.046	-1.549	-2.175	-2.421	-2.486
1.60	-0.004	-0.021	-0.043	-0.086	-0.173	-0.261	-0.350	-0.440	-0.531	-0.667	-0.894	-1.318	-1.904	-2.177	-2.285
1.70	-0.004	-0.019	-0.038	-0.076	-0.153	-0.231	-0.309	-0.387	-0.446	-0.583	-0.777	-1.139	-1.672	-1.953	-2.091
1.80	-0.003	-0.017	-0.034	-0.068	-0.137	-0.206	-0.275	-0.344	-0.413	-0.515	-0.683	-0.996	-1.476	-1.751	-1.908
1.90	-0.003	-0.015	-0.031	-0.062	-0.123	-0.185	-0.246	-0.307	-0.368	-0.458	-0.606	-0.880	-1.309	-1.571	-1.736
2.00	-0.003	-0.014	-0.028	-0.056	-0.111	-0.167	-0.222	-0.276	-0.330	-0.411	-0.541	-0.782	-1.167	-1.411	-1.577
2.20	-0.002	-0.012	-0.023	-0.046	-0.092	-0.137	-0.182	-0.226	-0.269	-0.334	-0.437	-0.629	-0.937	-1.143	-1.295
2.40	-0.002	-0.010	-0.019	-0.038	-0.076	-0.114	-0.150	-0.187	-0.222	-0.275	-0.359	-0.513	-0.761	-0.929	-1.058
2.60	-0.002	-0.008	-0.016	-0.032	-0.064	-0.095	-0.125	-0.155	-0.185	-0.228	-0.297	-0.422	-0.621	-0.756	-0.858
2.80	-0.001	-0.007	-0.014	-0.027	-0.054	-0.080	-0.105	-0.130	-0.154	-0.190	-0.246	-0.348	-0.508	-0.614	-0.689
3.00	-0.001	-0.006	-0.011	-0.023	-0.045	-0.067	-0.088	-0.109	-0.129	-0.159	-0.205	-0.288	-0.415	-0.495	-0.545
3.50	-0.001	-0.004	-0.007	-0.015	-0.029	-0.043	-0.056	-0.069	-0.081	-0.099	-0.127	-0.174	-0.239	-0.270	-0.264
4.00	0.000	-0.002	-0.005	-0.009	-0.017	-0.026	-0.033	-0.041	-0.048	-0.058	-0.072	-0.095	-0.116	-0.110	-0.061

Table D.4: h^{R1}/RT_c liquid values

T_r	P_r = 0.0100	0.0500	0.1000	0.2000	0.4000	0.6000	0.8000	1.0000	1.2000	1.5000	2.0000	3.0000	5.0000	7.0000	10.0000
0.30	-11.098	-11.096	-11.095	-11.091	-11.083	-11.076	-11.069	-11.062	-11.055	-11.044	-11.027	-10.992	-10.935	-10.872	-10.781
0.35	-10.656	-10.655	-10.654	-10.653	-10.650	-10.646	-10.643	-10.640	-10.637	-10.632	-10.624	-10.609	-10.581	-10.554	-10.529
0.40	-10.121	-10.121	-10.121	-10.120	-10.121	-10.121	-10.121	-10.121	-10.121	-10.121	-10.122	-10.123	-10.128	-10.135	-10.150
0.45	-9.515	-9.515	-9.516	-9.517	-9.519	-9.521	-9.523	-9.525	-9.527	-9.531	-9.537	-9.549	-9.576	-9.611	-9.663
0.50	-8.868	-8.869	-8.870	-8.872	-8.876	-8.880	-8.884	-8.888	-8.892	-8.899	-8.909	-8.932	-8.978	-9.030	-9.111
0.55	-0.080	-8.211	-8.212	-8.215	-8.221	-8.226	-8.232	-8.238	-8.243	-8.252	-8.267	-8.298	-8.360	-8.425	-8.531
0.60	-0.059	-7.568	-7.570	-7.573	-7.579	-7.585	-7.591	-7.596	-7.603	-7.614	-7.632	-7.669	-7.745	-7.824	-7.950
0.65	-0.045	-0.247	-6.949	-6.952	-6.959	-6.966	-6.973	-6.980	-6.987	-6.997	-7.017	-7.059	-7.147	-7.239	-7.381
0.70	-0.034	-0.185	-0.415	-6.360	-6.367	-6.373	-6.381	-6.388	-6.395	-6.407	-6.429	-6.475	-6.574	-6.677	-6.837
0.75	-0.027	-0.142	-0.306	-5.796	-5.802	-5.809	-5.816	-5.824	-5.832	-5.845	-5.868	-5.918	-6.027	-6.142	-6.318
0.80	-0.021	-0.110	-0.234	-0.542	-5.266	-5.271	-5.278	-5.285	-5.293	-5.306	-5.330	-5.385	-5.506	-5.632	-5.824
0.85	-0.017	-0.087	-0.182	-0.401	-4.753	-4.754	-4.758	-4.763	-4.771	-4.784	-4.810	-4.872	-5.000	-5.149	-5.358
0.90	-0.014	-0.070	-0.144	-0.308	-0.751	-4.254	-4.248	-4.249	-4.255	-4.268	-4.298	-4.371	-4.530	-4.688	-4.916
0.93	-0.012	-0.061	-0.126	-0.265	-0.612	-1.236	-3.942	-3.934	-3.937	-3.951	-3.987	-4.073	-4.251	-4.422	-4.662
0.95	-0.011	-0.056	-0.115	-0.241	-0.542	-0.994	-3.737	-3.712	-3.713	-3.730	-3.773	-3.873	-4.068	-4.248	-4.497
0.97	-0.010	-0.052	-0.105	-0.219	-0.483	-0.837	-1.616	-3.470	-3.467	-3.492	-3.551	-3.670	-3.885	-4.077	-4.336
0.98	-0.010	-0.050	-0.101	-0.209	-0.457	-0.776	-1.324	-3.332	-3.327	-3.363	-3.434	-3.568	-3.795	-3.992	-4.257
0.99	-0.009	-0.048	-0.097	-0.200	-0.433	-0.722	-1.154	-3.164	-3.164	-3.223	-3.313	-3.464	-3.705	-3.909	-4.178
1.00	-0.009	-0.046	-0.093	-0.191	-0.410	-0.675	-1.034	-2.471	-2.952	-3.065	-3.186	-3.358	-3.615	-3.825	-4.100
1.01	-0.009	-0.044	-0.089	-0.183	-0.389	-0.632	-0.940	-1.375	-2.595	-2.880	-3.051	-3.251	-3.525	-3.742	-4.023
1.02	-0.008	-0.042	-0.085	-0.175	-0.370	-0.594	-0.863	-1.180	-1.723	-2.650	-2.906	-3.142	-3.435	-3.661	-3.947
1.05	-0.007	-0.037	-0.075	-0.153	-0.318	-0.498	-0.691	-0.877	-0.878	-1.496	-2.381	-2.800	-3.167	-3.418	-3.722
1.10	-0.006	-0.030	-0.061	-0.123	-0.251	-0.381	-0.507	-0.617	-0.673	-0.617	-1.261	-2.167	-2.720	-3.023	-3.362
1.15	-0.005	-0.025	-0.050	-0.099	-0.199	-0.296	-0.385	-0.459	-0.503	-0.487	-0.604	-1.497	-2.275	-2.641	-3.019
1.20	-0.004	-0.020	-0.040	-0.080	-0.158	-0.232	-0.297	-0.349	-0.381	-0.381	-0.361	-0.934	-1.840	-2.273	-2.692
1.30	-0.003	-0.013	-0.026	-0.052	-0.100	-0.142	-0.177	-0.203	-0.218	-0.218	-0.178	-0.300	-1.066	-1.592	-2.086
1.40	-0.002	-0.008	-0.016	-0.032	-0.060	-0.083	-0.100	-0.111	-0.115	-0.128	-0.070	-0.044	-0.504	-1.012	-1.547
1.50	-0.001	-0.005	-0.009	-0.018	-0.032	-0.042	-0.048	-0.049	-0.046	-0.032	0.008	0.078	-0.142	-0.556	-1.080
1.60	0.000	-0.002	-0.004	-0.007	-0.012	-0.013	-0.011	-0.005	0.004	0.023	0.065	0.151	0.082	-0.217	-0.689
1.70	0.000	0.000	0.000	0.000	0.003	0.009	0.017	0.027	0.040	0.063	0.109	0.202	0.223	0.028	-0.369
1.80	0.000	0.001	0.003	0.006	0.015	0.025	0.037	0.051	0.067	0.094	0.143	0.241	0.317	0.203	-0.112
1.90	0.001	0.003	0.005	0.011	0.023	0.037	0.053	0.070	0.088	0.117	0.169	0.271	0.381	0.330	0.092
2.00	0.001	0.003	0.007	0.015	0.030	0.047	0.065	0.085	0.105	0.136	0.190	0.295	0.428	0.424	0.255
2.20	0.001	0.005	0.010	0.020	0.040	0.062	0.083	0.106	0.128	0.163	0.221	0.331	0.493	0.551	0.489
2.40	0.001	0.006	0.012	0.023	0.047	0.071	0.095	0.120	0.144	0.181	0.242	0.356	0.535	0.631	0.645
2.60	0.001	0.006	0.013	0.026	0.052	0.078	0.104	0.130	0.156	0.194	0.257	0.376	0.567	0.687	0.754
2.80	0.001	0.007	0.014	0.028	0.055	0.082	0.110	0.137	0.164	0.204	0.269	0.391	0.591	0.729	0.836
3.00	0.001	0.007	0.014	0.029	0.058	0.086	0.114	0.142	0.170	0.211	0.278	0.403	0.611	0.763	0.899
3.50	0.002	0.008	0.016	0.031	0.062	0.092	0.122	0.152	0.181	0.224	0.294	0.425	0.650	0.827	1.015
4.00	0.002	0.008	0.016	0.032	0.064	0.096	0.127	0.158	0.188	0.233	0.306	0.442	0.680	0.874	1.097

Appendix D - Lee-Kesler Tables (adapted from AIChE Journal 21, 515 - 522, 1975)

Table D.5: s^{R0}/R liquid values

	P_r													
T_r	0.0100	0.0500	0.1000	0.2000	0.4000	0.6000	0.8000	1.0000	1.2000	1.5000	2.0000	3.0000	5.0000	7.0000	10.0000
0.30	-11.614	-10.008	-9.319	-8.635	-7.961	-7.574	-7.304	-7.099	-6.935	-6.740	-6.497	-6.180	-5.847	-5.683	-5.578
0.35	-11.185	-9.579	-8.890	-8.205	-7.529	-7.140	-6.869	-6.663	-6.497	-6.299	-6.052	-5.728	-5.376	-5.194	-5.060
0.40	-10.802	-9.196	-8.506	-7.821	-7.144	-6.755	-6.483	-6.275	-6.109	-5.909	-5.660	-5.330	-4.967	-4.772	-4.619
0.45	-10.453	-8.847	-8.157	-7.472	-6.794	-6.404	-6.132	-5.924	-5.757	-5.557	-5.306	-4.974	-4.603	-4.401	-4.234
0.50	-10.137	-8.531	-7.841	-7.156	-6.479	-6.089	-5.816	-5.608	-5.441	-5.240	-4.989	-4.656	-4.282	-4.074	-3.899
0.55	-0.038	-8.245	-7.555	-6.870	-6.193	-5.803	-5.531	-5.324	-5.157	-4.956	-4.706	-4.373	-3.998	-3.788	-3.607
0.60	-0.029	-7.983	-7.294	-6.610	-5.933	-5.544	-5.273	-5.066	-4.900	-4.700	-4.451	-4.120	-3.747	-3.537	-3.353
0.65	-0.023	-0.122	-7.052	-6.368	-5.694	-5.306	-5.036	-4.830	-4.665	-4.467	-4.220	-3.892	-3.523	-3.315	-3.131
0.70	-0.018	-0.096	-0.206	-6.140	-5.467	-5.082	-4.814	-4.610	-4.446	-4.250	-4.007	-3.684	-3.322	-3.117	-2.935
0.75	-0.015	-0.078	-0.164	-5.917	-5.248	-4.866	-4.600	-4.399	-4.238	-4.045	-3.807	-3.491	-3.138	-2.939	-2.761
0.80	-0.013	-0.064	-0.134	-0.294	-5.026	-4.694	-4.388	-4.191	-4.034	-3.846	-3.615	-3.310	-2.970	-2.777	-2.605
0.85	-0.011	-0.054	-0.111	-0.239	-4.785	-4.418	-4.166	-3.976	-3.825	-3.646	-3.425	-3.135	-2.812	-2.629	-2.463
0.90	-0.009	-0.046	-0.094	-0.199	-0.463	-4.145	-3.912	-3.738	-3.599	-3.434	-3.231	-2.964	-2.663	-2.491	-2.334
0.93	-0.008	-0.042	-0.085	-0.179	-0.408	-0.750	-3.723	-3.569	-3.444	-3.295	-3.108	-2.860	-2.577	-2.412	-2.262
0.95	-0.008	-0.039	-0.080	-0.168	-0.377	-0.671	-3.556	-3.433	-3.326	-3.193	-3.023	-2.790	-2.520	-2.362	-2.215
0.97	-0.007	-0.037	-0.075	-0.157	-0.350	-0.607	-1.056	-3.259	-3.188	-3.081	-2.932	-2.719	-2.463	-2.312	-2.170
0.98	-0.007	-0.036	-0.073	-0.153	-0.337	-0.580	-0.971	-3.142	-3.106	-3.019	-2.884	-2.682	-2.436	-2.287	-2.148
0.99	-0.007	-0.035	-0.071	-0.148	-0.326	-0.555	-0.903	-2.972	-3.010	-2.953	-2.835	-2.646	-2.408	-2.263	-2.126
1.00	-0.007	-0.034	-0.069	-0.144	-0.315	-0.532	-0.847	-2.178	-2.893	-2.879	-2.784	-2.609	-2.380	-2.239	-2.105
1.01	-0.007	-0.033	-0.067	-0.139	-0.304	-0.510	-0.799	-1.391	-2.736	-2.798	-2.730	-2.571	-2.352	-2.215	-2.083
1.02	-0.006	-0.032	-0.065	-0.135	-0.294	-0.491	-0.757	-1.225	-2.495	-2.706	-2.673	-2.533	-2.325	-2.191	-2.062
1.05	-0.006	-0.030	-0.060	-0.124	-0.267	-0.439	-0.656	-0.965	-1.523	-2.328	-2.483	-2.415	-2.242	-2.121	-2.001
1.10	-0.005	-0.026	-0.053	-0.108	-0.230	-0.371	-0.537	-0.742	-1.012	-1.557	-2.081	-2.202	-2.104	-2.007	-1.903
1.15	-0.005	-0.023	-0.047	-0.096	-0.201	-0.319	-0.452	-0.607	-0.790	-1.126	-1.649	-1.968	-1.966	-1.897	-1.810
1.20	-0.004	-0.021	-0.042	-0.085	-0.177	-0.277	-0.389	-0.512	-0.651	-0.890	-1.308	-1.727	-1.827	-1.789	-1.722
1.30	-0.003	-0.017	-0.033	-0.068	-0.140	-0.217	-0.298	-0.385	-0.478	-0.628	-0.891	-1.299	-1.554	-1.581	-1.556
1.40	-0.003	-0.014	-0.027	-0.056	-0.114	-0.174	-0.237	-0.303	-0.375	-0.478	-0.663	-0.990	-1.303	-1.386	-1.402
1.50	-0.002	-0.011	-0.023	-0.046	-0.094	-0.143	-0.194	-0.246	-0.299	-0.381	-0.520	-0.777	-1.088	-1.208	-1.260
1.60	-0.002	-0.010	-0.019	-0.039	-0.079	-0.120	-0.162	-0.204	-0.247	-0.312	-0.421	-0.628	-0.913	-1.050	-1.130
1.70	-0.002	-0.008	-0.017	-0.033	-0.067	-0.102	-0.137	-0.172	-0.208	-0.261	-0.350	-0.519	-0.773	-0.915	-1.013
1.80	-0.001	-0.007	-0.014	-0.029	-0.058	-0.088	-0.117	-0.147	-0.177	-0.222	-0.296	-0.438	-0.661	-0.799	-0.908
1.90	-0.001	-0.006	-0.013	-0.025	-0.051	-0.076	-0.102	-0.127	-0.153	-0.191	-0.255	-0.375	-0.570	-0.702	-0.815
2.00	-0.001	-0.006	-0.011	-0.022	-0.044	-0.067	-0.089	-0.111	-0.134	-0.167	-0.221	-0.325	-0.497	-0.620	-0.733
2.20	-0.001	-0.004	-0.009	-0.018	-0.035	-0.053	-0.070	-0.087	-0.105	-0.130	-0.172	-0.251	-0.388	-0.492	-0.599
2.40	-0.001	-0.004	-0.007	-0.014	-0.028	-0.042	-0.056	-0.070	-0.084	-0.104	-0.138	-0.201	-0.311	-0.399	-0.496
2.60	-0.001	-0.003	-0.006	-0.012	-0.023	-0.035	-0.046	-0.058	-0.069	-0.086	-0.113	-0.164	-0.255	-0.329	-0.416
2.80	0.000	-0.002	-0.005	-0.010	-0.020	-0.029	-0.039	-0.048	-0.058	-0.072	-0.094	-0.137	-0.213	-0.277	-0.353
3.00	0.000	-0.002	-0.004	-0.008	-0.017	-0.025	-0.033	-0.041	-0.049	-0.061	-0.080	-0.116	-0.181	-0.236	-0.303
3.50	0.000	-0.001	-0.003	-0.006	-0.012	-0.017	-0.023	-0.029	-0.034	-0.042	-0.056	-0.081	-0.126	-0.166	-0.216
4.00	0.000	-0.001	-0.002	-0.004	-0.009	-0.013	-0.017	-0.021	-0.025	-0.031	-0.041	-0.059	-0.093	-0.123	-0.162

Table D.6: s^{R1}/R liquid values

	P_r													
T_r	0.0100	0.0500	0.1000	0.2000	0.4000	0.6000	0.8000	1.0000	1.2000	1.5000	2.0000	3.0000	5.0000	7.0000	10.0000
0.30	-16.782	-16.774	-16.764	-16.744	-16.705	-16.665	-16.626	-16.586	-16.547	-16.488	-16.390	-16.195	-15.837	-15.468	-14.925
0.35	-15.413	-15.408	-15.401	-15.387	-15.359	-15.333	-15.305	-15.278	-15.251	-15.211	-15.144	-15.011	-14.751	-14.496	-14.153
0.40	-13.990	-13.986	-13.981	-13.972	-13.953	-13.934	-13.915	-13.896	-13.877	-13.849	-13.803	-13.714	-13.541	-13.376	-13.144
0.45	-12.564	-12.561	-12.558	-12.551	-12.537	-12.523	-12.509	-12.496	-12.482	-12.462	-12.430	-12.367	-12.248	-12.145	-11.999
0.50	-11.202	-11.200	-11.197	-11.092	-11.082	-11.172	-11.162	-11.153	-11.143	-11.129	-11.107	-11.063	-10.985	-10.920	-10.836
0.55	-0.115	-9.948	-9.946	-9.942	-9.935	-9.928	-9.921	-9.914	-9.907	-9.897	-9.882	-9.853	-9.806	-9.769	-9.732
0.60	-0.078	-8.828	-8.826	-8.823	-8.817	-8.811	-8.806	-8.799	-8.794	-8.787	-8.777	-8.760	-8.736	-8.723	-8.720
0.65	-0.055	-0.309	-7.832	-7.829	-7.824	-7.819	-7.815	-7.810	-7.807	-7.801	-7.794	-7.784	-7.779	-7.785	-7.811
0.70	-0.040	-0.216	-0.491	-6.951	-6.945	-6.941	-6.937	-6.933	-6.930	-6.926	-6.922	-6.919	-6.929	-6.952	-7.002
0.75	-0.029	-0.156	-0.340	-6.173	-6.167	-6.162	-6.158	-6.155	-6.152	-6.149	-6.147	-6.149	-6.174	-6.213	-6.285
0.80	-0.022	-0.116	-0.246	-0.578	-5.475	-5.468	-5.462	-5.458	-5.455	-5.453	-5.452	-5.461	-5.501	-5.555	-5.648
0.85	-0.017	-0.088	-0.183	-0.400	-4.853	-4.841	-4.832	-4.826	-4.822	-4.820	-4.822	-4.839	-4.898	-4.969	-5.082
0.90	-0.013	-0.068	-0.140	-0.301	-0.744	-4.269	-4.249	-4.238	-4.232	-4.230	-4.236	-4.267	-4.351	-4.442	-4.578
0.93	-0.011	-0.058	-0.120	-0.254	-0.593	-1.219	-3.914	-3.894	-3.885	-3.884	-3.896	-3.941	-4.046	-4.151	-4.300
0.95	-0.010	-0.053	-0.109	-0.228	-0.517	-0.961	-3.697	-3.658	-3.647	-3.648	-3.669	-3.728	-3.851	-3.966	-4.125
0.97	-0.010	-0.048	-0.099	-0.206	-0.456	-0.797	-1.570	-3.406	-3.391	-3.401	-3.437	-3.517	-3.661	-3.788	-3.957
0.98	-0.009	-0.046	-0.094	-0.196	-0.429	-0.734	-1.270	-3.264	-3.247	-3.268	-3.318	-3.412	-3.569	-3.701	-3.875
0.99	-0.009	-0.044	-0.090	-0.186	-0.405	-0.680	-1.098	-3.093	-3.082	-3.126	-3.195	-3.306	-3.477	-3.616	-3.796
1.00	-0.008	-0.042	-0.086	-0.177	-0.382	-0.632	-0.977	-2.399	-2.868	-2.967	-3.067	-3.200	-3.387	-3.532	-3.717
1.01	-0.008	-0.040	-0.082	-0.169	-0.361	-0.590	-0.883	-1.306	-2.513	-2.784	-2.933	-3.094	-3.297	-3.450	-3.640
1.02	-0.008	-0.039	-0.078	-0.161	-0.342	-0.552	-0.807	-1.113	-1.655	-2.557	-2.790	-2.986	-3.209	-3.369	-3.565
1.05	-0.007	-0.034	-0.069	-0.140	-0.292	-0.460	-0.642	-0.820	-0.831	-1.443	-2.283	-2.655	-2.949	-3.134	-3.348
1.10	-0.005	-0.028	-0.055	-0.112	-0.229	-0.350	-0.470	-0.577	-0.640	-0.618	-1.241	-2.067	-2.534	-2.767	-3.013
1.15	-0.005	-0.023	-0.045	-0.091	-0.183	-0.275	-0.361	-0.437	-0.489	-0.502	-0.654	-1.471	-2.138	-2.428	-2.708
1.20	-0.004	-0.019	-0.037	-0.075	-0.149	-0.220	-0.286	-0.343	-0.385	-0.412	-0.447	-0.991	-1.767	-2.115	-2.430
1.30	-0.003	-0.013	-0.026	-0.052	-0.102	-0.148	-0.190	-0.226	-0.254	-0.282	-0.300	-0.481	-1.147	-1.569	-1.944
1.40	-0.002	-0.010	-0.019	-0.037	-0.072	-0.104	-0.133	-0.158	-0.178	-0.200	-0.220	-0.290	-0.730	-1.138	-1.544
1.50	-0.001	-0.007	-0.014	-0.027	-0.053	-0.076	-0.097	-0.115	-0.130	-0.147	-0.166	-0.206	-0.479	-0.823	-1.222
1.60	-0.001	-0.005	-0.011	-0.021	-0.040	-0.057	-0.073	-0.086	-0.098	-0.112	-0.129	-0.159	-0.334	-0.604	-0.969
1.70	-0.001	-0.004	-0.008	-0.016	-0.031	-0.044	-0.056	-0.067	-0.076	-0.087	-0.102	-0.127	-0.248	-0.456	-0.775
1.80	-0.001	-0.003	-0.006	-0.013	-0.024	-0.035	-0.044	-0.053	-0.060	-0.070	-0.083	-0.105	-0.195	-0.355	-0.628
1.90	-0.001	-0.003	-0.005	-0.010	-0.019	-0.028	-0.036	-0.043	-0.049	-0.057	-0.069	-0.089	-0.160	-0.286	-0.518
2.00	0.000	-0.002	-0.004	-0.008	-0.016	-0.023	-0.029	-0.035	-0.040	-0.048	-0.058	-0.077	-0.136	-0.238	-0.434
2.20	0.000	-0.001	-0.003	-0.006	-0.011	-0.016	-0.021	-0.025	-0.029	-0.035	-0.043	-0.060	-0.105	-0.178	-0.322
2.40	0.000	-0.001	-0.002	-0.004	-0.008	-0.012	-0.015	-0.019	-0.022	-0.027	-0.034	-0.048	-0.086	-0.143	-0.254
2.60	0.000	-0.001	-0.002	-0.003	-0.006	-0.009	-0.012	-0.015	-0.018	-0.021	-0.028	-0.041	-0.074	-0.120	-0.210
2.80	0.000	-0.001	-0.001	-0.003	-0.005	-0.008	-0.010	-0.012	-0.014	-0.018	-0.023	-0.025	-0.065	-0.104	-0.180
3.00	0.000	-0.001	-0.001	-0.002	-0.004	-0.006	-0.008	-0.010	-0.012	-0.015	-0.020	-0.031	-0.058	-0.093	-0.158
3.50	0.000	0.000	-0.001	-0.001	-0.003	-0.004	-0.006	-0.007	-0.009	-0.011	-0.015	-0.024	-0.046	-0.073	-0.122
4.00	0.000	0.000	-0.001	-0.001	-0.002	-0.003	-0.005	-0.006	-0.007	-0.009	-0.012	-0.020	-0.038	-0.060	-0.100

Appendix D - Lee-Kesler Tables (adapted from AIChE Journal 21, 515 - 522, 1975)

Table D.7: φ^0 liquid values

T_r	P_r 0.0100	0.0500	0.1000	0.2000	0.4000	0.6000	0.8000	1.0000	1.2000	1.5000	2.0000	3.0000	5.0000	7.0000	10.0000
0.30	0.0002	0.0000	0.0000	0.0000	0.0000	0.0000	0.0000	0.0000	0.0000	0.0000	0.0000	0.0000	0.0000	0.0000	0.0000
0.35	0.0034	0.0007	0.0003	0.0002	0.0001	0.0001	0.0001	0.0000	0.0000	0.0000	0.0000	0.0000	0.0000	0.0000	0.0000
0.40	0.0272	0.0055	0.0028	0.0014	0.0007	0.0005	0.0004	0.0003	0.0003	0.0003	0.0002	0.0002	0.0002	0.0002	0.0003
0.45	0.1321	0.0266	0.0135	0.0069	0.0036	0.0025	0.0020	0.0016	0.0014	0.0012	0.0010	0.0008	0.0008	0.0009	0.0012
0.50	0.4529	0.0912	0.0461	0.0235	0.0122	0.0085	0.0067	0.0055	0.0048	0.0041	0.0034	0.0028	0.0025	0.0027	0.0034
0.55	0.9817	0.2432	0.1227	0.0625	0.0325	0.0225	0.0176	0.0146	0.0127	0.0107	0.0089	0.0072	0.0063	0.0066	0.0080
0.60	0.9840	0.5383	0.2716	0.1384	0.0718	0.0497	0.0386	0.0321	0.0277	0.0234	0.0193	0.0154	0.0132	0.0135	0.0160
0.65	0.9886	0.9419	0.5212	0.2655	0.1374	0.0948	0.0738	0.0611	0.0527	0.0445	0.0364	0.0289	0.0244	0.0245	0.0282
0.70	0.9908	0.9528	0.9057	0.4560	0.2360	0.1626	0.1262	0.1045	0.0902	0.0759	0.0619	0.0488	0.0406	0.0402	0.0453
0.75	0.9931	0.9616	0.9226	0.7178	0.3715	0.2559	0.1982	0.1641	0.1413	0.1188	0.0966	0.0757	0.0625	0.0610	0.0673
0.80	0.9931	0.9683	0.9354	0.8730	0.5445	0.3750	0.2904	0.2404	0.2065	0.1738	0.1409	0.1102	0.0899	0.0867	0.0942
0.85	0.9954	0.9727	0.9462	0.8933	0.7534	0.5188	0.4018	0.3319	0.2858	0.2399	0.1945	0.1517	0.1227	0.1175	0.1256
0.90	0.9954	0.9772	0.9550	0.9099	0.8204	0.6823	0.5297	0.4375	0.3767	0.3162	0.2564	0.1995	0.1607	0.1524	0.1611
0.93	0.9954	0.9795	0.9594	0.9183	0.8375	0.7551	0.6109	0.5058	0.4355	0.3656	0.2972	0.2307	0.1854	0.1754	0.1841
0.95	0.9954	0.9817	0.9616	0.9226	0.8472	0.7709	0.6668	0.5521	0.4764	0.3999	0.3251	0.2523	0.2028	0.1910	0.2000
0.97	0.9954	0.9817	0.9638	0.9268	0.8570	0.7852	0.7112	0.5984	0.5164	0.4345	0.3532	0.2748	0.2203	0.2075	0.2163
0.98	0.9954	0.9817	0.9638	0.9290	0.8610	0.7925	0.7211	0.6223	0.5370	0.4529	0.3681	0.2864	0.2296	0.2158	0.2244
0.99	0.9977	0.9840	0.9661	0.9311	0.8650	0.7980	0.7295	0.6442	0.5572	0.4699	0.3828	0.2978	0.2388	0.2244	0.2328
1.00	0.9977	0.9840	0.9661	0.9333	0.8690	0.8035	0.7379	0.6668	0.5781	0.4875	0.3972	0.3097	0.2483	0.2328	0.2415
1.01	0.9977	0.9840	0.9683	0.9354	0.8730	0.8110	0.7464	0.6792	0.5970	0.5047	0.4121	0.3214	0.2576	0.2415	0.2500
1.02	0.9977	0.9840	0.9683	0.9376	0.8770	0.8166	0.7551	0.6902	0.6166	0.5224	0.4266	0.3334	0.2673	0.2506	0.2582
1.05	0.9977	0.9863	0.9705	0.9441	0.8872	0.8318	0.7762	0.7194	0.6607	0.5728	0.4710	0.3690	0.2958	0.2773	0.2844
1.10	0.9977	0.9886	0.9750	0.9506	0.9016	0.8531	0.8072	0.7586	0.7112	0.6412	0.5408	0.4285	0.3451	0.3228	0.3296
1.15	0.9977	0.9886	0.9795	0.9572	0.9141	0.8730	0.8318	0.7907	0.7499	0.6918	0.6026	0.4875	0.3954	0.3690	0.3750
1.20	0.9977	0.9908	0.9817	0.9616	0.9247	0.8892	0.8531	0.8166	0.7834	0.7328	0.6546	0.5420	0.4446	0.4150	0.4198
1.30	0.9977	0.9931	0.9863	0.9705	0.9419	0.9141	0.8872	0.8590	0.8318	0.7943	0.7345	0.6383	0.5383	0.5058	0.5093
1.40	0.9977	0.9931	0.9886	0.9772	0.9550	0.9333	0.9120	0.8892	0.8690	0.8395	0.7925	0.7145	0.6237	0.5902	0.5943
1.50	1.0000	0.9954	0.9908	0.9817	0.9638	0.9462	0.9290	0.9141	0.8974	0.8730	0.8375	0.7745	0.6966	0.6668	0.6714
1.60	1.0000	0.9954	0.9931	0.9863	0.9727	0.9572	0.9441	0.9311	0.9183	0.8995	0.8710	0.8222	0.7586	0.7328	0.7430
1.70	1.0000	0.9977	0.9954	0.9886	0.9772	0.9661	0.9550	0.9462	0.9354	0.9204	0.8995	0.8610	0.8091	0.7907	0.8054
1.80	1.0000	0.9977	0.9954	0.9908	0.9817	0.9727	0.9661	0.9572	0.9484	0.9376	0.9204	0.8913	0.8531	0.8414	0.8590
1.90	1.0000	0.9977	0.9954	0.9931	0.9863	0.9795	0.9727	0.9661	0.9594	0.9506	0.9376	0.9162	0.8872	0.8831	0.9057
2.00	1.0000	0.9977	0.9977	0.9954	0.9886	0.9840	0.9795	0.9727	0.9683	0.9616	0.9528	0.9354	0.9183	0.9183	0.9462
2.20	1.0000	1.0000	0.9977	0.9977	0.9931	0.9908	0.9886	0.9840	0.9817	0.9795	0.9727	0.9661	0.9616	0.9727	1.0093
2.40	1.0000	1.0000	1.0000	0.9977	0.9977	0.9954	0.9931	0.9931	0.9908	0.9908	0.9886	0.9863	0.9931	1.0116	1.0568
2.60	1.0000	1.0000	1.0000	1.0000	1.0000	0.9977	0.9977	0.9977	0.9977	0.9977	0.9977	1.0023	1.0162	1.0399	1.0889
2.80	1.0000	1.0000	1.0000	1.0000	1.0000	1.0000	1.0023	1.0023	1.0023	1.0046	1.0069	1.0116	1.0328	1.0593	1.1117
3.00	1.0000	1.0000	1.0000	1.0000	1.0023	1.0023	1.0046	1.0046	1.0069	1.0069	1.0116	1.0209	1.0423	1.0740	1.1298
3.50	1.0000	1.0000	1.0000	1.0023	1.0023	1.0046	1.0069	1.0093	1.0116	1.0139	1.0186	1.0304	1.0593	1.0914	1.1508
4.00	1.0000	1.0000	1.0000	1.0023	1.0046	1.0069	1.0093	1.0116	1.0139	1.0162	1.0233	1.0375	1.0666	1.0990	1.1588

Table D.8: φ^1 liquid values

T_r	P_r 0.0100	0.0500	0.1000	0.2000	0.4000	0.6000	0.8000	1.0000	1.2000	1.5000	2.0000	3.0000	5.0000	7.0000	10.0000
0.30	0.0000	0.0000	0.0000	0.0000	0.0000	0.0000	0.0000	0.0000	0.0000	0.0000	0.0000	0.0000	0.0000	0.0000	0.0000
0.35	0.0000	0.0000	0.0000	0.0000	0.0000	0.0000	0.0000	0.0000	0.0000	0.0000	0.0000	0.0000	0.0000	0.0000	0.0000
0.40	0.0000	0.0000	0.0000	0.0000	0.0000	0.0000	0.0000	0.0000	0.0000	0.0000	0.0000	0.0000	0.0000	0.0000	0.0000
0.45	0.0002	0.0002	0.0002	0.0002	0.0002	0.0002	0.0002	0.0002	0.0002	0.0002	0.0002	0.0001	0.0001	0.0001	0.0001
0.50	0.0014	0.0014	0.0014	0.0014	0.0014	0.0014	0.0013	0.0013	0.0013	0.0013	0.0012	0.0011	0.0009	0.0008	0.0006
0.55	0.9705	0.0069	0.0068	0.0068	0.0066	0.0065	0.0064	0.0063	0.0062	0.0061	0.0058	0.0053	0.0045	0.0039	0.0031
0.60	0.9795	0.0227	0.0226	0.0223	0.0220	0.0216	0.0213	0.0210	0.0207	0.0202	0.0194	0.0179	0.0154	0.0133	0.0108
0.65	0.9863	0.9311	0.0572	0.0568	0.0559	0.0551	0.0543	0.0536	0.0527	0.0516	0.0497	0.0461	0.0401	0.0350	0.0289
0.70	0.9908	0.9528	0.9036	0.1182	0.1163	0.1147	0.1131	0.1117	0.1102	0.1079	0.1040	0.0970	0.0851	0.0752	0.0629
0.75	0.9931	0.9683	0.9332	0.2112	0.2078	0.2050	0.2022	0.1995	0.1972	0.1932	0.1871	0.1754	0.1552	0.1387	0.1178
0.80	0.9954	0.9772	0.9550	0.9057	0.3302	0.3257	0.3212	0.3170	0.3133	0.3076	0.2978	0.2812	0.2512	0.2265	0.1954
0.85	0.9977	0.9863	0.9705	0.9375	0.4774	0.4708	0.4654	0.4592	0.4539	0.4457	0.4325	0.4093	0.3698	0.3365	0.2951
0.90	0.9977	0.9908	0.9795	0.9594	0.9141	0.6323	0.6250	0.6166	0.6095	0.5998	0.5834	0.5546	0.5058	0.4645	0.4130
0.93	0.9977	0.9931	0.9840	0.9705	0.9354	0.8953	0.7227	0.7145	0.7063	0.6950	0.6761	0.6457	0.5916	0.5470	0.4898
0.95	0.9977	0.9931	0.9885	0.9750	0.9484	0.9183	0.7888	0.7798	0.7691	0.7568	0.7379	0.7063	0.6501	0.6026	0.5432
0.97	1.0000	0.9954	0.9908	0.9795	0.9594	0.9354	0.9078	0.8414	0.8318	0.8185	0.7998	0.7656	0.7096	0.6607	0.5984
0.98	1.0000	0.9954	0.9908	0.9817	0.9638	0.9440	0.9225	0.8730	0.8630	0.8492	0.8298	0.7962	0.7379	0.6887	0.6266
0.99	1.0000	0.9954	0.9931	0.9840	0.9683	0.9528	0.9332	0.9036	0.8913	0.8790	0.8590	0.8241	0.7674	0.7178	0.6546
1.00	1.0000	0.9977	0.9931	0.9863	0.9727	0.9594	0.9440	0.9311	0.9204	0.9078	0.8872	0.8531	0.7962	0.7464	0.6823
1.01	1.0000	0.9977	0.9931	0.9885	0.9772	0.9638	0.9528	0.9462	0.9462	0.9333	0.9162	0.8831	0.8241	0.7745	0.7096
1.02	1.0000	0.9977	0.9954	0.9908	0.9795	0.9705	0.9616	0.9572	0.9661	0.9594	0.9419	0.9099	0.8531	0.8035	0.7379
1.05	1.0000	0.9977	0.9977	0.9954	0.9885	0.9863	0.9840	0.9840	0.9954	1.0186	1.0162	0.9886	0.9354	0.8872	0.8222
1.10	1.0000	1.0000	1.0000	1.0000	1.0023	1.0046	1.0093	1.0162	1.0280	1.0593	1.0990	1.1015	1.0617	1.0186	0.9572
1.15	1.0000	1.0000	1.0023	1.0046	1.0116	1.0186	1.0257	1.0375	1.0520	1.0814	1.1376	1.1858	1.1722	1.1403	1.0864
1.20	1.0000	1.0023	1.0046	1.0069	1.0163	1.0280	1.0399	1.0544	1.0691	1.0990	1.1588	1.2388	1.2647	1.2474	1.2050
1.30	1.0000	1.0023	1.0069	1.0116	1.0257	1.0399	1.0544	1.0715	1.0914	1.1194	1.1776	1.2853	1.3868	1.4125	1.4061
1.40	1.0000	1.0046	1.0069	1.0139	1.0304	1.0471	1.0642	1.0814	1.0990	1.1298	1.1858	1.2942	1.4488	1.5171	1.5524
1.50	1.0000	1.0046	1.0069	1.0163	1.0328	1.0496	1.0666	1.0864	1.1041	1.1350	1.1858	1.2942	1.4689	1.5740	1.6520
1.60	1.0000	1.0046	1.0069	1.0163	1.0328	1.0496	1.0691	1.0864	1.1041	1.1350	1.1858	1.2883	1.4689	1.5996	1.7140
1.70	1.0000	1.0046	1.0093	1.0163	1.0328	1.0496	1.0691	1.0864	1.1041	1.1324	1.1803	1.2794	1.4622	1.6033	1.7458
1.80	1.0000	1.0046	1.0069	1.0163	1.0328	1.0496	1.0666	1.0839	1.1015	1.1298	1.1749	1.2706	1.4488	1.5959	1.7620
1.90	1.0000	1.0046	1.0069	1.0163	1.0328	1.0496	1.0666	1.0814	1.0990	1.1272	1.1695	1.2618	1.4355	1.5849	1.7620
2.00	1.0000	1.0046	1.0069	1.0163	1.0304	1.0471	1.0642	1.0814	1.0965	1.1220	1.1641	1.2503	1.4191	1.5704	1.7539
2.20	1.0000	1.0046	1.0069	1.0139	1.0304	1.0447	1.0593	1.0765	1.0914	1.1143	1.1535	1.2331	1.3900	1.5346	1.7219
2.40	1.0000	1.0046	1.0069	1.0139	1.0280	1.0423	1.0568	1.0715	1.0864	1.1066	1.1429	1.2190	1.3614	1.4997	1.6866
2.60	1.0000	1.0023	1.0069	1.0139	1.0257	1.0399	1.0544	1.0666	1.0814	1.1015	1.1350	1.2023	1.3397	1.4689	1.6482
2.80	1.0000	1.0023	1.0069	1.0116	1.0257	1.0375	1.0496	1.0641	1.0765	1.0940	1.1272	1.1912	1.3183	1.4388	1.6144
3.00	1.0000	1.0023	1.0069	1.0116	1.0233	1.0352	1.0471	1.0593	1.0715	1.0889	1.1194	1.1803	1.3002	1.4158	1.5813
3.50	1.0000	1.0023	1.0046	1.0023	1.0209	1.0304	1.0423	1.0520	1.0617	1.0789	1.1041	1.1561	1.2618	1.3614	1.5101
4.00	1.0000	1.0023	1.0046	1.0093	1.0186	1.0280	1.0375	1.0471	1.0544	1.0691	1.0914	1.1403	1.2303	1.3213	1.4555

Appendix E: Antoine constants for vapor pressure of selected components

Values for ω are computed. Better values may exist in the open literature.

Paraffins — constants for $\ln(P^{sat} \text{ in kPa}) = A - B/[T(K)+C]$

Name	Formula	MW (g/mole)	A	B	C	T_{min} (K)	T_{max} (K)	T_c (K)	P_c (kPa)	ω
methane	CH_4	16.0430	13.9599	1072.62	3.004	95.1	190.6	190.6	4599	0.015
ethane	C_2H_6	30.0700	14.0358	1636.35	-11.307	122.5	305.1	305.3	4872	0.103
propane	C_3H_8	44.0970	14.2186	2101.35	-12.509	189.2	369.9	369.9	4251	0.157
n-butane	C_4H_{10}	58.1240	14.2315	2434.94	-19.866	213.1	425.1	425.1	3796	0.206
i-butane	C_4H_{10}	58.1240	14.0873	2291.61	-19.857	203.8	407.8	407.8	3629	0.189
n-pentane	C_5H_{12}	72.1510	14.2706	2730.90	-26.806	242.8	469.3	469.7	3369	0.258
isopentane	C_5H_{12}	72.1510	14.1921	2651.95	-24.500	235.6	460.4	460.4	3380	0.234
neopentane	C_5H_{12}	72.1510	14.1540	2552.02	-15.206	257.0	433.7	433.7	3196	0.200
n-hexane	C_6H_{14}	86.1780	14.3314	3004.30	-33.111	269.6	507.8	507.8	3042	0.307
isohexane	C_6H_{14}	86.1780	14.2553	2916.11	-31.318	262.2	497.7	497.7	3043	0.287
3-methylpentane	C_6H_{14}	86.1780	14.2678	2958.19	-30.412	264.5	504.6	504.6	3123	0.278
2,2-dimethylbutane	C_6H_{14}	86.1780	14.1273	2823.26	-26.490	252.3	489.1	489.1	3102	0.238
2,3-dimethylbutane	C_6H_{14}	86.1780	14.1655	2893.12	-28.734	259.5	500.2	500.2	3133	0.254
n-heptane	C_7H_{16}	100.2050	14.3748	3237.52	-40.249	294.3	540.1	541.2	2774	0.353
n-octane	C_8H_{18}	114.2320	14.3928	3431.68	-48.223	317.0	569.3	569.6	2507	0.401
n-nonane	C_9H_{20}	128.2590	14.4750	3659.61	-53.276	333.3	594.6	594.6	2280	0.451
n-decane	$C_{10}H_{22}$	142.2860	14.4359	3767.76	-64.075	325.4	617.7	617.7	2101	0.496

Olefins and unsaturates — constants for $\ln(P^{sat} \text{ in kPa}) = A - B/[T(K)+C]$

Name	Formula	MW (g/mole)	A	B	C	T_{min} (K)	T_{max} (K)	T_c (K)	P_c (kPa)	ω
ethylene	C_2H_4	28.0540	14.0885	1530.15	-8.150	138.2	282.0	282.4	5042	0.091
propylene	C_3H_6	42.0810	14.1480	1996.74	-16.512	173.1	363.6	364.2	4555	0.151
1-butene	C_4H_8	56.1080	14.2207	2371.23	-20.439	208.8	419.3	419.3	4010	0.198
cis-2-butene	C_4H_8	56.1080	14.2930	2458.70	-23.259	217.5	435.8	435.8	4236	0.209
trans-2-butene	C_4H_8	56.1080	14.3371	2478.74	-19.514	214.6	428.3	428.6	4019	0.215
isobutene	C_4H_8	56.1080	14.2268	2363.29	-20.691	208.4	418.1	418.1	4020	0.199
1,3 butadiene	C_4H_6	54.0920	14.3038	2416.48	-19.589	210.3	425.1	425.1	4270	0.194
1-pentene	C_5H_{10}	70.1350	14.3031	2707.19	-24.046	237.6	464.7	464.7	3547	0.241
1-hexene	C_6H_{12}	84.1620	14.3819	2991.80	-30.588	265.1	504.1	504.1	3210	0.292
1-heptene	C_7H_{14}	98.1890	14.3447	3192.69	-39.048	285.8	537.5	537.5	2852	0.338
1-octene	C_8H_{16}	112.2160	14.7027	3631.61	-35.034	312.9	566.6	566.6	2676	0.404
styrene	C_8H_8	104.1520	14.5590	3739.12	-43.032	332.3	635.2	635.2	3880	0.309
acetylene	C_2H_2	26.0380	14.9080	1865.61	-7.066	189.2	308.4	308.4	6206	0.199
methylacetylene	C_3H_4	40.0650	14.4245	2182.59	-27.133	170.0	402.4	402.7	5658	0.209

Aromatics and Cyclics — constants for $\ln(P^{sat} \text{ in kPa}) = A - B/[T(K)+C]$

Name	Formula	MW (g/mole)	A	B	C	T_{min} (K)	T_{max} (K)	T_c (K)	P_c (kPa)	ω
cyclopentane	C_5H_{10}	70.1350	14.3549	2887.91	-26.343	253.3	511.0	511.7	4510	0.199
cyclohexane	C_6H_{12}	84.1620	14.1996	3063.83	-34.723	279.8	553.2	553.6	4075	0.215
cyclohexene	C_6H_{10}	82.1460	14.3640	3169.41	-31.601	276.3	560.5	560.5	4420	0.223
benzene	C_6H_6	78.1140	14.3578	3087.10	-37.016	278.7	561.6	562.0	4906	0.217
toluene	C_7H_8	92.1410	14.5128	3451.60	-35.692	299.2	591.8	591.7	4126	0.273
o-xylene	C_8H_{10}	106.1680	14.6047	3773.32	-40.394	331.2	630.4	630.4	3746	0.318
m-xylene	C_8H_{10}	106.1680	14.6244	3734.82	-39.649	327.1	616.9	616.9	3540	0.334
p-xylene	C_8H_{10}	106.1680	14.6316	3752.14	-37.493	326.1	616.2	616.2	3524	0.330
ethylbenzene	C_8H_{10}	106.1680	14.4918	3631.28	-42.184	319.8	617.1	617.1	3616	0.310
phenol	C_6H_6O	94.1130	14.9343	3925.39	-75.351	386.7	694.3	694.3	5540	0.409

Appendix E: Antoine constants for vapor pressure of selected components

Oxygen Containing Organics

constants for $\ln(P^{sat}$ in kPa$)=A-B/[T(K)+C]$

Name	Formula	MW (g/mole)	A	B	C	T_{min} (K)	T_{max} (K)	T_c (K)	P_c (kPa)	ω
methanol	CH_3OH	32.0420	16.6878	3716.27	-29.846	288.3	513.4	513.4	8220	0.565
ethanol	C_2H_5OH	46.0690	16.0810	3338.53	-59.953	302.2	513.9	515.4	6300	0.635
ethylene oxide	C_2H_4O	44.0530	14.8213	2632.30	-25.737	238.3	469.0	469.0	7220	0.200
ethylene glycol	$C_2H_6O_2$	62.0680	15.5897	4116.92	-95.541	405.8	719.0	719.3	8200	0.526
1-propanol	C_3H_7OH	60.0960	15.0149	2865.97	-94.167	319.5	536.7	536.7	5120	0.609
2-propanol	C_3H_7OH	60.0960	15.0984	2772.37	-90.461	295.8	508.3	508.3	4750	0.657
1-butanol	C_4H_9OH	74.1230	14.6620	2871.74	-104.604	324.5	561.9	561.9	4420	0.597
2-butanol	C_4H_9OH	74.1230	14.5903	2707.75	-101.153	310.0	533.2	533.2	4240	0.613
acetone	C_3H_6O	58.0800	14.8771	3087.69	-28.820	262.0	507.5	508.1	4690	0.313
diethylether	$C_4H_{10}O$	74.1230	14.3944	2700.14	-31.842	243.4	466.8	466.8	3649	0.287
methylethylketone	C_4H_8O	72.1070	14.7327	3210.46	-35.865	280.9	536.5	536.5	4172	0.327
formaldehyde	CH_2O	30.0260	14.7929	2367.78	-21.044	211.0	416.5	416.5	6830	0.211
acetaldehyde	C_2H_4O	44.0530	14.7934	2848.98	-14.948	232.3	462.0	462.1	4800	0.267
acetic acid	$C_2H_4O_2$	60.0520	16.3798	4538.75	-5.963	328.9	593.0	593.0	5786	0.467

Miscellaneous Hydrocarbons

constants for $\ln(P^{sat}$ in kPa$)=A-B/[T(K)+C]$

Name	Formula	MW (g/mole)	A	B	C	T_{min} (K)	T_{max} (K)	T_c (K)	P_c (kPa)	ω
methylamine	CH_5N	31.0580	15.1174	2479.97	-30.872	224.8	430.6	430.6	7610	0.297
ethylamine	C_2H_7N	45.0850	14.5161	2426.43	-44.839	243.7	456.5	456.5	5640	0.283
acetonitrile	C_2H_3N	41.0530	15.6472	3900.54	-1.783	294.6	545.4	545.4	4866	0.350
methylchloride	CH_3Cl	50.4850	14.6540	2346.20	-15.619	205.8	416.2	416.2	6714	0.158
dichloromethane	CH_2Cl_2	84.9270	14.8171	2930.28	-26.170	260.6	501.4	508.0	6360	0.232
trichloromethane	$CHCl_3$	119.3690	14.8810	3232.90	-20.011	277.4	536.1	536.1	5600	0.238
carbon tetrachloride	CCl_4	153.8110	14.2869	3090.47	-30.993	275.4	556.4	556.5	4580	0.200
fluoromethane	CH_3F	34.0330	14.8879	1927.31	-7.557	160.9	317.3	317.3	5906	0.207
difluoromethane	CH_2F_2	52.0230	15.0640	2153.74	-15.686	184.7	351.3	351.3	5783	0.283
R-134a	$C_2H_2F_4$	102.0320	14.6714	2213.27	-27.279	197.0	374.1	374.2	4059	0.333

Inorganics

constants for $\ln(P^{sat}$ in kPa$)=A-B/[T(K)+C]$

Name	Formula	MW (g/mole)	A	B	C	T_{min} (K)	T_{max} (K)	T_c (K)	P_c (kPa)	ω
water	H_2O	18.0150	16.8116	4245.68	-25.008	354.5	647.1	647.1	22064	0.351
hydrogen	H_2	2.0160	12.2470	195.49	5.210	13.9	33.1	33.1	1296	-0.217
oxygen	O_2	31.9990	13.9412	835.35	-0.799	72.7	154.6	154.6	5046	0.026
nitrogen	N_2	28.0140	13.6275	691.98	-0.679	63.4	125.9	126.2	3396	0.041
carbon dioxide	CO_2*	44.0090	20.2910	2997.23	-3.494	150.0	303.7	304.1	7377	0.272
carbon monoxide	CO	28.0140	13.6758	725.96	-1.613	68.2	132.9	132.9	3498	0.053
helium	He	4.0030	10.5973	37.26	2.007	2.5	5.2	5.2	228	-0.377
argon	Ar	39.9480	14.0827	872.17	4.929	83.8	150.7	150.7	4863	0.001
ammonia	NH_3	17.0310	15.6376	2447.19	-17.977	211.8	405.4	405.5	11359	0.261
chlorine	Cl_2	70.9000	14.5517	2258.80	-12.250	197.0	416.9	416.9	7980	0.091
hydrogen chloride	HCl	36.4580	14.8883	1872.88	-6.112	159.1	324.7	324.7	8260	0.129
hydrogen sulfide	H_2S	34.0860	14.9418	2169.88	-2.229	212.6	372.5	373.1	8999	0.104
hydrogen cyanide	HCN	27.0260	16.2288	3663.72	16.250	259.9	456.6	456.7	4970	0.385
carbon disulfide	CS_2	76.1310	15.4247	3658.30	17.774	262.2	552.2	552.2	8430	0.157
sulfur dioxide	SO_2	64.0580	15.0567	2456.33	-28.145	221.0	430.1	430.6	7880	0.261
sulfur trioxide	SO_3	80.0630	14.2879	2054.31	-103.564	333.0	483.0	491.5	8491	0.434
nitric oxide	NO	30.0060	16.3145	1182.38	-20.265	109.8	177.0	177.0	6430	0.678
nitrogen dioxide	NO_2	46.0060	12.3245	1244.84	-131.930	217.6	293.6	431.0	10128	0.838
nitrous oxide	N_2O	44.0130	14.8486	1794.87	-9.075	182.4	309.2	309.5	7245	0.166
dinitrogen tetroxide	N_2O_4	92.0100	21.2697	5956.82	63.334	262.8	431.3	431.3	10230	0.856

* sublimes

Appendix F: Enthalpies and Free Energies of Formation

Enthalpies and Free Energies of Formation at 298.15 K and 1 bar.

Paraffins

Name	Formula	MW (g/mole)	State	$\Delta H_f°$ (J/mole)	$\Delta G_f°$ (J/mole)	State	$\Delta H_f°$ (J/mole)	$\Delta G_f°$ (J/mole)
methane	CH_4	16.0430	gas	-74900	-50600			
ethane	C_2H_6	30.0700	gas	-84500	-33000			
propane	C_3H_8	44.0970	gas	-104000	-23000			
n-butane	C_4H_{10}	58.1240	gas	-127200	-17000			
i-butane	C_4H_{10}	58.1240	gas	-135600	-21000			
n-pentane	C_5H_{12}	72.1510	gas	-146400	-8400	liq	-173200	-9500
isopentane	C_5H_{12}	72.1510	gas	-154400	-14800	liq	-179900	-15200
neopentane	C_5H_{12}	72.1510	gas	-166000	-15200			
n-hexane	C_6H_{14}	86.1780	gas	-167200	-300	liq	-198800	-4400
isohexane	C_6H_{14}	86.1780	gas	-174300	-2800			
3-methylpentane	C_6H_{14}	86.1780	gas	-171600	-2800			
2,2-dimethylbutane	C_6H_{14}	86.1780	gas	-185600	2480			
2,3-dimethylbutane	C_6H_{14}	86.1780	gas	-177800	-5240			
n-heptane	C_7H_{16}	100.2050	gas	-187800	8060			
n-octane	C_8H_{18}	114.2320	gas	-208700	16480			
n-nonane	C_9H_{20}	128.2590	gas	-228300	24900			
n-decane	$C_{10}H_{22}$	142.2860	gas	-249700	33320			

Olefins and unsaturates

Name	Formula	MW (g/mole)	State	$\Delta H_f°$ (J/mole)	$\Delta G_f°$ (J/mole)	State	$\Delta H_f°$ (J/mole)	$\Delta G_f°$ (J/mole)
ethylene	C_2H_4	28.0540	gas	52467	68421			
propylene	C_3H_6	42.0810	gas	20400	62700			
1-butene	C_4H_8	56.1080	gas	-100	71300			
cis-2-butene	C_4H_8	56.1080	gas	-7000	65900			
trans-2-butene	C_4H_8	56.1080	gas	-11200	63000			
isobutene	C_4H_8	56.1080	gas	-16900	58100			
1,3 butadiene	C_4H_6	54.0920	gas	110200	150700			
1-pentene	C_5H_{10}	70.1350	gas	-29300	79060			
1-hexene	C_6H_{12}	84.1620	gas	-44800	87480			
1-heptene	C_7H_{14}	98.1890	gas	-63000	95900			
1-octene	C_8H_{16}	112.2160	gas	-82930	104320			
styrene	C_8H_8	104.1520	gas	-15100	216730			
acetylene	C_2H_2	26.0380	gas	227400	209300			
methylacetylene	C_3H_4	40.0650	gas	185400	197450			

Aromatics and Cyclics

Name	Formula	MW (g/mole)	State	$\Delta H_f°$ (J/mole)	$\Delta G_f°$ (J/mole)	State	$\Delta H_f°$ (J/mole)	$\Delta G_f°$ (J/mole)
cyclopentane	C_5H_{10}	70.1350	gas	-77400	38600	liq	-105900	36400
cyclohexane	C_6H_{12}	84.1620	gas	-123000	31800	liq	-156200	26700
cyclohexene	C_6H_{10}	82.1460	gas	-5400	106900	liq	-38800	101600
benzene	C_6H_6	78.1140	gas	82800	129700	liq	49000	124300
toluene	C_7H_8	92.1410	gas	50000	122000	liq	12000	113800
o-xylene	C_8H_{10}	106.1680				liq	-24400	110500
m-xylene	C_8H_{10}	106.1680				liq	-25400	107700
p-xylene	C_8H_{10}	106.1680				liq	-24400	110000
ethylbenzene	C_8H_{10}	106.1680				liq	-12500	119700
phenol	C_6H_6O	94.1130	gas	-96400	-32900			

Appendix F: Enthalpies and Free Energies of Formation

Oxygen Containing Organics

Name	Formula	MW (g/mole)	State	ΔH_f° (J/mole)	ΔG_f° (J/mole)	State	ΔH_f° (J/mole)	ΔG_f° (J/mole)
methanol	CH_3OH	32.0420	gas	-201200	-162500	liq	-238600	-166200
ethanol	C_2H_5OH	46.0690	gas	-234800	-168300	liq	-277000	-174100
ethylene oxide	C_2H_4O	44.0530	gas	-52600	-13100	liq	-77400	-11400
ethylene glycol	$C_2H_6O_2$	62.0680	gas	-394400	-307680			
1-propanol	C_3H_7OH	60.0960	gas	-257500	-163000	liq	-304600	-170700
2-propanol	$C3H7OH$	60.0960	gas	-272600	-173600	liq	-318000	-180400
1-butanol	C_4H_9OH	74.1230	gas	-274400	-150700	liq	-325800	-161100
2-butanol	C_4H_9OH	74.1230	gas	-292300	-167300			
acetone	C_3H_6O	58.0800	gas	-217600	-153100	liq	-248100	-155400
diethylether	$C_4H_{10}O$	74.1230	gas	-252200	-122300	liq	-279500	-122900
methylethylketone	C_4H_8O	72.1070	gas	-258700	-135300			
formaldehyde	CH_2O	30.0260	gas	-115900	-109900			
acetaldehyde	C_2H_4O	44.0530	gas	-166400	-133300			
acetic acid	$C_2H_4O_2$	60.0520	gas	-435400	-299780			

Miscellaneous Hydrocarbons

Name	Formula	MW (g/mole)	State	ΔH_f° (J/mole)	ΔG_f° (J/mole)	State	ΔH_f° (J/mole)	ΔG_f° (J/mole)
methylamine	CH_5N	31.0580	gas	-23000	32300			
ethylamine	C_2H_7N	45.0850	gas	-46000	37300			
acetonitrile	C_2H_3N	41.0530	gas	87900	105600	liq	53100	98900
methylchloride	CH_3Cl	50.4850	gas	-86300	-62900			
dichloromethane	CH_2Cl_2	84.9270	gas	-95400	-68900			
trichloromethane	$CHCl_3$	119.3690	gas	-101300	-68500	liq	-132200	-71800
carbon tetrachloride	CCl_4	153.8110	gas	-95800	-53600			
fluoromethane	CH_3F	34.0330	gas	-233900	-210000			
difluoromethane	CH_2F_2	52.0230	gas	-452210	-432080			
R-134a	$C_2H_2F_4$	102.0320	gas	-879600	-818160			

Inorganics

Name	Formula	MW (g/mole)	State	ΔH_f° (J/mole)	ΔG_f° (J/mole)	State	ΔH_f° (J/mole)	ΔG_f° (J/mole)
water	H_2O	18.0150	gas	-241800	-228400	liq	-285800	-237200
hydrogen	H_2	2.0160	gas	0	0			
oxygen	O_2	31.9990	gas	0	0			
nitrogen	N_2	28.0140	gas	0	0			
carbon dioxide	CO_2	44.0090	gas	-393300	-394600			
carbon monoxide	CO	28.0140	gas	-110500	-137200			
helium	He	4.0030	gas	0	0			
argon	Ar	39.9480	gas	0	0			
ammonia	NH_3	17.0310	gas	-45600	-16300			
chlorine	Cl_2	70.9000	gas	0	0			
hydrogen chloride	HCl	36.4580	gas	-92500	-95400			
hydrogen sulfide	H_2S	34.0860	gas	-20100	-33100			
hydrogen cyanide	HCN	27.0260	gas	130500	120100			
carbon disulfide	CS_2	76.1310	gas	117200	66900			
sulfur dioxide	SO_2	64.0580	gas	-296200	-300400			
sulfur trioxide	SO_3	80.0630	gas	-395400	-370300			
nitric oxide	NO	30.0060	gas	90400	86600			
nitrogen dioxide	NO_2	46.0060	gas	33900	51900			
nitrous oxide	N_2O	44.0130	gas	81600	103800			
dinitrogen tetroxide	N_2O_4	92.0100	gas	9600	98300			

Appendix G - Steam Tables

Appendix G.1 - Saturated Steam Temperature Increment - SI Units

T °C	P kPa	v^L m³/kg	v^V m³/kg	h^L kJ/kg	h^V kJ/kg	s^L kJ/kg-K	s^V kJ/kg-K
0.01	0.6117	0.0010002	205.990	0.0006	2500.9	0.000000	9.1555
5	0.8726	0.0010001	147.011	21.020	2510.1	0.076254	9.0248
10	1.2282	0.0010004	106.303	42.021	2519.2	0.15109	8.8998
15	1.7058	0.0010009	77.876	62.981	2528.3	0.22446	8.7803
20	2.3393	0.0010018	57.757	83.914	2537.4	0.29648	8.6660
25	3.1699	0.0010030	43.337	104.83	2546.5	0.36722	8.5566
30	4.2470	0.0010044	32.879	125.73	2555.5	0.43675	8.4520
35	5.6290	0.0010060	25.205	146.63	2564.5	0.50513	8.3517
40	7.3849	0.0010079	19.515	167.53	2573.5	0.57240	8.2555
45	9.5950	0.0010099	15.252	188.43	2582.4	0.63861	8.1633
50	12.352	0.0010121	12.027	209.34	2591.3	0.70381	8.0748
55	15.762	0.0010145	9.5639	230.26	2600.1	0.76802	7.9898
60	19.946	0.0010171	7.6669	251.18	2608.8	0.83129	7.9081
65	25.042	0.0010199	6.1935	272.12	2617.5	0.89365	7.8296
70	31.201	0.0010228	5.0396	293.07	2626.1	0.95513	7.7540
75	38.595	0.0010258	4.1290	314.03	2634.6	1.0158	7.6812
80	47.414	0.0010291	3.4052	335.01	2643.0	1.0756	7.6111
85	57.867	0.0010324	2.8258	356.01	2651.3	1.1346	7.5434
90	70.182	0.0010359	2.3590	377.04	2659.5	1.1929	7.4781
95	84.608	0.0010396	1.9806	398.09	2667.6	1.2504	7.4151
100	101.42	0.0010435	1.6718	419.17	2675.6	1.3072	7.3541
105	120.90	0.0010474	1.4184	440.27	2683.4	1.3633	7.2952
110	143.38	0.0010516	1.2093	461.42	2691.1	1.4188	7.2381
115	169.18	0.0010559	1.0358	482.59	2698.6	1.4737	7.1828
120	198.67	0.0010603	0.89119	503.81	2705.9	1.5279	7.1291
125	232.24	0.0010649	0.77000	525.07	2713.1	1.5816	7.0770
130	270.28	0.0010697	0.66800	546.38	2720.1	1.6346	7.0264
135	313.23	0.0010746	0.58173	567.74	2726.9	1.6872	6.9772
140	361.54	0.0010798	0.50847	589.16	2733.4	1.7392	6.9293
145	415.68	0.0010850	0.44597	610.64	2739.8	1.7907	6.8826
150	476.16	0.0010905	0.39245	632.18	2745.9	1.8418	6.8371
155	543.50	0.0010962	0.34646	653.79	2751.8	1.8924	6.7926
160	618.23	0.0011020	0.30679	675.47	2757.4	1.9426	6.7491
165	700.93	0.0011080	0.27243	697.24	2762.8	1.9923	6.7066
170	792.19	0.0011143	0.24259	719.08	2767.9	2.0417	6.6650
175	892.60	0.0011207	0.21658	741.02	2772.7	2.0906	6.6241
180	1002.8	0.0011274	0.19384	763.05	2777.2	2.1392	6.5840
185	1123.5	0.0011343	0.17390	785.19	2781.4	2.1875	6.5447

Appendix G.1 - Saturated Steam Temperature Increment - English Units

T °F	P psia	v^L ft³/lbm	v^V ft³/lbm	h^L BTU/lbm	h^V BTU/lbm	s^L BTU/lbm-°F	s^V BTU/lbm-°F
32.02	0.08872	0.016022	3299.5	0.002279	1075.2	0.000004	2.1867
35	0.09998	0.016020	2945.4	3.0039	1076.5	0.006090	2.1762
40	0.12173	0.016020	2443.3	8.0323	1078.7	0.016204	2.1589
50	0.17814	0.016024	1702.8	18.066	1083.1	0.036086	2.1257
60	0.25640	0.016035	1206.0	28.078	1087.4	0.055539	2.0940
70	0.36336	0.016052	867.08	38.075	1091.8	0.074594	2.0639
80	0.50747	0.016074	632.39	48.065	1096.1	0.093277	2.0353
90	0.69904	0.016100	467.40	58.050	1100.4	0.11161	2.0079
100	0.95051	0.016131	349.83	68.033	1104.7	0.12961	1.9819
110	1.2767	0.016166	264.96	78.017	1109.0	0.14729	1.9570
120	1.6950	0.016205	202.95	88.003	1113.2	0.16466	1.9333
130	2.2259	0.016247	157.09	97.993	1117.4	0.18175	1.9106
140	2.8930	0.016293	122.82	107.99	1121.6	0.19855	1.8888
150	3.7232	0.016342	96.927	117.99	1125.7	0.21509	1.8680
160	4.7472	0.016394	77.184	128.00	1129.8	0.23136	1.8481
170	5.9998	0.016449	61.981	138.01	1133.9	0.24739	1.8290
180	7.5195	0.016508	50.168	148.04	1137.9	0.26319	1.8106
190	9.3496	0.016569	40.917	158.08	1141.8	0.27875	1.7930
200	11.538	0.016633	33.609	168.13	1145.7	0.29409	1.7760
210	14.136	0.016701	27.795	178.19	1149.5	0.30922	1.7597
220	17.201	0.016771	23.133	188.27	1153.3	0.32415	1.7440
230	20.795	0.016845	19.371	198.37	1156.9	0.33888	1.7288
240	24.986	0.016921	16.314	208.49	1160.5	0.35343	1.7141
250	29.844	0.017001	13.815	218.63	1164.0	0.36779	1.7000
260	35.447	0.017084	11.759	228.79	1167.4	0.38199	1.6863
270	41.878	0.017170	10.058	238.98	1170.7	0.39602	1.6730
280	49.222	0.017259	8.6430	249.20	1173.9	0.40990	1.6601
290	57.574	0.017352	7.4599	259.45	1177.0	0.42362	1.6476
300	67.029	0.017448	6.4658	269.73	1180.0	0.43721	1.6354
310	77.691	0.017548	5.6262	280.05	1182.8	0.45065	1.6236
320	89.667	0.017652	4.9142	290.40	1185.5	0.46397	1.6120
330	103.07	0.017760	4.3074	300.80	1188.0	0.47717	1.6007
340	118.02	0.017872	3.7883	311.24	1190.5	0.49025	1.5897
350	134.63	0.017988	3.3425	321.73	1192.7	0.50322	1.5789
360	153.03	0.018108	2.9580	332.28	1194.8	0.51608	1.5684
370	173.36	0.018233	2.6252	342.88	1196.7	0.52885	1.5580
380	195.74	0.018363	2.3361	353.54	1198.5	0.54153	1.5478
390	220.33	0.018498	2.0842	364.26	1200.1	0.55413	1.5378

Appendix G.1 - Saturated Steam Temperature Increment - SI Units

T °C	P kPa	v^L m³/kg	v^V m³/kg	h^L kJ/kg	h^V kJ/kg	s^L kJ/kg-K	s^V kJ/kg-K
190	1255.2	0.0011414	0.15636	807.43	2785.3	2.2355	6.5059
195	1398.8	0.0011489	0.14089	829.79	2788.8	2.2832	6.4678
200	1554.9	0.0011565	0.12721	852.27	2792.0	2.3305	6.4302
205	1724.3	0.0011645	0.11508	874.88	2794.8	2.3777	6.3930
210	1907.7	0.0011727	0.10429	897.63	2797.3	2.4245	6.3563
215	2105.8	0.0011813	0.094679	920.53	2799.3	2.4712	6.3200
220	2319.6	0.0011902	0.086096	943.58	2800.9	2.5177	6.2840
225	2549.7	0.0011994	0.078401	966.80	2802.1	2.5640	6.2483
230	2797.1	0.0012090	0.071505	990.19	2802.9	2.6101	6.2128
235	3062.5	0.0012190	0.065300	1013.8	2803.2	2.6561	6.1775
240	3346.9	0.0012295	0.059705	1037.6	2803.0	2.7020	6.1423
245	3651.2	0.0012404	0.054654	1061.5	2802.2	2.7478	6.1072
250	3976.2	0.0012517	0.050083	1085.8	2800.9	2.7935	6.0721
255	4322.9	0.0012636	0.045939	1110.2	2799.1	2.8392	6.0369
260	4692.3	0.0012761	0.042173	1135.0	2796.6	2.8849	6.0016
265	5085.3	0.0012892	0.038746	1160.0	2793.5	2.9307	5.9661
270	5503.0	0.0013030	0.035621	1185.3	2789.7	2.9765	5.9304
275	5946.4	0.0013175	0.032765	1210.9	2785.2	3.0224	5.8944
280	6416.6	0.0013328	0.030152	1236.9	2779.9	3.0685	5.8579
285	6914.7	0.0013491	0.027756	1263.2	2773.7	3.1147	5.8209
290	7441.8	0.0013663	0.025555	1290.0	2766.7	3.1612	5.7834
295	7999.1	0.0013846	0.023529	1317.3	2758.7	3.2080	5.7451
300	8587.9	0.0014042	0.021660	1345.0	2749.6	3.2552	5.7059
305	9209.4	0.0014252	0.019933	1373.3	2739.4	3.3028	5.6657
310	9865.1	0.0014479	0.018335	1402.2	2727.9	3.3510	5.6244
315	10556	0.0014724	0.016851	1431.8	2715.1	3.3998	5.5816
320	11284	0.0014990	0.015471	1462.2	2700.6	3.4494	5.5372
325	12051	0.0015283	0.014183	1493.5	2684.3	3.5000	5.4908
330	12858	0.0015606	0.012979	1525.9	2666.0	3.5518	5.4422
335	13707	0.0015967	0.011847	1559.5	2645.4	3.6050	5.3906
340	14601	0.0016375	0.010781	1594.5	2621.8	3.6601	5.3356
345	15541	0.0016846	0.0097694	1631.5	2594.9	3.7176	5.2762
350	16529	0.0017400	0.0088020	1670.9	2563.6	3.7784	5.2110
355	17570	0.0018079	0.0078684	1713.7	2526.6	3.8439	5.1380
360	18666	0.0018954	0.0069493	1761.7	2481.5	3.9167	5.0536
365	19821	0.0020172	0.0060114	1817.8	2422.9	4.0014	4.9497
370	21044	0.0022152	0.0049544	1890.7	2334.5	4.1112	4.8012
373.93	22060	0.0029378	0.0032953	2054.8	2116.6	4.3615	4.4570

Appendix G.1 - Saturated Steam Temperature Increment - English Units

T °F	P psia	v^L ft³/lbm	v^V ft³/lbm	h^L BTU/lbm	h^V BTU/lbm	s^L BTU/lbm-°F	s^V BTU/lbm-°F
400	247.26	0.018639	1.8638	375.05	1201.4	0.56665	1.5279
410	276.68	0.018785	1.6706	385.91	1202.6	0.57909	1.5182
420	308.76	0.018938	1.5006	396.85	1203.6	0.59147	1.5085
430	343.64	0.019097	1.3505	407.88	1204.3	0.60380	1.4990
440	381.48	0.019263	1.2177	418.99	1204.8	0.61607	1.4895
450	422.46	0.019437	1.0999	430.20	1205.1	0.62830	1.4802
460	466.75	0.019619	0.9950	441.51	1205.1	0.64049	1.4708
470	514.52	0.019809	0.90155	452.93	1204.9	0.65265	1.4615
480	565.95	0.020010	0.81793	464.47	1204.3	0.66479	1.4522
490	621.23	0.020220	0.74294	476.14	1203.5	0.67693	1.4428
500	680.55	0.020442	0.67554	487.94	1202.3	0.68906	1.4335
510	744.11	0.020675	0.61485	499.90	1200.8	0.70120	1.4240
520	812.10	0.020923	0.56007	512.01	1198.9	0.71336	1.4146
530	884.74	0.021185	0.51049	524.30	1196.7	0.72555	1.4050
540	962.24	0.021464	0.46553	536.78	1194.0	0.73779	1.3952
550	1044.8	0.021761	0.42466	549.47	1190.9	0.75010	1.3853
560	1132.7	0.022080	0.38742	562.40	1187.2	0.76248	1.3753
570	1226.2	0.022422	0.35341	575.58	1183.0	0.77497	1.3649
580	1325.5	0.022792	0.32227	589.05	1178.3	0.78759	1.3543
590	1430.8	0.023193	0.29369	602.84	1172.8	0.80036	1.3434
600	1542.5	0.023631	0.26739	617.01	1166.6	0.81334	1.3320
610	1660.9	0.024113	0.24311	631.60	1159.6	0.82655	1.3201
620	1786.2	0.024648	0.22062	646.68	1151.5	0.84006	1.3077
630	1918.9	0.025247	0.19970	662.35	1142.4	0.85394	1.2945
640	2059.2	0.025930	0.18017	678.73	1131.8	0.86830	1.2803
650	2207.8	0.026720	0.16180	696.01	1119.7	0.88328	1.2651
660	2364.9	0.027661	0.14439	714.48	1105.3	0.89914	1.2482
670	2531.2	0.028826	0.12768	734.63	1088.2	0.91628	1.2292
680	2707.3	0.030362	0.11132	757.38	1066.8	0.93548	1.2070
690	2894.0	0.032581	0.094563	784.50	1038.3	0.95822	1.1790
700	3093.0	0.036653	0.074794	822.45	990.99	0.98998	1.1353
705	3199.8	0.048464	0.051122	887.07	905.74	1.0449	1.0609

adapted from NIST data

The reference state for all portions of the steam table is u (internal energy) and s (entropy) are both zero for the saturated liquid at the triple point of water.

Appendix G.2 - Saturated Steam Pressure Increment - SI Units

T °C	P kPa	v^L m³/kg	v^V m³/kg	h^L kJ/kg	h^V kJ/kg	s^L kJ/kg-K	s^V kJ/kg-K
0.01	0.61165	0.001000	205.99	0.000612	2500.9	0.000000	9.1555
1.88	0.7	0.001000	181.22	7.8902	2504.3	0.028784	9.1058
3.76	0.8	0.001000	159.64	15.809	2507.8	0.057479	9.0567
5.44	0.9	0.001000	142.76	22.889	2510.9	0.082967	9.0135
6.97	1	0.001000	129.18	29.299	2513.7	0.10591	8.9749
17.50	2	0.001001	66.988	73.428	2532.9	0.26056	8.7226
24.08	3	0.001003	45.654	100.98	2544.8	0.35429	8.5764
28.96	4	0.001004	34.791	121.39	2553.7	0.42239	8.4734
32.87	5	0.001005	28.185	137.75	2560.7	0.47620	8.3938
36.16	6	0.001006	23.733	151.48	2566.6	0.52082	8.3290
39.00	7	0.001008	20.525	163.35	2571.7	0.55903	8.2745
41.51	8	0.001008	18.099	173.84	2576.2	0.59249	8.2273
43.76	9	0.001009	16.199	183.25	2580.2	0.6223	8.1858
45.81	10	0.001010	14.670	191.81	2583.9	0.6492	8.1488
60.06	20	0.001017	7.6482	251.42	2608.9	0.8320	7.9072
69.10	30	0.001022	5.2285	289.27	2624.5	0.9441	7.7675
75.86	40	0.001026	3.9930	317.62	2636.1	1.0261	7.6690
81.32	50	0.001030	3.2400	340.54	2645.2	1.0912	7.5930
85.93	60	0.001033	2.7317	359.91	2652.9	1.1454	7.5311
89.93	70	0.001036	2.3648	376.75	2659.4	1.1921	7.4790
93.49	80	0.001038	2.0871	391.71	2665.2	1.2330	7.4339
96.69	90	0.001041	1.8694	405.20	2670.3	1.2696	7.3943
99.61	100	0.001043	1.6939	417.50	2674.9	1.3028	7.3588
120.21	200	0.001061	0.88566	504.70	2706.2	1.5302	7.1269
133.52	300	0.001073	0.60577	561.43	2724.9	1.6717	6.9916
143.61	400	0.001084	0.46238	604.65	2738.1	1.7765	6.8955
151.83	500	0.001093	0.37481	640.09	2748.1	1.8604	6.8207
158.83	600	0.001101	0.31559	670.38	2756.1	1.9308	6.7592

Appendix G.2 - Saturated Steam Pressure Increment - English Units

T °F	P psia	v^L ft³/lbm	v^V ft³/lbm	h^L BTU/lbm	h^V BTU/lbm	s^L BTU/lbm-°F	s^V BTU/lbm-°F
32.02	0.08872	0.016022	3299.5	0.002279	1075.2	0.000004	2.1867
35.01	0.1	0.016020	2944.9	3.0089	1076.5	0.006101	2.1762
53.13	0.2	0.016027	1525.9	21.203	1084.4	0.042223	2.1156
64.45	0.3	0.016042	1039.4	32.529	1089.4	0.064068	2.0805
72.83	0.4	0.016058	791.83	40.906	1093.0	0.079924	2.0557
79.55	0.5	0.016073	641.31	47.612	1095.9	0.092439	2.0365
85.18	0.6	0.016087	539.87	53.235	1098.3	0.10281	2.0210
90.04	0.7	0.016100	466.79	58.093	1100.4	0.11169	2.0078
94.34	0.8	0.016113	411.56	62.382	1102.3	0.11946	1.9965
98.19	0.9	0.016125	368.31	66.229	1103.9	0.12638	1.9865
101.69	1	0.016137	333.50	69.722	1105.4	0.13262	1.9776
126.03	2	0.016230	173.71	94.022	1115.7	0.17499	1.9195
141.42	3	0.016300	118.70	109.40	1122.2	0.20091	1.8858
152.91	4	0.016357	90.629	120.90	1126.9	0.21985	1.8621
162.18	5	0.016406	73.519	130.18	1130.7	0.23489	1.8438
170.00	6	0.016449	61.977	138.02	1133.9	0.24740	1.8290
176.79	7	0.016489	53.648	144.82	1136.6	0.25814	1.8164
182.81	8	0.016525	47.344	150.86	1139.0	0.26758	1.8056
188.22	9	0.016558	42.402	156.30	1141.1	0.27600	1.7961
193.16	10	0.016589	38.422	161.25	1143.1	0.28362	1.7876
227.92	20	0.016829	20.091	196.27	1156.2	0.33583	1.7319
250.30	30	0.017003	13.747	218.93	1164.1	0.36822	1.6995
267.22	40	0.017146	10.500	236.14	1169.8	0.39213	1.6766
280.99	50	0.017268	8.5172	250.21	1174.2	0.41126	1.6588
292.68	60	0.017378	7.1762	262.20	1177.8	0.42728	1.6443
302.91	70	0.017477	6.2069	272.72	1180.8	0.44113	1.6319
312.02	80	0.017569	5.4732	282.13	1183.3	0.45335	1.6212
320.26	90	0.017655	4.8969	290.67	1185.6	0.46432	1.6117

Appendix G.2 - Saturated Steam Pressure Increment - SI Units

T	P	v^L	v^V	h^L	h^V	s^L	s^V
°C	kPa	m³/kg	m³/kg	kJ/kg	kJ/kg	kJ/kg-K	kJ/kg-K
164.95	700	0.001108	0.27278	697.00	2762.8	1.9918	6.7071
170.41	800	0.001115	0.24034	720.86	2768.3	2.0457	6.6616
175.35	900	0.001121	0.21489	742.56	2773.0	2.0940	6.6213
179.88	1000	0.001127	0.19436	762.52	2777.1	2.1381	6.5850
212.38	2000	0.001177	0.09958	908.50	2798.3	2.4468	6.3390
233.85	3000	0.001217	0.06666	1008.3	2803.2	2.6455	6.1856
250.35	4000	0.001253	0.049776	1087.5	2800.8	2.7968	6.0696
263.94	5000	0.001286	0.039446	1154.6	2794.2	2.9210	5.9737
275.58	6000	0.001319	0.032449	1213.9	2784.6	3.0278	5.8901
285.83	7000	0.001352	0.027379	1267.7	2772.6	3.1224	5.8148
295.01	8000	0.001385	0.023526	1317.3	2758.7	3.2081	5.7450
303.34	9000	0.001418	0.020490	1363.9	2742.9	3.2870	5.6791
311.00	10000	0.001453	0.018030	1408.1	2725.5	3.3606	5.6160
318.08	11000	0.001489	0.015990	1450.4	2706.3	3.4303	5.5545
324.68	12000	0.001526	0.014264	1491.5	2685.4	3.4967	5.4939
330.85	13000	0.001566	0.012780	1531.5	2662.7	3.5608	5.4336
336.67	14000	0.001610	0.011485	1571.0	2637.9	3.6232	5.3727
342.16	15000	0.001657	0.010338	1610.2	2610.7	3.6846	5.3106
347.35	16000	0.001709	0.0093093	1649.7	2580.8	3.7457	5.2463
352.29	17000	0.001769	0.0083710	1690.0	2547.5	3.8077	5.1787
356.99	18000	0.001840	0.0075019	1732.1	2509.8	3.8718	5.1061
361.47	19000	0.001927	0.0066774	1777.2	2466.0	3.9401	5.0256
365.75	20000	0.002040	0.0058651	1827.2	2412.3	4.0156	4.9314
369.83	21000	0.002206	0.0049960	1887.6	2338.6	4.1064	4.8079
373.71	22000	0.002704	0.0036475	2011.3	2173.1	4.2945	4.5446
373.93	22060	0.002938	0.0032953	2054.8	2116.6	4.3615	4.4570

adapted from NIST data

The reference state for all portions of the steam table is u (internal energy) and s (entropy) are both zero for the saturated liquid at the triple point of water.

Appendix G.2 - Saturated Steam Pressure Increment - English Units

T	P	v^L	v^V	h^L	h^V	s^L	s^V
°F	psia	ft³/lbm	ft³/lbm	BTU/lbm	BTU/lbm	BTU/lbm-°F	BTU/lbm-°F
327.81	100	0.017736	4.4326	298.51	1187.5	0.47428	1.6032
381.80	200	0.018387	2.2882	355.46	1198.8	0.54381	1.5460
417.35	300	0.018897	1.5435	393.95	1203.3	0.58820	1.5111
444.62	400	0.019342	1.1616	424.15	1205.0	0.62172	1.4852
467.04	500	0.019752	0.92816	449.54	1205.0	0.64905	1.4642
486.24	600	0.020140	0.77018	471.74	1203.8	0.67237	1.4463
503.13	700	0.020513	0.65587	491.67	1201.9	0.69286	1.4305
518.27	800	0.020879	0.56918	509.90	1199.3	0.71125	1.4162
532.02	900	0.021240	0.50105	526.81	1196.2	0.72802	1.4030
544.65	1000	0.021600	0.44605	542.66	1192.6	0.74350	1.3907
556.35	1100	0.021961	0.40062	557.65	1188.6	0.75795	1.3790
567.26	1200	0.022325	0.36244	571.94	1184.2	0.77154	1.3678
577.49	1300	0.022696	0.32983	585.64	1179.5	0.78441	1.3570
587.14	1400	0.023074	0.30163	598.86	1174.4	0.79669	1.3465
596.26	1500	0.023462	0.27697	611.67	1169.0	0.80846	1.3363
604.93	1600	0.023862	0.25519	624.14	1163.2	0.81981	1.3262
613.18	1700	0.024277	0.23577	636.34	1157.1	0.83081	1.3162
621.07	1800	0.024708	0.21832	648.32	1150.6	0.84152	1.3063
628.61	1900	0.025160	0.20252	660.14	1143.7	0.85198	1.2963
635.85	2000	0.025635	0.18813	671.83	1136.4	0.86227	1.2863
642.80	2100	0.026139	0.17491	683.47	1128.6	0.87242	1.2762
649.49	2200	0.026677	0.16271	695.11	1120.3	0.88250	1.2659
655.94	2300	0.027257	0.15136	706.81	1111.4	0.89257	1.2553
662.16	2400	0.027890	0.14073	718.67	1101.9	0.90271	1.2443
668.17	2500	0.028590	0.13071	730.78	1091.6	0.91302	1.2329
673.98	2600	0.029380	0.12116	743.30	1080.3	0.92361	1.2209
679.60	2700	0.030289	0.11198	756.40	1067.8	0.93466	1.2080
685.04	2800	0.031363	0.10300	770.33	1053.7	0.94636	1.1939
690.31	2900	0.032667	0.094020	785.44	1037.3	0.95901	1.1780
695.41	3000	0.034339	0.084660	802.48	1016.9	0.97326	1.1589
700.34	3100	0.036891	0.073959	824.30	988.54	0.99154	1.1331
705.10	3200	0.048464	0.051122	890.07	902.42	1.0474	1.0580

Appendix G.3 - Superheated Steam - SI Units

	P = 10 kPa (45.806 °C)				P = 50 kPa (81.317 °C)			
	v	u	h	s	v	u	h	s
T(°C)	m³/kg	kJ/kg	kJ/kg	kJ/kg-K	m³/kg	kJ/kg	kJ/kg	kJ/kg-K
(sat)	14.670	2437.2	2583.9	8.1488	3.2400	2483.2	2645.2	7.5930
50	14.867	2443.3	2592.0	8.1741	-	-	-	-
100	17.196	2515.5	2687.5	8.4489	3.4187	2511.5	2682.4	7.6953
150	19.513	2587.9	2783.0	8.6892	3.8897	2585.7	2780.2	7.9413
200	21.826	2661.3	2879.6	8.9049	4.3562	2660.0	2877.8	8.1592
250	24.136	2736.1	2977.4	9.1015	4.8206	2735.1	2976.1	8.3568
300	26.446	2812.3	3076.7	9.2827	5.2840	2811.6	3075.8	8.5386
350	28.755	2890.0	3177.5	9.4513	5.7469	2889.4	3176.8	8.7076
400	31.063	2969.3	3279.9	9.6094	6.2094	2968.9	3279.3	8.8659
450	33.371	3050.3	3384.0	9.7584	6.6717	3049.9	3383.5	9.0151
500	35.680	3132.9	3489.7	9.8998	7.1338	3132.6	3489.3	9.1566
550	37.988	3217.2	3597.1	10.0340	7.5957	3217.0	3596.8	9.2913
600	40.296	3303.3	3706.3	10.1630	8.0576	3303.1	3706.0	9.4201
650	42.603	3391.2	3817.2	10.2870	8.5195	3391.0	3816.9	9.5436
700	44.911	3480.8	3929.9	10.4060	8.9812	3480.6	3929.7	9.6625
750	47.219	3572.2	4044.4	10.5200	9.4430	3572.0	4044.2	9.7773
800	49.527	3665.3	4160.6	10.6310	9.9047	3665.2	4160.4	9.8882
850	51.835	3760.3	4278.6	10.7390	10.366	3760.1	4278.5	9.9957
900	54.142	3856.9	4398.3	10.8430	10.828	3856.8	4398.2	10.1000
950	56.450	3955.2	4519.7	10.9440	11.290	3955.1	4519.6	10.2010
1000	58.758	4055.2	4642.8	11.0430	11.751	4055.1	4642.7	10.3000
1050	61.065	4156.8	4767.5	11.1390	12.213	4156.8	4767.4	10.3960
1100	63.373	4260.0	4893.7	11.2330	12.674	4259.9	4893.7	10.4900
1150	65.681	4364.7	5021.5	11.3240	13.136	4364.6	5021.4	10.5810
1200	67.988	4470.9	5150.7	11.4130	13.598	4470.8	5150.7	10.6700
1250	70.296	4578.4	5281.4	11.5000	14.059	4578.4	5281.3	10.7580
1300	72.604	4687.4	5413.4	11.5860	14.521	4687.3	5413.3	10.8430

Appendix G.3 - Superheated Steam - English Units

	P = 1 psia (101.69 °F)				P = 5 psia (162.18 °F)			
	v	u	h	s	v	u	h	s
T(°F)	ft³/lbm	BTU/lbm	BTU/lbm	BTU/lbm-°F	ft³/lbm	BTU/lbm	BTU/lbm	BTU/lbm-°F
(sat)	333.50	1043.7	1105.4	1.9776	73.521	1062.7	1130.7	1.8438
200	392.52	1077.5	1150.1	2.0510	78.153	1076.2	1148.5	1.8716
240	416.44	1091.2	1168.3	2.0777	83.009	1090.3	1167.1	1.8989
280	440.33	1105.0	1186.5	2.1031	87.838	1104.3	1185.6	1.9247
320	464.20	1118.9	1204.8	2.1272	92.650	1118.4	1204.1	1.9490
360	488.07	1132.9	1223.3	2.1502	97.452	1132.5	1222.6	1.9722
400	511.92	1147.0	1241.8	2.1723	102.25	1146.7	1241.3	1.9944
440	535.77	1161.3	1260.4	2.1935	107.03	1160.9	1260.0	2.0157
500	571.54	1182.8	1288.6	2.2238	114.21	1182.6	1288.2	2.0461
600	631.14	1219.4	1336.2	2.2710	126.15	1219.2	1335.9	2.0934
700	690.73	1256.8	1384.6	2.3146	138.09	1256.7	1384.4	2.1371
800	750.31	1295.1	1433.9	2.3554	150.02	1294.9	1433.7	2.1779
900	809.89	1334.2	1484.1	2.3937	161.94	1334.1	1483.9	2.2162
1000	869.47	1374.2	1535.1	2.4299	173.86	1374.2	1535.0	2.2525
1100	929.04	1415.2	1587.1	2.4644	185.78	1415.1	1587.0	2.2869
1200	988.62	1457.1	1640.0	2.4973	197.70	1457.0	1640.0	2.3198
1300	1048.2	1499.9	1693.9	2.5288	209.62	1499.9	1693.8	2.3513
1400	1107.8	1543.7	1748.7	2.5591	221.54	1543.7	1748.6	2.3816
1500	1167.3	1588.4	1804.5	2.5883	233.46	1588.4	1804.4	2.4108

Appendix G.3 - Superheated Steam - SI Units

	P = 100 kPa (99.606 °C)				P = 200 kPa (120.21 °C)			
	v	u	h	s	v	u	h	s
T(°C)	m³/kg	kJ/kg	kJ/kg	kJ/kg-K	m³/kg	kJ/kg	kJ/kg	kJ/kg-K
(sat)	1.6939	2505.6	2674.9	7.3588	0.88568	2529.1	2706.2	7.1269
150	1.9367	2582.9	2776.6	7.6148	0.95986	2577.1	2769.1	7.2810
200	2.1724	2658.2	2875.5	7.8356	1.0805	2654.6	2870.7	7.5081
250	2.4062	2733.9	2974.5	8.0346	1.1989	2731.4	2971.2	7.7100
300	2.6388	2810.6	3074.5	8.2172	1.3162	2808.8	3072.1	7.8941
350	2.8710	2888.7	3175.8	8.3866	1.4330	2887.3	3173.9	8.0644
400	3.1027	2968.3	3278.6	8.5452	1.5493	2967.1	3277.0	8.2236
450	3.3342	3049.4	3382.8	8.6946	1.6655	3048.5	3381.6	8.3734
500	3.5655	3132.2	3488.7	8.8361	1.7814	3131.4	3487.7	8.5152
550	3.7968	3216.6	3596.3	8.9709	1.8973	3215.9	3595.4	8.6502
600	4.0279	3302.8	3705.6	9.0998	2.0130	3302.2	3704.8	8.7792
650	4.2590	3390.7	3816.6	9.2234	2.1287	3390.2	3815.9	8.9030
700	4.4900	3480.4	3929.4	9.3424	2.2443	3479.9	3928.8	9.0220
750	4.7209	3571.8	4043.9	9.4572	2.3599	3571.4	4043.4	9.1369
800	4.9519	3665.0	4160.2	9.5681	2.4755	3664.7	4159.8	9.2479
850	5.1828	3760.0	4278.2	9.6757	2.5910	3759.6	4277.8	9.3555
900	5.4137	3856.6	4398.0	9.7800	2.7066	3856.3	4397.6	9.4598
950	5.6446	3955.0	4519.5	9.8813	2.8221	3954.7	4519.1	9.5612
1000	5.8754	4055.0	4642.6	9.9800	2.9375	4054.8	4642.3	9.6599
1050	6.1063	4156.6	4767.3	10.0760	3.0530	4156.4	4767.0	9.7560
1100	6.3371	4259.8	4893.5	10.1700	3.1685	4259.6	4893.3	9.8497
1150	6.5680	4364.5	5021.3	10.2610	3.2839	4364.3	5021.1	9.9411
1200	6.7988	4470.7	5150.6	10.3500	3.3994	4470.5	5150.4	10.0300
1250	7.0296	4578.3	5281.2	10.4380	3.5148	4578.1	5281.1	10.1180
1300	7.2604	4687.2	5413.2	10.5230	3.6302	4687.0	5413.1	10.2030

Appendix G.3 - Superheated Steam - English Units

	P = 10 psia (193.16 °F)				P = 14.696 psia (211.95 °F)			
	v	u	h	s	v	u	h	s
T(°F)	ft³/lbm	BTU/lbm	BTU/lbm	BTU/lbm-°F	ft³/lbm	BTU/lbm	BTU/lbm	BTU/lbm-°F
(sat)	38.421	1072.0	1143.1	1.7876	26.802	1077.4	1150.3	1.7566
200	38.849	1074.5	1146.4	1.7926	-	-	-	-
240	41.326	1089.0	1165.5	1.8208	28.004	1087.9	1164.0	1.7766
280	43.774	1103.4	1184.4	1.8470	29.692	1102.5	1183.2	1.8033
320	46.205	1117.6	1203.1	1.8717	31.363	1116.9	1202.2	1.8283
360	48.624	1131.9	1221.8	1.8951	33.021	1131.3	1221.1	1.8519
400	51.035	1146.2	1240.6	1.9174	34.671	1145.7	1240.0	1.8744
440	53.440	1160.5	1259.4	1.9388	36.315	1160.1	1258.9	1.8959
500	57.041	1182.2	1287.8	1.9693	38.774	1181.9	1287.4	1.9266
600	63.029	1219.0	1335.6	2.0167	42.858	1218.7	1335.3	1.9741
700	69.007	1256.5	1384.2	2.0605	46.933	1256.3	1383.9	2.0179
800	74.979	1294.8	1433.5	2.1014	51.002	1294.6	1433.3	2.0588
900	80.948	1334.0	1483.8	2.1397	55.067	1333.9	1483.6	2.0972
1000	86.913	1374.1	1534.9	2.1760	59.129	1374.0	1534.8	2.1335
1100	92.877	1415.0	1586.9	2.2105	63.189	1415.0	1586.8	2.1680
1200	98.839	1457.0	1639.9	2.2434	67.248	1456.9	1639.8	2.2009
1300	104.80	1499.8	1693.7	2.2749	71.306	1499.8	1693.7	2.2324
1400	110.76	1543.6	1748.6	2.3052	75.363	1543.6	1748.5	2.2627
1500	116.72	1588.3	1804.3	2.3344	79.420	1588.3	1804.3	2.2919

Appendix G.3 - Superheated Steam - SI Units

	P = 300 kPa (133.52 °C)				P = 400 kPa (143.61 °C)			
	v	u	h	s	v	u	h	s
T(°C)	m³/kg	kJ/kg	kJ/kg	kJ/kg-K	m³/kg	kJ/kg	kJ/kg	kJ/kg-K
(sat)	0.60576	2543.2	2724.9	6.9916	0.46238	2553.1	2738.1	6.8955
150	0.63401	2571.0	2761.2	7.0791	0.47088	2564.4	2752.8	6.9306
200	0.71642	2651.0	2865.9	7.3131	0.53433	2647.2	2860.9	7.1723
250	0.79644	2728.9	2967.9	7.5180	0.59520	2726.4	2964.5	7.3804
300	0.87534	2807.0	3069.6	7.7037	0.65489	2805.1	3067.1	7.5677
350	0.95363	2885.9	3172.0	7.8750	0.71396	2884.4	3170.0	7.7399
400	1.0315	2966.0	3275.5	8.0347	0.77264	2964.9	3273.9	7.9002
450	1.1092	3047.5	3380.3	8.1849	0.83109	3046.6	3379.0	8.0508
500	1.1867	3130.6	3486.6	8.3271	0.88936	3129.8	3485.5	8.1933
550	1.2641	3215.3	3594.5	8.4623	0.94751	3214.6	3593.6	8.3287
600	1.3414	3301.6	3704.0	8.5914	1.0056	3301.0	3703.2	8.4580
650	1.4186	3389.7	3815.3	8.7153	1.0636	3389.1	3814.6	8.5820
700	1.4958	3479.5	3928.2	8.8344	1.1215	3479.0	3927.6	8.7012
750	1.5729	3571.0	4042.9	8.9494	1.1794	3570.6	4042.4	8.8162
800	1.6500	3664.3	4159.3	9.0604	1.2373	3663.9	4158.8	8.9273
850	1.7271	3759.3	4277.4	9.1680	1.2951	3759.0	4277.0	9.0350
900	1.8042	3856.0	4397.3	9.2724	1.3530	3855.7	4396.9	9.1394
950	1.8812	3954.4	4518.8	9.3739	1.4108	3954.2	4518.5	9.2409
1000	1.9582	4054.5	4642.0	9.4726	1.4686	4054.3	4641.7	9.3396
1050	2.0352	4156.2	4766.7	9.5687	1.5264	4155.9	4766.5	9.4357
1100	2.1122	4259.4	4893.1	9.6624	1.5841	4259.2	4892.8	9.5295
1150	2.1892	4364.1	5020.9	9.7538	1.6419	4363.9	5020.7	9.6209
1200	2.2662	4470.3	5150.2	9.8431	1.6997	4470.1	5150.0	9.7102
1250	2.3432	4577.9	5280.9	9.9303	1.7574	4577.8	5280.7	9.7975
1300	2.4202	4686.9	5412.9	10.0160	1.8152	4686.7	5412.8	9.8828

Appendix G.3 - Superheated Steam - English Units

	P = 20 psia (227.92 °F)				P = 40 psia (267.22 °F)			
	v	u	h	s	v	u	h	s
T(°F)	ft³/lbm	BTU/lbm	BTU/lbm	BTU/lbm-°F	ft³/lbm	BTU/lbm	BTU/lbm	BTU/lbm-°F
(sat)	20.091	1081.8	1156.2	1.7319	10.500	1092.1	1169.8	1.6766
240	20.478	1086.5	1162.3	1.7406	-	-	-	-
280	21.739	1101.4	1181.9	1.7679	10.713	1097.3	1176.6	1.6858
320	22.980	1116.1	1201.2	1.7933	11.363	1112.9	1197.0	1.7128
360	24.209	1130.6	1220.2	1.8172	11.999	1128.1	1216.9	1.7377
400	25.429	1145.1	1239.3	1.8398	12.625	1143.1	1236.5	1.7610
440	26.644	1159.7	1258.3	1.8614	13.244	1157.9	1256.0	1.7831
500	28.458	1181.5	1286.9	1.8922	14.165	1180.2	1285.0	1.8144
600	31.467	1218.5	1334.9	1.9399	15.686	1217.5	1333.6	1.8625
700	34.467	1256.1	1383.6	1.9838	17.197	1255.3	1382.6	1.9067
800	37.461	1294.5	1433.1	2.0247	18.702	1293.9	1432.3	1.9478
900	40.451	1333.7	1483.4	2.0631	20.202	1333.2	1482.8	1.9864
1000	43.438	1373.8	1534.6	2.0994	21.700	1373.4	1534.0	2.0227
1100	46.423	1414.9	1586.7	2.1339	23.196	1414.5	1586.2	2.0573
1200	49.407	1456.8	1639.7	2.1669	24.691	1456.5	1639.3	2.0903
1300	52.390	1499.7	1693.6	2.1984	26.185	1499.4	1693.2	2.1219
1400	55.372	1543.5	1748.4	2.2287	27.678	1543.3	1748.1	2.1522
1500	58.354	1588.2	1804.2	2.2579	29.170	1588.0	1803.9	2.1814
1600	61.335	1633.9	1860.9	2.2862	30.662	1633.7	1860.7	2.2097

Appendix G.3 - Superheated Steam - SI Units

	P = 500 kPa (151.83 °C)				P = 600 kPa (158.83 °C)			
T(°C)	v m³/kg	u kJ/kg	h kJ/kg	s kJ/kg-K	v m³/kg	u kJ/kg	h kJ/kg	s kJ/kg-K
(sat)	0.37481	2560.7	2748.1	6.8207	0.31558	2566.8	2756.1	6.7592
200	0.42503	2643.3	2855.8	7.0610	0.35212	2639.3	2850.6	6.9683
250	0.47443	2723.8	2961.0	7.2724	0.39390	2721.2	2957.6	7.1832
300	0.52261	2803.2	3064.6	7.4614	0.43442	2801.4	3062.0	7.3740
350	0.57015	2883.0	3168.1	7.6346	0.47427	2881.6	3166.1	7.5481
400	0.61730	2963.7	3272.3	7.7955	0.51374	2962.5	3270.8	7.7097
450	0.66421	3045.6	3377.7	7.9465	0.55296	3044.7	3376.5	7.8611
500	0.71094	3129.0	3484.5	8.0892	0.59200	3128.2	3483.4	8.0041
550	0.75756	3213.9	3592.7	8.2249	0.63093	3213.2	3591.8	8.1399
600	0.80409	3300.4	3702.5	8.3543	0.66976	3299.8	3701.7	8.2695
650	0.85055	3388.6	3813.9	8.4784	0.70853	3388.1	3813.2	8.3937
700	0.89696	3478.5	3927.0	8.5977	0.74725	3478.1	3926.4	8.5131
750	0.94332	3570.2	4041.8	8.7128	0.78592	3569.8	4041.3	8.6283
800	0.98966	3663.6	4158.4	8.8240	0.82457	3663.2	4157.9	8.7395
850	1.0360	3758.6	4276.6	8.9317	0.86319	3758.3	4276.2	8.8472
900	1.0823	3855.4	4396.6	9.0362	0.90178	3855.1	4396.2	8.9518
950	1.1285	3953.9	4518.2	9.1377	0.94037	3953.6	4517.8	9.0533
1000	1.1748	4054.0	4641.4	9.2364	0.97893	4053.7	4641.1	9.1521
1050	1.2210	4155.7	4766.2	9.3326	1.0175	4155.5	4766.0	9.2482
1100	1.2673	4259.0	4892.6	9.4263	1.0560	4258.7	4892.4	9.3420
1150	1.3135	4363.7	5020.5	9.5178	1.0946	4363.5	5020.3	9.4335
1200	1.3597	4470.0	5149.8	9.6071	1.1331	4469.8	5149.6	9.5228
1250	1.4059	4577.6	5280.5	9.6944	1.1716	4577.4	5280.4	9.6101
1300	1.4521	4686.6	5412.6	9.7797	1.2101	4686.4	5412.5	9.6954

Appendix G.3 - Superheated Steam - English Units

	P = 60 psia (292.68 °F)				P = 80 psia (312.02 °F)			
T(°F)	v ft³/lbm	u BTU/lbm	h BTU/lbm	s BTU/lbm-°F	v ft³/lbm	u BTU/lbm	h BTU/lbm	s BTU/lbm-°F
(sat)	7.1761	1098.1	1177.8	1.6443	5.4730	1102.3	1183.3	1.6212
320	7.4863	1109.5	1192.7	1.6637	5.5440	1105.9	1187.9	1.6271
360	7.9259	1125.5	1213.5	1.6897	5.8875	1122.7	1209.9	1.6545
400	8.3548	1140.9	1233.7	1.7138	6.2186	1138.7	1230.8	1.6795
440	8.7766	1156.1	1253.6	1.7364	6.5420	1154.3	1251.2	1.7026
500	9.4004	1178.8	1283.1	1.7682	7.0176	1177.3	1281.2	1.7350
600	10.426	1216.5	1332.2	1.8168	7.7951	1215.4	1330.8	1.7842
700	11.440	1254.5	1381.6	1.8613	8.5616	1253.8	1380.5	1.8290
800	12.448	1293.3	1431.5	1.9026	9.3217	1292.6	1430.6	1.8704
900	13.453	1332.7	1482.1	1.9413	10.078	1332.2	1481.4	1.9092
1000	14.454	1373.0	1533.5	1.9778	10.831	1372.6	1532.9	1.9458
1100	15.454	1414.2	1585.7	2.0124	11.583	1413.8	1585.3	1.9804
1200	16.452	1456.2	1638.9	2.0454	12.333	1455.9	1638.5	2.0135
1300	17.450	1499.1	1692.9	2.0770	13.082	1498.9	1692.5	2.0451
1400	18.446	1543.0	1747.8	2.1074	13.831	1542.8	1747.5	2.0755
1500	19.442	1587.8	1803.7	2.1366	14.578	1587.6	1803.4	2.1048
1600	20.438	1633.5	1860.4	2.1649	15.326	1633.3	1860.2	2.1331
1700	21.433	1680.1	1918.1	2.1922	16.073	1680.0	1917.9	2.1604
1800	22.428	1727.6	1976.6	2.2187	16.819	1727.5	1976.5	2.1869
1900	23.422	1776.0	2036.0	2.2444	17.565	1775.8	2035.9	2.2126
2000	24.417	1825.1	2096.2	2.2694	18.312	1825.0	2096.1	2.2376

Appendix G.3 - Superheated Steam - SI Units

	P = 800 kPa (170.41 °C)				P = 1000 kPa (179.88 °C)			
	v	u	h	s	v	u	h	s
T(°C)	m³/kg	kJ/kg	kJ/kg	kJ/kg-K	m³/kg	kJ/kg	kJ/kg	kJ/kg-K
(sat)	0.24034	2576.0	2768.3	6.6616	0.19436	2582.7	2777.1	6.5850
200	0.26088	2631.0	2839.7	6.8176	0.20602	2622.2	2828.3	6.6955
250	0.29320	2715.9	2950.4	7.0401	0.23275	2710.4	2943.1	6.9265
300	0.32416	2797.5	3056.9	7.2345	0.25799	2793.6	3051.6	7.1246
350	0.35442	2878.6	3162.2	7.4106	0.28250	2875.7	3158.2	7.3029
400	0.38428	2960.2	3267.6	7.5734	0.30661	2957.9	3264.5	7.4669
450	0.41389	3042.8	3373.9	7.7257	0.33045	3040.9	3371.3	7.6200
500	0.44332	3126.6	3481.3	7.8692	0.35411	3125.0	3479.1	7.7641
550	0.47263	3211.9	3590.0	8.0054	0.37766	3210.5	3588.1	7.9008
600	0.50185	3298.7	3700.1	8.1354	0.40111	3297.5	3698.6	8.0310
650	0.53101	3387.1	3811.9	8.2598	0.42449	3386.0	3810.5	8.1557
700	0.56011	3477.2	3925.3	8.3794	0.44783	3476.2	3924.1	8.2755
750	0.58917	3569.0	4040.3	8.4947	0.47112	3568.1	4039.3	8.3909
800	0.61820	3662.4	4157.0	8.6061	0.49438	3661.7	4156.1	8.5024
850	0.64721	3757.6	4275.4	8.7139	0.51762	3757.0	4274.6	8.6103
900	0.67619	3854.5	4395.5	8.8185	0.54083	3853.9	4394.8	8.7150
950	0.70515	3953.1	4517.2	8.9201	0.56403	3952.5	4516.5	8.8166
1000	0.73411	4053.2	4640.5	9.0189	0.58721	4052.7	4639.9	8.9155
1050	0.76304	4155.0	4765.4	9.1151	0.61038	4154.5	4764.9	9.0118
1100	0.79197	4258.3	4891.9	9.2089	0.63354	4257.9	4891.4	9.1056
1150	0.82089	4363.1	5019.8	9.3004	0.65669	4362.7	5019.4	9.1972
1200	0.84980	4469.4	5149.2	9.3898	0.67983	4469.0	5148.9	9.2866
1250	0.87871	4577.1	5280.0	9.4771	0.70297	4576.7	5279.7	9.3739
1300	0.90760	4686.1	5412.2	9.5625	0.72610	4685.8	5411.9	9.4593

Appendix G.3 - Superheated Steam - English Units

	P = 100 psia (327.81 °F)				P = 150 psia (358.42 °F)			
	v	u	h	s	v	u	h	s
T(°F)	ft³/lbm	BTU/lbm	BTU/lbm	BTU/lbm-°F	ft³/lbm	BTU/lbm	BTU/lbm	BTU/lbm-°F
(sat)	4.4326	1105.5	1187.5	1.6032	3.0150	1110.8	1194.5	1.5700
350	4.5926	1115.5	1200.4	1.6194	-	-	-	-
400	4.9359	1136.4	1227.7	1.6521	3.2223	1130.3	1219.7	1.6001
450	5.2658	1156.4	1253.8	1.6816	3.4559	1151.7	1247.6	1.6317
500	5.5876	1175.9	1279.3	1.7089	3.6798	1172.1	1274.3	1.6602
550	5.9040	1195.2	1304.4	1.7344	3.8978	1192.1	1300.3	1.6866
600	6.2166	1214.4	1329.4	1.7586	4.1116	1211.8	1325.9	1.7114
700	6.8344	1253.0	1379.5	1.8037	4.5313	1251.0	1376.8	1.7573
800	7.4457	1292.0	1429.8	1.8453	4.9443	1290.5	1427.7	1.7994
900	8.0530	1331.7	1480.7	1.8842	5.3531	1330.5	1479.0	1.8386
1000	8.6575	1372.2	1532.4	1.9209	5.7591	1371.1	1531.0	1.8755
1100	9.2602	1413.4	1584.8	1.9556	6.1632	1412.5	1583.6	1.9103
1200	9.8614	1455.6	1638.1	1.9887	6.5659	1454.8	1637.1	1.9436
1300	10.462	1498.6	1692.2	2.0204	6.9676	1497.9	1691.3	1.9753
1400	11.061	1542.5	1747.2	2.0508	7.3685	1542.0	1746.5	2.0058
1500	11.660	1587.4	1803.2	2.0801	7.7688	1586.9	1802.5	2.0351
1600	12.258	1633.1	1860.0	2.1084	8.1686	1632.7	1859.4	2.0635
1700	12.856	1679.8	1917.7	2.1357	8.5681	1679.4	1917.2	2.0908
1800	13.454	1727.3	1976.3	2.1622	8.9672	1726.9	1975.8	2.1174
1900	14.051	1775.7	2035.7	2.1880	9.3660	1775.3	2035.3	2.1431
2000	14.649	1824.9	2096.0	2.2130	9.7647	1824.6	2095.6	2.1682

Appendix G.3 - Superheated Steam - SI Units

P = 1200 kPa (187.96 °C)

T(°C)	v m³/kg	u kJ/kg	h kJ/kg	s kJ/kg-K
(sat)	0.16326	2587.8	2783.7	6.5217
200	0.16934	2612.9	2816.1	6.5909
250	0.19241	2704.7	2935.6	6.8313
300	0.21386	2789.7	3046.3	7.0335
350	0.23455	2872.7	3154.2	7.2139
400	0.25482	2955.5	3261.3	7.3793
450	0.27482	3038.9	3368.7	7.5332
500	0.29464	3123.4	3476.9	7.6779
550	0.31434	3209.1	3586.3	7.8150
600	0.33394	3296.3	3697.0	7.9455
650	0.35348	3385.0	3809.2	8.0704
700	0.37297	3475.3	3922.9	8.1904
750	0.39242	3567.3	4038.2	8.3060
800	0.41184	3661.0	4155.2	8.4176
850	0.43123	3756.3	4273.8	8.5256
900	0.45059	3853.3	4394.0	8.6303
950	0.46994	3952.0	4515.9	8.7320
1000	0.48928	4052.2	4639.4	8.8310
1050	0.50860	4154.1	4764.4	8.9273
1100	0.52792	4257.5	4891.0	9.0212
1150	0.54722	4362.3	5019.0	9.1128
1200	0.56652	4468.7	5148.5	9.2022
1250	0.58581	4576.4	5279.3	9.2895
1300	0.60509	4685.4	5411.5	9.3749

P = 1400 kPa (195.04 °C)

v m³/kg	u kJ/kg	h kJ/kg	s kJ/kg-K
0.14078	2591.8	2788.8	6.4675
0.14303	2602.7	2803.0	6.4975
0.16356	2698.9	2927.9	6.7488
0.18232	2785.7	3040.9	6.9552
0.20029	2869.7	3150.1	7.1379
0.21782	2953.1	3258.1	7.3046
0.23508	3037.0	3366.1	7.4594
0.25216	3121.8	3474.8	7.6047
0.26911	3207.7	3584.5	7.7422
0.28597	3295.1	3695.4	7.8730
0.30276	3384.0	3807.8	7.9982
0.31951	3474.4	3921.7	8.1183
0.33621	3566.5	4037.2	8.2340
0.35287	3660.2	4154.3	8.3457
0.36952	3755.6	4273.0	8.4538
0.38614	3852.7	4393.3	8.5587
0.40274	3951.4	4515.2	8.6604
0.41933	4051.7	4638.8	8.7594
0.43591	4153.6	4763.9	8.8558
0.45247	4257.0	4890.5	8.9497
0.46903	4361.9	5018.6	9.0413
0.48558	4468.3	5148.1	9.1308
0.50212	4576.0	5279.0	9.2182
0.51866	4685.1	5411.2	9.3036

Appendix G.3 - Superheated Steam - English Units

P = 200 psia (381.80 °F)

T(°F)	v ft³/lbm	u BTU/lbm	h BTU/lbm	s BTU/lbm-°F
(sat)	2.2882	1114.1	1198.8	1.5460
400	2.3615	1123.5	1210.9	1.5602
450	2.5488	1146.7	1241.0	1.5943
500	2.7247	1168.2	1269.0	1.6243
550	2.8939	1188.9	1296.0	1.6517
600	3.0586	1209.1	1322.3	1.6771
700	3.3795	1249.0	1374.1	1.7238
800	3.6934	1288.9	1425.6	1.7664
900	4.0030	1329.2	1477.3	1.8060
1000	4.3098	1370.1	1529.6	1.8430
1100	4.6147	1411.6	1582.4	1.8781
1200	4.9182	1454.0	1636.1	1.9114
1300	5.2206	1497.3	1690.5	1.9432
1400	5.5222	1541.4	1745.7	1.9738
1500	5.8232	1586.3	1801.9	2.0031
1600	6.1238	1632.2	1858.8	2.0315
1700	6.4239	1678.9	1916.7	2.0589
1800	6.7237	1726.5	1975.4	2.0855
1900	7.0233	1775.0	2034.9	2.1113
2000	7.3227	1824.2	2095.3	2.1363

P = 300 psia (417.35 °F)

v ft³/lbm	u BTU/lbm	h BTU/lbm	s BTU/lbm-°F
1.5435	1117.7	1203.3	1.5111
-	-	-	-
1.6369	1135.6	1226.4	1.5370
1.7670	1159.8	1257.9	1.5706
1.8884	1182.1	1287.0	1.6002
2.0046	1203.5	1314.8	1.6271
2.2273	1244.9	1368.6	1.6756
2.4424	1285.7	1421.3	1.7192
2.6529	1326.6	1473.9	1.7594
2.8605	1367.9	1526.7	1.7969
3.0662	1409.8	1580.1	1.8322
3.2704	1452.5	1634.0	1.8657
3.4735	1495.9	1688.7	1.8978
3.6759	1540.2	1744.2	1.9284
3.8776	1585.3	1800.5	1.9579
4.0789	1631.3	1857.7	1.9864
4.2798	1678.1	1915.7	2.0138
4.4803	1725.8	1974.5	2.0405
4.6806	1774.3	2034.1	2.0663
4.8807	1823.6	2094.6	2.0914

Appendix G.3 - Superheated Steam - SI Units

	P = 1600 kPa (201.37 °C)				P = 1800 kPa (207.11 °C)			
T(°C)	v m³/kg	u kJ/kg	h kJ/kg	s kJ/kg-K	v m³/kg	u kJ/kg	h kJ/kg	s kJ/kg-K
(sat)	0.12374	2594.8	2792.8	6.4199	0.11037	2597.2	2795.9	6.3775
250	0.16356	2698.9	2927.9	6.7488	0.14190	2692.9	2919.9	6.6753
300	0.18232	2785.7	3040.9	6.9552	0.15866	2781.6	3035.4	6.8863
350	0.20029	2869.7	3150.1	7.1379	0.17459	2866.6	3146.0	7.0713
400	0.21782	2953.1	3258.1	7.3046	0.19007	2950.7	3254.9	7.2394
450	0.23508	3037.0	3366.1	7.4594	0.20527	3035.0	3363.5	7.3950
500	0.25216	3121.8	3474.8	7.6047	0.22029	3120.1	3472.6	7.5409
550	0.26911	3207.7	3584.5	7.7422	0.23519	3206.3	3582.6	7.6788
600	0.28597	3295.1	3695.4	7.8730	0.24999	3293.9	3693.9	7.8100
650	0.30276	3384.0	3807.8	7.9982	0.26472	3382.9	3806.5	7.9354
700	0.31951	3474.4	3921.7	8.1183	0.27940	3473.5	3920.5	8.0557
750	0.33621	3566.5	4037.2	8.2340	0.29404	3565.7	4036.1	8.1716
800	0.35287	3660.2	4154.3	8.3457	0.30865	3659.5	4153.3	8.2834
850	0.36952	3755.6	4273.0	8.4538	0.32323	3755.0	4272.2	8.3916
900	0.38614	3852.7	4393.3	8.5587	0.33780	3852.1	4392.6	8.4965
950	0.40274	3951.4	4515.2	8.6604	0.35234	3950.9	4514.6	8.5984
1000	0.41933	4051.7	4638.8	8.7594	0.36687	4051.2	4638.2	8.6974
1050	0.43591	4153.6	4763.9	8.8558	0.38138	4153.1	4763.4	8.7938
1100	0.45247	4257.0	4890.5	8.9497	0.39589	4256.6	4890.0	8.8878
1150	0.46903	4361.9	5018.6	9.0413	0.41038	4361.5	5018.2	8.9794
1200	0.48558	4468.3	5148.1	9.1308	0.42487	4467.9	5147.7	9.0689
1250	0.50212	4576.0	5279.0	9.2182	0.43936	4575.7	5278.7	9.1563
1300	0.51866	4685.1	5411.2	9.3036	0.45383	4684.8	5410.9	9.2417
1350	0.46830	4795.2	5544.5	9.3253	0.41628	4794.9	5544.2	9.2708

Appendix G.3 - Superheated Steam - English Units

	P = 400 psia (444.62 °F)				P = 600 psia (486.24 °F)			
T(°F)	v ft³/lbm	u BTU/lbm	h BTU/lbm	s BTU/lbm-°F	v ft³/lbm	u BTU/lbm	h BTU/lbm	s BTU/lbm-°F
(sat)	1.1616	1119.0	1205.0	1.4852	0.77015	1118.3000	1203.8000	1.4463
450	1.1747	1122.5	1209.4	1.4901	-	-	-	-
500	1.2851	1150.4	1245.6	1.5288	0.79526	1128.2	1216.5	1.4596
550	1.3840	1174.9	1277.3	1.5611	0.87542	1158.6	1255.8	1.4996
600	1.4765	1197.6	1306.9	1.5897	0.94605	1184.9	1289.9	1.5326
700	1.6507	1240.7	1362.9	1.6402	1.0732	1231.9	1351.0	1.5877
800	1.8166	1282.5	1416.9	1.6849	1.1904	1275.8	1408.0	1.6348
900	1.9777	1324.0	1470.4	1.7257	1.3023	1318.7	1463.3	1.6771
1000	2.1358	1365.8	1523.9	1.7637	1.4110	1361.4	1518.1	1.7160
1100	2.2919	1408.0	1577.7	1.7993	1.5175	1404.4	1572.8	1.7523
1200	2.4465	1450.9	1632.0	1.8331	1.6225	1447.8	1627.9	1.7865
1300	2.6000	1494.6	1687.0	1.8653	1.7264	1491.8	1683.5	1.8190
1400	2.7527	1539.0	1742.7	1.8961	1.8296	1536.6	1739.7	1.8501
1500	2.9048	1584.2	1799.2	1.9257	1.9320	1582.1	1796.6	1.8799
1600	3.0565	1630.3	1856.5	1.9542	2.0340	1628.4	1854.2	1.9086
1700	3.2077	1677.2	1914.7	1.9817	2.1356	1675.5	1912.6	1.9362
1800	3.3586	1725.0	1973.6	2.0084	2.2369	1723.4	1971.8	1.9630
1900	3.5093	1773.6	2033.3	2.0343	2.3379	1772.2	2031.7	1.9890
2000	3.6597	1823.0	2093.8	2.0594	2.4387	1821.7	2092.4	2.0142

Appendix G.3 - Superheated Steam - SI Units

	P = 2000 kPa (212.38 °C)				P = 2500 kPa (223.95 °C)			
T(°C)	v m³/kg	u kJ/kg	h kJ/kg	s kJ/kg-K	v m³/kg	u kJ/kg	h kJ/kg	s kJ/kg-K
(sat)	0.099585	2599.1	2798.3	6.339	0.07995	2602.1	2801.9	6.2558
250	0.11150	2680.2	2903.2	6.5475	0.08705	2663.3	2880.9	6.4107
300	0.12551	2773.2	3024.2	6.7684	0.09894	2762.2	3009.6	6.6459
350	0.13860	2860.5	3137.7	6.9583	0.10979	2852.5	3127.0	6.8424
400	0.15121	2945.9	3248.3	7.1292	0.12012	2939.8	3240.1	7.0170
450	0.16354	3031.1	3358.2	7.2866	0.13015	3026.2	3351.6	7.1767
500	0.17568	3116.9	3468.2	7.4337	0.13999	3112.8	3462.7	7.3254
550	0.18770	3203.6	3579.0	7.5725	0.14970	3200.1	3574.3	7.4653
600	0.19961	3291.5	3690.7	7.7043	0.15931	3288.5	3686.8	7.5979
650	0.21146	3380.8	3803.8	7.8302	0.16886	3378.2	3800.4	7.7243
700	0.22326	3471.6	3918.2	7.9509	0.17835	3469.3	3915.2	7.8455
750	0.23502	3564.0	4034.1	8.0670	0.18780	3562.0	4031.5	7.9620
800	0.24674	3658.0	4151.5	8.1790	0.19721	3656.2	4149.2	8.0743
850	0.25844	3753.6	4270.5	8.2874	0.20660	3752.0	4268.5	8.1830
900	0.27012	3850.9	4391.1	8.3925	0.21597	3849.4	4389.3	8.2882
950	0.28178	3949.8	4513.3	8.4945	0.22532	3948.4	4511.7	8.3904
1000	0.29342	4050.2	4637.0	8.5936	0.23466	4048.9	4635.6	8.4896
1050	0.30505	4152.2	4762.3	8.6901	0.24399	4151.0	4761.0	8.5863
1100	0.31667	4255.7	4889.1	8.7842	0.25330	4254.7	4887.9	8.6804
1150	0.32828	4360.7	5017.3	8.8759	0.26260	4359.7	5016.2	8.7722
1200	0.33989	4467.2	5147.0	8.9654	0.27190	4466.2	5146.0	8.8618
1250	0.35149	4575.0	5278.0	9.0529	0.28119	4574.1	5277.1	8.9493
1300	0.36308	4684.1	5410.3	9.1384	0.29047	4683.3	5409.5	9.0349
1350	0.37466	4794.6	5543.9	9.222	0.29975	4793.8	5543.2	9.1185

Appendix G.3 - Superheated Steam - English Units

	P = 800 psia (518.27 °F)				P = 1000 psia (544.65 °F)			
T(°F)	v ft³/lbm	u BTU/lbm	h BTU/lbm	s BTU/lbm-°F	v ft³/lbm	u BTU/lbm	h BTU/lbm	s BTU/lbm-°F
(sat)	0.56917	1115.0	1199.3	1.4162	0.4461	1110.1	1192.6	1.3907
550	0.61586	1139.3	1230.5	1.4476	0.4538	1115.2	1199.2	1.3972
600	0.67799	1170.5	1270.9	1.4867	0.5143	1154.1	1249.3	1.4457
650	0.73279	1197.6	1306.0	1.5191	0.5641	1185.1	1289.4	1.4827
600	0.67799	1170.5	1270.9	1.4867	0.5143	1154.1	1249.3	1.4457
700	0.78330	1222.4	1338.4	1.5476	0.6084	1212.4	1325.0	1.5140
800	0.87678	1268.9	1398.7	1.5975	0.6882	1261.7	1389.0	1.5671
900	0.96433	1313.3	1456.0	1.6414	0.7614	1307.7	1448.6	1.6126
1000	1.0484	1357.0	1512.2	1.6812	0.8308	1352.5	1506.2	1.6535
1100	1.1302	1400.7	1568.0	1.7182	0.8978	1396.9	1563.0	1.6912
1200	1.2105	1444.6	1623.8	1.7529	0.9633	1441.4	1619.7	1.7264
1300	1.2897	1489.1	1680.0	1.7857	1.0276	1486.3	1676.5	1.7596
1400	1.3680	1534.2	1736.7	1.8171	1.0910	1531.8	1733.6	1.7912
1500	1.4456	1580.0	1794.0	1.8471	1.1538	1577.8	1791.3	1.8214
1600	1.5228	1626.5	1851.9	1.8759	1.2161	1624.6	1849.6	1.8504
1700	1.5996	1673.8	1910.6	1.9037	1.2780	1672.1	1908.6	1.8784
1800	1.6761	1721.9	1970.0	1.9306	1.3396	1720.3	1968.2	1.9053
1900	1.7523	1770.7	2030.2	1.9567	1.4009	1769.3	2028.6	1.9315
2000	1.8282	1820.4	2091.0	1.9819	1.4619	1819.1	2089.6	1.9568

Appendix G.3 - Superheated Steam - SI Units

P = 3000 kPa (233.85 °C)

T(°C)	v m³/kg	u kJ/kg	h kJ/kg	s kJ/kg-K
(sat)	0.06666	2603.2	2803.2	6.1856
250	0.07063	2644.7	2856.5	6.2893
300	0.08118	2750.8	2994.3	6.5412
350	0.09056	2844.4	3116.1	6.7449
400	0.09938	2933.5	3231.7	6.9234
450	0.10789	3021.2	3344.8	7.0856
500	0.11620	3108.6	3457.2	7.2359
550	0.12437	3196.6	3569.7	7.3768
600	0.13245	3285.5	3682.8	7.5103
650	0.14045	3375.6	3796.9	7.6373
700	0.14841	3467.0	3912.2	7.7590
750	0.15632	3559.9	4028.9	7.8758
800	0.16420	3654.3	4146.9	7.9885
850	0.17205	3750.3	4266.5	8.0973
900	0.17988	3847.9	4387.5	8.2028
950	0.18769	3947.0	4510.1	8.3051
1000	0.19549	4047.7	4634.1	8.4045
1050	0.20327	4149.9	4759.7	8.5012
1100	0.21105	4253.6	4886.7	8.5955
1150	0.21882	4358.7	5015.2	8.6874
1200	0.22657	4465.3	5145.3	8.7770
1250	0.23433	4573.3	5276.2	8.8646
1300	0.24207	4682.5	5408.8	8.9502
1350	0.24981	4793.1	5542.5	9.0339
1400	0.25755	4904.9	5677.5	9.1158
1450	0.26528	5017.8	5813.7	9.196
1500	0.27301	5131.9	5950.9	9.2745

P = 4000 kPa (250.35 °C)

T(°C)	v m³/kg	u kJ/kg	h kJ/kg	s kJ/kg-K
(sat)	0.049776	2601.7	2800.8	6.0696
250	-	-	-	-
300	0.05887	2726.2	2961.7	6.3639
350	0.06647	2827.4	3093.3	6.5843
400	0.07343	2920.7	3214.5	6.7714
450	0.08004	3011.0	3331.2	6.9386
500	0.08644	3100.3	3446.0	7.0922
550	0.09270	3189.5	3560.3	7.2355
600	0.09886	3279.4	3674.9	7.3705
650	0.10494	3370.3	3790.1	7.4988
700	0.11098	3462.4	3906.3	7.6214
750	0.11697	3555.8	4023.6	7.7390
800	0.12292	3650.6	4142.3	7.8523
850	0.12885	3747.0	4262.4	7.9616
900	0.13476	3844.8	4383.9	8.0674
950	0.14065	3944.2	4506.8	8.1701
1000	0.14652	4045.1	4631.2	8.2697
1050	0.15239	4147.5	4757.1	8.3667
1100	0.15824	4251.4	4884.4	8.4611
1150	0.16408	4356.7	5013.1	8.5532
1200	0.16992	4463.5	5143.1	8.6430
1250	0.17575	4571.5	5274.5	8.7307
1300	0.18157	4680.9	5407.2	8.8164
1350	0.18739	4791.6	5541.1	8.9002
1400	0.19320	4903.5	5676.3	8.9822
1450	0.19901	5016.5	5812.5	9.0625
1500	0.20481	5130.7	5949.9	9.1411

Appendix G.3 - Superheated Steam - English Units

P = 1200 psia (567.26 °F)

T(°F)	v ft³/lbm	u BTU/lbm	h BTU/lbm	s BTU/lbm-°F
(sat)	0.36243	1103.8	1184.2	1.3678
600	0.40202	1134.9	1224.2	1.4061
700	0.49098	1201.5	1310.6	1.4842
800	0.56214	1254.1	1379.0	1.5408
900	0.62588	1302.0	1441.0	1.5882
1000	0.68560	1347.9	1500.1	1.6302
1100	0.74284	1393.1	1558.1	1.6686
1200	0.79842	1438.2	1615.5	1.7043
1300	0.85283	1483.5	1672.9	1.7379
1400	0.90636	1529.3	1730.6	1.7698
1500	0.95925	1575.7	1788.7	1.8002
1600	1.01160	1622.7	1847.3	1.8294
1700	1.06360	1670.4	1906.5	1.8575
1800	1.11520	1718.8	1966.4	1.8846
1900	1.16660	1767.9	2027.0	1.9108
2000	1.21780	1817.8	2088.2	1.9362
2100	1.26870	1868.4	2150.1	1.9609
2200	1.3195	1919.7	2212.7	1.9848
2300	1.3702	1971.6	2275.9	2.0082
2400	1.4208	2024.3	2339.8	2.0309
2500	1.4713	2077.6	2404.3	2.0531
2600	1.5217	2131.5	2469.4	2.0747
2700	1.5720	2186.0	2535.0	2.0958
2800	1.6222	2241.0	2601.2	2.1165
2900	1.6724	2296.6	2668.0	2.1366
3000	1.7226	2352.7	2735.2	2.1563

P = 1400 psia (587.14 °F)

T(°F)	v ft³/lbm	u BTU/lbm	h BTU/lbm	s BTU/lbm-°F
(sat)	0.30163	1096.3	1174.4	1.3465
600	0.31778	1111.5	1193.8	1.3649
700	0.40622	1189.8	1295.1	1.4566
800	0.47177	1246.3	1368.5	1.5174
900	0.52896	1296.1	1433.1	1.5668
1000	0.58184	1343.2	1494.0	1.6100
1100	0.63211	1389.3	1553.0	1.6491
1200	0.68066	1435.0	1611.3	1.6853
1300	0.72801	1480.8	1669.4	1.7193
1400	0.77447	1526.9	1727.5	1.7515
1500	0.82028	1573.5	1786.0	1.7821
1600	0.86556	1620.8	1845.0	1.8115
1700	0.91044	1668.6	1904.5	1.8397
1800	0.95499	1717.2	1964.6	1.8669
1900	0.99928	1766.5	2025.4	1.8932
2000	1.04330	1816.5	2086.8	1.9187
2100	1.08720	1867.2	2148.9	1.9434
2200	1.1309	1918.6	2211.6	1.9674
2300	1.1745	1970.6	2274.9	1.9908
2400	1.218	2023.4	2338.9	2.0136
2500	1.2614	2076.7	2403.5	2.0358
2600	1.3047	2130.7	2468.7	2.0575
2700	1.3479	2185.2	2534.4	2.0786
2800	1.3911	2240.3	2600.7	2.0992
2900	1.4342	2295.9	2667.5	2.1194
3000	1.4772	2352.1	2734.8	2.1392

Appendix G.3 - Superheated Steam - SI Units

P = 5000 kPa (263.94 °C)

T(°C)	v m³/kg	u kJ/kg	h kJ/kg	s kJ/kg-K
(sat)	0.039446	2597.0	2794.2	5.9737
300	0.045346	2699.0	2925.7	6.2110
350	0.051969	2809.5	3069.3	6.4516
400	0.057837	2907.5	3196.7	6.6483
450	0.063323	3000.6	3317.2	6.8210
500	0.068583	3091.7	3434.7	6.9781
550	0.073694	3182.4	3550.9	7.1237
600	0.078704	3273.3	3666.8	7.2605
650	0.083639	3365.0	3783.2	7.3901
700	0.088518	3457.7	3900.3	7.5136
750	0.093355	3551.6	4018.4	7.6320
800	0.098158	3646.9	4137.7	7.7458
850	0.10293	3743.6	4258.3	7.8556
900	0.10769	3841.8	4380.2	7.9618
950	0.11242	3941.5	4503.6	8.0648
1000	0.11715	4042.6	4628.3	8.1648
1050	0.12185	4145.2	4754.5	8.2620
1100	0.12655	4249.3	4882.0	8.3566
1150	0.13124	4354.8	5011.0	8.4488
1200	0.13592	4461.6	5141.2	8.5388
1250	0.14060	4569.8	5272.8	8.6266
1300	0.14527	4679.3	5405.7	8.7124
1350	0.14993	4790.1	5539.7	8.7963
1400	0.15459	4902.0	5675.0	8.8784
1450	0.15925	5015.2	5811.4	8.9587
1500	0.16390	5129.4	5948.9	9.0374

P = 6000 kPa (275.58 °C)

T(°C)	v m³/kg	u kJ/kg	h kJ/kg	s kJ/kg-K
(sat)	0.03245	2589.9	2784.6	5.8901
300	0.03619	2668.4	2885.5	6.0703
350	0.04225	2790.4	3043.9	6.3357
400	0.04742	2893.7	3178.2	6.5432
450	0.05217	2989.9	3302.9	6.7219
500	0.05667	3083.1	3423.1	6.8826
550	0.06102	3175.2	3541.3	7.0307
600	0.06527	3267.2	3658.7	7.1693
650	0.06943	3359.6	3776.2	7.3001
700	0.07355	3453.0	3894.3	7.4246
750	0.07761	3547.5	4013.2	7.5438
800	0.08165	3643.2	4133.1	7.6582
850	0.08566	3740.3	4254.2	7.7685
900	0.08964	3838.8	4376.6	7.8751
950	0.09361	3938.7	4500.3	7.9784
1000	0.09756	4040.1	4625.4	8.0786
1050	0.10150	4142.9	4751.9	8.1760
1100	0.10543	4247.1	4879.7	8.2709
1150	0.10935	4352.8	5008.9	8.3632
1200	0.11326	4459.8	5139.3	8.4534
1250	0.11717	4568.1	5271.1	8.5413
1300	0.12107	4677.7	5404.1	8.6272
1350	0.12496	4788.6	5538.3	8.7112
1400	0.12885	4900.6	5673.7	8.7934
1450	0.13274	5013.9	5810.3	8.8738
1500	0.13662	5128.2	5947.9	8.9525

Appendix G.3 - Superheated Steam - English Units

P = 1600 psia (604.93 °F)

T(°F)	v ft³/lbm	u BTU/lbm	h BTU/lbm	s BTU/lbm-°F
(sat)	0.25518	1087.7	1163.2	1.3262
700	0.34179	1177.1	1278.3	1.4302
800	0.40369	1238.1	1357.6	1.4959
900	0.45615	1290.1	1425.1	1.5475
1000	0.50396	1338.5	1487.7	1.5920
1100	0.54903	1385.4	1547.9	1.6319
1200	0.59233	1431.7	1607.1	1.6686
1300	0.63439	1477.9	1665.8	1.7030
1400	0.67556	1524.4	1724.5	1.7354
1500	0.71605	1571.4	1783.4	1.7663
1600	0.75603	1618.8	1842.7	1.7958
1700	0.79559	1666.9	1902.5	1.8242
1800	0.83482	1715.7	1962.8	1.8515
1900	0.87379	1765.1	2023.8	1.8779
2000	0.91253	1815.2	2085.4	1.9034
2100	0.95110	1866.0	2147.6	1.9282
2200	0.98950	1917.5	2210.5	1.9523
2300	1.0278	1969.6	2273.9	1.9757
2400	1.0659	2022.4	2338.0	1.9986
2500	1.1040	2075.8	2402.7	2.0208
2600	1.1420	2129.9	2468.0	2.0425
2700	1.1799	2184.4	2533.8	2.0636
2800	1.2177	2239.6	2600.1	2.0843
2900	1.2555	2295.3	2667.0	2.1045
3000	1.2932	2351.5	2734.4	2.1243

P = 1800 psia (621.07 °F)

T(°F)	v ft³/lbm	u BTU/lbm	h BTU/lbm	s BTU/lbm-°F
(sat)	0.21832	1077.9	1150.6	1.3063
700	0.29075	1163.1	1260.0	1.4044
800	0.35047	1229.5	1346.2	1.4759
900	0.39940	1283.9	1416.9	1.5299
1000	0.44333	1333.7	1481.3	1.5756
1100	0.48439	1381.4	1542.8	1.6164
1200	0.52361	1428.4	1602.8	1.6537
1300	0.56158	1475.1	1662.2	1.6884
1400	0.59862	1522.0	1721.4	1.7211
1500	0.63499	1569.2	1780.7	1.7522
1600	0.67083	1616.9	1840.3	1.7819
1700	0.70626	1665.2	1900.4	1.8104
1800	0.74136	1714.1	1961.0	1.8378
1900	0.77619	1763.7	2022.2	1.8643
2000	0.81080	1813.9	2084.0	1.8899
2100	0.84522	1864.8	2146.4	1.9148
2200	0.87949	1916.4	2209.3	1.9389
2300	0.91363	1968.6	2272.9	1.9624
2400	0.94765	2021.5	2337.1	1.9852
2500	0.98157	2075.0	2401.9	2.0075
2600	1.01540	2129.0	2467.3	2.0292
2700	1.04920	2183.7	2533.2	2.0504
2800	1.08290	2238.9	2599.6	2.0711
2900	1.11650	2294.6	2666.5	2.0913
3000	1.15010	2350.9	2734.0	2.1111

Appendix G.3 - Superheated Steam - SI Units

	P = 8000 kPa (295.01 °C)				P = 10000 kPa (311.00 °C)			
	v	u	h	s	v	u	h	s
T(°C)	m³/kg	kJ/kg	kJ/kg	kJ/kg-K	m³/kg	kJ/kg	kJ/kg	kJ/kg-K
(sat)	0.023526	2570.5	2758.7	5.7450	0.018030	2545.2	2725.5	5.6160
300	0.024279	2592.3	2786.5	5.7937	-	-	-	-
350	0.029975	2748.3	2988.1	6.1321	0.022440	2699.6	2924.0	5.9459
400	0.034344	2864.6	3139.4	6.3658	0.026436	2833.1	3097.4	6.2141
450	0.038194	2967.8	3273.3	6.5579	0.029782	2944.5	3242.3	6.4219
500	0.041767	3065.4	3399.5	6.7266	0.032811	3047.0	3375.1	6.5995
550	0.045172	3160.5	3521.8	6.8799	0.035654	3145.4	3502.0	6.7585
600	0.048463	3254.7	3642.4	7.0221	0.038378	3242.0	3625.8	6.9045
650	0.051675	3348.9	3762.3	7.1556	0.041018	3337.9	3748.1	7.0408
700	0.054828	3443.6	3882.2	7.2821	0.043597	3434.0	3870.0	7.1693
750	0.057937	3539.1	4002.6	7.4028	0.046131	3530.7	3992.0	7.2916
800	0.061011	3635.7	4123.8	7.5184	0.048629	3628.2	4114.5	7.4085
850	0.064057	3733.5	4246.0	7.6297	0.051099	3726.8	4237.8	7.5207
900	0.067082	3832.6	4369.3	7.7371	0.053547	3826.5	4362.0	7.6290
950	0.070088	3933.1	4493.8	7.8411	0.055976	3927.5	4487.3	7.7335
1000	0.073079	4035.0	4619.6	7.9419	0.058390	4029.9	4613.8	7.8349
1050	0.076057	4138.2	4746.7	8.0397	0.060792	4133.5	4741.4	7.9332
1100	0.079025	4242.8	4875.0	8.1350	0.063183	4238.5	4870.3	8.0288
1150	0.081983	4348.8	5004.6	8.2277	0.065564	4344.8	5000.4	8.1219
1200	0.084934	4456.1	5135.5	8.3181	0.067938	4452.3	5131.7	8.2126
1250	0.087878	4564.6	5267.7	8.4063	0.070305	4561.2	5264.2	8.3010
1300	0.090816	4674.5	5401.0	8.4924	0.072667	4671.3	5397.9	8.3874
1350	0.093749	4785.6	5535.6	8.5766	0.075023	4782.6	5532.8	8.4718
1400	0.096678	4897.8	5671.2	8.6589	0.077374	4895.0	5668.7	8.5543
1450	0.099602	5011.2	5808.0	8.7395	0.079722	5008.6	5805.8	8.6350
1500	0.10252	5125.7	5945.9	8.8184	0.082066	5123.2	5943.9	8.7140

adapted from NIST data

Appendix G.3 - Superheated Steam - English Units

	P = 2000 psia (635.85 °F)				P = 2500 psia (668.17 °F)			
	v	u	h	s	v	u	h	s
T(°F)	ft³/lbm	BTU/lbm	BTU/lbm	BTU/lbm-°F	ft³/lbm	BTU/lbm	BTU/lbm	BTU/lbm-°F
(sat)	0.18813	1066.8	1136.4	1.2863	0.13071	1031.1	1091.6	1.2329
700	0.24894	1147.6	1239.7	1.3783	0.16849	1098.4	1176.3	1.3072
800	0.30763	1220.5	1334.3	1.4568	0.22949	1195.9	1302.0	1.4116
900	0.35390	1277.5	1408.5	1.5135	0.27165	1260.7	1386.4	1.4762
1000	0.39479	1328.7	1474.8	1.5606	0.30726	1316.1	1458.2	1.5271
1100	0.43266	1377.5	1537.6	1.6022	0.33949	1367.3	1524.4	1.5710
1200	0.46864	1425.1	1598.5	1.6400	0.36966	1416.6	1587.6	1.6103
1300	0.50332	1472.3	1658.5	1.6752	0.39847	1465.1	1649.4	1.6465
1400	0.53708	1519.5	1718.3	1.7082	0.42631	1513.3	1710.5	1.6802
1500	0.57015	1567.0	1778.0	1.7395	0.45344	1561.6	1771.3	1.7121
1600	0.60269	1615.0	1838.0	1.7693	0.48004	1610.1	1832.2	1.7424
1700	0.63481	1663.5	1898.4	1.7980	0.50621	1659.1	1893.3	1.7714
1800	0.66660	1712.5	1959.2	1.8255	0.53204	1708.6	1954.8	1.7992
1900	0.69812	1762.2	2020.6	1.8521	0.55761	1758.7	2016.7	1.8260
2000	0.72942	1812.6	2082.6	1.8778	0.58295	1809.4	2079.1	1.8519
2100	0.76053	1863.6	2145.1	1.9027	0.60810	1860.7	2142.0	1.8770
2200	0.79149	1915.3	2208.2	1.9269	0.63310	1912.6	2205.5	1.9013
2300	0.82231	1967.6	2272.0	1.9504	0.65796	1965.1	2269.5	1.9249
2400	0.85303	2020.6	2336.3	1.9733	0.68272	2018.2	2334.1	1.9479
2500	0.88364	2074.1	2401.1	1.9956	0.70737	2071.9	2399.2	1.9703
2600	0.91417	2128.2	2466.6	2.0173	0.73194	2126.2	2464.8	1.9921
2700	0.94463	2182.9	2532.6	2.0386	0.75644	2181.1	2531.0	2.0134
2800	0.97502	2238.2	2599.1	2.0593	0.78087	2236.4	2597.7	2.0342
2900	1.0054	2294.0	2666.1	2.0795	0.80525	2292.3	2664.9	2.0545
3000	1.0356	2350.3	2733.6	2.0993	0.82957	2348.5	2732.5	2.0743

The reference state for all portions of the steam table is u (internal energy) and s (entropy) are both zero for the saturated liquid at the triple point of water.

Appendix G.4 - Compressed Liquid Water - SI Units

P = 1000 kPa (179.88 °C)

T(°C)	v m³/kg	u kJ/kg	h kJ/kg	s kJ/kg-K
10	0.001000	42.00	43.00	0.1510
20	0.001001	83.85	84.85	0.2963
30	0.001004	125.64	126.64	0.4365
40	0.001007	167.41	168.41	0.5720
50	0.001012	209.18	210.19	0.7034
60	0.001017	250.99	252.00	0.8308
70	0.001022	292.84	293.86	0.9546
80	0.001029	334.74	335.77	1.0750
90	0.001036	376.72	377.76	1.1922
100	0.001043	418.80	419.84	1.3065
110	0.001051	460.99	462.04	1.4181
120	0.001060	503.32	504.38	1.5272
130	0.001069	545.81	546.88	1.6339
140	0.001079	588.50	589.58	1.7386
150	0.001090	631.41	632.50	1.8412
160	0.001102	674.59	675.70	1.9421
170	0.001114	718.08	719.20	2.0414

P = 2000 kPa (212.38 °C)

T(°C)	v m³/kg	u kJ/kg	h kJ/kg	s kJ/kg-K
10	0.000999	41.97	43.97	0.15091
20	0.001001	83.79	85.79	0.29607
30	0.001004	125.55	127.55	0.43615
40	0.001007	167.28	169.30	0.57163
50	0.001011	209.03	211.06	0.70289
60	0.001016	250.81	252.84	0.83024
70	0.001022	292.63	294.68	0.95396
80	0.001028	334.51	336.57	1.07430
90	0.001035	376.46	378.53	1.19150
100	0.001043	418.51	420.59	1.30570
110	0.001051	460.67	462.77	1.41730
120	0.001059	502.96	505.08	1.52630
130	0.001069	545.42	547.55	1.63300
140	0.001079	588.07	590.22	1.73750
150	0.001090	630.94	633.12	1.84010
160	0.001101	674.08	676.28	1.94010
170	0.001113	717.52	719.74	2.04010
180	0.001127	761.30	763.56	2.13790
190	0.001141	805.49	807.77	2.23440
200	0.001156	850.14	852.45	2.32980
210	0.001173	895.32	897.66	2.42440

P = 5000 kPa (263.94 °C)

T(°C)	v m³/kg	u kJ/kg	h kJ/kg	s kJ/kg-K
20	0.001000	83.61	88.61	0.2954
40	0.001006	166.92	171.95	0.5705
60	0.001015	250.29	255.36	0.8287
80	0.001027	333.82	338.95	1.0723
100	0.001041	417.64	422.85	1.3034
120	0.001058	501.90	507.19	1.5236
140	0.001077	586.79	592.18	1.7344
160	0.001099	672.55	678.04	1.9374
180	0.001124	759.46	765.08	2.1338
200	0.001153	847.91	853.68	2.3251
220	0.001187	938.39	944.32	2.5127
240	0.001227	1031.60	1037.70	2.6983
260	0.001276	1128.50	1134.90	2.8841

P = 10000 kPa (311.00 °C)

T(°C)	v m³/kg	u kJ/kg	h kJ/kg	s kJ/kg-K
20	0.000997	83.31	93.28	0.29435
40	0.001004	166.33	176.36	0.56851
60	0.001013	249.42	259.55	0.82602
80	0.001024	332.69	342.94	1.06910
100	0.001039	416.23	426.62	1.29960
120	0.001055	500.18	510.73	1.51910
140	0.001074	584.71	595.45	1.72930
160	0.001095	670.06	681.01	1.93150
180	0.001120	756.48	767.68	2.12710
200	0.001148	844.31	855.80	2.31740
220	0.001181	934.00	945.81	2.50370
240	0.001219	1026.10	1038.30	2.68760
260	0.001265	1121.60	1134.30	2.87100
280	0.001323	1221.80	1235.00	3.05650
300	0.001398	1329.40	1343.30	3.24880

Appendix G.4 - Compressed Liquid Water - English Units

P = 150 psia (358.42 °F)

T(°F)	v ft³/lbm	u BTU/lbm	h BTU/lbm	s BTU/lbm-°F
40	0.016012	8.03	8.48	0.01620
60	0.016027	28.06	28.50	0.05550
80	0.016066	48.03	48.47	0.09321
100	0.016124	67.98	68.43	0.12952
120	0.016198	87.93	88.38	0.16455
140	0.016285	107.90	108.35	0.19842
160	0.016387	127.89	128.35	0.23122
180	0.016500	147.92	148.38	0.26303
200	0.016626	167.98	168.44	0.29392
220	0.016764	188.10	188.57	0.32397
240	0.016914	208.29	208.76	0.35325
260	0.017076	228.55	229.03	0.38181
280	0.017252	248.92	249.39	0.40973
300	0.017442	269.40	269.88	0.43706
320	0.017647	290.02	290.51	0.46385
340	0.017869	310.80	311.29	0.49018

P = 300 psia (417.35 °F)

T(°F)	v ft³/lbm	u BTU/lbm	h BTU/lbm	s BTU/lbm-°F
40	0.016003	8.03	8.92	0.01620
60	0.016020	28.04	28.93	0.05546
80	0.016059	47.99	48.88	0.09314
100	0.016117	67.93	68.82	0.12942
120	0.016190	87.87	88.77	0.16444
140	0.016278	107.82	108.73	0.19829
160	0.016379	127.80	128.71	0.23107
180	0.016492	147.81	148.73	0.26286
200	0.016618	167.86	168.79	0.29374
220	0.016755	187.97	188.90	0.32377
240	0.016905	208.13	209.07	0.35303
260	0.017067	228.38	229.33	0.38158
280	0.017242	248.73	249.69	0.40947
300	0.017431	269.19	270.16	0.43678
320	0.017635	289.79	290.77	0.46356
340	0.017856	310.54	311.54	0.48986
360	0.018094	331.49	332.49	0.51575
380	0.018352	352.65	353.67	0.54127
400	0.018633	374.07	375.11	0.56650

P = 750 psia (510.89 °F)

T(°F)	v ft³/lbm	u BTU/lbm	h BTU/lbm	s BTU/lbm-°F
40	0.015979	8.03	10.24	0.01619
80	0.016037	47.88	50.11	0.09294
120	0.016168	87.68	89.92	0.16410
160	0.016356	127.53	129.80	0.23063
200	0.016593	167.51	169.81	0.29319
240	0.016878	207.69	210.03	0.35238
280	0.017212	248.17	250.56	0.40872
320	0.017600	289.11	291.55	0.46268
360	0.018051	330.65	333.16	0.51472
400	0.018580	373.04	375.62	0.56530
440	0.019209	416.57	419.24	0.61488
480	0.019973	461.70	464.47	0.66407

P = 1500 psia (596.26 °F)

T(°F)	v ft³/lbm	u BTU/lbm	h BTU/lbm	s BTU/lbm-°F
40	0.015939	8.01	12.44	0.01615
80	0.016000	47.70	52.14	0.09259
120	0.016132	87.36	91.84	0.16354
160	0.016319	127.08	131.61	0.22989
200	0.016553	166.92	171.52	0.29229
240	0.016834	206.95	211.62	0.35132
280	0.017162	247.27	252.03	0.40748
320	0.017542	288.00	292.87	0.46125
360	0.017983	329.30	334.29	0.51306
400	0.018496	371.37	376.51	0.56334
440	0.019102	414.49	419.79	0.61254
480	0.019832	459.03	464.53	0.66119
520	0.020739	505.60	511.36	0.70999
560	0.021928	555.32	561.41	0.76005

Appendix G.4 - Compressed Liquid Water - SI Units

		P = 25000 kPa				P = 50000 kPa		
	v	u	h	s	v	u	h	s
T(°C)	m³/kg	kJ/kg	kJ/kg	kJ/kg-K	m³/kg	kJ/kg	kJ/kg	kJ/kg-K
20	0.000991	82.41	107.18	0.2909	0.000980	80.93	129.95	0.28454
100	0.001031	412.17	437.95	1.2883	0.001020	405.93	456.94	1.27050
200	0.001135	834.24	862.61	2.2956	0.001115	819.45	875.19	2.26280
300	0.001346	1297.60	1331.30	3.1919	0.001288	1259.60	1324.00	3.12180
400	0.006005	2428.50	2578.60	5.1400	0.001731	1787.80	1874.40	4.00290
500	0.011143	2887.30	3165.90	5.9642	0.003890	2528.10	2722.60	5.17620
600	0.014140	3140.00	3493.50	6.3637	0.006108	2947.10	3252.50	5.82450
700	0.016643	3359.90	3776.00	6.6702	0.007717	3228.70	3614.60	6.21780
800	0.018922	3570.70	4043.80	6.9322	0.009072	3472.20	3925.80	6.52250
900	0.021075	3780.20	4307.10	7.1668	0.010296	3702.00	4216.80	6.78190
1000	0.023150	3991.50	4570.20	7.3820	0.011441	3927.30	4499.40	7.01310
1100	0.025172	4206.00	4835.40	7.5825	0.012534	4152.20	4778.90	7.22440
1200	0.027157	4424.60	5103.50	7.7710	0.013590	4378.60	5058.10	7.42070
1300	0.029115	4647.20	5375.10	7.9493	0.014620	4607.40	5338.40	7.60480
1400	0.031052	4873.90	5650.30	8.1189	0.015631	4839.20	5620.80	7.77880
1500	0.032974	5104.70	5929.00	8.2807	0.016626	5074.10	5905.40	7.94400

Appendix G.4 - Compressed Liquid Water - English Units

		P = 5000 psia				P = 10000 psia		
	v	u	h	s	v	u	h	s
T(°F)	ft³/lbm	BTU/lbm	BTU/lbm	BTU/lbm-°F	ft³/lbm	BTU/lbm	BTU/lbm	BTU/lbm-°F
40	0.015762	7.92	22.50	0.01574	0.015530	7.65	36.39	0.01457
200	0.016375	164.35	179.51	0.28824	0.016145	161.08	190.96	0.28280
400	0.018145	364.35	381.14	0.55493	0.017730	355.96	388.77	0.54450
600	0.021943	584.41	604.72	0.78804	0.020657	561.97	600.20	0.76523
800	0.059365	986.87	1041.80	1.15820	0.027646	800.24	851.40	0.98158
1000	0.131280	1244.00	1365.50	1.40040	0.049495	1081.90	1173.50	1.21870
1200	0.171850	1372.10	1531.10	1.50710	0.075585	1277.30	1417.20	1.37590
1400	0.205080	1481.40	1671.10	1.58680	0.096242	1416.10	1594.20	1.47680
1600	0.235050	1585.60	1803.10	1.65420	0.113860	1536.40	1747.10	1.55500
1800	0.263200	1689.00	1932.50	1.71420	0.129790	1650.00	1890.20	1.62130
2000	0.290230	1793.20	2061.70	1.76900	0.144670	1761.20	2028.90	1.68010
2200	0.316510	1899.00	2191.80	1.81980	0.158850	1872.20	2166.20	1.73380
2400	0.342260	2006.60	2323.30	1.86750	0.172550	1983.80	2303.10	1.78340

adapted from NIST data

The reference state for all portions of the steam table is u (internal energy) and s (entropy) are both zero for the saturated liquid at the triple point of water.

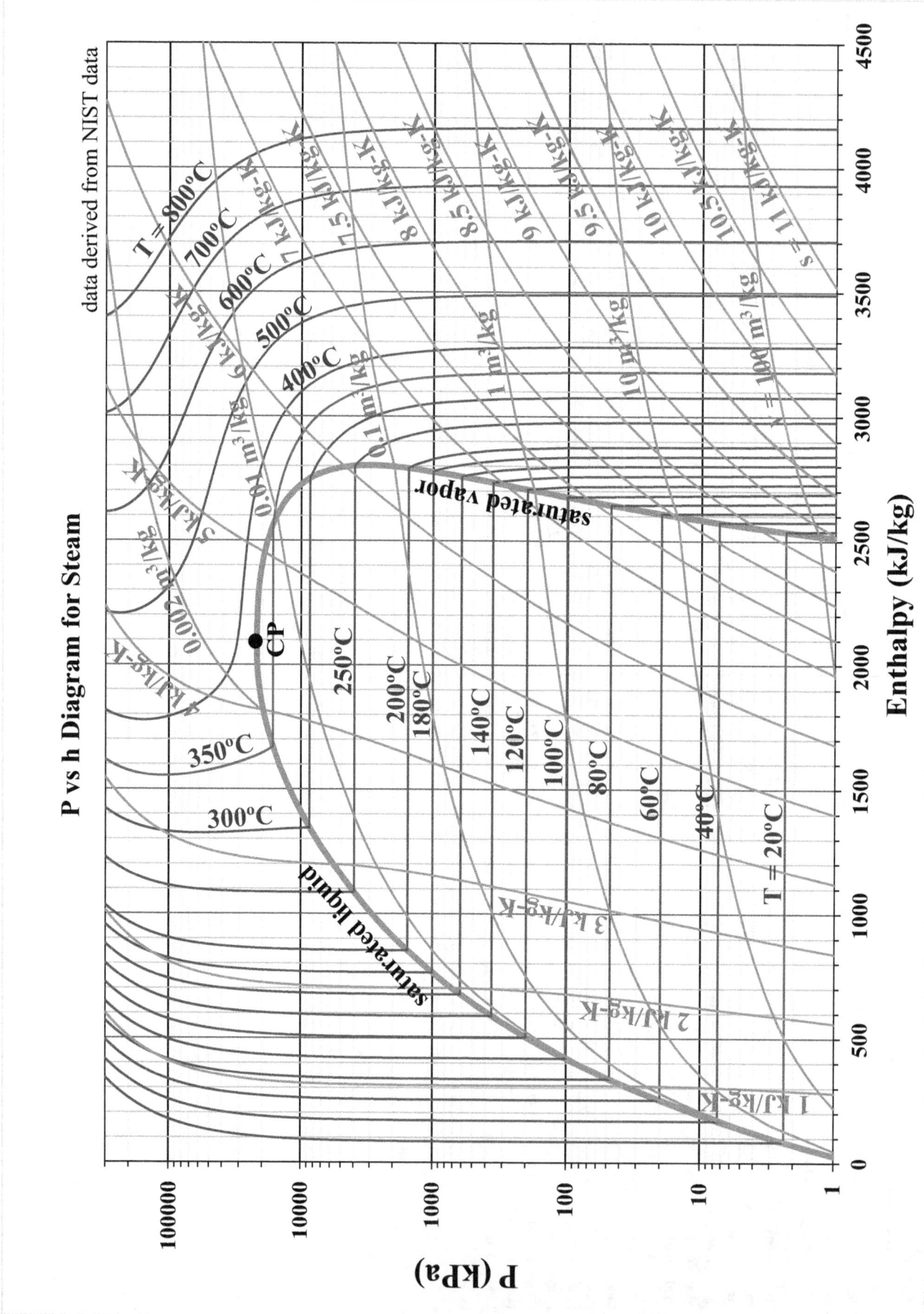

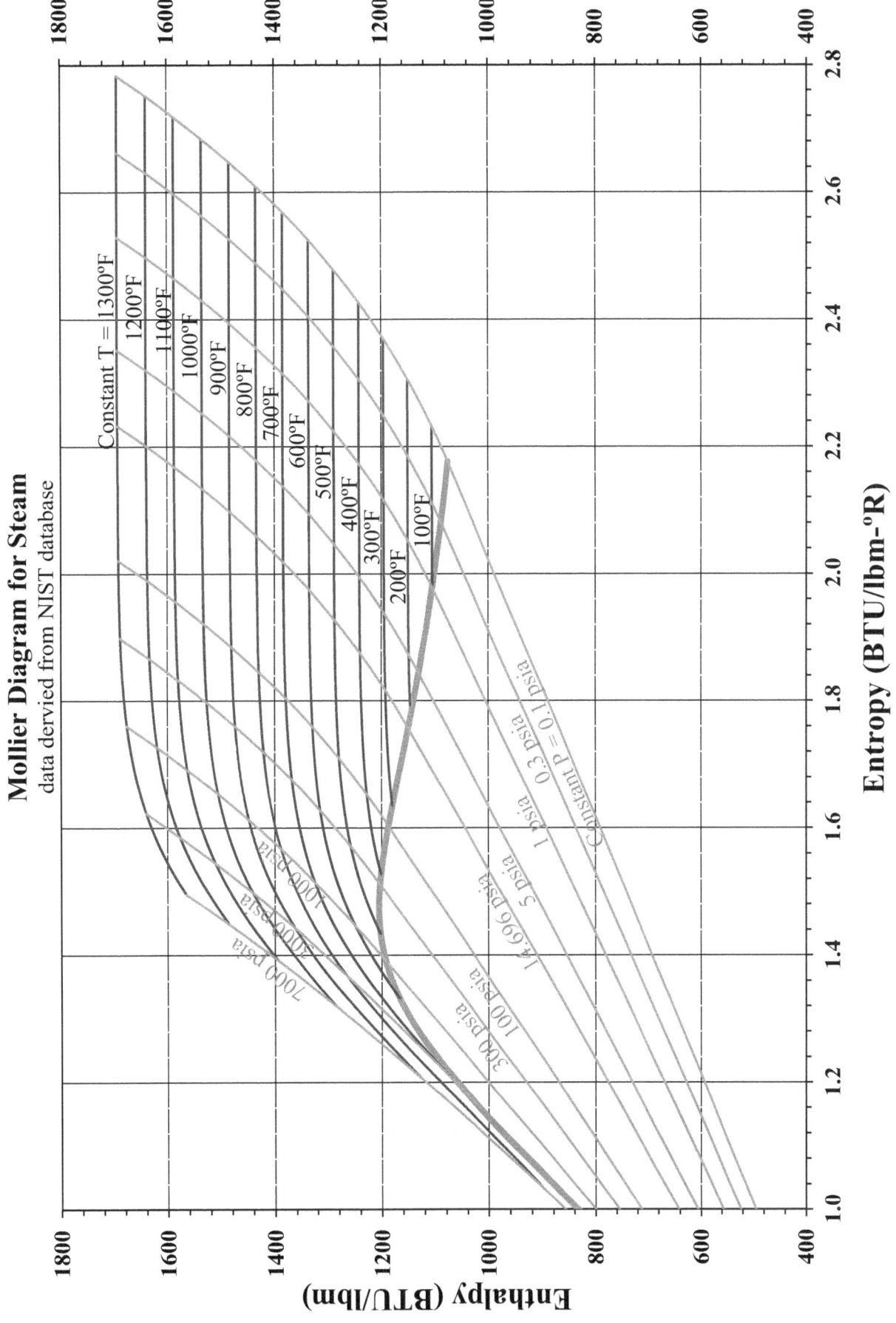

Appendix G - Steam Tables

T-S Diagram for Steam — data derived from NIST database

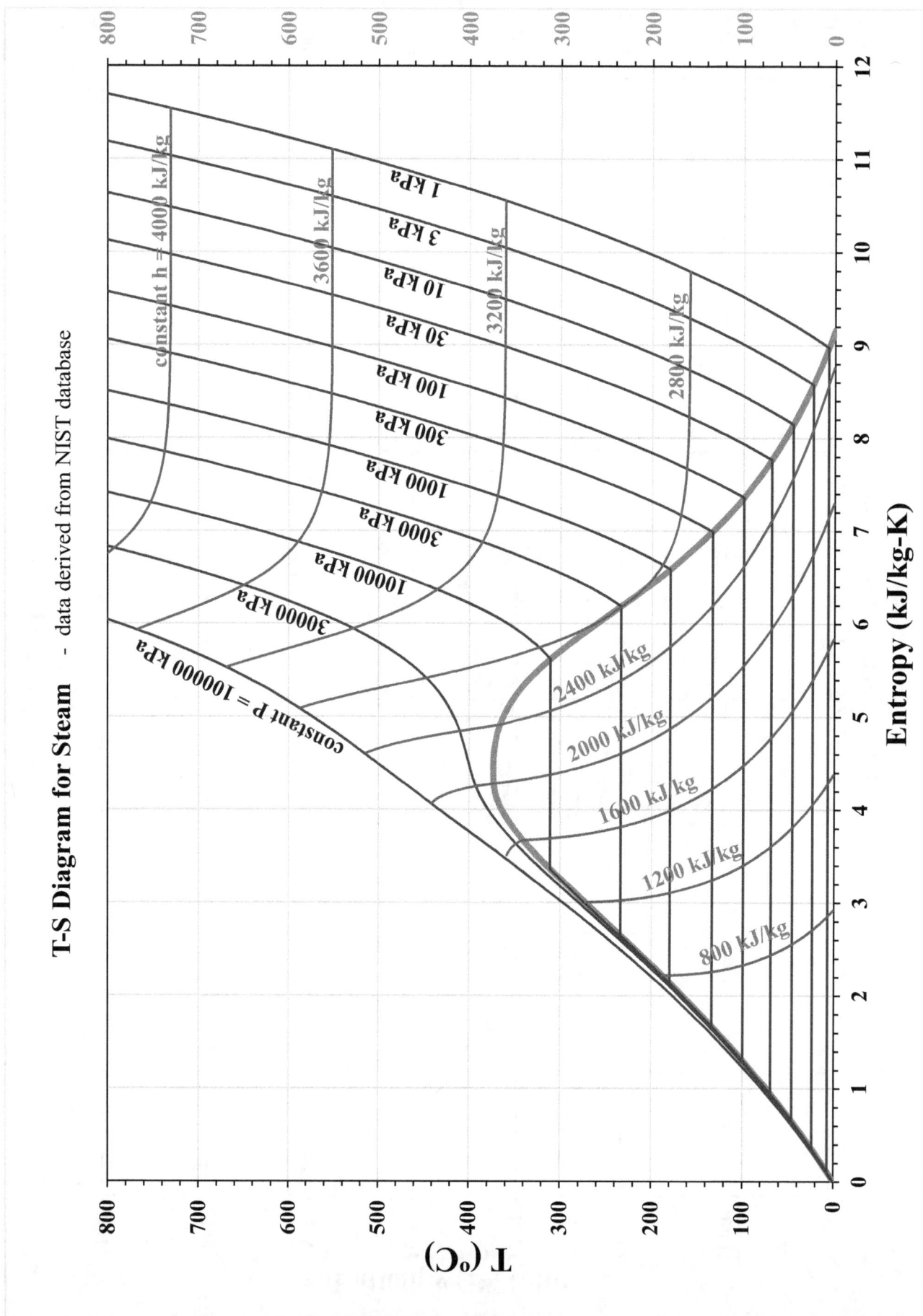

Appendix H.1 - Saturated R-134a (1,1,1,2-tetrafuoroethane)

Temperature Entry

T	P	v^L	v^V	u^L	u^V	h^L	h^V	s^L	s^V
°C	kPa	m³/kg	m³/kg	kJ/kg	kJ/kg	kJ/kg	kJ/kg	kJ/kg-K	kJ/kg-K
-100	0.56	0.000632	25.19300	75.36	322.76	75.36	336.85	0.4354	1.9456
-95	0.94	0.000637	15.43500	81.29	325.29	81.29	339.78	0.4691	1.9201
-90	1.52	0.000643	9.76980	87.23	327.87	87.23	342.76	0.5020	1.8972
-85	2.40	0.000648	6.37070	93.18	330.49	93.18	345.77	0.5341	1.8766
-80	3.67	0.000654	4.26820	99.16	333.15	99.16	348.83	0.5654	1.8580
-75	5.48	0.000660	2.93120	105.16	335.85	105.17	351.91	0.5961	1.8414
-70	7.98	0.000666	2.05900	111.19	338.59	111.20	355.02	0.6262	1.8264
-65	11.38	0.000672	1.47650	117.26	341.35	117.26	358.16	0.6557	1.8130
-60	15.91	0.000678	1.07900	123.35	344.15	123.36	361.31	0.6846	1.8010
-55	21.83	0.000685	0.80236	129.48	346.96	129.50	364.48	0.7131	1.7902
-50	29.45	0.000691	0.60620	135.65	349.80	135.67	367.65	0.7410	1.7806
-45	39.12	0.000698	0.46473	141.86	352.65	141.89	370.83	0.7685	1.7720
-40	51.21	0.000705	0.36108	148.11	355.51	148.14	374.00	0.7956	1.7643
-35	66.14	0.000713	0.28402	154.40	358.38	154.44	377.17	0.8223	1.7575
-30	84.38	0.000720	0.22594	160.73	361.25	160.79	380.32	0.8486	1.7515
-25	106.40	0.000728	0.18162	167.11	364.12	167.19	383.45	0.8746	1.7461
-20	132.73	0.000736	0.14739	173.54	366.99	173.64	386.55	0.9003	1.7413
-15	163.94	0.000745	0.12067	180.02	369.85	180.14	389.63	0.9256	1.7371
-10	200.60	0.000754	0.09959	186.55	372.69	186.70	392.66	0.9507	1.7333
-5	243.34	0.000763	0.08280	193.13	375.51	193.32	395.66	0.9754	1.7300
0	292.80	0.000772	0.06931	199.77	378.31	200.00	398.60	1.0000	1.7271
5	349.66	0.000782	0.05837	206.48	381.08	206.75	401.49	1.0243	1.7245
10	414.61	0.000793	0.04944	213.25	383.82	213.58	404.32	1.0485	1.7221
15	488.37	0.000804	0.04209	220.09	386.52	220.48	407.07	1.0724	1.7200
20	571.71	0.000816	0.03600	227.00	389.17	227.47	409.75	1.0962	1.7180
25	665.38	0.000829	0.03091	233.99	391.77	234.55	412.33	1.1199	1.7162
30	770.20	0.000842	0.02664	241.07	394.30	241.72	414.82	1.1435	1.7145
35	886.98	0.000857	0.02303	248.25	396.76	249.01	417.19	1.1670	1.7128
40	1016.60	0.000872	0.01997	255.52	399.13	256.41	419.43	1.1905	1.7111
45	1159.90	0.000889	0.01734	262.91	401.40	263.94	421.52	1.2139	1.7092
50	1317.90	0.000907	0.01509	270.43	403.55	271.62	423.44	1.2375	1.7072
55	1491.50	0.000927	0.01314	278.09	405.55	279.47	425.15	1.2611	1.7050
60	1681.80	0.000950	0.01144	285.91	407.38	287.50	426.63	1.2848	1.7024
65	1889.80	0.000975	0.00996	293.92	408.99	295.76	427.82	1.3088	1.6993
70	2116.80	0.001004	0.00865	302.16	410.33	304.28	428.65	1.3332	1.6956
75	2364.10	0.001037	0.00749	310.68	411.32	313.13	429.03	1.3580	1.6909
80	2633.20	0.001077	0.00645	319.55	411.83	322.39	428.81	1.3836	1.6850
85	2925.80	0.001127	0.00550	328.93	411.67	332.22	427.76	1.4104	1.6771
90	3244.20	0.001194	0.00461	339.06	410.45	342.93	425.42	1.4390	1.6662
95	3591.20	0.001294	0.00374	350.60	407.23	355.25	420.67	1.4715	1.6492
100	3972.40	0.001536	0.00268	367.20	397.03	373.30	407.68	1.5188	1.6109
101.06	4059.10	0.001954	0.00195	381.71	381.71	389.64	389.64	1.5621	1.5621

Appendix H.1 - Saturated R-134a (1,1,1,2-tetrafuoroethane)

Pressure Entry

P kPa	T °C	v^L m³/kg	v^V m³/kg	u^L kJ/kg	u^V kJ/kg	h^L kJ/kg	h^V kJ/kg	s^L kJ/kg-K	s^V kJ/kg-K
1	-94.37	0.000638	14.54300	82.04	325.61	82.04	340.16	0.4733	1.9171
2	-87.04	0.000646	7.56150	90.74	329.41	90.75	344.54	0.5211	1.8848
3	-82.41	0.000651	5.16170	96.27	331.86	96.27	347.35	0.5504	1.8667
4	-78.96	0.000655	3.93830	100.41	333.71	100.41	349.47	0.5719	1.8544
5	-76.17	0.000658	3.19340	103.76	335.22	103.76	351.19	0.5890	1.8451
6	-73.82	0.000661	2.69100	106.59	336.50	106.59	352.64	0.6033	1.8377
7	-71.78	0.000664	2.32860	109.05	337.61	109.05	353.91	0.6156	1.8315
8	-69.97	0.000666	2.05450	111.23	338.61	111.24	355.04	0.6264	1.8263
9	-68.34	0.000668	1.83970	113.20	339.50	113.21	356.06	0.6361	1.8218
10	-66.86	0.000670	1.66670	115.00	340.32	115.01	356.99	0.6448	1.8178
20	-56.41	0.000683	0.87081	127.75	346.17	127.77	363.58	0.7051	1.7931
30	-49.68	0.000692	0.59580	136.05	349.98	136.07	367.85	0.7428	1.7800
40	-44.60	0.000699	0.45511	142.36	352.88	142.39	371.09	0.7707	1.7713
50	-40.45	0.000705	0.36925	147.54	355.25	147.57	373.71	0.7932	1.7650
60	-36.94	0.000710	0.31123	151.96	357.27	152.00	375.94	0.8120	1.7601
70	-33.86	0.000714	0.26930	155.84	359.04	155.89	377.89	0.8284	1.7561
80	-31.12	0.000719	0.23755	159.31	360.61	159.37	379.62	0.8428	1.7528
90	-28.63	0.000722	0.21264	162.47	362.04	162.54	381.18	0.8558	1.7499
100	-26.36	0.000726	0.19256	165.37	363.34	165.44	382.60	0.8676	1.7475
200	-10.08	0.000753	0.09988	186.45	372.64	186.60	392.62	0.9503	1.7334
300	0.67	0.000774	0.06770	200.67	378.68	200.90	398.99	1.0033	1.7267
400	8.93	0.000791	0.05121	211.79	383.24	212.11	403.72	1.0433	1.7226
500	15.74	0.000806	0.04112	221.10	386.91	221.50	407.47	1.0759	1.7197
600	21.57	0.000820	0.03430	229.19	389.99	229.68	410.57	1.1037	1.7175
700	26.71	0.000833	0.02937	236.41	392.64	236.99	413.20	1.1280	1.7156
800	31.33	0.000846	0.02562	242.97	394.96	243.65	415.46	1.1497	1.7140
900	35.53	0.000858	0.02269	249.01	397.01	249.78	417.43	1.1695	1.7126
1000	39.39	0.000870	0.02032	254.63	398.85	255.50	419.16	1.1876	1.7113
1200	46.32	0.000894	0.01672	264.87	401.98	265.95	422.04	1.2201	1.7087
1400	52.42	0.000917	0.01411	274.12	404.54	275.40	424.30	1.2489	1.7062
1600	57.91	0.000940	0.01213	282.61	406.64	284.11	426.04	1.2748	1.7036
1800	62.90	0.000964	0.01056	290.52	408.35	292.26	427.36	1.2987	1.7007
2000	67.48	0.000989	0.00929	297.98	409.70	299.95	428.28	1.3209	1.6976
2200	71.73	0.001015	0.00824	305.07	410.72	307.30	428.84	1.3417	1.6941
2400	75.69	0.001042	0.00734	311.88	411.42	314.38	429.04	1.3615	1.6902
2600	79.41	0.001072	0.00657	318.47	411.80	321.26	428.88	1.3805	1.6858
2800	82.90	0.001105	0.00589	324.92	411.84	328.01	428.33	1.3990	1.6807
3000	86.20	0.001141	0.00528	331.28	411.49	334.70	427.34	1.4171	1.6748
3200	89.33	0.001183	0.00473	337.64	410.70	341.43	425.83	1.4350	1.6679
3400	92.30	0.001234	0.00422	344.12	409.32	348.31	423.65	1.4533	1.6595
3600	95.12	0.001297	0.00372	350.91	407.11	355.58	420.50	1.4724	1.6487
3800	97.80	0.001389	0.00321	358.50	403.41	363.77	415.63	1.4939	1.6336
4000	100.34	0.001580	0.00256	369.25	395.13	375.57	405.38	1.5247	1.6046
4059.10	101.06	0.001954	0.00195	381.71	381.71	389.64	389.64	1.5621	1.5621

adapted from NIST data   The reference state for all R-134a data is h = 200 kJ/kg and s = 1 kJ/kg-K for the saturated liquid at 0°C. This choice is common and results in h > 0 for all commonly encountered states.

Appendix H.2 - Superheated R-134a (1,1,1,2-tetrafuoroethane)

T(°C)	P = 100 kPa (-26.36°C)				P = 200 kPa (-10.08°C)			
	v m³/kg	u kJ/kg	h kJ/kg	s kJ/kg-K	v m³/kg	u kJ/kg	h kJ/kg	s kJ/kg-K
sat	0.19256	363.34	382.60	1.7475	0.09988	372.64	392.62	1.7334
-20	0.19841	367.81	387.65	1.7677	-	-	-	-
-10	0.20743	374.89	395.64	1.7986	0.09992	372.70	392.68	1.7337
0	0.21630	382.10	403.73	1.8288	0.10481	380.23	401.20	1.7654
10	0.22506	389.45	411.95	1.8584	0.10955	387.81	409.73	1.7961
20	0.23373	396.94	420.31	1.8874	0.11419	395.49	418.33	1.8259
30	0.24233	404.59	428.82	1.9159	0.11874	403.29	427.04	1.8551
40	0.25088	412.40	437.49	1.9441	0.12323	411.22	435.87	1.8838
50	0.25938	420.37	446.30	1.9718	0.12766	419.29	444.83	1.9120
60	0.26783	428.49	455.28	1.9991	0.13206	427.51	453.92	1.9397
70	0.27626	436.78	464.41	2.0261	0.13642	435.88	463.16	1.9670
80	0.28466	445.23	473.70	2.0528	0.14074	444.39	472.54	1.9939
90	0.29303	453.84	483.14	2.0792	0.14505	453.06	482.07	2.0206
100	0.30138	462.61	492.74	2.1053	0.14933	461.88	491.75	2.0468
110	0.30971	471.53	502.50	2.1311	0.15359	470.86	501.57	2.0728
120	0.31803	480.61	512.42	2.1566	0.15783	479.98	511.55	2.0985
130	0.32633	489.85	522.49	2.1819	0.16206	489.26	521.67	2.1239
140	0.33462	499.25	532.71	2.2070	0.16628	498.69	531.94	2.1491
150	0.34290	508.80	543.09	2.2318	0.17048	508.27	542.37	2.1740
160	0.35117	518.50	553.62	2.2564	0.17468	518.00	552.94	2.1987
170	0.35943	528.36	564.30	2.2807	0.17887	527.88	563.66	2.2232
180	0.36768	538.36	575.13	2.3049	0.18304	537.91	574.52	2.2474

T(°C)	P = 300 kPa (0.67 °C)				P = 400 kPa (8.93 °C)			
	v m³/kg	u kJ/kg	h kJ/kg	s kJ/kg-K	v m³/kg	u kJ/kg	h kJ/kg	s kJ/kg-K
sat	0.06770	378.68	398.99	1.7267	0.05121	383.24	403.72	1.7226
10	0.07093	386.06	407.34	1.7567	0.05151	384.12	404.72	1.7261
20	0.07425	393.96	416.24	1.7876	0.05421	392.32	414.01	1.7584
30	0.07748	401.93	425.17	1.8175	0.05680	400.50	423.22	1.7893
40	0.08063	410.00	434.19	1.8468	0.05929	408.73	432.45	1.8192
50	0.08372	418.19	443.30	1.8755	0.06172	417.05	441.74	1.8484
60	0.08677	426.50	452.53	1.9036	0.06410	425.47	451.11	1.8770
70	0.08978	434.95	461.89	1.9312	0.06644	434.01	460.58	1.9050
80	0.09276	443.54	471.37	1.9585	0.06875	442.67	470.17	1.9325
90	0.09571	452.27	480.99	1.9853	0.07102	451.47	479.88	1.9596
100	0.09863	461.15	490.74	2.0118	0.07328	460.41	489.72	1.9864
110	0.10154	470.17	500.63	2.0380	0.07551	469.48	499.68	2.0127
120	0.10443	479.34	510.67	2.0638	0.07772	478.69	509.78	2.0387
130	0.10730	488.66	520.85	2.0894	0.07991	488.05	520.02	2.0645
140	0.11016	498.12	531.17	2.1147	0.08210	497.55	530.39	2.0899
150	0.11301	507.74	541.64	2.1397	0.08427	507.20	540.90	2.1150
160	0.11585	517.50	552.25	2.1645	0.08643	516.99	551.56	2.1399
170	0.11868	527.40	563.01	2.1891	0.08858	526.92	562.35	2.1645
180	0.12150	537.46	573.91	2.2134	0.09072	537.00	573.29	2.1889

Appendix H.2 - Superheated R-134a (1,1,1,2-tetrafuoroethane)

T(°C)	P = 500 kPa (15.74 °C)				P = 600 kPa (21.57 °C)			
	v m³/kg	u kJ/kg	h kJ/kg	s kJ/kg-K	v m³/kg	u kJ/kg	h kJ/kg	s kJ/kg-K
sat	0.04112	386.91	407.47	1.7197	0.03430	389.99	410.57	1.7175
20	0.04212	390.55	411.61	1.7339	-	-	-	-
30	0.04434	398.99	421.16	1.7659	0.03598	397.37	418.96	1.7455
40	0.04646	407.40	430.63	1.7967	0.03787	406.01	428.73	1.7772
50	0.04850	415.86	440.11	1.8265	0.03966	414.63	438.43	1.8077
60	0.05049	424.40	449.64	1.8555	0.04139	423.30	448.13	1.8373
70	0.05243	433.04	459.25	1.8839	0.04307	432.04	457.88	1.8661
80	0.05433	441.78	468.95	1.9118	0.04471	440.87	467.70	1.8943
90	0.05621	450.65	478.75	1.9392	0.04632	449.82	477.61	1.9220
100	0.05805	459.65	488.67	1.9661	0.04790	458.88	487.62	1.9492
110	0.05988	468.78	498.72	1.9927	0.04946	468.06	497.74	1.9759
120	0.06169	478.04	508.88	2.0189	0.05100	477.37	507.97	2.0023
130	0.06348	487.44	519.18	2.0447	0.05252	486.82	518.33	2.0283
140	0.06526	496.98	529.60	2.0703	0.05403	496.39	528.81	2.0540
150	0.06702	506.65	540.17	2.0955	0.05552	506.11	539.42	2.0794
160	0.06878	516.47	550.86	2.1205	0.05701	515.96	550.16	2.1045
170	0.07052	526.44	561.70	2.1453	0.05848	525.95	561.03	2.1293
180	0.07226	536.54	572.67	2.1697	0.05995	536.08	572.04	2.1539

T(°C)	P = 800 kPa (31.33 °C)				P = 1000 kPa (39.39 °C)			
	v m³/kg	u kJ/kg	h kJ/kg	s kJ/kg-K	v m³/kg	u kJ/kg	h kJ/kg	s kJ/kg-K
sat	0.02562	394.96	415.46	1.7140	0.02032	398.85	419.16	1.7113
40	0.02704	402.96	424.59	1.7436	0.02041	399.45	419.86	1.7135
50	0.02855	412.00	434.84	1.7758	0.02180	409.09	430.88	1.7482
60	0.02997	420.97	444.95	1.8067	0.02307	418.46	441.53	1.7806
70	0.03134	429.96	455.03	1.8364	0.02426	427.74	452.00	1.8116
80	0.03266	438.99	465.12	1.8654	0.02540	437.00	462.40	1.8414
90	0.03394	448.10	475.25	1.8937	0.02649	446.30	472.79	1.8705
100	0.03519	457.30	485.45	1.9214	0.02755	455.65	483.20	1.8988
110	0.03642	466.60	495.73	1.9486	0.02858	465.09	493.67	1.9264
120	0.03763	476.01	506.12	1.9754	0.02959	474.62	504.21	1.9536
130	0.03881	485.55	516.60	2.0017	0.03058	484.25	514.83	1.9803
140	0.03999	495.21	527.20	2.0277	0.03155	494.00	525.55	2.0065
150	0.04114	504.99	537.91	2.0533	0.03251	503.86	536.37	2.0324
160	0.04229	514.91	548.74	2.0786	0.03346	513.84	547.30	2.0579
170	0.04343	524.96	559.70	2.1036	0.03439	523.95	558.34	2.0831
180	0.04455	535.14	570.78	2.1283	0.03532	534.19	569.51	2.1080

Appendix H.2 - Superheated R-134a (1,1,1,2-tetrafuoroethane)

T(°C)	P = 2000 kPa (67.48 °C)				P = 4000 kPa (100.34 °C)			
	v m^3/kg	u kJ/kg	h kJ/kg	s kJ/kg-K	v m^3/kg	u kJ/kg	h kJ/kg	s kJ/kg-K
sat	0.00929	409.70	428.28	1.6976	0.00256	395.13	405.38	1.6046
70	0.00957	412.91	432.06	1.7086	-	-	-	-
80	0.01054	424.70	445.77	1.7480	-	-	-	-
90	0.01136	435.66	458.38	1.7833	-	-	-	-
100	0.01211	446.24	470.45	1.8160	-	-	-	-
110	0.01279	456.63	482.21	1.8471	0.00427	429.17	446.26	1.7131
120	0.01344	466.93	493.80	1.8770	0.00499	445.13	465.10	1.7616
130	0.01405	477.21	505.31	1.9059	0.00555	458.71	480.90	1.8013
140	0.01464	487.49	516.77	1.9340	0.00602	471.28	495.37	1.8368
150	0.01521	497.82	528.24	1.9614	0.00645	483.32	509.12	1.8697
160	0.01576	508.21	539.74	1.9883	0.00684	495.06	522.43	1.9008
170	0.01630	518.68	551.28	2.0146	0.00721	506.63	535.47	1.9305
180	0.01683	529.23	562.89	2.0405	0.00755	518.11	548.33	1.9592

T(°C)	P = 6000 kPa				P = 10000 kPa			
	v m^3/kg	u kJ/kg	h kJ/kg	s kJ/kg-K	v m^3/kg	u kJ/kg	h kJ/kg	s kJ/kg-K
110	0.00131	367.51	375.36	1.5167	0.00110	352.04	363.03	1.4723
120	0.00169	395.70	405.85	1.5952	0.00118	368.93	380.69	1.5178
130	0.00240	426.08	440.51	1.6823	0.00127	386.55	399.29	1.5645
140	0.00299	447.23	465.18	1.7428	0.00140	404.85	418.86	1.6125
150	0.00344	463.84	484.50	1.7891	0.00156	423.51	439.14	1.6610
160	0.00382	478.49	501.42	1.8286	0.00176	441.95	459.51	1.7085
170	0.00416	492.10	517.03	1.8642	0.00196	459.61	479.24	1.7536
180	0.00446	505.12	531.86	1.8973	0.00217	476.22	497.95	1.7953

adapted from NIST data

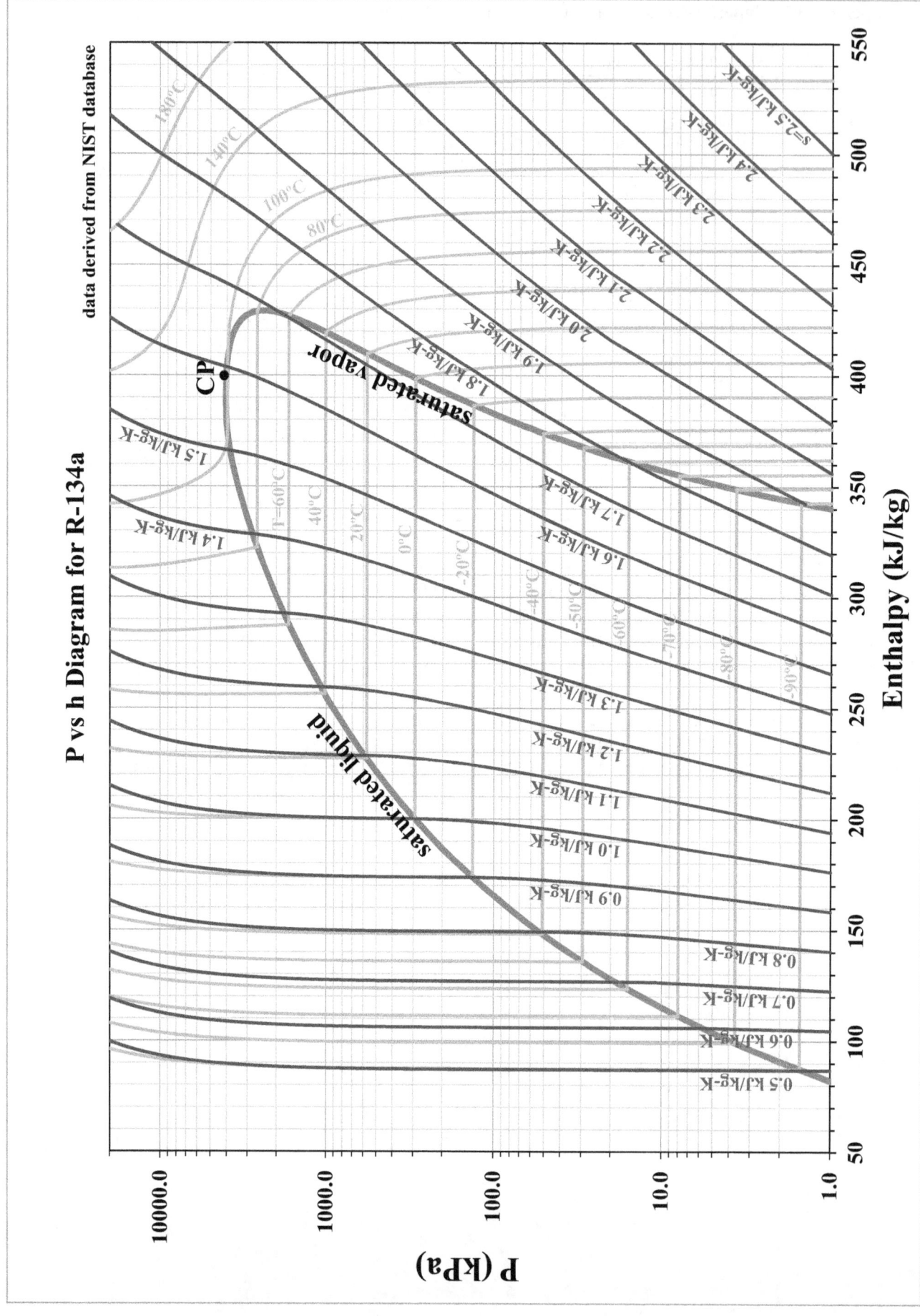

Appendix I.1 - Saturated R-23 (trifuoromethane)

temperature entry

T	P	v^L	v^V	u^L	u^V	h^L	h^V	s^L	s^V
°C	kPa	m^3/kg	m^3/kg	kJ/kg	kJ/kg	kJ/kg	kJ/kg	kJ/kg-K	kJ/kg-K
-150	0.14	0.000593	105.95000	2.54	277.11	2.54	291.73	-0.0190	2.3293
-145	0.30	0.000599	51.14600	8.55	278.99	8.55	294.18	0.0288	2.2578
-140	0.60	0.000605	26.33000	14.53	280.86	14.53	296.65	0.0746	2.1934
-135	1.14	0.000612	14.34600	20.51	282.74	20.51	299.11	0.1187	2.1353
-130	2.06	0.000618	8.21940	26.48	284.63	26.48	301.56	0.1612	2.0828
-125	3.55	0.000625	4.92400	32.46	286.51	32.46	304.01	0.2022	2.0352
-120	5.88	0.000632	3.06950	38.45	288.40	38.45	306.45	0.2420	1.9919
-115	9.38	0.000639	1.98260	44.45	290.27	44.45	308.87	0.2805	1.9525
-110	14.47	0.000646	1.32190	50.46	292.14	50.47	311.27	0.3180	1.9165
-105	21.67	0.000653	0.90675	56.49	294.00	56.50	313.64	0.3544	1.8836
-100	31.57	0.000661	0.63807	62.54	295.84	62.56	315.98	0.3898	1.8534
-95	44.88	0.000669	0.45941	68.62	297.65	68.65	318.27	0.4244	1.8256
-90	62.39	0.000678	0.33766	74.72	299.44	74.76	320.51	0.4582	1.8000
-85	85.01	0.000686	0.25282	80.85	301.20	80.91	322.69	0.4912	1.7763
-80	113.70	0.000695	0.19248	87.03	302.92	87.10	324.81	0.5236	1.7543
-75	149.55	0.000705	0.14875	93.24	304.60	93.35	326.84	0.5554	1.7338
-70	193.70	0.000715	0.11652	99.50	306.22	99.64	328.79	0.5866	1.7146
-65	247.37	0.000725	0.09239	105.82	307.79	106.00	330.65	0.6174	1.6966
-60	311.88	0.000736	0.07406	112.20	309.30	112.43	332.40	0.6476	1.6796
-55	388.59	0.000748	0.05995	118.64	310.73	118.93	334.02	0.6775	1.6635
-50	478.92	0.000760	0.04895	125.16	312.08	125.53	335.52	0.7071	1.6482
-45	584.39	0.000773	0.04028	131.77	313.33	132.22	336.86	0.7364	1.6334
-40	706.53	0.000788	0.03336	138.47	314.47	139.02	338.04	0.7655	1.6191
-35	846.97	0.000803	0.02780	145.27	315.49	145.95	339.04	0.7945	1.6052
-30	1007.40	0.000819	0.02328	152.20	316.36	153.03	339.82	0.8233	1.5915
-25	1189.50	0.000837	0.01958	159.27	317.07	160.26	340.36	0.8521	1.5779
-20	1395.30	0.000857	0.01651	166.49	317.57	167.68	340.62	0.8810	1.5642
-15	1626.50	0.000879	0.01396	173.89	317.84	175.32	340.55	0.9101	1.5502
-10	1885.30	0.000904	0.01181	181.51	317.83	183.21	340.09	0.9395	1.5357
-5	2173.90	0.000933	0.00999	189.39	317.45	191.42	339.16	0.9694	1.5204
0	2494.70	0.000966	0.00843	197.59	316.62	200.00	337.64	1.0000	1.5039
5	2850.30	0.001006	0.00708	206.21	315.19	209.08	335.36	1.0317	1.4856
10	3243.80	0.001056	0.00589	215.41	312.91	218.84	332.01	1.0650	1.4647
15	3679.10	0.001123	0.00482	225.51	309.33	229.64	327.06	1.1011	1.4392
20	4161.00	0.001225	0.00381	237.27	303.33	242.36	319.17	1.1430	1.4050
25	4698.60	0.001470	0.00263	255.03	289.18	261.94	301.55	1.2067	1.3396
26.143	4831.70	0.001899	0.00190	271.79	271.79	280.97	280.97	1.2697	1.2697

adapted from NIST data   The reference state for all R-23 properties is h = 0 and s = 0 for the saturated liquid at its triple point which is -155°C.

Appendix I.1 - Saturated R-23 (trifluoromethane)

pressure entry

P	T	v^L	v^V	u^L	u^V	h^L	h^V	s^L	s^V
kPa	°C	m³/kg	m³/kg	kJ/kg	kJ/kg	kJ/kg	kJ/kg	kJ/kg-K	kJ/kg-K
1	-136.06	0.000610	16.23900	19.24	282.35	19.24	298.59	0.1095	2.1471
2	-130.26	0.000618	8.45220	26.17	284.53	26.17	301.44	0.1590	2.0854
3	-126.60	0.000623	5.77330	30.55	285.91	30.55	303.23	0.1893	2.0499
4	-123.86	0.000626	4.40670	33.82	286.94	33.83	304.57	0.2114	2.0250
5	-121.66	0.000629	3.57450	36.46	287.77	36.47	305.65	0.2290	2.0058
10	-114.29	0.000640	1.86770	45.30	290.54	45.31	309.22	0.2859	1.9472
15	-109.57	0.000647	1.27840	50.98	292.30	50.99	311.48	0.3211	1.9136
20	-106.02	0.000652	0.97713	55.26	293.62	55.27	313.16	0.3470	1.8901
25	-103.14	0.000656	0.79333	58.74	294.68	58.76	314.52	0.3677	1.8721
30	-100.70	0.000660	0.66915	61.70	295.58	61.72	315.66	0.3849	1.8575
40	-96.67	0.000667	0.51153	66.58	297.05	66.61	317.51	0.4129	1.8347
50	-93.40	0.000672	0.41530	70.57	298.23	70.60	319.00	0.4353	1.8172
60	-90.61	0.000677	0.35024	73.97	299.23	74.01	320.24	0.4541	1.8030
70	-88.18	0.000681	0.30323	76.95	300.09	77.00	321.31	0.4703	1.7911
80	-86.01	0.000685	0.26761	79.62	300.85	79.67	322.26	0.4847	1.7809
90	-84.04	0.000688	0.23966	82.03	301.53	82.09	323.10	0.4975	1.7720
100	-82.25	0.000691	0.21712	84.25	302.15	84.32	323.87	0.5092	1.7640
110	-80.59	0.000694	0.19855	86.30	302.72	86.38	324.56	0.5199	1.7568
120	-79.04	0.000697	0.18298	88.22	303.25	88.30	325.20	0.5298	1.7502
130	-77.59	0.000700	0.16972	90.01	303.73	90.10	325.80	0.5390	1.7442
140	-76.23	0.000703	0.15829	91.71	304.19	91.80	326.35	0.5476	1.7387
160	-73.72	0.000707	0.13958	94.84	305.02	94.95	327.35	0.5634	1.7288
180	-71.45	0.000712	0.12489	97.69	305.76	97.81	328.24	0.5776	1.7200
200	-69.36	0.000716	0.11304	100.31	306.43	100.45	329.04	0.5906	1.7123
300	-60.86	0.000734	0.07686	111.10	309.04	111.32	332.10	0.6425	1.6825
400	-54.32	0.000750	0.05829	119.52	310.92	119.82	334.23	0.6816	1.6614
500	-48.94	0.000763	0.04693	126.56	312.35	126.94	335.82	0.7134	1.6450
600	-44.32	0.000775	0.03924	132.67	313.49	133.14	337.03	0.7404	1.6314
700	-40.25	0.000787	0.03368	138.13	314.42	138.68	337.99	0.7641	1.6198
800	-36.60	0.000798	0.02945	143.09	315.18	143.72	338.74	0.7852	1.6096
900	-33.27	0.000808	0.02614	147.65	315.81	148.38	339.33	0.8044	1.6005
1000	-30.22	0.000819	0.02346	151.90	316.33	152.72	339.79	0.8221	1.5921
1200	-24.73	0.000838	0.01940	159.65	317.10	160.66	340.38	0.8537	1.5771
1400	-19.89	0.000858	0.01645	166.64	317.58	167.84	340.62	0.8817	1.5639
1600	-15.55	0.000877	0.01422	173.07	317.83	174.48	340.57	0.9070	1.5517
1800	-11.59	0.000896	0.01245	179.06	317.87	180.68	340.28	0.9302	1.5404
2000	-7.95	0.000915	0.01103	184.71	317.72	186.54	339.77	0.9517	1.5295
3000	6.96	0.001024	0.00659	209.74	314.42	212.81	334.19	1.0445	1.4778
4000	18.39	0.001186	0.00413	233.19	305.67	237.94	322.20	1.1285	1.4175
4831.7	26.14	0.001899	0.00190	271.79	271.79	280.97	280.97	1.2697	1.2697

Appendix I.2 - Superheated R-23 (trifuoromethane)

	P = 50 kPa (-93.40 °C)				P = 75 kPa (-87.06 °C)			
	v	u	h	s	v	u	h	s
T(°C)	m³/kg	kJ/kg	kJ/kg	kJ/kg-K	m³/kg	kJ/kg	kJ/kg	kJ/kg-K
sat	0.41530	298.23	319.00	1.8172	0.28427	300.48	321.80	1.7859
-90	0.42421	299.97	321.18	1.8292	-	-	-	-
-80	0.44989	305.00	327.50	1.8628	0.29682	304.22	326.48	1.8105
-70	0.47501	310.00	333.75	1.8944	0.31409	309.39	332.95	1.8432
-60	0.49979	315.03	340.02	1.9245	0.33099	314.54	339.37	1.8740
-50	0.52435	320.13	346.35	1.9535	0.34765	319.73	345.81	1.9036
-40	0.54874	325.34	352.77	1.9817	0.36413	325.00	352.31	1.9321
-30	0.57302	330.66	359.31	2.0091	0.38050	330.36	358.90	1.9597
-20	0.59721	336.10	365.96	2.0359	0.39678	335.84	365.60	1.9867
-10	0.62134	341.67	372.74	2.0622	0.41299	341.44	372.42	2.0132
0	0.64541	347.38	379.65	2.0880	0.42916	347.18	379.36	2.0391
10	0.66945	353.23	386.70	2.1133	0.44528	353.04	386.44	2.0645
20	0.69345	359.22	393.89	2.1383	0.46136	359.05	393.65	2.0895
30	0.71742	365.35	401.22	2.1628	0.47742	365.19	400.99	2.1142
40	0.74137	371.62	408.69	2.1871	0.49345	371.47	408.48	2.1384
50	0.76530	378.03	416.30	2.2110	0.50947	377.89	416.10	2.1624
60	0.78922	384.58	424.05	2.2346	0.52546	384.45	423.86	2.1861
70	0.81312	391.28	431.93	2.2579	0.54145	391.16	431.77	2.2094
80	0.83700	398.11	439.96	2.2810	0.55742	398.00	439.80	2.2325
90	0.86088	405.08	448.13	2.3038	0.57337	404.97	447.98	2.2553
100	0.88475	412.19	456.43	2.3263	0.58932	412.09	456.29	2.2779
110	0.90860	419.43	464.86	2.3486	0.60526	419.34	464.73	2.3002
120	0.93245	426.81	473.43	2.3707	0.62120	426.72	473.31	2.3223
130	0.95630	434.32	482.13	2.3926	0.63712	434.23	482.01	2.3442
140	0.98013	441.95	490.96	2.4142	0.65304	441.87	490.85	2.3658
150	1.00400	449.72	499.91	2.4356	0.66896	449.63	499.81	2.3873
160	1.02780	457.60	508.99	2.4568	0.68487	457.52	508.89	2.4085
170	1.05160	465.61	518.19	2.4778	0.70077	465.54	518.09	2.4295
180	1.07540	473.74	527.51	2.4986	0.71667	473.67	527.42	2.4503
190	1.09920	481.98	536.95	2.5192	0.73257	481.92	536.86	2.4709
200	1.12300	490.35	546.50	2.5396	0.74846	490.28	546.41	2.4913

Appendix I.2 - Superheated R-23 (trifuoromethane)

	P = 100 kPa (-82.25 °C)				P = 200 kPa (-69.36°C)			
	v	u	h	s	v	u	h	s
T(°C)	m³/kg	kJ/kg	kJ/kg	kJ/kg-K	m³/kg	kJ/kg	kJ/kg	kJ/kg-K
sat	0.21712	302.15	323.87	1.7640	0.11304	306.43	329.04	1.7123
-80	0.22021	303.39	325.41	1.7720	-	-	-	-
-70	0.23359	308.77	332.13	1.8059	-	-	-	-
-60	0.24656	314.05	338.70	1.8375	0.11976	311.94	335.89	1.7452
-50	0.25928	319.33	345.25	1.8676	0.12663	317.62	342.95	1.7775
-40	0.27182	324.65	351.84	1.8964	0.13328	323.23	349.89	1.8079
-30	0.28423	330.07	358.49	1.9244	0.13979	328.85	356.81	1.8370
-20	0.29656	335.58	365.24	1.9515	0.14620	334.52	363.76	1.8650
-10	0.30882	341.21	372.09	1.9781	0.15253	340.27	370.78	1.8922
0	0.32102	346.97	379.07	2.0041	0.15881	346.12	377.89	1.9187
10	0.33319	352.85	386.17	2.0297	0.16504	352.09	385.10	1.9446
20	0.34531	358.87	393.40	2.0548	0.17123	358.17	392.42	1.9700
30	0.35741	365.03	400.77	2.0795	0.17740	364.38	399.86	1.9950
40	0.36949	371.32	408.27	2.1038	0.18354	370.72	407.43	2.0196
50	0.38155	377.75	415.91	2.1278	0.18966	377.20	415.13	2.0438
60	0.39358	384.33	423.68	2.1515	0.19576	383.80	422.96	2.0676
70	0.40561	391.03	431.60	2.1749	0.20185	390.54	430.92	2.0912
80	0.41762	397.88	439.64	2.1980	0.20793	397.42	439.01	2.1144
90	0.42962	404.87	447.83	2.2209	0.21399	404.43	447.23	2.1374
100	0.44161	411.98	456.15	2.2435	0.22004	411.57	455.58	2.1600
110	0.45359	419.24	464.60	2.2658	0.22609	418.84	464.06	2.1825
120	0.46557	426.62	473.18	2.2879	0.23213	426.25	472.67	2.2047
130	0.47754	434.14	481.89	2.3098	0.23816	433.78	481.41	2.2266
140	0.48950	441.78	490.73	2.3315	0.24418	441.44	490.28	2.2483
150	0.50145	449.55	499.70	2.3529	0.25020	449.22	499.26	2.2698
160	0.51340	457.45	508.79	2.3741	0.25621	457.13	508.37	2.2911
170	0.52535	465.46	518.00	2.3952	0.26222	465.16	517.60	2.3122
180	0.53729	473.59	527.32	2.4160	0.26822	473.30	526.95	2.3330
190	0.54923	481.85	536.77	2.4366	0.27422	481.57	536.41	2.3537
200	0.56116	490.21	546.33	2.4570	0.28022	489.94	545.98	2.3741

Appendix I.2 - Superheated R-23 (trifuoromethane)

	P = 300 kPa (-60.86 °C)				P = 400 kPa (-54.32 °C)			
	v	u	h	s	v	u	h	s
T(°C)	m³/kg	kJ/kg	kJ/kg	kJ/kg-K	m³/kg	kJ/kg	kJ/kg	kJ/kg-K
sat	0.07686	309.04	332.10	1.6825	0.05829	310.92	334.23	1.6614
-60	0.07731	309.59	332.79	1.6857	-	-	-	-
-50	0.08230	315.79	340.48	1.7209	0.06003	313.79	337.80	1.6775
-40	0.08703	321.73	347.84	1.7532	0.06384	320.13	345.67	1.7120
-30	0.09159	327.58	355.06	1.7836	0.06745	326.26	353.24	1.7438
-20	0.09604	333.42	362.24	1.8125	0.07094	332.29	360.66	1.7738
-10	0.10041	339.31	369.43	1.8403	0.07433	338.31	368.05	1.8024
0	0.10472	345.26	376.68	1.8674	0.07766	344.38	375.44	1.8299
10	0.10898	351.31	384.00	1.8937	0.08093	350.52	382.89	1.8567
20	0.11320	357.46	391.42	1.9195	0.08417	356.74	390.41	1.8828
30	0.11739	363.73	398.95	1.9447	0.08737	363.07	398.02	1.9083
40	0.12155	370.12	406.59	1.9695	0.09055	369.51	405.73	1.9334
50	0.12569	376.63	414.34	1.9939	0.09371	376.07	413.55	1.9580
60	0.12982	383.28	422.22	2.0179	0.09685	382.75	421.49	1.9821
70	0.13393	390.05	430.23	2.0416	0.09997	389.55	429.54	2.0060
80	0.13803	396.95	438.36	2.0649	0.10308	396.49	437.72	2.0294
90	0.14211	403.99	446.62	2.0880	0.10617	403.55	446.02	2.0526
100	0.14619	411.15	455.01	2.1108	0.10926	410.74	454.44	2.0755
110	0.15025	418.45	463.52	2.1333	0.11234	418.05	462.98	2.0981
120	0.15431	425.87	472.16	2.1556	0.11540	425.49	471.65	2.1204
130	0.15836	433.42	480.93	2.1776	0.11847	433.06	480.45	2.1425
140	0.16241	441.10	489.82	2.1993	0.12152	440.75	489.36	2.1644
150	0.16645	448.89	498.83	2.2209	0.12457	448.56	498.39	2.1860
160	0.17048	456.81	507.96	2.2422	0.12762	456.50	507.55	2.2073
170	0.17451	464.85	517.21	2.2633	0.13066	464.55	516.81	2.2285
180	0.17853	473.01	526.57	2.2842	0.13369	472.72	526.20	2.2494
190	0.18256	481.29	536.05	2.3049	0.13672	481.00	535.69	2.2702
200	0.18657	489.67	545.64	2.3254	0.13975	489.40	545.30	2.2907

Appendix I.2 - Superheated R-23 (trifuoromethane)

T(°C)	P = 600 kPa (-44.32 °C)				P = 800 kPa (-36.60 °C)			
	v m³/kg	u kJ/kg	h kJ/kg	s kJ/kg-K	v m³/kg	u kJ/kg	h kJ/kg	s kJ/kg-K
sat	0.07686	309.04	332.10	1.6825	0.02945	315.18	338.74	1.6096
-40	0.04049	316.59	340.89	1.6481	-	-	-	-
-30	0.04321	323.39	349.32	1.6835	0.03098	320.18	344.96	1.6356
-20	0.04577	329.88	357.34	1.7159	0.03311	327.27	353.75	1.6710
-10	0.04821	336.24	365.16	1.7462	0.03509	334.03	362.10	1.7033
0	0.05057	342.55	372.89	1.7750	0.03699	340.63	370.22	1.7336
10	0.05287	348.88	380.60	1.8027	0.03881	347.18	378.23	1.7624
20	0.05513	355.26	388.34	1.8296	0.04059	353.73	386.20	1.7901
30	0.05735	361.72	396.13	1.8557	0.04232	360.33	394.19	1.8169
40	0.05954	368.27	403.99	1.8812	0.04403	366.99	402.22	1.8429
50	0.06171	374.91	411.94	1.9062	0.04571	373.74	410.31	1.8684
60	0.06387	381.67	419.99	1.9307	0.04737	380.58	418.48	1.8933
70	0.06600	388.55	428.15	1.9549	0.04901	387.53	426.74	1.9177
80	0.06812	395.54	436.41	1.9786	0.05064	394.58	435.10	1.9417
90	0.07023	402.66	444.79	2.0020	0.05226	401.75	443.56	1.9653
100	0.07233	409.89	453.29	2.0251	0.05386	409.04	452.13	1.9886
110	0.07442	417.25	461.90	2.0479	0.05546	416.44	460.81	2.0116
120	0.07650	424.73	470.63	2.0703	0.05704	423.97	469.60	2.0342
130	0.07857	432.33	479.48	2.0926	0.05862	431.60	478.50	2.0566
140	0.08064	440.06	488.44	2.1145	0.06019	439.36	487.52	2.0787
150	0.08270	447.90	497.52	2.1362	0.06176	447.24	496.64	2.1005
160	0.08475	455.86	506.71	2.1577	0.06332	455.23	505.88	2.1221
170	0.08680	463.94	516.02	2.1790	0.06488	463.33	515.23	2.1434
180	0.08885	472.14	525.44	2.2000	0.06643	471.55	524.69	2.1645
190	0.09089	480.44	534.98	2.2208	0.06797	479.88	534.26	2.1854
200	0.09293	488.86	544.62	2.2414	0.06952	488.32	543.93	2.2061

Appendix I.2 - Superheated R-23 (trifuoromethane)

	P = 1000 kPa (-30.22 °C)				P = 1500 kPa (-17.67 °C)			
	v	u	h	s	v	u	h	s
T(°C)	m³/kg	kJ/kg	kJ/kg	kJ/kg-K	m³/kg	kJ/kg	kJ/kg	kJ/kg-K
sat	0.02346	316.33	339.79	1.5921	0.01526	317.73	340.63	1.5577
-30	0.02351	316.51	340.01	1.5930	-	-	-	-
-20	0.02543	324.39	349.82	1.6326	-	-	-	-
-10	0.02718	331.65	358.83	1.6675	0.01644	324.73	349.38	1.5914
0	0.02880	338.60	367.41	1.6995	0.01778	332.92	359.59	1.6295
10	0.03035	345.40	375.76	1.7295	0.01900	340.56	369.06	1.6636
20	0.03185	352.15	384.00	1.7581	0.02014	347.92	378.13	1.6951
30	0.03330	358.90	392.20	1.7856	0.02122	355.13	386.96	1.7247
40	0.03471	365.69	400.40	1.8122	0.02226	362.28	395.67	1.7530
50	0.03610	372.54	408.64	1.8381	0.02327	369.42	404.33	1.7802
60	0.03747	379.47	416.94	1.8634	0.02425	376.60	412.98	1.8065
70	0.03882	386.49	425.31	1.8882	0.02521	383.83	421.65	1.8322
80	0.04015	393.62	433.77	1.9125	0.02616	391.13	430.37	1.8572
90	0.04147	400.84	442.31	1.9363	0.02708	398.52	439.14	1.8817
100	0.04278	408.18	450.96	1.9598	0.02800	405.99	447.99	1.9058
110	0.04408	415.63	459.71	1.9830	0.02891	413.56	456.92	1.9294
120	0.04537	423.19	468.56	2.0058	0.02980	421.24	465.94	1.9526
130	0.04665	430.87	477.52	2.0283	0.03069	429.01	475.05	1.9755
140	0.04793	438.66	486.59	2.0505	0.03157	436.89	484.25	1.9981
150	0.04920	446.57	495.76	2.0724	0.03245	444.88	493.55	2.0203
160	0.05046	454.59	505.05	2.0941	0.03332	452.98	502.95	2.0422
170	0.05172	462.72	514.44	2.1155	0.03418	461.18	512.45	2.0639
180	0.05297	470.96	523.93	2.1367	0.03504	469.48	522.04	2.0853
190	0.05423	479.31	533.54	2.1577	0.03590	477.89	531.73	2.1065
200	0.05547	487.77	543.24	2.1784	0.03675	486.41	541.53	2.1274

Appendix I.2 - Superheated R-23 (trifuoromethane)

T(°C)	P = 2000 kPa (-7.95 °C)				P = 4000 kPa (100.34 °C)			
	v m³/kg	u kJ/kg	h kJ/kg	s kJ/kg-K	v m³/kg	u kJ/kg	h kJ/kg	s kJ/kg-K
sat	0.01103	317.72	339.77	1.5295	0.00413	305.67	322.20	1.4175
0	0.01209	325.97	350.14	1.5681	-	-	-	-
10	0.01322	334.96	361.39	1.6085	-	-	-	-
20	0.01422	343.18	371.62	1.6440	0.00445	310.48	328.26	1.4383
30	0.01514	351.01	381.28	1.6765	0.00561	328.08	350.51	1.5130
40	0.01600	358.62	390.63	1.7068	0.00639	340.10	365.65	1.5622
50	0.01683	366.12	399.79	1.7356	0.00703	350.34	378.46	1.6025
60	0.01763	373.59	408.84	1.7632	0.00760	359.73	390.13	1.6380
70	0.01840	381.06	417.86	1.7898	0.00812	368.65	401.12	1.6705
80	0.01915	388.57	426.87	1.8157	0.00861	377.29	411.71	1.7010
90	0.01989	396.12	435.90	1.8409	0.00907	385.77	422.04	1.7298
100	0.02061	403.75	444.97	1.8656	0.00951	394.17	432.20	1.7574
110	0.02132	411.45	454.09	1.8897	0.00993	402.53	442.27	1.7840
120	0.02202	419.24	463.28	1.9134	0.01035	410.89	452.28	1.8098
130	0.02271	427.12	472.55	1.9367	0.01075	419.27	462.26	1.8349
140	0.02340	435.10	481.89	1.9596	0.01114	427.69	472.25	1.8594
150	0.02408	443.17	491.32	1.9821	0.01153	436.15	482.25	1.8833
160	0.02475	451.35	500.84	2.0043	0.01190	444.68	492.29	1.9067
170	0.02541	459.62	510.44	2.0263	0.01227	453.27	502.36	1.9297
180	0.02608	467.99	520.14	2.0479	0.01264	461.93	512.49	1.9523
190	0.02673	476.46	529.93	2.0693	0.01300	470.67	522.67	1.9746
200	0.02739	485.03	539.81	2.0904	0.01336	479.49	532.92	1.9964

Appendix I.2 - Superheated R-23 (trifuoromethane)

T(°C)	P = 6000 kPa				P = 10000 kPa			
	v	u	h	s	v	u	h	s
	m³/kg	kJ/kg	kJ/kg	kJ/kg-K	m³/kg	kJ/kg	kJ/kg	kJ/kg-K
-150	0.00059	1.71	5.26	-0.0258	0.00059	1.18	7.07	-0.0302
-140	0.00060	13.57	17.18	0.0673	0.00060	12.95	18.96	0.0626
-130	0.00062	25.38	29.07	0.1535	0.00061	24.69	30.82	0.1485
-120	0.00063	37.21	40.98	0.2338	0.00063	36.42	42.68	0.2286
-110	0.00064	49.06	52.91	0.3093	0.00064	48.17	54.57	0.3037
-100	0.00066	60.96	64.90	0.3806	0.00065	59.97	66.50	0.3747
-90	0.00067	72.93	76.97	0.4484	0.00067	71.81	78.49	0.4421
-80	0.00069	85.01	89.14	0.5131	0.00068	83.74	90.58	0.5063
-70	0.00071	97.22	101.46	0.5752	0.00070	95.77	102.79	0.5679
-60	0.00073	109.60	113.96	0.6353	0.00072	107.95	115.15	0.6273
-50	0.00075	122.20	126.69	0.6937	0.00074	120.30	127.70	0.6849
-40	0.00077	135.08	139.71	0.7508	0.00076	132.86	140.49	0.7409
-30	0.00080	148.31	153.11	0.8070	0.00079	145.69	153.56	0.7958
-20	0.00083	161.99	166.98	0.8629	0.00082	158.83	166.99	0.8499
-10	0.00087	176.26	181.48	0.9191	0.00085	172.36	180.84	0.9036
0	0.00092	191.34	196.85	0.9764	0.00089	186.37	195.23	0.9573
10	0.00098	207.66	213.57	1.0365	0.00093	200.97	210.30	1.0114
20	0.00108	226.24	232.75	1.1030	0.00099	216.36	226.28	1.0669
30	0.00131	251.33	259.20	1.1916	0.00107	232.81	243.51	1.1247
40	0.00258	302.61	318.07	1.3826	0.00118	250.79	262.60	1.1866
50	0.00352	326.60	347.75	1.4760	0.00135	270.87	284.40	1.2551
60	0.00414	341.35	366.17	1.5322	0.00162	292.73	308.94	1.3299
70	0.00463	353.34	381.12	1.5764	0.00196	313.63	333.27	1.4018
80	0.00506	364.03	394.38	1.6145	0.00231	331.40	354.50	1.4629
90	0.00545	374.00	406.69	1.6489	0.00263	346.37	372.67	1.5136
100	0.00581	383.55	418.39	1.6807	0.00292	359.50	388.68	1.5571
110	0.00614	392.83	429.69	1.7106	0.00318	371.45	403.27	1.5957
120	0.00647	401.94	440.73	1.7390	0.00343	382.64	416.91	1.6309
130	0.00677	410.95	451.59	1.7663	0.00366	393.32	429.88	1.6634
140	0.00707	419.91	462.34	1.7926	0.00387	403.65	442.39	1.6941
150	0.00736	428.85	473.00	1.8181	0.00408	413.75	454.55	1.7232
160	0.00764	437.79	483.62	1.8429	0.00428	423.68	466.47	1.7510
170	0.00791	446.75	494.22	1.8671	0.00447	433.49	478.21	1.7778
180	0.00818	455.75	504.82	1.8908	0.00466	443.24	489.82	1.8037
190	0.00844	464.79	515.43	1.9140	0.00484	452.95	501.33	1.8288
200	0.00870	473.87	526.07	1.9367	0.00501	462.63	512.77	1.8533

adapted from NIST data

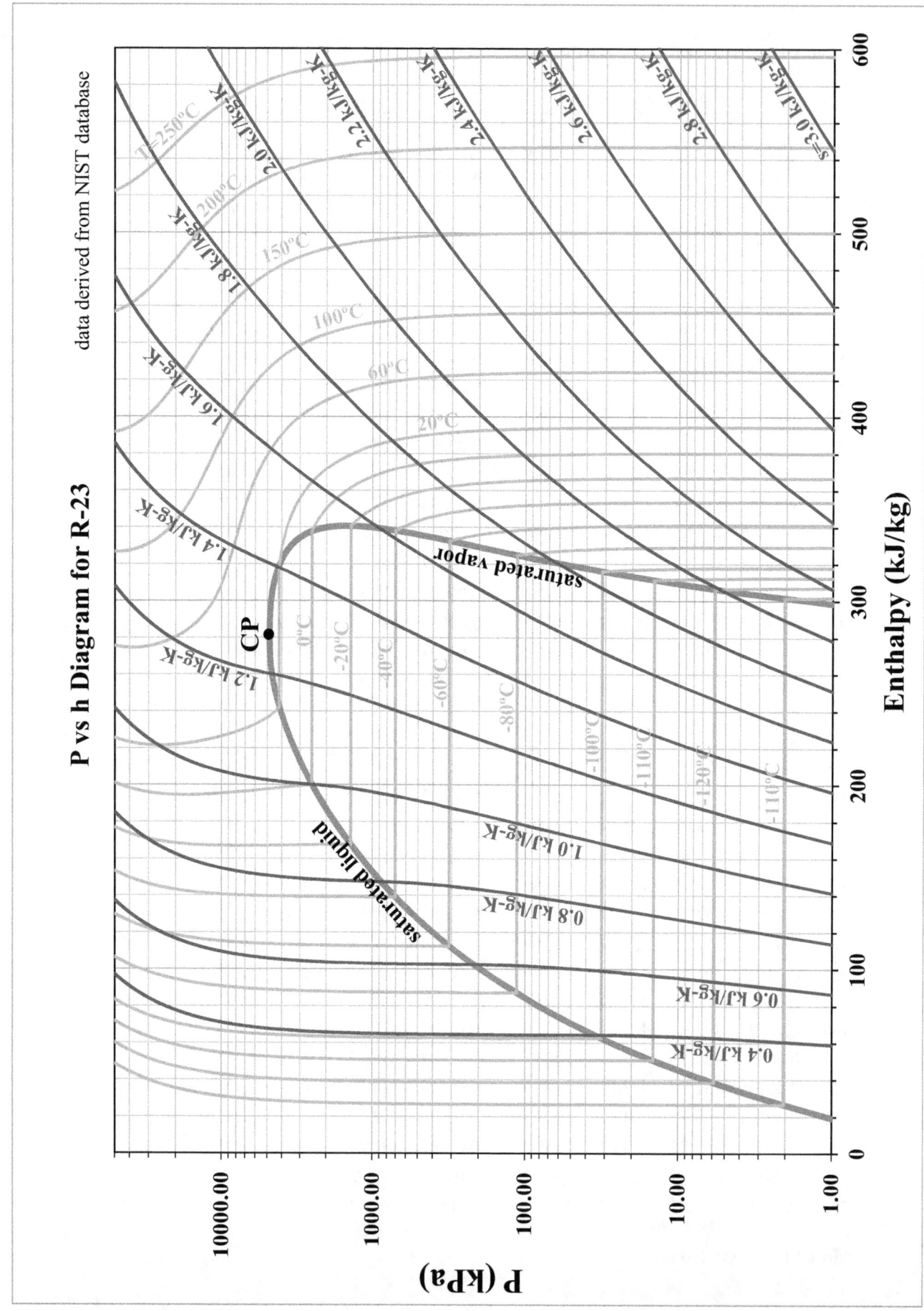

Appendix J - Properties of Ammonia

Appendix J.1 - Saturated Ammonia (NH₃) - SI Units

Temperature Entry

T °C	P kPa	v^L m³/kg	v^V m³/kg	h^L kJ/kg	h^V kJ/kg	s^L kJ/kg-K	s^V kJ/kg-K
-70	10.90	0.001379	9.04280	33.07	1502.90	0.1659	7.4012
-65	15.57	0.001390	6.47130	54.76	1511.90	0.2713	7.2716
-60	21.84	0.001402	4.71600	76.54	1520.70	0.3746	7.1498
-55	30.09	0.001413	3.49490	98.41	1529.30	0.4760	7.0350
-50	40.78	0.001425	2.63050	120.38	1537.70	0.5755	6.9268
-45	54.43	0.001437	2.00830	142.45	1545.80	0.6733	6.8245
-40	71.63	0.001450	1.55380	164.62	1553.80	0.7693	6.7276
-35	93.04	0.001463	1.21690	186.88	1561.50	0.8636	6.6358
-30	119.38	0.001476	0.96390	209.25	1569.00	0.9564	6.5486
-25	151.42	0.001490	0.77157	231.71	1576.20	1.0476	6.4656
-20	190.03	0.001504	0.62363	254.28	1583.10	1.1374	6.3865
-15	236.11	0.001519	0.50862	276.95	1589.70	1.2259	6.3110
-10	290.64	0.001534	0.41828	299.74	1595.90	1.3130	6.2387
-5	354.66	0.001550	0.34666	322.64	1601.90	1.3988	6.1694
0	429.25	0.001566	0.28935	345.67	1607.40	1.4835	6.1028
5	515.56	0.001583	0.24312	368.85	1612.60	1.5671	6.0387
10	614.79	0.001601	0.20552	392.17	1617.40	1.6496	5.9768
15	728.19	0.001619	0.17471	415.65	1621.70	1.7312	5.9168
20	857.04	0.001638	0.14930	439.32	1625.60	1.8119	5.8586
25	1002.7	0.001659	0.12819	463.17	1629.00	1.8918	5.8019
30	1166.5	0.001680	0.11055	487.25	1631.80	1.9709	5.7466
35	1350.0	0.001702	0.09571	511.56	1634.10	2.0494	5.6923
40	1554.5	0.001725	0.08317	536.12	1635.80	2.1274	5.6390
45	1781.7	0.001750	0.07251	560.97	1636.80	2.2049	5.5863
50	2033.0	0.001776	0.06339	586.14	1637.10	2.2820	5.5342
55	2310.0	0.001804	0.05557	611.65	1636.60	2.3588	5.4823
60	2614.5	0.001834	0.04882	637.55	1635.30	2.4354	5.4305
65	2948.1	0.001866	0.04297	663.88	1633.20	2.5120	5.3785
70	3312.5	0.001900	0.03787	690.69	1630.00	2.5887	5.3260
75	3709.6	0.001937	0.03342	718.05	1625.70	2.6657	5.2728
80	4141.3	0.001977	0.02951	746.04	1620.20	2.7431	5.2185
85	4609.5	0.002022	0.02605	774.76	1613.40	2.8212	5.1628
90	5116.4	0.002071	0.02299	804.33	1605.00	2.9003	5.1051
95	5664.2	0.002126	0.02027	834.91	1594.80	2.9808	5.0448
100	6255.1	0.002189	0.01782	866.74	1582.40	3.0633	4.9812
110	7577.5	0.002349	0.01361	935.60	1549.10	3.2374	4.8385
120	9108.8	0.002597	0.01001	1016.50	1496.70	3.4360	4.6573
130	10888.0	0.003190	0.00638	1136.20	1384.10	3.7236	4.3384
132.41	11363.0	0.004287	0.00429	1247.80	1247.80	3.9952	3.9952

adapted from NIST data

Appendix J.1 - Saturated Ammonia (NH₃) - English Units

T °F	P psia	v^L ft³/lbm	v^V ft³/lbm	h^L BTU/lbm	h^V BTU/lbm	s^L BTU/lbm-°F	s^V BTU/lbm-°F
-100	1.2314	0.021977	183.030	8.0290	643.97	0.02255	1.7907
-90	1.8568	0.022171	124.540	18.370	648.29	0.05091	1.7549
-80	2.7316	0.022370	86.763	28.758	652.53	0.07862	1.7216
-70	3.9290	0.022574	61.754	39.195	656.68	0.10575	1.6904
-60	5.5352	0.022784	44.825	49.685	660.72	0.13231	1.6612
-50	7.6505	0.023000	33.127	60.226	664.65	0.15834	1.6337
-40	10.390	0.023222	24.889	70.820	668.47	0.18386	1.6079
-30	13.882	0.023452	18.984	81.466	672.15	0.20889	1.5836
-20	18.271	0.023690	14.682	92.164	675.69	0.23346	1.5607
-10	23.716	0.023937	11.501	102.92	679.09	0.25758	1.5389
0	30.389	0.024193	9.115	113.72	682.34	0.28129	1.5183
10	38.478	0.024458	7.301	124.59	685.41	0.30459	1.4987
20	48.181	0.024735	5.9068	135.51	688.31	0.32752	1.4800
30	59.712	0.025023	4.8220	146.51	691.02	0.35009	1.4621
40	73.295	0.025324	3.9692	157.57	693.53	0.37233	1.4450
50	89.168	0.025638	3.2921	168.71	695.82	0.39426	1.4285
60	107.58	0.025968	2.7495	179.95	697.89	0.41591	1.4126
70	128.78	0.026314	2.3110	191.27	699.70	0.43731	1.3972
80	153.05	0.026678	1.9536	202.70	701.24	0.45847	1.3823
90	180.65	0.027061	1.6602	214.25	702.50	0.47942	1.3677
100	211.89	0.027467	1.4175	225.93	703.44	0.50019	1.3534
110	247.05	0.027898	1.2154	237.76	704.05	0.52081	1.3393
120	286.44	0.028357	1.0460	249.74	704.29	0.54131	1.3255
130	330.38	0.028848	0.9031	261.91	704.13	0.56172	1.3117
140	379.20	0.029375	0.7820	274.28	703.54	0.58208	1.2979
150	433.23	0.029944	0.6786	286.88	702.47	0.60243	1.2841
160	492.82	0.030561	0.5900	299.74	700.87	0.62280	1.2701
170	558.33	0.031237	0.5135	312.90	698.68	0.64327	1.2559
180	630.16	0.031983	0.4472	326.40	695.81	0.66389	1.2414
190	708.69	0.032814	0.3894	340.33	692.18	0.68476	1.2263
200	794.36	0.033752	0.3386	354.75	687.65	0.70598	1.2106
210	887.62	0.034830	0.2938	369.79	682.04	0.72771	1.1940
220	988.99	0.036095	0.2539	385.62	675.10	0.75018	1.1761
230	1099.0	0.037625	0.2180	402.51	666.43	0.77376	1.1564
240	1218.4	0.039563	0.1852	420.91	655.35	0.79903	1.1341
250	1347.9	0.042201	0.1543	441.74	640.49	0.82719	1.1073
260	1488.6	0.046375	0.1235	467.33	618.23	0.86137	1.0710
270	1642.6	0.060657	0.0790	518.31	558.54	0.92961	0.9847
270.33	1648.0	0.065986	0.0716	530.86	543.11	0.94672	0.9635

The reference state for all ammonia properties is h = 0 and s = 0 for the saturated liquid at its triple point which is -77.66°C.

Appendix J.1 - Saturated Ammonia (NH$_3$) - SI Units
Pressure Entry

T °C	P kPa	v^L m³/kg	v^V m³/kg	h^L kJ/kg	h^V kJ/kg	s^L kJ/kg-K	s^V kJ/kg-K
-71.16	10	0.001377	9.80040	28.05	1500.80	0.1411	7.4324
-61.33	20	0.001399	5.12030	70.75	1518.30	0.3474	7.1814
-55.05	30	0.001413	3.50430	98.21	1529.20	0.4751	7.0360
-50.32	40	0.001424	2.67810	118.95	1537.10	0.5691	6.9336
-46.50	50	0.001434	2.17410	135.84	1543.40	0.6442	6.8545
-43.25	60	0.001442	1.83360	150.18	1548.70	0.7070	6.7901
-40.43	70	0.001449	1.58760	162.71	1553.10	0.7611	6.7357
-37.92	80	0.001455	1.40140	173.88	1557.10	0.8088	6.6888
-35.65	90	0.001461	1.25530	183.99	1560.50	0.8515	6.6475
-33.58	100	0.001466	1.13750	193.24	1563.70	0.8902	6.6105
-25.20	150	0.001489	0.77841	230.80	1575.90	1.0440	6.4689
-18.84	200	0.001507	0.59441	259.52	1584.60	1.1580	6.3687
-13.65	250	0.001523	0.48200	283.11	1591.40	1.2496	6.2911
-9.22	300	0.001536	0.40597	303.31	1596.90	1.3265	6.2277
-5.34	350	0.001548	0.35101	321.09	1601.50	1.3930	6.1740
-1.87	400	0.001560	0.30936	337.03	1605.40	1.4519	6.1275
1.27	450	0.001570	0.27668	351.54	1608.80	1.5048	6.0863
4.15	500	0.001580	0.25031	364.89	1611.80	1.5529	6.0494
6.81	550	0.001589	0.22859	377.29	1614.40	1.5971	6.0160
9.3	600	0.001598	0.21036	388.88	1616.80	1.6380	5.9854
11.6	650	0.001607	0.19485	399.78	1618.90	1.6762	5.9571
13.8	700	0.001615	0.18147	410.08	1620.80	1.7120	5.9308
15.9	750	0.001622	0.16982	419.86	1622.50	1.7456	5.9063
17.9	800	0.001630	0.15958	429.18	1624.00	1.7775	5.8833
19.7	850	0.001637	0.15050	438.09	1625.40	1.8077	5.8616
21.5	900	0.001644	0.14239	446.64	1626.70	1.8365	5.8410
23.3	950	0.001651	0.13511	454.85	1627.90	1.8640	5.8215
24.9	1000	0.001658	0.12853	462.76	1628.90	1.8904	5.8029
38.7	1500	0.001719	0.08619	529.80	1635.40	2.1074	5.6526
49.4	2000	0.001773	0.06446	582.95	1637.10	2.2723	5.5407
58.2	2500	0.001823	0.05117	628.04	1635.90	2.4074	5.4494
65.7	3000	0.001871	0.04217	667.81	1632.80	2.5234	5.3707
72.4	3500	0.001917	0.03565	703.83	1628.10	2.6258	5.3004
78.4	4000	0.001964	0.03070	737.06	1622.10	2.7184	5.2360
83.9	4500	0.002011	0.02680	768.18	1615.10	2.8034	5.1756
88.9	5000	0.002060	0.02364	797.66	1607.00	2.8826	5.1182
93.5	5500	0.002109	0.02103	825.87	1598.00	2.9572	5.0627
97.9	6000	0.002162	0.01882	853.12	1587.90	3.0282	5.0085
102.0	6500	0.002216	0.01693	879.67	1576.90	3.0964	4.9550
105.8	7000	0.002275	0.01528	905.75	1564.70	3.1626	4.9014
109.5	7500	0.002339	0.01382	931.60	1551.30	3.2274	4.8471
112.9	8000	0.002408	0.01252	957.47	1536.40	3.2916	4.7913
116.2	8500	0.002486	0.01133	983.65	1519.90	3.3560	4.7332
119.3	9000	0.002575	0.01024	1010.50	1501.10	3.4215	4.6714
122.3	9500	0.002680	0.00922	1038.60	1479.40	3.4895	4.6040
125.2	10000	0.002811	0.00823	1068.90	1453.40	3.5623	4.5275
132.41	11363.0	0.004287	0.00429	1247.80	1247.80	3.9952	3.9952

Appendix J.1 - Saturated Ammonia (NH$_3$) - English Units
Pressure Entry

T °F	P psia	v^L ft³/lbm	v^V ft³/lbm	h^L BTU/lbm	h^V BTU/lbm	s^L BTU/lbm-°F	s^V BTU/lbm-°F
-41.28	10	0.023194	25.7940	69.46	667.98	0.18062	1.6112
-16.59	20	0.023773	13.4910	95.82	676.87	0.24173	1.5531
-0.53	30	0.024179	9.2256	113.15	682.17	0.28004	1.5194
11.69	40	0.024504	7.0396	126.43	685.92	0.30849	1.4954
21.69	50	0.024783	5.7037	137.37	688.78	0.33137	1.4769
30.23	60	0.025030	4.8001	146.76	691.08	0.35060	1.4617
37.72	70	0.025254	4.1467	155.04	692.98	0.36728	1.4488
44.41	80	0.025461	3.6516	162.48	694.57	0.38204	1.4376
50.49	90	0.025654	3.2630	169.26	695.93	0.39532	1.4277
56.06	100	0.025836	2.9495	175.51	697.10	0.40741	1.4188
66.03	120	0.026174	2.4744	186.76	699.01	0.42883	1.4033
74.79	140	0.026486	2.1309	196.73	700.47	0.44747	1.3900
82.64	160	0.026777	1.8705	205.74	701.60	0.46402	1.3784
89.78	180	0.027053	1.6661	213.99	702.47	0.47895	1.3680
96.33	200	0.027316	1.5012	221.64	703.13	0.49260	1.3586
102.41	220	0.027569	1.3653	228.77	703.62	0.50518	1.3500
108.09	240	0.027814	1.2513	235.48	703.96	0.51688	1.3420
113.41	260	0.028052	1.1542	241.83	704.17	0.52782	1.3346
118.44	280	0.028283	1.0705	247.86	704.28	0.53812	1.3276
123.20	300	0.028511	0.9976	253.62	704.29	0.54786	1.3210
127.73	320	0.028733	0.9335	259.13	704.21	0.55710	1.3148
132.05	340	0.028953	0.8766	264.43	704.05	0.56591	1.3089
136.19	360	0.029169	0.8259	269.54	703.82	0.57433	1.3032
140.16	380	0.029383	0.7803	274.47	703.53	0.58240	1.2977
143.97	400	0.029595	0.7390	279.25	703.18	0.59016	1.2924
147.64	420	0.029805	0.7016	283.88	702.77	0.59763	1.2874
151.19	440	0.030014	0.6674	288.39	702.31	0.60484	1.2824
154.61	460	0.030222	0.6361	292.77	701.80	0.61182	1.2777
157.93	480	0.030429	0.6073	297.05	701.25	0.61858	1.2730
161.14	500	0.030636	0.5807	301.23	700.65	0.62514	1.2685
168.78	550	0.031151	0.5223	311.27	698.98	0.64076	1.2577
175.91	600	0.031669	0.4732	320.84	697.07	0.65544	1.2474
182.61	650	0.032191	0.4314	330.00	694.94	0.66931	1.2375
188.94	700	0.032721	0.3952	338.82	692.61	0.68252	1.2280
194.93	750	0.033261	0.3636	347.37	690.07	0.69516	1.2187
200.63	800	0.033816	0.3357	355.68	687.33	0.70733	1.2096
206.07	850	0.034387	0.3108	363.79	684.39	0.71909	1.2007
211.27	900	0.034979	0.2885	371.74	681.24	0.73051	1.1918
216.25	950	0.035595	0.2684	379.57	677.88	0.74165	1.1830
221.04	1000	0.036240	0.2500	387.31	674.29	0.75257	1.1742
230.09	1100	0.037640	0.2177	402.66	666.35	0.77396	1.1563
238.51	1200	0.039240	0.1899	418.05	657.19	0.79512	1.1376
246.40	1300	0.041139	0.1653	433.87	646.42	0.81659	1.1176
253.80	1400	0.043520	0.1428	450.67	633.27	0.83916	1.0951
260.77	1500	0.046826	0.1209	469.68	615.93	0.86450	1.0675
267.33	1600	0.052881	0.0964	495.39	587.64	0.89875	1.0256
270.33	1648	0.065986	0.0716	530.86	543.11	0.94672	0.9635

Appendix J.2 - Superheated Ammonia (NH$_3$) - SI Units

	P = 100 kPa (-33.58°C)				P = 200 kPa (-18.84°C)			
	v	u	h	s	v	u	h	s
T(°C)	m³/kg	kJ/kg	kJ/kg	kJ/kg-K	m³/kg	kJ/kg	kJ/kg	kJ/kg-K
sat	1.1375	1449.9	1563.7	6.6105	0.5944	1465.8	1584.6	6.3687
-30	1.1568	1456.2	1571.8	6.6443	-	-	-	-
-20	1.2098	1473.3	1594.3	6.7348	-	-	-	-
-10	1.2620	1490.2	1616.4	6.8204	0.6191	1482.1	1605.9	6.4510
0	1.3135	1506.9	1638.2	6.9019	0.6464	1500.1	1629.4	6.5384
10	1.3646	1523.5	1659.9	6.9799	0.6731	1517.7	1652.3	6.6209
20	1.4153	1540.0	1681.6	7.0550	0.6995	1535.0	1674.9	6.6995
30	1.4656	1556.6	1703.2	7.1276	0.7255	1552.3	1697.4	6.7747
40	1.5158	1573.3	1724.8	7.1978	0.7512	1569.4	1719.7	6.8471
50	1.5657	1590.0	1746.6	7.2661	0.7768	1586.6	1741.9	6.9171
60	1.6155	1606.9	1768.4	7.3326	0.8022	1603.7	1764.2	6.9849
70	1.6652	1623.8	1790.3	7.3974	0.8275	1621.0	1786.5	7.0509
80	1.7148	1640.9	1812.3	7.4607	0.8527	1638.3	1808.9	7.1151
90	1.7643	1658.1	1834.5	7.5227	0.8778	1655.8	1831.3	7.1779
100	1.8137	1675.5	1856.9	7.5834	0.9028	1673.3	1853.9	7.2392
110	1.8631	1693.1	1879.4	7.6429	0.9278	1691.1	1876.6	7.2993
120	1.9124	1710.8	1902.0	7.7013	0.9527	1708.9	1899.5	7.3581
130	1.9616	1728.7	1924.9	7.7586	0.9775	1727.0	1922.5	7.4159
140	2.0109	1746.8	1947.9	7.8150	1.0024	1745.2	1945.6	7.4727
150	2.0601	1765.1	1971.1	7.8704	1.0271	1763.5	1969.0	7.5284
160	2.1092	1783.5	1994.4	7.9250	1.0519	1782.1	1992.4	7.5833
170	2.1584	1802.1	2018.0	7.9787	1.0766	1800.8	2016.1	7.6373
180	2.2075	1820.9	2041.7	8.0317	1.1013	1819.6	2039.9	7.6904
190	2.2566	1839.9	2065.6	8.0838	1.1260	1838.7	2063.9	7.7428
200	2.3057	1859.1	2089.7	8.1353	1.1507	1857.9	2088.1	7.7944

continued...

Appendix J.2 - Superheated Ammonia (NH$_3$) - English Units

	P = 5 psia (-63.03 °F)				P = 10 psia (-41.28°F)			
	v	u	h	s	v	u	h	s
T(°F)	ft³/lbm	BTU/lbm	BTU/lbm	BTU/lbm-°F	ft³/lbm	BTU/lbm	BTU/lbm	BTU/lbm-°F
sat	49.294	613.87	659.51	1.6698	25.794	620.22	667.98	1.6112
-60	49.697	615.04	661.05	1.6737	-	-	-	-
-55	50.359	616.97	663.60	1.6800	-	-	-	-
-50	51.019	618.90	666.14	1.6862	-	-	-	-
-45	51.678	620.82	668.66	1.6924	-	-	-	-
-40	52.334	622.73	671.19	1.6984	25.881	620.74	668.67	1.6128
-35	52.989	624.64	673.70	1.7044	26.221	622.76	671.31	1.6191
-30	53.642	626.55	676.22	1.7103	26.560	624.77	673.95	1.6252
-25	54.293	628.45	678.72	1.7161	26.897	626.77	676.57	1.6313
-20	54.944	630.36	681.23	1.7218	27.233	628.76	679.18	1.6373
-15	55.593	632.26	683.73	1.7275	27.568	630.74	681.79	1.6431
-10	56.240	634.16	686.23	1.7330	27.901	632.71	684.38	1.6489
-5	56.887	636.05	688.72	1.7386	28.233	634.68	686.96	1.6547
0	57.533	637.95	691.22	1.7440	28.564	636.65	689.54	1.6603
10	58.822	641.75	696.21	1.7548	29.223	640.57	694.68	1.6714
20	60.108	645.55	701.20	1.7653	29.880	644.47	699.80	1.6822
30	61.391	649.36	706.20	1.7756	30.533	648.37	704.91	1.6927
40	62.672	653.17	711.20	1.7857	31.184	652.27	710.02	1.7030
50	63.950	657.00	716.21	1.7956	31.833	656.17	715.11	1.7131
60	65.227	660.84	721.23	1.8054	32.480	660.07	720.21	1.7230
70	66.503	664.69	726.26	1.8150	33.125	663.98	725.32	1.7328
80	67.777	668.55	731.30	1.8244	33.769	667.89	730.43	1.7423
90	69.049	672.43	736.37	1.8337	34.412	671.82	735.54	1.7517
100	70.321	676.33	741.44	1.8429	35.054	675.76	740.67	1.7610
110	71.592	680.25	746.54	1.8519	35.694	679.72	745.81	1.7701

Appendix J.2 - Superheated Ammonia (NH$_3$) - SI Units

	P = 100 kPa (-33.58°C)				P = 200 kPa (-18.84°C)			
T(°C)	v m^3/kg	u kJ/kg	h kJ/kg	s kJ/kg-K	v m^3/kg	u kJ/kg	h kJ/kg	s kJ/kg-K
210	2.3547	1878.5	2113.9	8.1860	1.1753	1877.4	2112.4	7.8453
220	2.4038	1898.0	2138.4	8.2361	1.2000	1896.9	2136.9	7.8956
230	2.4528	1917.7	2163.0	8.2856	1.2246	1916.7	2161.6	7.9452
240	2.5018	1937.6	2187.8	8.3344	1.2492	1936.7	2186.5	7.9941
250	2.5508	1957.7	2212.8	8.3826	1.2738	1956.8	2211.6	8.0425
260	2.5998	1978.0	2238.0	8.4303	1.2984	1977.1	2236.8	8.0902
270	2.6488	1998.5	2263.4	8.4774	1.3229	1997.6	2262.2	8.1375
280	2.6978	2019.1	2288.9	8.5240	1.3475	2018.3	2287.8	8.1841
290	2.7467	2039.9	2314.6	8.5701	1.3720	2039.2	2313.6	8.2303
300	2.7957	2061.0	2340.5	8.6157	1.3966	2060.2	2339.5	8.2760
310	2.8446	2082.1	2366.6	8.6608	1.4211	2081.4	2365.6	8.3212
320	2.8936	2103.5	2392.9	8.7055	1.4457	2102.8	2392.0	8.3659
330	2.9425	2125.1	2419.3	8.7497	1.4702	2124.4	2418.4	8.4102
340	2.9914	2146.8	2446.0	8.7935	1.4947	2146.2	2445.1	8.4540
350	3.0404	2168.7	2472.8	8.8369	1.5192	2168.1	2471.9	8.4975
360	3.0893	2190.8	2499.8	8.8798	1.5437	2190.2	2499.0	8.5405
370	3.1382	2213.1	2526.9	8.9224	1.5682	2212.5	2526.2	8.5831
380	3.1871	2235.6	2554.3	8.9646	1.5927	2235.0	2553.5	8.6253
390	3.2360	2258.2	2581.8	9.0064	1.6172	2257.6	2581.1	8.6672
400	3.2849	2281.0	2609.5	9.0479	1.6417	2280.5	2608.8	8.7087
410	3.3338	2304.0	2637.4	9.0890	1.6662	2303.5	2636.7	8.7498
420	3.3827	2327.2	2665.5	9.1298	1.6907	2326.7	2664.8	8.7907
430	3.4316	2350.5	2693.7	9.1702	1.7152	2350.0	2693.1	8.8311
440	3.4805	2374.1	2722.1	9.2104	1.7397	2373.6	2721.5	8.8713
450	3.5294	2397.8	2750.7	9.2502	1.7641	2397.3	2750.1	8.9111

Appendix J.2 - Superheated Ammonia (NH$_3$) - English Units

	P = 5 psia (-63.03 °F)				P = 10 psia (-41.28°F)			
T(°F)	v ft^3/lbm	u BTU/lbm	h BTU/lbm	s BTU/lbm-°F	v ft^3/lbm	u BTU/lbm	h BTU/lbm	s BTU/lbm-°F
120	72.862	684.19	751.65	1.8608	36.334	683.69	750.97	1.7790
130	74.131	688.15	756.78	1.8696	36.974	687.67	756.14	1.7879
140	75.400	692.12	761.93	1.8782	37.612	691.68	761.32	1.7966
150	76.668	696.12	767.11	1.8868	38.250	695.70	766.53	1.8052
160	77.936	700.14	772.30	1.8952	38.888	699.74	771.75	1.8137
170	79.203	704.19	777.52	1.9036	39.525	703.80	776.99	1.8221
180	80.469	708.25	782.76	1.9118	40.161	707.89	782.26	1.8304
190	81.736	712.34	788.02	1.9200	40.797	711.99	787.54	1.8386
200	83.001	716.45	793.30	1.9281	41.433	716.12	792.84	1.8467
220	85.532	724.75	803.94	1.9439	42.704	724.44	803.52	1.8626
240	88.062	733.14	814.67	1.9595	43.973	732.86	814.28	1.8782
260	90.591	741.63	825.50	1.9748	45.242	741.37	825.14	1.8935
280	93.119	750.21	836.43	1.9897	46.509	749.97	836.09	1.9085
300	95.646	758.90	847.45	2.0045	47.776	758.67	847.14	1.9233
320	98.173	767.68	858.58	2.0189	49.043	767.47	858.28	1.9378
340	100.70	776.57	869.80	2.0331	50.309	776.37	869.53	1.9520
360	103.22	785.55	881.12	2.0471	51.574	785.36	880.86	1.9660
380	105.75	794.63	892.54	2.0609	52.839	794.46	892.30	1.9798
400	108.28	803.81	904.06	2.0744	54.104	803.65	903.83	1.9934
450	114.59	827.20	933.29	2.1075	57.264	827.06	933.09	2.0264
500	120.90	851.21	963.14	2.1394	60.423	851.08	962.97	2.0584
550	127.20	875.83	993.60	2.1704	63.581	875.71	993.45	2.0894
600	133.51	901.06	1024.70	2.2004	66.737	900.95	1024.5	2.1194
650	139.82	926.90	1056.30	2.2296	69.893	926.80	1056.2	2.1486
700	146.12	953.34	1088.60	2.2580	73.048	953.25	1088.5	2.1771
750	152.43	980.37	1121.50	2.2858	76.202	980.29	1121.4	2.2048
800	158.73	1008.0	1155.0	2.3129	79.356	1007.9	1154.9	2.2320

Appendix J.2 - Superheated Ammonia (NH$_3$) - SI Units

	P = 300 kPa (-9.22°C)				P = 400 kPa (-1.87°C)			
	v	u	h	s	v	u	h	s
T(°C)	m³/kg	kJ/kg	kJ/kg	kJ/kg-K	m³/kg	kJ/kg	kJ/kg	kJ/kg-K
sat	0.4060	1475.1	1596.9	6.2277	0.3094	1481.7	1605.4	6.1275
0	0.4238	1493.0	1620.1	6.3140	0.3122	1485.5	1610.4	6.1457
10	0.4425	1511.7	1644.4	6.4015	0.3270	1505.2	1636.2	6.2386
20	0.4607	1529.9	1668.1	6.4838	0.3413	1524.6	1661.1	6.3250
30	0.4787	1547.8	1691.4	6.5619	0.3552	1543.2	1685.3	6.4061
40	0.4963	1565.5	1714.4	6.6366	0.3688	1561.5	1709.0	6.4832
50	0.5138	1583.1	1737.2	6.7083	0.3822	1579.5	1732.4	6.5568
60	0.5311	1600.6	1759.9	6.7776	0.3955	1597.5	1755.6	6.6275
70	0.5482	1618.2	1782.6	6.8447	0.4086	1615.3	1778.7	6.6958
80	0.5653	1635.8	1805.3	6.9099	0.4216	1633.2	1801.8	6.7620
90	0.5823	1653.4	1828.1	6.9734	0.4345	1651.0	1824.8	6.8264
100	0.5991	1671.2	1850.9	7.0354	0.4473	1669.0	1847.9	6.8891
110	0.6160	1689.1	1873.9	7.0961	0.4601	1687.0	1871.1	6.9503
120	0.6328	1707.1	1896.9	7.1554	0.4728	1705.2	1894.3	7.0102
130	0.6495	1725.2	1920.1	7.2136	0.4854	1723.5	1917.6	7.0688
140	0.6662	1743.5	1943.4	7.2707	0.4981	1741.9	1941.1	7.1263
150	0.6828	1762.0	1966.8	7.3268	0.5107	1760.4	1964.7	7.1827
160	0.6994	1780.6	1990.4	7.3820	0.5232	1779.2	1988.4	7.2382
170	0.7160	1799.4	2014.2	7.4362	0.5357	1798.0	2012.3	7.2927
180	0.7326	1818.3	2038.1	7.4896	0.5482	1817.1	2036.3	7.3463
190	0.7492	1837.5	2062.2	7.5422	0.5607	1836.3	2060.5	7.3991
200	0.7657	1856.8	2086.5	7.5940	0.5732	1855.6	2084.9	7.4511

continued...

Appendix J.2 - Superheated Ammonia (NH$_3$) - English Units

	P = 25 psia (-7.915°F)				P = 50 psia (21.69°F)			
	v	u	h	s	v	u	h	s
T(°F)	ft³/lbm	BTU/lbm	BTU/lbm	BTU/lbm-°F	ft³/lbm	BTU/lbm	BTU/lbm	BTU/lbm-°F
sat	10.946	629.11	679.78	1.5345	5.7037	635.98	688.78	1.4769
-5	11.030	630.39	681.45	1.5382	-	-	-	-
0	11.174	632.57	684.30	1.5444	-	-	-	-
10	11.457	636.88	689.92	1.5565	-	-	-	-
20	11.737	641.13	695.47	1.5682	-	-	-	-
30	12.013	645.33	700.95	1.5795	5.8306	639.93	693.91	1.4875
40	12.287	649.49	706.37	1.5905	5.9805	644.59	699.96	1.4997
50	12.559	653.62	711.76	1.6012	6.1276	649.15	705.88	1.5114
60	12.828	657.72	717.11	1.6116	6.2725	653.63	711.70	1.5227
70	13.096	661.81	722.44	1.6217	6.4154	658.04	717.44	1.5337
80	13.363	665.89	727.75	1.6317	6.5566	662.41	723.11	1.5443
90	13.628	669.95	733.04	1.6414	6.6964	666.73	728.73	1.5546
100	13.892	674.02	738.33	1.6509	6.8349	671.02	734.30	1.5646
110	14.155	678.08	743.61	1.6603	6.9724	675.29	739.84	1.5744
120	14.417	682.15	748.89	1.6695	7.1089	679.54	745.35	1.5840
130	14.678	686.23	754.18	1.6785	7.2445	683.77	750.85	1.5934
140	14.939	690.32	759.47	1.6874	7.3794	688.00	756.32	1.6026
150	15.199	694.41	764.77	1.6962	7.5136	692.23	761.79	1.6117
160	15.458	698.52	770.08	1.7048	7.6472	696.45	767.26	1.6206
170	15.717	702.64	775.40	1.7133	7.7804	700.68	772.72	1.6293
180	15.976	706.78	780.74	1.7217	7.9130	704.92	778.18	1.6379
190	16.234	710.94	786.09	1.7300	8.0452	709.16	783.65	1.6464

Appendix J.2 - Superheated Ammonia (NH$_3$) - SI Units

	P = 300 kPa (-9.22°C)				P = 400 kPa (-1.87°C)			
	v	u	h	s	v	u	h	s
T(°C)	m³/kg	kJ/kg	kJ/kg	kJ/kg-K	m³/kg	kJ/kg	kJ/kg	kJ/kg-K
210	0.7822	1876.2	2110.9	7.6451	0.5856	1875.1	2109.4	7.5023
220	0.7987	1895.9	2135.5	7.6955	0.5981	1894.8	2134.1	7.5529
230	0.8152	1915.7	2160.3	7.7452	0.6105	1914.7	2158.9	7.6027
240	0.8316	1935.7	2185.2	7.7943	0.6229	1934.8	2183.9	7.6520
250	0.8481	1955.9	2210.3	7.8428	0.6353	1955.0	2209.1	7.7005
260	0.8645	1976.2	2235.6	7.8906	0.6476	1975.4	2234.4	7.7485
270	0.8810	1996.8	2261.1	7.9380	0.6600	1995.9	2259.9	7.7959
280	0.8974	2017.5	2286.7	7.9847	0.6724	2016.7	2285.6	7.8428
290	0.9138	2038.4	2312.5	8.0310	0.6847	2037.6	2311.5	7.8891
300	0.9302	2059.5	2338.5	8.0767	0.6971	2058.7	2337.5	7.9350
310	0.9466	2080.7	2364.7	8.1220	0.7094	2080.0	2363.7	7.9803
320	0.9630	2102.1	2391.0	8.1668	0.7217	2101.4	2390.1	8.0252
330	0.9794	2123.7	2417.6	8.2111	0.7340	2123.1	2416.7	8.0695
340	0.9958	2145.5	2444.2	8.2550	0.7463	2144.9	2443.4	8.1135
350	1.0122	2167.5	2471.1	8.2985	0.7587	2166.8	2470.3	8.1570
360	1.0285	2189.6	2498.2	8.3416	0.7710	2189.0	2497.4	8.2001
370	1.0449	2211.9	2525.4	8.3842	0.7833	2211.3	2524.6	8.2428
380	1.0613	2234.4	2552.8	8.4265	0.7956	2233.8	2552.1	8.2852
390	1.0776	2257.1	2580.4	8.4684	0.8078	2256.5	2579.7	8.3271
400	1.0940	2279.9	2608.1	8.5099	0.8201	2279.4	2607.4	8.3687
410	1.1103	2303.0	2636.1	8.5511	0.8324	2302.4	2635.4	8.4099
420	1.1267	2326.2	2664.2	8.5920	0.8447	2325.6	2663.5	8.4508
430	1.1430	2349.5	2692.4	8.6325	0.8570	2349.0	2691.8	8.4913
440	1.1594	2373.1	2720.9	8.6727	0.8692	2372.6	2720.3	8.5315
450	1.1757	2396.8	2749.5	8.7125	0.8815	2396.3	2748.9	8.5714

Appendix J.2 - Superheated Ammonia (NH$_3$) - English Units

	P = 25 psia (-7.915°F)				P = 50 psia (21.69°F)			
	v	u	h	s	v	u	h	s
T(°F)	ft³/lbm	BTU/lbm	BTU/lbm	BTU/lbm-°F	ft³/lbm	BTU/lbm	BTU/lbm	BTU/lbm-°F
200	16.492	715.11	791.46	1.7382	8.1771	713.42	789.13	1.6548
220	17.006	723.52	802.25	1.7543	8.4397	721.97	800.11	1.6712
240	17.519	732.01	813.11	1.7701	8.7012	730.58	811.14	1.6872
260	18.032	740.58	824.05	1.7855	8.9617	739.25	822.23	1.7028
280	18.544	749.24	835.08	1.8006	9.2214	748.01	833.39	1.7181
300	19.055	757.99	846.20	1.8155	9.4803	756.84	844.62	1.7331
320	19.565	766.83	857.40	1.8300	9.7386	765.76	855.92	1.7478
340	20.075	775.77	868.70	1.8443	9.9964	774.76	867.31	1.7622
360	20.584	784.80	880.09	1.8584	10.2540	783.85	878.78	1.7763
380	21.093	793.92	891.57	1.8722	10.5110	793.03	890.34	1.7903
400	21.601	803.14	903.14	1.8858	10.7670	802.30	901.99	1.8040
450	22.871	826.61	932.49	1.9190	11.4070	825.87	931.49	1.8373
500	24.140	850.69	962.44	1.9511	12.0450	850.03	961.55	1.8695
550	25.407	875.36	992.98	1.9821	12.6820	874.78	992.20	1.9006
600	26.673	900.64	1024.1	2.0122	13.3180	900.11	1023.4	1.9308
650	27.938	926.51	1055.8	2.0414	13.9530	926.04	1055.2	1.9601
700	29.203	952.99	1088.2	2.0699	14.5880	952.55	1087.6	1.9887
750	30.467	980.05	1121.1	2.0977	15.2220	979.65	1120.6	2.0165
800	31.730	1007.7	1154.6	2.1249	15.8550	1007.3	1154.1	2.0437

Appendix J.2 - Superheated Ammonia (NH$_3$) - SI Units

	P = 500 kPa (4.15 °C)				P = 600 kPa (9.30 °C)			
	v	u	h	s	v	u	h	s
T(°C)	m³/kg	kJ/kg	kJ/kg	kJ/kg-K	m³/kg	kJ/kg	kJ/kg	kJ/kg-K
sat	0.25031	1486.6	1611.8	6.0494	0.21036	1490.5	1616.8	5.9854
10	0.25753	1498.9	1627.7	6.1062	0.21111	1492.1	1618.8	5.9924
20	0.26947	1519.1	1653.8	6.1969	0.22152	1513.4	1646.3	6.0880
30	0.28100	1538.5	1679.0	6.2814	0.23150	1533.6	1672.5	6.1760
40	0.29224	1557.4	1703.5	6.3609	0.24115	1553.2	1697.9	6.2583
50	0.30324	1575.9	1727.5	6.4365	0.25056	1572.2	1722.6	6.3359
60	0.31407	1594.2	1751.3	6.5088	0.25978	1591.0	1746.8	6.4099
70	0.32475	1612.4	1774.8	6.5784	0.26885	1609.5	1770.8	6.4807
80	0.33532	1630.5	1798.2	6.6456	0.27780	1627.9	1794.6	6.5490
90	0.34579	1648.6	1821.5	6.7108	0.28666	1646.2	1818.2	6.6151
100	0.35619	1666.8	1844.9	6.7742	0.29544	1664.6	1841.8	6.6792
110	0.36652	1685.0	1868.3	6.8360	0.30414	1682.9	1865.4	6.7416
120	0.37679	1703.3	1891.7	6.8964	0.31279	1701.4	1889.1	6.8025
130	0.38702	1721.7	1915.2	6.9555	0.32140	1719.9	1912.8	6.8620
140	0.39720	1740.2	1938.8	7.0134	0.32996	1738.6	1936.5	6.9203
150	0.40735	1758.9	1962.6	7.0701	0.33848	1757.3	1960.4	6.9774
160	0.41747	1777.7	1986.4	7.1259	0.34697	1776.2	1984.4	7.0334
170	0.42756	1796.6	2010.4	7.1806	0.35544	1795.3	2008.5	7.0885
180	0.43762	1815.8	2034.6	7.2345	0.36387	1814.4	2032.8	7.1426
190	0.44766	1835.0	2058.8	7.2875	0.37229	1833.8	2057.1	7.1958
200	0.45769	1854.4	2083.3	7.3397	0.38068	1853.1	2081.7	7.2482

continued...

Appendix J.2 - Superheated Ammonia (NH$_3$) - English Units

	P = 75 psia (41.15°F)				P = 100 psia (56.06°F)			
	v	u	h	s	v	u	h	s
T(°F)	ft³/lbm	BTU/lbm	BTU/lbm	BTU/lbm-°F	ft³/lbm	BTU/lbm	BTU/lbm	BTU/lbm-°F
sat	3.8833	639.88	693.81	1.4430	2.9495	642.48	697.10	1.4188
50	3.9779	644.35	699.60	1.4545	-	-	-	-
60	4.0822	649.27	705.96	1.4669	2.9828	644.60	699.83	1.4241
70	4.1844	654.06	712.17	1.4787	3.0653	649.83	706.59	1.4369
80	4.2846	658.75	718.25	1.4901	3.1456	654.89	713.14	1.4492
90	4.3831	663.36	724.23	1.5010	3.2240	659.82	719.52	1.4609
100	4.4803	667.90	730.12	1.5117	3.3009	664.64	725.76	1.4722
110	4.5762	672.38	735.94	1.5220	3.3765	669.37	731.89	1.483
120	4.6711	676.83	741.70	1.5320	3.4508	674.02	737.92	1.4935
130	4.7651	681.24	747.41	1.5418	3.5242	678.62	743.88	1.5037
140	4.8583	685.62	753.09	1.5513	3.5967	683.17	749.77	1.5136
150	4.9508	689.98	758.74	1.5607	3.6685	687.68	755.61	1.5233
160	5.0426	694.33	764.37	1.5698	3.7395	692.17	761.41	1.5327
170	5.1339	698.68	769.98	1.5788	3.8100	696.63	767.18	1.5419
180	5.2247	703.01	775.58	1.5876	3.8800	701.07	772.92	1.551
190	5.3150	707.35	781.17	1.5963	3.9494	705.51	778.64	1.5599
200	5.4050	711.69	786.76	1.6048	4.0185	709.94	784.35	1.5686
220	5.5838	720.39	797.94	1.6215	4.1555	718.79	795.74	1.5856
240	5.7614	729.13	809.14	1.6378	4.2912	727.66	807.13	1.6021
260	5.9380	737.92	820.38	1.6536	4.4259	736.57	818.52	1.6182
280	6.1137	746.77	831.67	1.6691	4.5597	745.51	829.95	1.6338

Appendix J.2 - Superheated Ammonia (NH$_3$) - SI Units

	P = 500 kPa (4.15 °C)				P = 600 kPa (9.30 °C)			
	v	u	h	s	v	u	h	s
T(°C)	m³/kg	kJ/kg	kJ/kg	kJ/kg-K	m³/kg	kJ/kg	kJ/kg	kJ/kg-K
210	0.46769	1874.0	2107.9	7.3911	0.38906	1872.9	2106.3	7.2998
220	0.47768	1893.8	2132.6	7.4418	0.39742	1892.7	2131.2	7.3506
230	0.48765	1913.7	2157.5	7.4918	0.40577	1912.7	2156.1	7.4008
240	0.49761	1933.8	2182.6	7.5411	0.41410	1932.8	2181.3	7.4502
250	0.50756	1954.0	2207.8	7.5898	0.42242	1953.1	2206.6	7.4991
260	0.51749	1974.5	2233.2	7.6379	0.43073	1973.6	2232.0	7.5473
270	0.52742	1995.1	2258.8	7.6854	0.43903	1994.2	2257.7	7.5949
280	0.53733	2015.9	2284.5	7.7324	0.44731	2015.1	2283.4	7.6419
290	0.54724	2036.8	2310.4	7.7788	0.45559	2036.0	2309.4	7.6884
300	0.55714	2057.9	2336.5	7.8247	0.46387	2057.2	2335.5	7.7344
310	0.56703	2079.2	2362.8	7.8701	0.47213	2078.5	2361.8	7.7799
320	0.57692	2100.7	2389.2	7.9150	0.48039	2100.0	2388.3	7.8248
330	0.58680	2122.4	2415.8	7.9595	0.48864	2121.7	2414.9	7.8694
340	0.59667	2144.2	2442.5	8.0035	0.49689	2143.6	2441.7	7.9134
350	0.60654	2166.2	2469.5	8.0471	0.50513	2165.6	2468.7	7.9570
360	0.61640	2188.4	2496.6	8.0902	0.51336	2187.8	2495.8	8.0003
370	0.62626	2210.7	2523.9	8.1330	0.52160	2210.1	2523.1	8.0431
380	0.63611	2233.3	2551.3	8.1753	0.52982	2232.7	2550.6	8.0855
390	0.64596	2256.0	2579.0	8.2173	0.53804	2255.4	2578.2	8.1275
400	0.65581	2278.9	2606.8	8.2589	0.54626	2278.3	2606.1	8.1691
410	0.66565	2301.9	2634.7	8.3002	0.55448	2301.4	2634.1	8.2104
420	0.67549	2325.1	2662.9	8.3411	0.56269	2324.6	2662.2	8.2514
430	0.68533	2348.5	2691.2	8.3817	0.57090	2348.0	2690.6	8.2919
440	0.69516	2372.1	2719.7	8.4219	0.57911	2371.6	2719.1	8.3322
450	0.70499	2395.9	2748.4	8.4618	0.58731	2395.4	2747.8	8.3722

Appendix J.2 - Superheated Ammonia (NH$_3$) - English Units

	P = 75 psia (41.15°F)				P = 100 psia (56.06°F)			
	v	u	h	s	v	u	h	s
T(°F)	ft³/lbm	BTU/lbm	BTU/lbm	BTU/lbm-°F	ft³/lbm	BTU/lbm	BTU/lbm	BTU/lbm-°F
300	6.2887	755.68	843.02	1.6842	4.6928	754.52	841.42	1.6491
320	6.4631	764.68	854.43	1.6991	4.8252	763.59	852.94	1.6641
340	6.6369	773.74	865.92	1.7136	4.9570	772.73	864.52	1.6788
360	6.8102	782.90	877.48	1.7279	5.0884	781.94	876.16	1.6931
380	6.9831	792.13	889.11	1.7419	5.2193	791.23	887.87	1.7073
400	7.1556	801.45	900.83	1.7557	5.3498	800.60	899.66	1.7211
420	7.3277	810.86	912.62	1.7693	5.4799	810.05	911.52	1.7348
440	7.4995	820.35	924.51	1.7826	5.6098	819.59	923.47	1.7482
460	7.6711	829.94	936.47	1.7958	5.7393	829.21	935.49	1.7614
480	7.8424	839.61	948.53	1.8087	5.8687	838.92	947.59	1.7744
500	8.0135	849.38	960.67	1.8215	5.9978	848.72	959.78	1.7873
520	8.1844	859.23	972.90	1.8341	6.1267	858.61	972.06	1.7999
540	8.3552	869.18	985.22	1.8466	6.2554	868.58	984.42	1.8124
560	8.5257	879.22	997.63	1.8589	6.3839	878.65	996.86	1.8247
580	8.6961	889.36	1010.1	1.8710	6.5123	888.81	1009.4	1.8369
600	8.8663	899.59	1022.7	1.8830	6.6406	899.06	1022.0	1.8489
620	9.0365	909.90	1035.4	1.8949	6.7687	909.40	1034.7	1.8608
640	9.2065	920.32	1048.2	1.9066	6.8967	919.83	1047.5	1.8726
660	9.3764	930.82	1061.0	1.9182	7.0246	930.35	1060.4	1.8842
680	9.5462	941.42	1074.0	1.9296	7.1524	940.97	1073.4	1.8957
700	9.7159	952.11	1087.0	1.9410	7.2801	951.67	1086.5	1.9071
720	9.8855	962.90	1100.2	1.9522	7.4077	962.47	1099.6	1.9183
740	10.055	973.77	1113.4	1.9634	7.5352	973.37	1112.9	1.9294
760	10.225	984.75	1126.7	1.9744	7.6627	984.35	1126.2	1.9405
780	10.394	995.81	1140.2	1.9853	7.7901	995.42	1139.7	1.9514
800	10.563	1007.0	1153.7	1.9961	7.9174	1006.6	1153.2	1.9622
820	10.733	1018.2	1167.3	2.0068	8.0447	1017.9	1166.8	1.9729
840	10.902	1029.6	1181.0	2.0174	8.1719	1029.2	1180.5	1.9836

Appendix J.2 - Superheated Ammonia (NH$_3$) - SI Units

	P = 800 kPa (17.86 °C)				P = 1000 kPa (24.91 °C)			
	v	u	h	s	v	u	h	s
T(°C)	m^3/kg	kJ/kg	kJ/kg	kJ/kg·K	m^3/kg	kJ/kg	kJ/kg	kJ/kg·K
sat	0.15958	1496.4	1624.0	5.5833	0.12853	1500.4	1628.9	5.8029
20	0.16137	1501.3	1630.4	5.9051	-	-	-	-
30	0.16946	1523.5	1659.0	6.0011	0.13205	1512.6	1644.6	5.8551
40	0.17718	1544.4	1686.2	6.0893	0.13867	1535.2	1673.9	5.9501
50	0.18463	1564.6	1712.3	6.1715	0.14497	1556.7	1701.7	6.0374
60	0.19186	1584.2	1737.7	6.2489	0.15104	1577.3	1728.3	6.1187
70	0.19893	1603.5	1762.6	6.3226	0.15693	1597.3	1754.2	6.1953
80	0.20588	1622.5	1787.2	6.3931	0.16268	1616.9	1779.6	6.2682
90	0.21272	1641.3	1811.5	6.4609	0.16832	1636.3	1804.6	6.3380
100	0.21947	1660.1	1835.6	6.5266	0.17387	1655.5	1829.3	6.4052
110	0.22616	1678.8	1859.7	6.5903	0.17935	1674.6	1853.9	6.4702
120	0.23278	1697.5	1883.8	6.6522	0.18476	1693.6	1878.4	6.5333
130	0.23936	1716.4	1907.9	6.7127	0.19012	1712.7	1902.9	6.5947
140	0.24589	1735.2	1931.9	6.7717	0.19544	1731.9	1927.3	6.6546
150	0.25238	1754.2	1956.1	6.8295	0.20072	1751.0	1951.8	6.7131
160	0.25884	1773.3	1980.4	6.8862	0.20596	1770.3	1976.3	6.7703
170	0.26528	1792.5	2004.7	6.9417	0.21118	1789.7	2000.9	6.8265
180	0.27168	1811.8	2029.2	6.9963	0.21636	1809.2	2025.5	6.8815
190	0.27807	1831.3	2053.7	7.0500	0.22153	1828.8	2050.3	6.9356
200	0.28443	1850.9	2078.4	7.1027	0.22667	1848.5	2075.2	6.9888

continued....

Appendix J.2 - Superheated Ammonia (NH$_3$) - English Units

	P = 125 psia (68.32°F)				P = 150 psia (78.81°F)			
	v	u	h	s	v	u	h	s
T(°F)	ft^3/lbm	BTU/lbm	BTU/lbm	BTU/lbm·°F	ft^3/lbm	BTU/lbm	BTU/lbm	BTU/lbm·°F
sat	2.3786	644.35	699.41	1.3998	1.9923	645.73	701.07	1.384
70	2.3905	645.30	700.63	1.4021	-	-	-	-
80	2.4595	650.79	707.73	1.4153	1.9996	646.43	701.97	1.386
90	2.5264	656.09	714.57	1.4279	2.0593	652.14	709.34	1.399
100	2.5915	661.22	721.21	1.4399	2.1169	657.63	716.43	1.412
110	2.6551	666.22	727.68	1.4513	2.1728	662.93	723.28	1.424
120	2.7174	671.11	734.01	1.4624	2.2273	668.08	729.95	1.436
130	2.7786	675.91	740.23	1.4730	2.2806	673.11	736.46	1.447
140	2.8389	680.65	746.36	1.4833	2.3328	678.04	742.84	1.458
150	2.8983	685.32	752.40	1.4933	2.3842	682.88	749.11	1.468
160	2.9570	689.94	758.39	1.5030	2.4347	687.66	755.29	1.478
170	3.0151	694.53	764.32	1.5125	2.4847	692.39	761.40	1.488
180	3.0726	699.09	770.21	1.5218	2.5340	697.07	767.45	1.497
190	3.1296	703.63	776.07	1.5309	2.5827	701.72	773.46	1.507
200	3.1862	708.16	781.91	1.5398	2.6310	706.34	779.42	1.516
220	3.2982	717.17	793.52	1.5572	2.7265	715.53	791.26	1.533
240	3.4089	726.18	805.09	1.5739	2.8205	724.68	803.02	1.551
260	3.5185	735.20	816.64	1.5902	2.9134	733.82	814.74	1.567
280	3.6272	744.25	828.21	1.6061	3.0054	742.98	826.45	1.583
300	3.7351	753.34	839.80	1.6215	3.0966	752.16	838.17	1.599

Appendix J.2 - Superheated Ammonia (NH₃) - SI Units

	P = 800 kPa (17.86 °C)				P = 1000 kPa (24.91 °C)			
T(°C)	v m³/kg	u kJ/kg	h kJ/kg	s kJ/kg-K	v m³/kg	u kJ/kg	h kJ/kg	s kJ/kg-K
210	0.29077	1870.7	2103.3	7.1547	0.23180	1868.4	2100.2	7.0411
220	0.29710	1890.6	2128.3	7.2058	0.23690	1888.4	2125.3	7.0926
230	0.30341	1910.7	2153.4	7.2563	0.24200	1908.6	2150.6	7.1433
240	0.30971	1930.9	2178.6	7.3060	0.24708	1928.9	2176.0	7.1933
250	0.31600	1951.3	2204.1	7.3551	0.25214	1949.4	2201.5	7.2426
260	0.32227	1971.8	2229.6	7.4035	0.25720	1970.0	2227.2	7.2912
270	0.32854	1992.5	2255.4	7.4513	0.26224	1990.8	2253.1	7.3392
280	0.33479	2013.4	2281.2	7.4985	0.26728	2011.8	2279.1	7.3866
290	0.34104	2034.5	2307.3	7.5452	0.27230	2032.9	2305.2	7.4334
300	0.34728	2055.7	2333.5	7.5913	0.27732	2054.2	2331.5	7.4797
310	0.35351	2077.1	2359.9	7.6369	0.28233	2075.6	2357.9	7.5255
320	0.35973	2098.6	2386.4	7.6820	0.28734	2097.2	2384.6	7.5707
330	0.36595	2120.3	2413.1	7.7267	0.29233	2119.0	2411.3	7.6155
340	0.37216	2142.2	2440.0	7.7708	0.29733	2140.9	2438.3	7.6598
350	0.37837	2164.3	2467.0	7.8146	0.30231	2163.0	2465.4	7.7036
360	0.38457	2186.5	2494.2	7.8579	0.30729	2185.3	2492.6	7.7470
370	0.39077	2209.0	2521.6	7.9008	0.31227	2207.8	2520.0	7.7900
380	0.39696	2231.5	2549.1	7.9432	0.31724	2230.4	2547.6	7.8325
390	0.40315	2254.3	2576.8	7.9853	0.32221	2253.2	2575.4	7.8747
400	0.40933	2277.2	2604.7	8.0271	0.32717	2276.1	2603.3	7.9165
410	0.41551	2300.3	2632.7	8.0684	0.33214	2299.3	2631.4	7.9579
420	0.42169	2323.6	2661.0	8.1094	0.33709	2322.6	2659.7	7.9990
430	0.42787	2347.0	2689.3	8.1501	0.34205	2346.0	2688.1	8.0397
440	0.43404	2370.7	2717.9	8.1904	0.34700	2369.7	2716.7	8.0801
450	0.44021	2394.4	2746.6	8.2304	0.35195	2393.5	2745.4	8.1202

Appendix J.2 - Superheated Ammonia (NH₃) - English Units

	P = 125 psia (68.32°F)				P = 150 psia (78.81°F)			
T(°F)	v ft³/lbm	u BTU/lbm	h BTU/lbm	s BTU/lbm-°F	v ft³/lbm	u BTU/lbm	h BTU/lbm	s BTU/lbm-°F
320	3.8424	762.49	851.43	1.6366	3.1871	761.39	849.91	1.614
340	3.9490	771.70	863.11	1.6514	3.2770	770.67	861.69	1.629
360	4.0552	780.97	874.84	1.6659	3.3664	780.01	873.51	1.644
380	4.1609	790.32	886.63	1.6801	3.4553	789.41	885.38	1.658
400	4.2662	799.74	898.49	1.6941	3.5439	798.88	897.32	1.672
420	4.3712	809.24	910.42	1.7078	3.6321	808.43	909.31	1.686
440	4.4759	818.82	922.42	1.7213	3.7199	818.05	921.38	1.699
460	4.5803	828.48	934.50	1.7346	3.8075	827.75	933.51	1.713
480	4.6844	838.23	946.66	1.7477	3.8949	837.54	945.72	1.726
500	4.7883	848.06	958.89	1.7605	3.9820	847.40	958.00	1.739
520	4.8920	857.98	971.21	1.7732	4.0689	857.35	970.36	1.751
540	4.9955	867.98	983.61	1.7858	4.1556	867.38	982.81	1.764
560	5.0988	878.08	996.10	1.7981	4.2421	877.50	995.33	1.776
580	5.2020	888.26	1008.7	1.8103	4.3285	887.70	1007.9	1.789
600	5.3051	898.53	1021.3	1.8224	4.4148	898.00	1020.6	1.801
620	5.4080	908.89	1034.1	1.8343	4.5009	908.38	1033.4	1.813
640	5.5108	919.34	1046.9	1.8461	4.5869	918.85	1046.3	1.824
660	5.6135	929.88	1059.8	1.8577	4.6728	929.41	1059.2	1.836
680	5.7161	940.51	1072.8	1.8692	4.7586	940.06	1072.2	1.848
700	5.8186	951.24	1085.9	1.8806	4.8443	950.80	1085.4	1.859
720	5.9210	962.05	1099.1	1.8919	4.9299	961.63	1098.6	1.870
740	6.0233	972.96	1112.4	1.9031	5.0154	972.54	1111.9	1.882
760	6.1256	983.95	1125.7	1.9141	5.1009	983.55	1125.2	1.893
780	6.2278	995.04	1139.2	1.9251	5.1862	994.65	1138.7	1.904
800	6.3299	1006.2	1152.7	1.9359	5.2716	1005.8	1152.3	1.914
820	6.4320	1017.5	1166.4	1.9466	5.3568	1017.1	1165.9	1.925
840	6.5340	1028.8	1180.1	1.9573	5.4421	1028.5	1179.7	1.936

Appendix J.2 - Superheated Ammonia (NH$_3$) - SI Units

	P = 2000 kPa (49.37 °C)				P = 3000 kPa (65.74 °C)			
	v	u	h	s	v	u	h	s
T(°C)	m³/kg	kJ/kg	kJ/kg	kJ/kg·K	m³/kg	kJ/kg	kJ/kg	kJ/kg·K
sat	0.0645	1508.2	1637.1	5.5407	0.0422	1506.2	1632.8	5.3707
50	0.0647	1510.0	1639.5	5.5481	-	-	-	-
60	0.0688	1537.7	1675.2	5.6571	-	-	-	-
70	0.0725	1563.0	1708.0	5.7539	0.0436	1520.5	1651.2	5.4248
80	0.0759	1586.7	1738.6	5.8420	0.0465	1551.0	1690.5	5.5379
90	0.0793	1609.3	1767.9	5.9236	0.0492	1578.4	1726.1	5.6371
100	0.0825	1631.1	1796.1	6.0003	0.0517	1603.9	1759.1	5.7268
110	0.0856	1652.4	1823.5	6.0729	0.0541	1628.1	1790.4	5.8096
120	0.0886	1673.3	1850.5	6.1423	0.0564	1651.3	1820.4	5.8870
130	0.0916	1693.9	1877.0	6.2090	0.0586	1673.8	1849.5	5.9601
140	0.0945	1714.4	1903.3	6.2734	0.0607	1695.9	1878.0	6.0299
150	0.0973	1734.8	1929.4	6.3358	0.0628	1717.7	1906.0	6.0968
160	0.1001	1755.1	1955.4	6.3965	0.0648	1739.2	1933.6	6.1612
170	0.1029	1775.4	1981.2	6.4555	0.0668	1760.5	1960.9	6.2236
180	0.1057	1795.7	2007.1	6.5132	0.0688	1781.8	1988.0	6.2842
190	0.1084	1816.1	2032.9	6.5696	0.0707	1803.0	2015.0	6.3431
200	0.1111	1836.5	2058.8	6.6248	0.0726	1824.1	2041.9	6.4005

continued...

Appendix J.2 - Superheated Ammonia (NH$_3$) - English Units

	P = 300 psia (123.20°F)				P = 500 psia (161.14°F)			
	v	u	h	s	v	u	h	s
T(°F)	ft³/lbm	BTU/lbm	BTU/lbm	BTU/lbm-°F	ft³/lbm	BTU/lbm	BTU/lbm	BTU/lbm-°F
sat	0.99762	648.87	704.29	1.3210	0.58066	646.89	700.65	1.2685
130	1.0223	653.61	710.40	1.3315	-	-	-	-
140	1.0569	660.23	718.94	1.3458	-	-	-	-
150	1.0900	666.50	727.05	1.3592	-	-	-	-
160	1.1217	672.50	734.81	1.3719	-	-	-	-
170	1.1523	678.28	742.29	1.3839	0.60435	654.44	710.39	1.2841
180	1.1820	683.89	749.55	1.3953	0.62913	662.27	720.52	1.3001
190	1.2109	689.35	756.62	1.4062	0.65231	669.56	729.96	1.3147
200	1.2392	694.69	763.53	1.4168	0.67426	676.43	738.86	1.3283
220	1.2941	705.10	776.99	1.4369	0.71533	689.27	755.50	1.3532
240	1.3472	715.24	790.08	1.4559	0.75364	701.26	771.03	1.3757
260	1.3989	725.21	802.92	1.4740	0.78993	712.67	785.81	1.3965
280	1.4495	735.06	815.58	1.4913	0.82470	723.70	800.06	1.4160
300	1.4991	744.85	828.13	1.5081	0.85827	734.46	813.93	1.4346
320	1.5480	754.60	840.59	1.5243	0.89087	745.04	827.52	1.4522
340	1.5962	764.33	853.00	1.5400	0.92268	755.48	840.91	1.4692
360	1.6438	774.08	865.39	1.5553	0.95383	765.84	854.16	1.4855

Appendix J.2 - Superheated Ammonia (NH₃) - SI Units

	P = 2000 kPa (49.37 °C)				P = 3000 kPa (65.74 °C)			
	v	u	h	s	v	u	h	s
T(°C)	m³/kg	kJ/kg	kJ/kg	kJ/kg·K	m³/kg	kJ/kg	kJ/kg	kJ/kg·K
210	0.1138	1857.0	2084.7	6.6789	0.0745	1845.3	2068.7	6.4566
220	0.1165	1877.6	2110.6	6.7321	0.0763	1866.5	2095.5	6.5115
230	0.1192	1898.3	2136.6	6.7843	0.0782	1887.7	2122.3	6.5653
240	0.1218	1919.1	2162.7	6.8356	0.0800	1909.1	2149.1	6.6180
250	0.1244	1940.0	2188.9	6.8861	0.0818	1930.4	2176.0	6.6698
260	0.1270	1961.0	2215.1	6.9359	0.0837	1951.9	2202.9	6.7208
270	0.1297	1982.2	2241.5	6.9849	0.0854	1973.5	2229.8	6.7708
280	0.1323	2003.5	2268.0	7.0333	0.0872	1995.1	2256.8	6.8202
290	0.1348	2025.0	2294.6	7.0809	0.0890	2016.9	2284.0	6.8687
300	0.1374	2046.5	2321.4	7.1280	0.0908	2038.8	2311.2	6.9166
310	0.1400	2068.3	2348.2	7.1745	0.0925	2060.9	2338.5	6.9638
320	0.1426	2090.1	2375.2	7.2204	0.0943	2083.0	2365.9	7.0104
330	0.1451	2112.2	2402.4	7.2658	0.0960	2105.3	2393.4	7.0564
340	0.1477	2134.3	2429.7	7.3106	0.0978	2127.7	2421.0	7.1019
350	0.1502	2156.7	2457.1	7.3550	0.0995	2150.3	2448.8	7.1468
360	0.1527	2179.2	2484.7	7.3989	0.1012	2173.0	2476.7	7.1912
370	0.1553	2201.8	2512.4	7.4423	0.1030	2195.8	2504.7	7.2351
380	0.1578	2224.6	2540.2	7.4853	0.1047	2218.8	2532.8	7.2785
390	0.1603	2247.6	2568.3	7.5279	0.1064	2241.9	2561.1	7.3215
400	0.1629	2270.7	2596.4	7.5700	0.1081	2265.2	2589.5	7.3640
410	0.1654	2294.0	2624.7	7.6118	0.1098	2288.7	2618.1	7.4061
420	0.1679	2317.4	2653.2	7.6532	0.1115	2312.3	2646.8	7.4478
430	0.1704	2341.0	2681.9	7.6942	0.1132	2336.0	2675.6	7.4892
440	0.1729	2364.8	2710.7	7.7349	0.1149	2359.9	2704.6	7.5301
450	0.1754	2388.8	2739.6	7.7752	0.1166	2384.0	2733.8	7.5707

Appendix J.2 - Superheated Ammonia (NH₃) - English Units

	P = 300 psia (123.20°F)				P = 500 psia (161.14°F)			
	v	u	h	s	v	u	h	s
T(°F)	ft³/lbm	BTU/lbm	BTU/lbm	BTU/lbm·°F	ft³/lbm	BTU/lbm	BTU/lbm	BTU/lbm·°F
380	1.6909	783.85	877.78	1.5702	0.98440	776.15	867.30	1.5014
400	1.7376	793.65	890.17	1.5848	1.0145	786.43	880.36	1.5167
420	1.7839	803.49	902.58	1.5991	1.0442	796.70	893.38	1.5317
440	1.8298	813.38	915.03	1.6131	1.0735	806.98	906.37	1.5463
460	1.8755	823.32	927.51	1.6268	1.1024	817.27	919.34	1.5606
480	1.9209	833.33	940.04	1.6403	1.1311	827.60	932.32	1.5745
500	1.9661	843.40	952.61	1.6535	1.1596	837.95	945.31	1.5882
520	2.0110	853.53	965.25	1.6665	1.1878	848.35	958.32	1.6016
540	2.0557	863.74	977.94	1.6794	1.2158	858.80	971.36	1.6148
560	2.1003	874.02	990.69	1.6920	1.2436	869.30	984.44	1.6278
580	2.1447	884.37	1003.5	1.7045	1.2712	879.86	997.56	1.6405
600	2.1890	894.80	1016.4	1.7167	1.2987	890.48	1010.7	1.6531
620	2.2331	905.31	1029.4	1.7288	1.3260	901.17	1023.9	1.6654
640	2.2771	915.90	1042.4	1.7408	1.3533	911.92	1037.2	1.6776
660	2.3210	926.57	1055.5	1.7526	1.3804	922.74	1050.5	1.6896
680	2.3648	937.32	1068.7	1.7643	1.4074	933.63	1063.9	1.7015
700	2.4085	948.15	1082.0	1.7758	1.4343	944.60	1077.4	1.7132
720	2.4521	959.07	1095.3	1.7872	1.4611	955.64	1090.9	1.7247
740	2.4956	970.08	1108.7	1.7985	1.4878	966.76	1104.5	1.7362
760	2.5391	981.16	1122.2	1.8097	1.5145	977.96	1118.2	1.7475
780	2.5825	992.34	1135.8	1.8207	1.5411	989.24	1131.9	1.7586
800	2.6258	1003.6	1149.5	1.8317	1.5676	1000.6	1145.7	1.7697
820	2.6691	1014.9	1163.2	1.8425	1.5941	1012.0	1159.6	1.7806
840	2.7123	1026.4	1177.1	1.8532	1.6205	1023.5	1173.6	1.7914

Appendix J.2 - Superheated Ammonia (NH₃) - SI Units

P = 5000 kPa (88.89 °C)

T(°C)	v m³/kg	u kJ/kg	h kJ/kg	s kJ/kg-K
sat	0.02364	1488.8	1607.0	5.1182
90	0.02395	1494.1	1613.8	5.1369
100	0.02636	1535.3	1667.1	5.2816
110	0.02840	1569.7	1711.7	5.3998
120	0.03023	1600.3	1751.5	5.5022
130	0.03191	1628.5	1788.0	5.5941
140	0.03348	1655.0	1822.4	5.6783
150	0.03498	1680.4	1855.3	5.7569
160	0.03641	1704.9	1886.9	5.8309
170	0.03779	1728.8	1917.8	5.9012
180	0.03913	1752.3	1947.9	5.9685
190	0.04044	1775.4	1977.6	6.0332
200	0.04171	1798.2	2006.8	6.0957
210	0.04296	1820.9	2035.8	6.1562
220	0.04419	1843.5	2064.5	6.2150
230	0.04540	1866.0	2093.0	6.2723
240	0.04659	1888.4	2121.4	6.3281
250	0.04777	1910.8	2149.6	6.3827
260	0.04893	1933.2	2177.8	6.4361
270	0.05008	1955.6	2206.0	6.4885
280	0.05122	1978.1	2234.2	6.5398
290	0.05235	2000.6	2262.3	6.5903
300	0.05347	2023.1	2290.5	6.6398
310	0.05458	2045.8	2318.7	6.6886
320	0.05568	2068.5	2346.9	6.7367
330	0.05678	2091.3	2375.2	6.7840
340	0.05786	2114.3	2403.6	6.8306
350	0.05895	2137.3	2432.0	6.8766
360	0.06003	2160.4	2460.6	6.9221
370	0.06110	2183.7	2489.2	6.9669
380	0.06217	2207.1	2517.9	7.0112
390	0.06323	2230.6	2546.7	7.0550
400	0.06429	2254.2	2575.7	7.0983
410	0.06535	2278.0	2604.7	7.1412
420	0.06640	2301.9	2633.9	7.1836
430	0.06745	2325.9	2663.2	7.2255
440	0.06850	2350.1	2692.6	7.2670
450	0.06954	2374.4	2722.1	7.3082

P = 6000 kPa (97.89 °C)

T(°C)	v m³/kg	u kJ/kg	h kJ/kg	s kJ/kg-K
sat	0.01882	1475.0	1587.9	5.0085
90	-	-	-	-
100	0.01938	1486.6	1602.9	5.0489
110	0.02162	1532.5	1662.2	5.2058
120	0.02346	1569.7	1710.5	5.3302
130	0.02508	1602.3	1752.8	5.4365
140	0.02657	1632.0	1791.4	5.5311
150	0.02794	1659.8	1827.5	5.6174
160	0.02925	1686.3	1861.8	5.6975
170	0.03049	1711.8	1894.8	5.7728
180	0.03169	1736.6	1926.7	5.8441
190	0.03285	1760.8	1957.9	5.9122
200	0.03397	1784.7	1988.5	5.9776
210	0.03507	1808.2	2018.7	6.0406
220	0.03614	1831.6	2048.4	6.1016
230	0.03719	1854.7	2077.9	6.1607
240	0.03823	1877.7	2107.1	6.2182
250	0.03924	1900.7	2136.2	6.2743
260	0.04025	1923.6	2165.1	6.3291
270	0.04124	1946.5	2193.9	6.3826
280	0.04221	1969.4	2222.6	6.4350
290	0.04318	1992.2	2251.3	6.4865
300	0.04414	2015.2	2280.0	6.5369
310	0.04509	2038.1	2308.7	6.5865
320	0.04603	2061.2	2337.4	6.6353
330	0.04696	2084.3	2366.1	6.6833
340	0.04789	2107.5	2394.8	6.7306
350	0.04881	2130.7	2423.6	6.7772
360	0.04973	2154.1	2452.5	6.8231
370	0.05064	2177.6	2481.4	6.8685
380	0.05155	2201.2	2510.4	6.9132
390	0.05245	2224.9	2539.5	6.9575
400	0.05334	2248.7	2568.7	7.0012
410	0.05424	2272.6	2598.0	7.0444
420	0.05513	2296.5	2627.4	7.0871
430	0.05602	2320.9	2657.0	7.1294
440	0.05690	2345.2	2686.6	7.1712
450	0.05778	2369.6	2716.3	7.2126

Appendix J.2 - Superheated Ammonia (NH₃) - English Units

P = 750 psia (194.93°F)

T(°F)	v ft³/lbm	u BTU/lbm	h BTU/lbm	s BTU/lbm-°F
sat	0.36355	639.58	690.07	1.2187
200	0.37554	645.32	697.47	1.2300
220	0.41612	664.51	722.30	1.2671
240	0.45036	680.50	743.05	1.2972
260	0.48091	694.72	761.51	1.3232
280	0.50904	707.85	778.54	1.3465
300	0.53543	720.24	794.60	1.3679
320	0.56050	732.13	809.97	1.3879
340	0.58456	743.67	824.85	1.4068
360	0.60779	754.95	839.36	1.4247
380	0.63034	766.05	853.59	1.4418
400	0.65232	777.01	867.61	1.4583
420	0.67382	787.89	881.47	1.4743
440	0.69491	798.70	895.21	1.4897
460	0.71563	809.48	908.86	1.5047
480	0.73604	820.23	922.45	1.5194
500	0.75617	830.98	935.99	1.5336
520	0.77606	841.73	949.51	1.5476
540	0.79572	852.50	963.01	1.5612
560	0.81518	863.30	976.51	1.5746
580	0.83447	874.13	990.02	1.5877
600	0.85360	885.00	1003.6	1.6006
620	0.87258	895.92	1017.1	1.6132
640	0.89143	906.88	1030.7	1.6257
660	0.91016	917.90	1044.3	1.6380
680	0.92878	928.98	1058.0	1.6501
700	0.94730	940.12	1071.7	1.6620
720	0.96573	951.32	1085.4	1.6738
740	0.98407	962.59	1099.3	1.6854
760	1.0023	973.93	1113.1	1.6969
780	1.0205	985.33	1127.1	1.7082
800	1.0387	996.81	1141.1	1.7194
820	1.0567	1008.4	1155.1	1.7305
840	1.0747	1020.0	1169.2	1.7414

P = 1000 psia (221.04°F)

T(°F)	v ft³/lbm	u BTU/lbm	h BTU/lbm	s BTU/lbm-°F
sat	0.25001	627.99	674.29	1.1742
200	-	-	-	-
220	-	-	-	-
240	0.28926	653.13	706.69	1.2211
260	0.32076	672.82	732.21	1.2571
280	0.34752	689.38	753.73	1.2866
300	0.37146	704.19	772.97	1.3123
320	0.39351	717.89	790.75	1.3354
340	0.41417	730.84	807.53	1.3567
360	0.43379	743.27	823.60	1.3765
380	0.45258	755.32	839.13	1.3952
400	0.47070	767.10	854.26	1.4130
420	0.48825	778.67	869.08	1.4301
440	0.50534	790.09	883.67	1.4465
460	0.52202	801.40	898.07	1.4623
480	0.53836	812.63	912.32	1.4776
500	0.55439	823.80	926.46	1.4925
520	0.57016	834.94	940.52	1.5070
540	0.58569	846.06	954.52	1.5212
560	0.60101	857.18	968.47	1.5350
580	0.61614	868.30	982.39	1.5485
600	0.63110	879.43	996.29	1.5618
620	0.64591	890.59	1010.2	1.5748
640	0.66059	901.78	1024.1	1.5875
660	0.67514	913.01	1038.0	1.6001
680	0.68958	924.28	1052.0	1.6124
700	0.70391	935.59	1065.9	1.6246
720	0.71815	946.96	1079.9	1.6365
740	0.73230	958.38	1094.0	1.6483
760	0.74637	969.86	1108.1	1.6600
780	0.76036	981.41	1122.2	1.6715
800	0.77429	993.01	1136.4	1.6828
820	0.78815	1004.7	1150.6	1.6940
840	0.80196	1016.4	1164.9	1.7051

Appendix J.2 - Superheated Ammonia (NH$_3$) - SI Units

	P = 8000 kPa (112.91 °C)				P = 10000 kPa (125.20 °C)			
T(°C)	v m³/kg	u kJ/kg	h kJ/kg	s kJ/kg-K	v m³/kg	u kJ/kg	h kJ/kg	s kJ/kg-K
sat	0.01252	1436.3	1536.4	4.7913	0.00823	1371.1	1453.4	4.5275
120	0.01433	1487.5	1602.2	4.9601	-	-	-	-
130	0.01618	1538.2	1667.6	5.1247	0.00994	1436.6	1536.0	4.7337
140	0.01768	1578.6	1720.0	5.2531	0.01194	1507.7	1627.1	4.9573
150	0.01899	1613.6	1765.6	5.3620	0.01338	1557.1	1691.0	5.1101
160	0.02018	1645.4	1806.9	5.4585	0.01459	1597.7	1743.6	5.2332
170	0.02129	1675.0	1845.3	5.5463	0.01566	1633.5	1790.1	5.3391
180	0.02233	1703.1	1881.7	5.6277	0.01663	1666.1	1832.4	5.4337
190	0.02331	1730.1	1916.6	5.7036	0.01754	1696.6	1872.6	5.5202
200	0.02426	1756.2	1950.3	5.7757	0.01840	1725.7	1909.6	5.6004
210	0.02517	1781.7	1983.1	5.8443	0.01921	1753.6	1945.7	5.6758
220	0.02606	1806.8	2015.2	5.9101	0.01999	1780.6	1980.5	5.7473
230	0.02692	1831.4	2046.8	5.9734	0.02074	1807.1	2014.5	5.8154
240	0.02776	1855.8	2077.9	6.0346	0.02147	1833.0	2047.7	5.8807
250	0.02858	1880.0	2108.6	6.0939	0.02218	1858.5	2080.2	5.9436
260	0.02939	1903.9	2139.0	6.1516	0.02287	1883.6	2112.3	6.0044
270	0.03018	1927.8	2169.2	6.2077	0.02354	1908.6	2144.0	6.0632
280	0.03096	1951.6	2199.2	6.2624	0.02421	1933.3	2175.4	6.1205
290	0.03173	1975.3	2229.1	6.3159	0.02486	1957.9	2206.5	6.1762
300	0.03248	1999.0	2258.8	6.3682	0.02549	1982.4	2237.3	6.2305
310	0.03323	2022.6	2288.5	6.4195	0.02612	2006.8	2268.0	6.2836
320	0.03397	2046.3	2318.1	6.4698	0.02674	2031.1	2298.6	6.3355
330	0.03470	2070.0	2347.6	6.5193	0.02735	2055.5	2329.0	6.3864
340	0.03543	2093.7	2377.2	6.5678	0.02796	2079.8	2359.4	6.4363
350	0.03615	2117.5	2406.7	6.6156	0.02856	2104.1	2389.7	6.4854
360	0.03686	2141.4	2436.3	6.6627	0.02915	2128.4	2419.9	6.5335
370	0.03757	2165.5	2465.8	6.7090	0.02974	2152.8	2450.2	6.5809
380	0.03827	2189.3	2495.5	6.7547	0.03032	2177.2	2480.4	6.6276
390	0.03897	2213.3	2525.1	6.7998	0.03090	2201.7	2510.7	6.6736
400	0.03967	2237.5	2554.9	6.8443	0.03147	2226.3	2541.0	6.7189
410	0.04036	2261.8	2584.7	6.8883	0.03204	2250.9	2571.3	6.7636
420	0.04105	2286.2	2614.5	6.9317	0.03260	2275.6	2601.7	6.8077
430	0.04173	2310.7	2644.5	6.9746	0.03316	2300.5	2632.1	6.8513
440	0.04241	2335.3	2674.6	7.0170	0.03372	2325.4	2662.6	6.8944
450	0.04309	2360.0	2704.7	7.0590	0.03428	2350.4	2693.1	6.9369

Appendix J.2 - Superheated Ammonia (NH$_3$) - English Units

	P = 1250 psia (242.52°F)				P = 1500 psia (260.77°F)			
T(°F)	v ft³/lbm	u BTU/lbm	h BTU/lbm	s BTU/lbm-°F	v ft³/lbm	u BTU/lbm	h BTU/lbm	s BTU/lbm-°F
sat	0.17726	611.02	652.05	1.1279	0.12094	582.34	615.93	1.0675
260	0.21618	643.10	693.14	1.1857	-	-	-	-
280	0.24611	666.74	723.70	1.2277	0.17153	635.81	683.45	1.1603
300	0.27034	685.54	748.12	1.2602	0.19957	662.82	718.25	1.2067
320	0.29151	701.89	769.36	1.2879	0.22158	683.47	745.01	1.2415
340	0.31071	716.76	788.68	1.3123	0.24053	701.09	767.90	1.2705
360	0.32853	730.66	806.71	1.3346	0.25756	716.93	788.47	1.2959
380	0.34531	743.89	823.82	1.3552	0.27326	731.63	807.53	1.3189
400	0.36128	756.63	840.25	1.3746	0.28797	745.53	825.52	1.3401
420	0.37659	769.01	856.18	1.3929	0.30191	758.86	842.72	1.3598
440	0.39137	781.12	871.71	1.4103	0.31524	771.77	859.33	1.3785
460	0.40570	793.03	886.94	1.4271	0.32806	784.35	875.47	1.3963
480	0.41965	804.79	901.92	1.4432	0.34046	796.69	891.26	1.4132
500	0.43327	816.43	916.72	1.4588	0.35250	808.84	906.75	1.4296
520	0.44660	827.98	931.36	1.4739	0.36423	820.85	922.02	1.4453
540	0.45967	839.48	945.88	1.4885	0.37570	832.74	937.10	1.4605
560	0.47253	850.93	960.30	1.5028	0.38692	844.56	952.03	1.4753
580	0.48518	862.35	974.66	1.5168	0.39794	856.31	966.84	1.4897
600	0.49766	873.77	988.96	1.5304	0.40877	868.02	981.56	1.5037
620	0.50998	885.18	1003.2	1.5437	0.41944	879.70	996.21	1.5174
640	0.52216	896.61	1017.5	1.5568	0.42996	891.37	1010.8	1.5308
660	0.53421	908.05	1031.7	1.5696	0.44034	903.04	1025.4	1.5439
680	0.54614	919.52	1045.9	1.5822	0.45060	914.72	1039.9	1.5568
700	0.55797	931.02	1060.2	1.5946	0.46076	926.41	1054.4	1.5694
720	0.56969	942.56	1074.4	1.6068	0.47081	938.13	1068.9	1.5818
740	0.58133	954.14	1088.7	1.6188	0.48077	949.88	1083.4	1.5940
760	0.59288	965.77	1103.0	1.6306	0.49065	961.66	1097.9	1.6060
780	0.60436	977.45	1117.3	1.6423	0.50045	973.48	1112.5	1.6179
800	0.61577	989.19	1131.7	1.6538	0.51018	985.35	1127.1	1.6295
820	0.62712	1001.0	1146.1	1.6652	0.51985	997.27	1141.7	1.6410
840	0.63840	1012.8	1160.6	1.6764	0.52945	1009.2	1156.3	1.6524

adapted from NIST data

The reference state for all ammonia properties is h = 0 and s = 0 for the saturated liquid at its triple point which is −77.66°C.

Appendix J - Properties of Ammonia

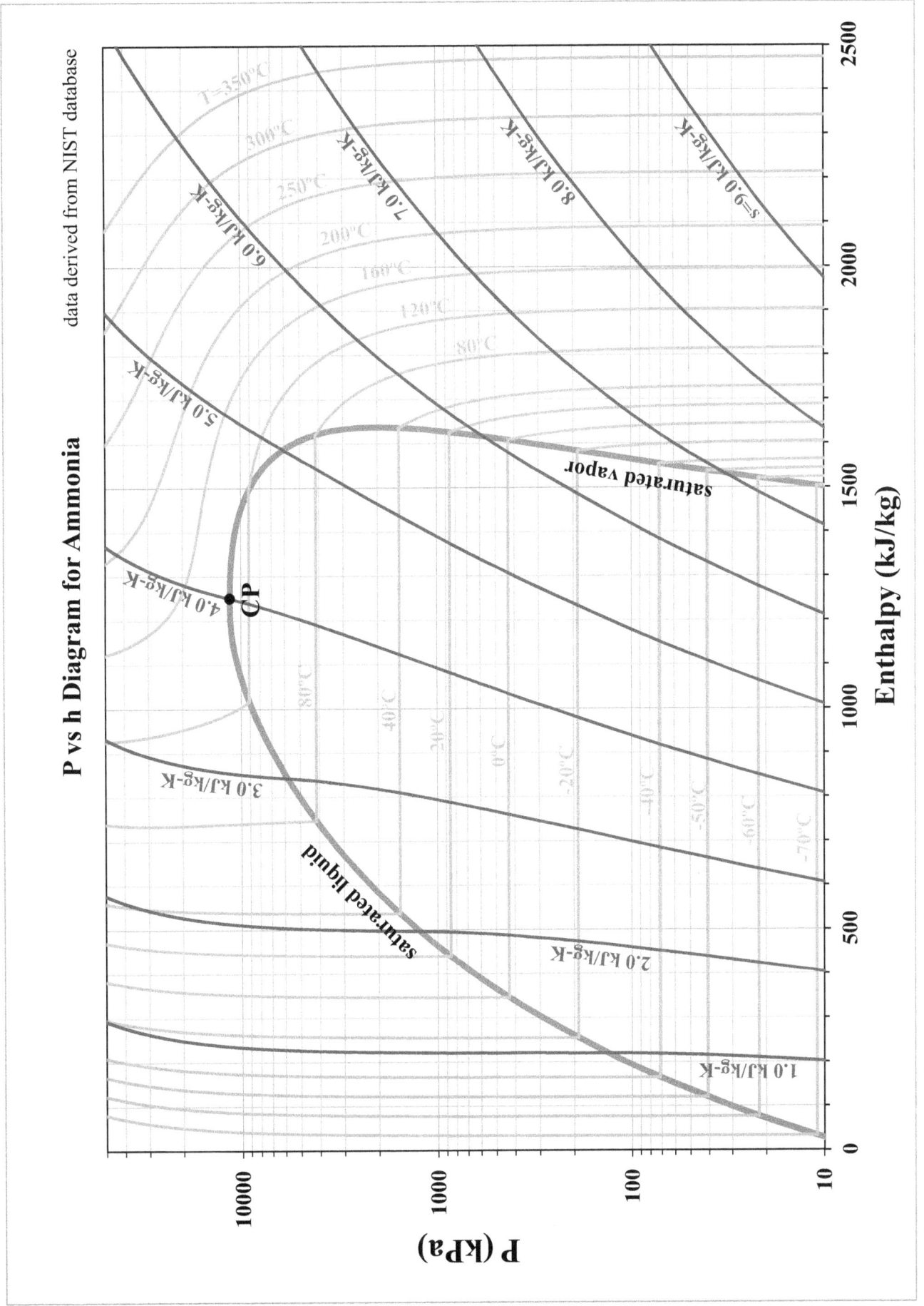

P vs h Diagram for Ammonia

Appendix K.1 - Saturated Nitrogen (N_2)

Temperature Entry

T °C	P kPa	v^L m³/kg	v^V m³/kg	u^L kJ/kg	u^V kJ/kg	h^L kJ/kg	h^V kJ/kg	s^L kJ/kg-K	s^V kJ/kg-K
-210	12.52	0.001153	1.48330	-150.75	46.21	-150.74	64.78	2.4256	5.8384
-208	17.86	0.001164	1.06950	-146.75	47.58	-146.73	66.68	2.4880	5.7636
-206	24.89	0.001176	0.78812	-142.74	48.92	-142.71	68.53	2.5486	5.6945
-204	33.97	0.001187	0.59228	-138.73	50.22	-138.69	70.34	2.6076	5.6304
-202	45.48	0.001200	0.45303	-134.70	51.50	-134.65	72.10	2.6649	5.5707
-200	59.84	0.001212	0.35208	-130.67	52.73	-130.59	73.80	2.7209	5.5151
-198	77.49	0.001226	0.27759	-126.62	53.93	-126.53	75.44	2.7754	5.4629
-196	98.90	0.001239	0.22172	-122.56	55.07	-122.44	77.00	2.8288	5.4139
-194	124.56	0.001253	0.17919	-118.49	56.17	-118.33	78.49	2.8809	5.3676
-192	154.97	0.001268	0.14636	-114.39	57.22	-114.20	79.90	2.9320	5.3238
-190	190.67	0.001284	0.12071	-110.28	58.20	-110.04	81.22	2.9821	5.2822
-188	232.19	0.001300	0.10042	-106.14	59.12	-105.84	82.44	3.0313	5.2425
-186	280.09	0.001317	0.08421	-101.98	59.98	-101.61	83.56	3.0797	5.2045
-184	334.92	0.001334	0.07111	-97.79	60.76	-97.35	84.58	3.1273	5.1679
-182	397.25	0.001353	0.06044	-93.57	61.46	-93.03	85.47	3.1742	5.1325
-180	467.67	0.001373	0.05167	-89.31	62.07	-88.66	86.24	3.2206	5.0982
-178	546.77	0.001394	0.04440	-85.00	62.59	-84.24	86.87	3.2664	5.0647
-176	635.14	0.001416	0.03833	-80.64	63.01	-79.74	87.36	3.3119	5.0319
-174	733.38	0.001440	0.03321	-76.23	63.32	-75.18	87.68	3.3570	4.9995
-172	842.12	0.001466	0.02888	-71.76	63.51	-70.52	87.83	3.4019	4.9674
-170	961.98	0.001493	0.02519	-67.20	63.55	-65.77	87.79	3.4467	4.9353
-168	1093.60	0.001523	0.02202	-62.56	63.45	-60.90	87.53	3.4915	4.9031
-166	1237.70	0.001556	0.01928	-57.82	63.17	-55.90	87.03	3.5366	4.8705
-164	1394.90	0.001592	0.01690	-52.96	62.69	-50.73	86.26	3.5820	4.8371
-162	1565.90	0.001633	0.01481	-47.94	61.97	-45.39	85.16	3.6281	4.8026
-160	1751.60	0.001679	0.01297	-42.75	60.96	-39.81	83.69	3.6751	4.7665
-158	1952.70	0.001732	0.01134	-37.32	59.61	-33.94	81.75	3.7235	4.7281
-156	2170.20	0.001795	0.00987	-31.59	57.80	-27.70	79.22	3.7739	4.6865
-154	2405.10	0.001872	0.00853	-25.45	55.38	-20.95	75.90	3.8274	4.6402
-152	2658.80	0.001971	0.00729	-18.68	52.06	-13.44	71.43	3.8858	4.5863
-150	2932.90	0.002113	0.00608	-10.81	47.18	-4.61	65.01	3.9535	4.5188
-148	3230.00	0.002377	0.00476	-0.09	38.45	7.59	53.83	4.0463	4.4158
-146.96	3395.80	0.002918	0.00292	135.47	135.47	145.38	145.38	1.3341	1.3341

Appendix K.1 - Saturated Nitrogen (N$_2$)

Pressure Entry

P	T	v^L	v^V	u^L	u^V	h^L	h^V	s^L	s^V
kPa	°C	m^3/kg	m^3/kg	kJ/kg	kJ/kg	kJ/kg	kJ/kg	kJ/kg-K	kJ/kg-K
20	-207.33	0.001168	0.96376	-145.41	48.03	-145.39	67.30	2.5084	5.7400
30	-204.82	0.001183	0.66396	-140.36	49.69	-140.33	69.61	2.5837	5.6560
40	-202.90	0.001194	0.50978	-136.51	50.93	-136.46	71.32	2.6394	5.5970
50	-201.32	0.001204	0.41529	-133.34	51.92	-133.28	72.68	2.6840	5.5515
60	-199.98	0.001213	0.35123	-130.63	52.74	-130.55	73.82	2.7214	5.5145
70	-198.80	0.001220	0.30481	-128.24	53.45	-128.16	74.79	2.7537	5.4834
80	-197.75	0.001227	0.26956	-126.10	54.08	-126.01	75.64	2.7823	5.4565
90	-196.79	0.001234	0.24185	-124.16	54.63	-124.05	76.40	2.8079	5.4328
100	-195.91	0.001240	0.21947	-122.37	55.13	-122.25	77.07	2.8312	5.4116
120	-194.33	0.001251	0.18547	-119.16	55.99	-119.01	78.25	2.8724	5.3751
140	-192.94	0.001261	0.16082	-116.33	56.73	-116.15	79.25	2.9081	5.3442
160	-191.70	0.001270	0.14209	-113.77	57.37	-113.57	80.10	2.9396	5.3174
180	-190.57	0.001279	0.12736	-111.45	57.93	-111.22	80.85	2.9680	5.2938
200	-189.52	0.001287	0.11545	-109.30	58.43	-109.04	81.52	2.9939	5.2726
300	-185.24	0.001323	0.07893	-100.40	60.28	-100.00	83.96	3.0978	5.1905
400	-181.92	0.001354	0.06005	-93.39	61.49	-92.85	85.50	3.1762	5.1311
500	-179.15	0.001382	0.04844	-87.49	62.31	-86.80	86.52	3.2400	5.0840
600	-176.77	0.001407	0.04054	-82.33	62.87	-81.48	87.19	3.2944	5.0444
700	-174.66	0.001432	0.03480	-77.69	63.23	-76.69	87.59	3.3422	5.0101
800	-172.75	0.001456	0.03043	-73.45	63.45	-72.28	87.80	3.3850	4.9794
900	-171.01	0.001479	0.02699	-69.51	63.55	-68.18	87.83	3.4241	4.9515
1000	-169.40	0.001502	0.02420	-65.83	63.54	-64.33	87.73	3.4601	4.9257
1500	-162.75	0.001617	0.01557	-49.84	62.27	-47.42	85.62	3.6107	4.8157
2000	-157.55	0.001745	0.01100	-36.07	59.24	-32.57	81.24	3.7346	4.7191
2500	-153.23	0.001907	0.00805	-22.95	54.24	-18.18	74.35	3.8490	4.6207
3000	-149.53	0.002158	0.00579	-8.70	45.66	-2.22	63.04	3.9717	4.4996
-146.96	3395.80	0.002918	0.00292	135.47	135.47	145.38	145.38	1.3341	1.3341

adapted from NIST data The reference state for all nitrogen properties is h = 0 and s = 0 for the saturated liquid at its normal boiling point which is -195.8°C.

Appendix K.2 - Superheated Nitrogen (N$_2$)

	P = 100 kPa (-195.91°C)				P = 200 kPa (-189.52°C)			
	v	u	h	s	v	u	h	s
T(°C)	m^3/kg	kJ/kg	kJ/kg	kJ/kg-K	m^3/kg	kJ/kg	kJ/kg	kJ/kg-K
sat	0.2195	177.14	199.09	2.5775	0.1155	180.4	203.5	2.4384
-190	0.2384	59.80	83.64	5.4936	-	-	-	-
-180	0.2698	67.56	94.54	5.6174	0.1314	66.21	92.49	5.3969
-170	0.3007	75.21	105.28	5.7269	0.1475	74.12	103.62	5.5105
-160	0.3313	82.79	115.92	5.8253	0.1633	81.88	114.54	5.6115
-150	0.3617	90.33	126.50	5.9149	0.1789	89.55	125.33	5.7029
-140	0.3919	97.85	137.03	5.9972	0.1943	97.17	136.02	5.7864
-130	0.4220	105.34	147.54	6.0733	0.2096	104.74	146.66	5.8634
-120	0.4521	112.82	158.03	6.1441	0.2248	112.28	157.24	5.9349
-110	0.4821	120.29	168.50	6.2104	0.2400	119.80	167.80	6.0016
-100	0.5121	127.76	178.96	6.2726	0.2551	127.31	178.33	6.0643
-90	0.5420	135.21	189.41	6.3313	0.2702	134.80	188.84	6.1233
-80	0.5719	142.67	199.85	6.3868	0.2852	142.29	199.33	6.1791
-70	0.6017	150.12	210.29	6.4395	0.3003	149.76	209.81	6.2320
-60	0.6316	157.56	220.72	6.4896	0.3153	157.23	220.29	6.2823
-50	0.6614	165.01	231.15	6.5374	0.3303	164.70	230.75	6.3303
-40	0.6912	172.45	241.57	6.5831	0.3452	172.16	241.20	6.3761
-30	0.7210	179.89	251.99	6.6268	0.3602	179.61	251.65	6.4200
-20	0.7508	187.33	262.41	6.6688	0.3751	187.07	262.09	6.4621
-10	0.7806	194.76	272.82	6.7092	0.3901	194.52	272.53	6.5025
0	0.8104	202.20	283.24	6.7480	0.4050	201.97	282.97	6.5414
10	0.8401	209.64	293.65	6.7855	0.4199	209.42	293.40	6.5789
20	0.8699	217.08	304.06	6.8216	0.4348	216.86	303.83	6.6151
30	0.8996	224.51	314.48	6.8565	0.4497	224.31	314.26	6.6501
40	0.9294	231.95	324.89	6.8903	0.4647	231.76	324.69	6.6840
50	0.9591	239.40	335.31	6.9231	0.4796	239.21	335.12	6.7168
60	0.9888	246.84	345.72	6.9548	0.4944	246.66	345.55	6.7486
70	1.0186	254.28	356.14	6.9856	0.5093	254.11	355.98	6.7794
80	1.0483	261.73	366.56	7.0156	0.5242	261.57	366.41	6.8094
90	1.0780	269.19	376.99	7.0447	0.5391	269.03	376.85	6.8385
100	1.1077	276.65	387.42	7.0730	0.5540	276.49	387.29	6.8669
110	1.1375	284.11	397.86	7.1006	0.5689	283.96	397.74	6.8945
120	1.1672	291.58	408.30	7.1275	0.5838	291.44	408.19	6.9214
130	1.1969	299.05	418.75	7.1538	0.5986	298.92	418.64	6.9477
140	1.2266	306.54	429.20	7.1794	0.6135	306.41	429.11	6.9733
150	1.2563	314.03	439.67	7.2044	0.6284	313.90	439.58	6.9984
160	1.2860	321.53	450.14	7.2289	0.6432	321.41	450.06	7.0228
170	1.3157	329.05	460.62	7.2528	0.6581	328.93	460.55	7.0468
180	1.3455	336.57	471.12	7.2762	0.6730	336.46	471.05	7.0702
190	1.3752	344.11	481.62	7.2991	0.6878	343.99	481.56	7.0932
200	1.4049	351.66	492.14	7.3216	0.7027	351.55	492.09	7.1156

continued next page…

Appendix K.2 - Superheated Nitrogen (N$_2$)

	P = 100 kPa (-195.91°C)				P = 200 kPa (-189.52°C)			
	v	u	h	s	v	u	h	s
T(°C)	m^3/kg	kJ/kg	kJ/kg	kJ/kg-K	m^3/kg	kJ/kg	kJ/kg	kJ/kg-K
210	1.4346	359.22	502.67	7.3436	0.7176	359.11	502.62	7.1377
220	1.4643	366.79	513.22	7.3652	0.7324	366.69	513.17	7.1593
230	1.4940	374.38	523.78	7.3864	0.7473	374.28	523.74	7.1805
240	1.5237	381.99	534.36	7.4073	0.7621	381.89	534.32	7.2013
250	1.5534	389.62	544.95	7.4277	0.7770	389.52	544.92	7.2218
260	1.5831	397.26	555.56	7.4478	0.7919	397.16	555.54	7.2419
270	1.6128	404.92	566.19	7.4675	0.8067	404.83	566.17	7.2617
280	1.6425	412.59	576.84	7.4870	0.8216	412.51	576.82	7.2811
290	1.6722	420.29	587.50	7.5061	0.8364	420.20	587.49	7.3002
300	1.7018	428.00	598.19	7.5249	0.8513	427.92	598.18	7.3190
310	1.7315	435.74	608.90	7.5434	0.8661	435.66	608.89	7.3375
320	1.7612	443.50	619.62	7.5616	0.8810	443.42	619.62	7.3558
330	1.7909	451.28	630.37	7.5796	0.8958	451.20	630.37	7.3738
340	1.8206	459.08	641.14	7.5973	0.9107	459.00	641.14	7.3915
350	1.8503	466.90	651.93	7.6148	0.9256	466.82	651.93	7.4089
360	1.8800	474.74	662.74	7.6320	0.9404	474.67	662.75	7.4261
370	1.9097	482.61	673.58	7.6490	0.9553	482.54	673.59	7.4431
380	1.9394	490.50	684.44	7.6657	0.9701	490.43	684.45	7.4599
390	1.9691	498.41	695.32	7.6823	0.9850	498.34	695.33	7.4764
400	1.9988	506.35	706.22	7.6986	0.9998	506.28	706.24	7.4928
410	2.0285	514.31	717.15	7.7147	1.0147	514.24	717.17	7.5089
420	2.0582	522.29	728.10	7.7306	1.0295	522.23	728.13	7.5248
430	2.0878	530.30	739.08	7.7463	1.0444	530.24	739.11	7.5405
440	2.1175	538.33	750.08	7.7619	1.0592	538.27	750.11	7.5561
450	2.1472	546.38	761.10	7.7772	1.0740	546.32	761.13	7.5714
460	2.1769	554.46	772.15	7.7924	1.0889	554.41	772.19	7.5866
470	2.2066	562.57	783.23	7.8074	1.1037	562.51	783.26	7.6016
480	2.2363	570.70	794.32	7.8222	1.1186	570.64	794.36	7.6164
490	2.2660	578.85	805.45	7.8369	1.1334	578.79	805.48	7.6311
500	2.2957	587.03	816.59	7.8514	1.1483	586.97	816.63	7.6456

Appendix K.2 - Superheated Nitrogen (N$_2$)

	P = 300 kPa (-185.24°C)				P = 400 kPa (-181.92°C)			
	v	u	h	s	v	u	h	s
T(°C)	m^3/kg	kJ/kg	kJ/kg	kJ/kg-K	m^3/kg	kJ/kg	kJ/kg	kJ/kg-K
sat	0.0789	60.28	83.96	5.1905	0.0600	61.49	85.50	5.1311
-180	0.0851	64.77	90.30	5.2605	0.0618	63.21	87.95	5.1576
-170	0.0964	72.99	101.90	5.3789	0.0708	71.80	100.10	5.2816
-160	0.1073	80.95	113.13	5.4828	0.0792	79.98	111.67	5.3887
-150	0.1179	88.76	124.13	5.5760	0.0874	87.95	122.91	5.4839
-140	0.1284	96.48	134.99	5.6608	0.0954	95.77	133.95	5.5701
-130	0.1388	104.13	145.76	5.7388	0.1034	103.51	144.85	5.6490
-120	0.1491	111.74	156.45	5.8110	0.1112	111.18	155.65	5.7220
-110	0.1593	119.31	167.09	5.8783	0.1189	118.81	166.38	5.7898
-100	0.1695	126.86	177.69	5.9413	0.1266	126.41	177.06	5.8533
-90	0.1796	134.39	188.27	6.0007	0.1343	133.97	187.69	5.9130
-80	0.1897	141.91	198.81	6.0568	0.1419	141.52	198.29	5.9694
-70	0.1998	149.41	209.34	6.1099	0.1495	149.05	208.86	6.0228
-60	0.2098	156.90	219.85	6.1604	0.1571	156.57	219.41	6.0735
-50	0.2199	164.39	230.35	6.2085	0.1647	164.08	229.95	6.1218
-40	0.2299	171.87	240.83	6.2545	0.1722	171.57	240.47	6.1679
-30	0.2399	179.34	251.31	6.2985	0.1798	179.06	250.97	6.2120
-20	0.2499	186.81	261.78	6.3407	0.1873	186.55	261.47	6.2543
-10	0.2599	194.27	272.24	6.3812	0.1948	194.03	271.95	6.2949
0	0.2699	201.73	282.70	6.4202	0.2023	201.50	282.43	6.3340
10	0.2799	209.19	293.15	6.4578	0.2098	208.97	292.90	6.3717
20	0.2898	216.65	303.60	6.4941	0.2173	216.44	303.37	6.4080
30	0.2998	224.11	314.05	6.5291	0.2248	223.91	313.83	6.4431
40	0.3097	231.57	324.49	6.5630	0.2323	231.37	324.29	6.4770
50	0.3197	239.02	334.93	6.5958	0.2398	238.84	334.75	6.5099
60	0.3296	246.48	345.38	6.6277	0.2473	246.31	345.20	6.5418
70	0.3396	253.94	355.82	6.6586	0.2547	253.77	355.66	6.5727
80	0.3495	261.41	366.26	6.6886	0.2622	261.24	366.12	6.6027
90	0.3595	268.87	376.71	6.7177	0.2697	268.71	376.57	6.6319
100	0.3694	276.34	387.16	6.7461	0.2771	276.19	387.03	6.6603
110	0.3793	283.82	397.62	6.7738	0.2846	283.67	397.50	6.6880
120	0.3893	291.30	408.08	6.8007	0.2920	291.15	407.97	6.7150
130	0.3992	298.78	418.54	6.8270	0.2995	298.65	418.44	6.7413
140	0.4091	306.28	429.01	6.8527	0.3069	306.14	428.92	6.7670
150	0.4191	313.78	439.49	6.8777	0.3144	313.65	439.40	6.7920
160	0.4290	321.29	449.98	6.9022	0.3218	321.16	449.90	6.8165
170	0.4389	328.81	460.47	6.9262	0.3293	328.69	460.40	6.8405
180	0.4488	336.34	470.98	6.9496	0.3367	336.22	470.91	6.8640
190	0.4587	343.88	481.50	6.9726	0.3442	343.77	481.44	6.8870
200	0.4686	351.44	492.03	6.9951	0.3516	351.33	491.97	6.9095

continued next page…

Appendix K.2 - Superheated Nitrogen (N$_2$)

	P = 300 kPa (-185.24°C)				P = 400 kPa (-181.92°C)			
	v	u	h	s	v	u	h	s
T(°C)	m^3/kg	kJ/kg	kJ/kg	kJ/kg-K	m^3/kg	kJ/kg	kJ/kg	kJ/kg-K
210	0.4786	359.01	502.57	7.0171	0.3591	358.90	502.52	6.9315
220	0.4885	366.59	513.13	7.0388	0.3665	366.48	513.09	6.9532
230	0.4984	374.18	523.70	7.0600	0.3739	374.08	523.66	6.9744
240	0.5083	381.80	534.29	7.0808	0.3814	381.70	534.25	6.9952
250	0.5182	389.43	544.89	7.1013	0.3888	389.33	544.86	7.0157
260	0.5281	397.07	555.51	7.1214	0.3963	396.98	555.48	7.0358
270	0.5380	404.74	566.15	7.1411	0.4037	404.65	566.12	7.0556
280	0.5479	412.42	576.80	7.1606	0.4111	412.33	576.78	7.0750
290	0.5579	420.12	587.47	7.1797	0.4186	420.03	587.46	7.0942
300	0.5678	427.84	598.17	7.1985	0.4260	427.76	598.16	7.1130
310	0.5777	435.58	608.88	7.2171	0.4334	435.50	608.87	7.1315
320	0.5876	443.34	619.61	7.2353	0.4409	443.26	619.61	7.1498
330	0.5975	451.12	630.37	7.2533	0.4483	451.05	630.37	7.1678
340	0.6074	458.93	641.14	7.2710	0.4557	458.85	641.14	7.1855
350	0.6173	466.75	651.94	7.2885	0.4632	466.68	651.94	7.2030
360	0.6272	474.60	662.76	7.3057	0.4706	474.53	662.77	7.2202
370	0.6371	482.47	673.60	7.3227	0.4780	482.40	673.61	7.2372
380	0.6470	490.36	684.46	7.3394	0.4855	490.29	684.48	7.2540
390	0.6569	498.28	695.35	7.3560	0.4929	498.21	695.37	7.2705
400	0.6668	506.22	706.26	7.3723	0.5003	506.15	706.28	7.2868
410	0.6767	514.18	717.19	7.3884	0.5078	514.11	717.21	7.3030
420	0.6866	522.16	728.15	7.4044	0.5152	522.10	728.17	7.3189
430	0.6965	530.17	739.13	7.4201	0.5226	530.11	739.16	7.3346
440	0.7064	538.21	750.14	7.4356	0.5300	538.15	750.16	7.3502
450	0.7163	546.27	761.16	7.4510	0.5375	546.21	761.19	7.3655
460	0.7262	554.35	772.22	7.4662	0.5449	554.29	772.25	7.3807
470	0.7361	562.46	783.29	7.4812	0.5523	562.40	783.33	7.3957
480	0.7460	570.59	794.40	7.4960	0.5598	570.53	794.43	7.4106
490	0.7559	578.74	805.52	7.5107	0.5672	578.69	805.56	7.4252
500	0.7658	586.92	816.67	7.5252	0.5746	586.87	816.71	7.4397

Appendix K.2 - Superheated Nitrogen (N$_2$)

T(°C)	\multicolumn{4}{c}{P = 500 kPa (-179.15 °C)}	\multicolumn{4}{c}{P = 600 kPa (-176.77°C)}
	v (m^3/kg)	u (kJ/kg)	h (kJ/kg)	s (kJ/kg-K)	v (m^3/kg)	u (kJ/kg)	h (kJ/kg)	s (kJ/kg-K)
sat	0.04844	184.32	208.54	2.2498	0.04054	62.87	87.19	5.0444
-170	0.05534	70.55	98.21	5.2027	0.04500	69.22	96.22	5.1351
-160	0.06237	78.99	110.17	5.3134	0.05111	77.95	108.62	5.2498
-150	0.06911	87.12	121.67	5.4108	0.05689	86.26	120.40	5.3496
-140	0.07566	95.06	132.89	5.4984	0.06247	94.33	131.82	5.4388
-130	0.08209	102.89	143.93	5.5784	0.06792	102.25	143.00	5.5198
-120	0.08843	110.63	154.84	5.6521	0.07327	110.07	154.03	5.5943
-110	0.09470	118.31	165.66	5.7205	0.07856	117.81	164.94	5.6633
-100	0.10093	125.95	176.41	5.7845	0.08379	125.49	175.77	5.7277
-90	0.10711	133.56	187.11	5.8445	0.08899	133.14	186.53	5.7881
-80	0.11326	141.14	197.77	5.9012	0.09415	140.75	197.24	5.8451
-70	0.11938	148.70	208.39	5.9548	0.09928	148.34	207.91	5.8989
-60	0.12548	156.24	218.98	6.0057	0.10439	155.91	218.54	5.9500
-50	0.13156	163.77	229.55	6.0542	0.10949	163.46	229.15	5.9986
-40	0.13763	171.28	240.10	6.1004	0.11457	170.99	239.73	6.0450
-30	0.14369	178.79	250.63	6.1446	0.11963	178.51	250.29	6.0894
-20	0.14973	186.29	261.15	6.1870	0.12469	186.03	260.84	6.1319
-10	0.15576	193.78	271.66	6.2278	0.12973	193.54	271.37	6.1727
0	0.16179	201.27	282.16	6.2669	0.13477	201.04	281.89	6.2120
10	0.16780	208.75	292.65	6.3046	0.13979	208.53	292.41	6.2498
20	0.17382	216.23	303.14	6.3410	0.14482	216.02	302.91	6.2862
30	0.17982	223.71	313.62	6.3762	0.14983	223.51	313.40	6.3214
40	0.18582	231.18	324.09	6.4102	0.15484	230.99	323.89	6.3555
50	0.19182	238.66	334.56	6.4431	0.15985	238.47	334.38	6.3884
60	0.19781	246.13	345.03	6.4750	0.16485	245.95	344.86	6.4204
70	0.20379	253.60	355.50	6.5060	0.16985	253.43	355.34	6.4514
80	0.20978	261.08	365.97	6.5360	0.17484	260.92	365.82	6.4815
90	0.21576	268.56	376.44	6.5653	0.17983	268.40	376.30	6.5107
100	0.22174	276.04	386.91	6.5937	0.18482	275.89	386.78	6.5392
110	0.22771	283.52	397.38	6.6214	0.18981	283.38	397.26	6.5669
120	0.23368	291.01	407.86	6.6484	0.19479	290.87	407.75	6.5939
130	0.23966	298.51	418.34	6.6747	0.19977	298.37	418.24	6.6203
140	0.24562	306.01	428.82	6.7004	0.20475	305.88	428.73	6.6460
150	0.25159	313.52	439.32	6.7255	0.20973	313.39	439.23	6.6711
160	0.25756	321.04	449.82	6.7500	0.21470	320.92	449.74	6.6956
170	0.26352	328.57	460.33	6.7740	0.21968	328.45	460.26	6.7196
180	0.26948	336.11	470.85	6.7975	0.22465	335.99	470.78	6.7431
190	0.27544	343.66	481.38	6.8205	0.22962	343.54	481.32	6.7661
200	0.28140	351.22	491.92	6.8430	0.23459	351.11	491.87	6.7887

continued next page…

Appendix K.2 - Superheated Nitrogen (N$_2$)

	P = 500 kPa (-179.15 °C)				P = 600 kPa (-176.77°C)			
	v	u	h	s	v	u	h	s
T(°C)	m^3/kg	kJ/kg	kJ/kg	kJ/kg-K	m^3/kg	kJ/kg	kJ/kg	kJ/kg-K
210	0.28736	358.79	502.47	6.8651	0.23956	358.69	502.42	6.8107
220	0.29332	366.38	513.04	6.8867	0.24453	366.28	513.00	6.8324
230	0.29927	373.98	523.62	6.9080	0.24950	373.88	523.58	6.8537
240	0.30523	381.60	534.22	6.9288	0.25446	381.51	534.18	6.8745
250	0.31118	389.24	544.83	6.9493	0.25943	389.14	544.80	6.8950
260	0.31714	396.89	555.46	6.9694	0.26439	396.80	555.43	6.9151
270	0.32309	404.56	566.10	6.9892	0.26936	404.47	566.08	6.9349
280	0.32904	412.24	576.76	7.0087	0.27432	412.16	576.75	6.9544
290	0.33499	419.95	587.45	7.0278	0.27928	419.86	587.43	6.9735
300	0.34095	427.67	598.15	7.0466	0.28424	427.59	598.14	6.9924
310	0.34690	435.42	608.87	7.0652	0.28920	435.34	608.86	7.0109
320	0.35284	443.18	619.60	7.0834	0.29416	443.10	619.60	7.0292
330	0.35879	450.97	630.37	7.1014	0.29912	450.89	630.36	7.0472
340	0.36474	458.78	641.15	7.1191	0.30408	458.70	641.15	7.0649
350	0.37069	466.60	651.95	7.1366	0.30904	466.53	651.95	7.0824
360	0.37664	474.45	662.77	7.1539	0.31400	474.38	662.78	7.0996
370	0.38259	482.33	673.62	7.1708	0.31896	482.26	673.63	7.1166
380	0.38853	490.22	684.49	7.1876	0.32391	490.16	684.50	7.1334
390	0.39448	498.14	695.38	7.2042	0.32887	498.08	695.40	7.1500
400	0.40042	506.08	706.30	7.2205	0.33383	506.02	706.32	7.1663
410	0.40637	514.05	717.24	7.2366	0.33878	513.99	717.26	7.1824
420	0.41231	522.04	728.20	7.2526	0.34374	521.98	728.22	7.1984
430	0.41826	530.05	739.18	7.2683	0.34869	529.99	739.21	7.2141
440	0.42420	538.09	750.19	7.2838	0.35365	538.03	750.22	7.2296
450	0.43015	546.15	761.22	7.2992	0.35860	546.09	761.25	7.2450
460	0.43609	554.23	772.28	7.3144	0.36356	554.18	772.31	7.2602
470	0.44204	562.34	783.36	7.3294	0.36851	562.29	783.40	7.2752
480	0.44798	570.48	794.47	7.3442	0.37347	570.42	794.50	7.2901
490	0.45392	578.63	805.60	7.3589	0.37842	578.58	805.63	7.3047
500	0.45987	586.82	816.75	7.3734	0.38337	586.76	816.79	7.3193

Appendix K.2 - Superheated Nitrogen (N$_2$)

T(°C)	P = 700 kPa (-174.66°C)				P = 800 kPa (-172.75°C)			
	v m^3/kg	u kJ/kg	h kJ/kg	s kJ/kg-K	v m^3/kg	u kJ/kg	h kJ/kg	s kJ/kg-K
sat	0.03480	63.23	87.59	5.0101	0.03043	63.5	87.8	4.9794
-170	0.03757	67.81	94.11	5.0748	0.03194	66.3	91.9	5.0193
-160	0.04305	76.88	107.01	5.1942	0.03697	75.76	105.34	5.1442
-150	0.04815	85.39	119.09	5.2966	0.04158	84.49	117.76	5.2494
-140	0.05304	93.59	130.72	5.3874	0.04596	92.84	129.61	5.3420
-130	0.05779	101.61	142.06	5.4695	0.05019	100.96	141.11	5.4253
-120	0.06244	109.50	153.21	5.5448	0.05432	108.93	152.38	5.5014
-110	0.06703	117.30	164.22	5.6144	0.05838	116.79	163.49	5.5716
-100	0.07156	125.03	175.12	5.6793	0.06238	124.57	174.47	5.6370
-90	0.07605	132.72	185.95	5.7401	0.06634	132.30	185.37	5.6982
-80	0.08050	140.36	196.71	5.7973	0.07026	139.98	196.19	5.7557
-70	0.08493	147.98	207.43	5.8514	0.07416	147.62	206.95	5.8100
-60	0.08933	155.57	218.11	5.9027	0.07804	155.24	217.67	5.8615
-50	0.09372	163.14	228.75	5.9515	0.08190	162.83	228.35	5.9105
-40	0.09809	170.70	239.36	5.9980	0.08574	170.41	239.00	5.9572
-30	0.10245	178.24	249.96	6.0425	0.08957	177.96	249.62	6.0018
-20	0.10680	185.77	260.53	6.0851	0.09338	185.51	260.22	6.0445
-10	0.11114	193.29	271.09	6.1260	0.09719	193.04	270.80	6.0855
0	0.11546	200.80	281.63	6.1654	0.10099	200.57	281.36	6.1249
10	0.11979	208.31	292.16	6.2032	0.10478	208.09	291.91	6.1628
20	0.12410	215.81	302.68	6.2397	0.10857	215.60	302.45	6.1994
30	0.12841	223.30	313.19	6.2750	0.11234	223.10	312.98	6.2347
40	0.13271	230.80	323.70	6.3091	0.11612	230.60	323.50	6.2688
50	0.13701	238.29	334.19	6.3421	0.11989	238.10	334.01	6.3019
60	0.14131	245.77	344.69	6.3741	0.12365	245.60	344.52	6.3339
70	0.14560	253.26	355.18	6.4051	0.12741	253.09	355.02	6.3650
80	0.14989	260.75	365.67	6.4352	0.13117	260.59	365.52	6.3951
90	0.15417	268.24	376.16	6.4645	0.13492	268.09	376.02	6.4245
100	0.15845	275.73	386.65	6.4930	0.13868	275.58	386.52	6.4530
110	0.16273	283.23	397.14	6.5208	0.14243	283.09	397.03	6.4808
120	0.16701	290.73	407.64	6.5478	0.14617	290.59	407.53	6.5078
130	0.17128	298.24	418.13	6.5742	0.14992	298.10	418.03	6.5342
140	0.17556	305.75	428.64	6.5999	0.15366	305.62	428.54	6.5600
150	0.17983	313.27	439.15	6.6250	0.15740	313.14	439.06	6.5851
160	0.18409	320.79	449.66	6.6496	0.16114	320.67	449.58	6.6097
170	0.18836	328.33	460.18	6.6736	0.16488	328.21	460.11	6.6337
180	0.19263	335.88	470.72	6.6971	0.16861	335.76	470.65	6.6572
190	0.19689	343.43	481.26	6.7201	0.17235	343.32	481.20	6.6803
200	0.20116	351.00	491.81	6.7427	0.17608	350.89	491.76	6.7028

continued next page…

Appendix K.2 - Superheated Nitrogen (N$_2$)

	P = 700 kPa (-174.66°C)				P = 800 kPa (-172.75°C)			
	v	u	h	s	v	u	h	s
T(°C)	m³/kg	kJ/kg	kJ/kg	kJ/kg-K	m³/kg	kJ/kg	kJ/kg	kJ/kg-K
210	0.20542	358.58	502.38	6.7648	0.17981	358.48	502.33	6.7249
220	0.20968	366.18	512.95	6.7864	0.18354	366.07	512.91	6.7466
230	0.21394	373.78	523.54	6.8077	0.18727	373.69	523.50	6.7679
240	0.21820	381.41	534.15	6.8286	0.19100	381.31	534.11	6.7888
250	0.22246	389.05	544.77	6.8491	0.19473	388.95	544.74	6.8093
260	0.22672	396.70	555.41	6.8692	0.19846	396.61	555.38	6.8294
270	0.23097	404.38	566.06	6.8890	0.20219	404.29	566.04	6.8492
280	0.23523	412.07	576.73	6.9085	0.20591	411.98	576.71	6.8687
290	0.23948	419.78	587.42	6.9276	0.20964	419.69	587.40	6.8878
300	0.24374	427.51	598.12	6.9465	0.21336	427.43	598.11	6.9067
310	0.24799	435.26	608.85	6.9650	0.21709	435.18	608.84	6.9253
320	0.25225	443.03	619.60	6.9833	0.22081	442.95	619.59	6.9435
330	0.25650	450.81	630.36	7.0013	0.22453	450.74	630.36	6.9615
340	0.26075	458.63	641.15	7.0190	0.22826	458.55	641.15	6.9793
350	0.26500	466.46	651.96	7.0365	0.23198	466.38	651.97	6.9968
360	0.26926	474.31	662.79	7.0538	0.23570	474.24	662.80	7.0140
370	0.27351	482.19	673.64	7.0708	0.23942	482.12	673.65	7.0310
380	0.27776	490.09	684.52	7.0875	0.24314	490.02	684.53	7.0478
390	0.28201	498.01	695.41	7.1041	0.24686	497.94	695.43	7.0644
400	0.28626	505.95	706.33	7.1204	0.25058	505.89	706.35	7.0807
410	0.29051	513.92	717.28	7.1366	0.25430	513.86	717.30	7.0969
420	0.29476	521.92	728.24	7.1525	0.25802	521.85	728.27	7.1128
430	0.29900	529.93	739.23	7.1683	0.26174	529.87	739.26	7.1285
440	0.30325	537.97	750.25	7.1838	0.26546	537.91	750.28	7.1441
450	0.30750	546.03	761.28	7.1992	0.26917	545.98	761.32	7.1595
460	0.31175	554.12	772.35	7.2144	0.27289	554.06	772.38	7.1747
470	0.31600	562.23	783.43	7.2294	0.27661	562.18	783.46	7.1897
480	0.32024	570.37	794.54	7.2442	0.28033	570.31	794.58	7.2045
490	0.32449	578.53	805.67	7.2589	0.28404	578.47	805.71	7.2192
500	0.32874	586.71	816.83	7.2734	0.28776	586.66	816.87	7.2337

Appendix K.2 - Superheated Nitrogen (N$_2$)

	P = 1000 kPa (-169.40°C)				P = 2000 kPa (-157.55°C)			
	v	u	h	s	v	u	h	s
T(°C)	m^3/kg	kJ/kg	kJ/kg	kJ/kg-K	m^3/kg	kJ/kg	kJ/kg	kJ/kg-K
sat	0.02420	63.54	87.73	4.9257	0.01100	59.24	81.24	4.7191
-160	0.02841	73.37	101.78	5.0555	-	-	-	-
-150	0.03236	82.62	114.98	5.1674	0.01351	70.95	97.97	4.8596
-140	0.03604	91.29	127.33	5.2638	0.01601	82.42	114.44	4.9884
-130	0.03955	99.63	139.18	5.3496	0.01817	92.36	128.69	5.0916
-120	0.04294	107.76	150.71	5.4275	0.02015	101.55	141.85	5.1806
-110	0.04627	115.75	162.02	5.4990	0.02203	110.31	154.36	5.2597
-100	0.04953	123.64	173.17	5.5654	0.02383	118.79	166.44	5.3316
-90	0.05275	131.44	184.19	5.6273	0.02558	127.07	178.22	5.3977
-80	0.05594	139.19	195.13	5.6854	0.02729	135.20	189.78	5.4591
-70	0.05909	146.90	205.99	5.7403	0.02896	143.23	201.16	5.5166
-60	0.06223	154.57	216.79	5.7922	0.03062	151.17	212.41	5.5707
-50	0.06534	162.21	227.55	5.8415	0.03225	159.05	223.55	5.6217
-40	0.06844	169.82	238.26	5.8884	0.03387	166.87	234.60	5.6702
-30	0.07153	177.41	248.94	5.9333	0.03547	174.64	245.58	5.7163
-20	0.07461	184.99	259.59	5.9762	0.03706	182.38	256.50	5.7603
-10	0.07767	192.55	270.22	6.0174	0.03864	190.09	267.37	5.8024
0	0.08073	200.10	280.83	6.0570	0.04021	197.77	278.19	5.8428
10	0.08378	207.64	291.42	6.0950	0.04178	205.42	288.98	5.8816
20	0.08682	215.17	301.99	6.1317	0.04334	213.06	299.73	5.9189
30	0.08986	222.70	312.55	6.1672	0.04489	220.69	310.46	5.9549
40	0.09289	230.22	323.10	6.2014	0.04643	228.29	321.16	5.9896
50	0.09591	237.73	333.64	6.2345	0.04798	235.89	331.84	6.0232
60	0.09894	245.24	344.18	6.2666	0.04951	243.48	342.51	6.0557
70	0.10195	252.75	354.71	6.2978	0.05105	251.06	353.16	6.0872
80	0.10497	260.26	365.23	6.3280	0.05258	258.64	363.79	6.1177
90	0.10798	267.77	375.75	6.3574	0.05410	266.21	374.42	6.1474
100	0.11099	275.28	386.27	6.3860	0.05563	273.78	385.03	6.1763
110	0.11400	282.79	396.79	6.4138	0.05715	281.34	395.65	6.2043
120	0.11700	290.31	407.31	6.4409	0.05867	288.91	406.25	6.2316
130	0.12000	297.83	417.83	6.4673	0.06019	296.48	416.86	6.2583
140	0.12301	305.35	428.36	6.4931	0.06171	304.05	427.46	6.2843
150	0.12600	312.89	438.89	6.5183	0.06322	311.62	438.06	6.3096
160	0.12900	320.43	449.43	6.5429	0.06473	319.20	448.67	6.3344
170	0.13200	327.97	459.97	6.5670	0.06624	326.79	459.27	6.3586
180	0.13499	335.53	470.52	6.5905	0.06775	334.38	469.89	6.3823
190	0.13798	343.10	481.08	6.6136	0.06926	341.98	480.50	6.4055
200	0.14097	350.67	491.65	6.6361	0.07077	349.60	491.13	6.4281

continued next page…

Appendix K.2 - Superheated Nitrogen (N_2)

	P = 1000 kPa (-169.40°C)				P = 2000 kPa (-157.55°C)			
	v	u	h	s	v	u	h	s
T(°C)	m³/kg	kJ/kg	kJ/kg	kJ/kg-K	m³/kg	kJ/kg	kJ/kg	kJ/kg-K
210	0.14396	358.27	502.23	6.6583	0.07227	357.22	501.76	6.4504
220	0.14695	365.87	512.82	6.6800	0.07378	364.85	512.41	6.4722
230	0.14994	373.49	523.43	6.7012	0.07528	372.50	523.06	6.4936
240	0.15293	381.12	534.05	6.7221	0.07678	380.16	533.72	6.5146
250	0.15592	388.77	544.68	6.7427	0.07829	387.83	544.40	6.5352
260	0.15890	396.43	555.33	6.7628	0.07979	395.52	555.09	6.5554
270	0.16189	404.11	566.00	6.7827	0.08129	403.22	565.80	6.5753
280	0.16487	411.81	576.68	6.8021	0.08279	410.94	576.52	6.5949
290	0.16785	419.52	587.38	6.8213	0.08429	418.68	587.26	6.6141
300	0.17084	427.26	598.10	6.8402	0.08578	426.44	598.01	6.6330
310	0.17382	435.02	608.83	6.8587	0.08728	434.22	608.78	6.6517
320	0.17680	442.79	619.59	6.8770	0.08878	442.01	619.57	6.6700
330	0.17978	450.59	630.36	6.8951	0.09028	449.83	630.38	6.6881
340	0.18276	458.40	641.16	6.9128	0.09177	457.66	641.20	6.7059
350	0.18574	466.24	651.98	6.9303	0.09327	465.52	652.05	6.7234
360	0.18872	474.10	662.82	6.9476	0.09476	473.39	662.92	6.7407
370	0.19170	481.98	673.68	6.9646	0.09626	481.29	673.81	6.7578
380	0.19468	489.88	684.56	6.9814	0.09775	489.21	684.71	6.7746
390	0.19765	497.81	695.47	6.9979	0.09925	497.15	695.64	6.7912
400	0.20063	505.76	706.39	7.0143	0.10074	505.12	706.60	6.8076
410	0.20361	513.73	717.34	7.0304	0.10223	513.10	717.57	6.8238
420	0.20659	521.73	728.32	7.0464	0.10373	521.11	728.57	6.8398
430	0.20956	529.75	739.31	7.0621	0.10522	529.15	739.58	6.8556
440	0.21254	537.79	750.33	7.0777	0.10671	537.20	750.63	6.8712
450	0.21552	545.86	761.38	7.0931	0.10820	545.28	761.69	6.8866
460	0.21849	553.95	772.44	7.1083	0.10969	553.39	772.78	6.9018
470	0.22147	562.07	783.53	7.1233	0.11119	561.52	783.89	6.9168
480	0.22444	570.21	794.65	7.1382	0.11268	569.67	795.02	6.9317
490	0.22742	578.37	805.79	7.1529	0.11417	577.84	806.18	6.9464
500	0.23039	586.56	816.95	7.1674	0.11566	586.04	817.36	6.9610

Appendix K.2 - Superheated Nitrogen (N$_2$)

T(°C)	P = 3000 kPa (-149.53°C)				P = 4000 kPa			
	v m^3/kg	u kJ/kg	h kJ/kg	s kJ/kg-K	v m^3/kg	u kJ/kg	h kJ/kg	s kJ/kg-K
sat	0.00579	45.66	63.04	4.4996	-	-	-	-
-140	0.00902	70.46	97.52	4.7705	0.00499	49.82	69.76	4.5103
-130	0.01092	83.67	116.44	4.9078	0.00719	72.82	101.58	4.7419
-120	0.01251	94.56	132.08	5.0134	0.00865	86.60	121.20	4.8745
-110	0.01393	104.40	146.19	5.1028	0.00988	97.96	137.47	4.9775
-100	0.01526	113.64	159.42	5.1815	0.01099	108.17	152.11	5.0647
-90	0.01653	122.49	172.08	5.2526	0.01202	117.71	165.78	5.1414
-80	0.01775	131.08	184.33	5.3177	0.01300	126.82	178.81	5.2107
-70	0.01894	139.47	196.28	5.3780	0.01394	135.62	191.38	5.2742
-60	0.02010	147.71	208.01	5.4344	0.01486	144.20	203.62	5.3330
-50	0.02124	155.84	219.56	5.4873	0.01575	152.61	215.59	5.3879
-40	0.02236	163.88	230.96	5.5373	0.01662	160.88	227.37	5.4395
-30	0.02347	171.85	242.25	5.5848	0.01748	169.05	238.97	5.4882
-20	0.02456	179.76	253.45	5.6299	0.01833	177.13	250.44	5.5345
-10	0.02565	187.62	264.56	5.6729	0.01916	185.15	261.80	5.5785
0	0.02672	195.43	275.60	5.7141	0.01999	193.10	273.07	5.6205
10	0.02779	203.21	286.59	5.7536	0.02081	201.00	284.25	5.6607
20	0.02886	210.96	297.52	5.7916	0.02163	208.86	295.36	5.6993
30	0.02991	218.68	308.41	5.8281	0.02243	216.69	306.42	5.7364
40	0.03096	226.38	319.27	5.8633	0.02324	224.48	317.42	5.7721
50	0.03201	234.06	330.09	5.8973	0.02403	232.25	328.38	5.8065
60	0.03305	241.73	340.88	5.9302	0.02483	239.99	339.30	5.8398
70	0.03409	249.38	351.65	5.9621	0.02562	247.72	350.18	5.8720
80	0.03512	257.02	362.39	5.9929	0.02640	255.42	361.04	5.9032
90	0.03616	264.66	373.12	6.0229	0.02719	263.12	371.87	5.9334
100	0.03719	272.28	383.84	6.0520	0.02797	270.80	382.68	5.9628
110	0.03821	279.91	394.54	6.0803	0.02875	278.48	393.47	5.9913
120	0.03924	287.52	405.23	6.1079	0.02953	286.15	404.25	6.0191
130	0.04026	295.14	415.91	6.1347	0.03030	293.81	415.01	6.0461
140	0.04128	302.75	426.59	6.1608	0.03107	301.47	425.76	6.0725
150	0.04230	310.37	437.27	6.1864	0.03184	309.13	436.50	6.0982
160	0.04332	317.99	447.94	6.2113	0.03261	316.79	447.24	6.1232
170	0.04433	325.62	458.61	6.2357	0.03338	324.46	457.98	6.1477
180	0.04535	333.25	469.28	6.2595	0.03415	332.12	468.71	6.1717
190	0.04636	340.88	479.96	6.2828	0.03491	339.79	479.44	6.1951
200	0.04737	348.53	490.64	6.3056	0.03568	347.47	490.17	6.2180

continued next page…

Appendix K.2 - Superheated Nitrogen (N_2)

T(°C)	P = 3000 kPa (-149.53°C)				P = 4000 kPa			
	v	u	h	s	v	u	h	s
	m³/kg	kJ/kg	kJ/kg	kJ/kg-K	m³/kg	kJ/kg	kJ/kg	kJ/kg-K
210	0.04838	356.18	501.32	6.3279	0.03644	355.15	500.91	6.2405
220	0.04939	363.84	512.01	6.3498	0.03720	362.85	511.65	6.2625
230	0.05040	371.52	522.71	6.3713	0.03796	370.55	522.40	6.2841
240	0.05141	379.20	533.42	6.3924	0.03872	378.26	533.15	6.3052
250	0.05241	386.90	544.15	6.4131	0.03948	385.99	543.91	6.3260
260	0.05342	394.62	554.88	6.4334	0.04024	393.73	554.68	6.3464
270	0.05443	402.35	565.62	6.4534	0.04100	401.48	565.47	6.3664
280	0.05543	410.09	576.38	6.4730	0.04176	409.24	576.26	6.3861
290	0.05643	417.85	587.15	6.4923	0.04251	417.03	587.07	6.4055
300	0.05744	425.63	597.94	6.5113	0.04327	424.83	597.89	6.4245
310	0.05844	433.43	608.75	6.5300	0.04402	432.64	608.73	6.4433
320	0.05944	441.24	619.57	6.5484	0.04478	440.48	619.58	6.4617
330	0.06045	449.07	630.41	6.5665	0.04553	448.33	630.45	6.4799
340	0.06145	456.93	641.26	6.5844	0.04629	456.20	641.34	6.4978
350	0.06245	464.80	652.14	6.6020	0.04704	464.09	652.24	6.5155
360	0.06345	472.69	663.03	6.6193	0.04779	472.00	663.16	6.5328
370	0.06445	480.61	673.95	6.6364	0.04854	479.93	674.10	6.5500
380	0.06545	488.54	684.88	6.6533	0.04930	487.88	685.06	6.5669
390	0.06645	496.50	695.84	6.6699	0.05005	495.85	696.04	6.5836
400	0.06744	504.48	706.81	6.6863	0.05080	503.85	707.04	6.6000
410	0.06844	512.48	717.81	6.7026	0.05155	511.86	718.06	6.6163
420	0.06944	520.51	728.83	6.7186	0.05230	519.90	729.10	6.6323
430	0.07044	528.55	739.87	6.7344	0.05305	527.96	740.16	6.6482
440	0.07144	536.62	750.93	6.7500	0.05380	536.04	751.25	6.6638
450	0.07243	544.71	762.01	6.7654	0.05455	544.15	762.35	6.6793
460	0.07343	552.83	773.12	6.7807	0.05530	552.28	773.48	6.6946
470	0.07443	560.97	784.25	6.7958	0.05605	560.43	784.63	6.7097
480	0.07542	569.13	795.40	6.8107	0.05680	568.60	795.79	6.7246
490	0.07642	577.32	806.58	6.8254	0.05755	576.80	806.99	6.7394
500	0.07742	585.53	817.77	6.8400	0.05830	585.02	818.20	6.7540

Appendix K.2 - Superheated Nitrogen (N$_2$)

	P = 6000 kPa				P = 8000 kPa			
	v	u	h	s	v	u	h	s
T(°C)	m^3/kg	kJ/kg	kJ/kg	kJ/kg-K	m^3/kg	kJ/kg	kJ/kg	kJ/kg-K
-140	0.00218	1.78	14.85	4.0541	0.00193	-8.93	6.54	3.9611
-130	0.00343	41.69	62.26	4.3971	0.00240	18.95	38.18	4.1899
-120	0.00481	67.19	96.03	4.6258	0.00317	46.63	72.02	4.4186
-110	0.00586	83.42	118.59	4.7688	0.00400	67.74	99.72	4.5940
-100	0.00675	96.33	136.85	4.8775	0.00473	83.81	121.68	4.7247
-90	0.00755	107.63	152.93	4.9678	0.00539	97.15	140.29	4.8293
-80	0.00829	117.98	167.71	5.0464	0.00600	108.93	156.90	4.9177
-70	0.00898	127.73	181.63	5.1167	0.00656	119.72	172.21	4.9950
-60	0.00965	137.05	194.95	5.1807	0.00710	129.85	186.62	5.0642
-50	0.01029	146.06	207.82	5.2397	0.00761	139.52	200.38	5.1273
-40	0.01092	154.84	220.34	5.2946	0.00810	148.83	213.64	5.1855
-30	0.01153	163.44	232.59	5.3460	0.00858	157.87	226.53	5.2396
-20	0.01212	171.89	244.62	5.3945	0.00905	166.71	239.11	5.2903
-10	0.01271	180.23	256.47	5.4404	0.00951	175.37	251.44	5.3381
0	0.01329	188.46	268.18	5.4841	0.00996	183.90	263.56	5.3833
10	0.01386	196.62	279.75	5.5257	0.01040	192.31	275.52	5.4263
20	0.01442	204.71	291.22	5.5655	0.01084	200.63	287.33	5.4673
30	0.01498	212.74	302.60	5.6037	0.01127	208.86	299.01	5.5065
40	0.01553	220.72	313.89	5.6403	0.01170	217.03	310.59	5.5440
50	0.01608	228.66	325.12	5.6756	0.01212	225.14	322.08	5.5802
60	0.01662	236.56	336.29	5.7097	0.01254	233.19	333.48	5.6149
70	0.01716	244.43	347.40	5.7425	0.01295	241.21	344.82	5.6484
80	0.01770	252.27	358.47	5.7743	0.01336	249.18	356.09	5.6808
90	0.01824	260.09	369.50	5.8051	0.01377	257.12	367.31	5.7121
100	0.01877	267.89	380.49	5.8350	0.01418	265.03	378.48	5.7425
110	0.01930	275.67	391.46	5.8640	0.01459	272.92	389.60	5.7719
120	0.01983	283.44	402.39	5.8922	0.01499	280.79	400.69	5.8005
130	0.02035	291.20	413.31	5.9196	0.01539	288.64	411.75	5.8283
140	0.02088	298.95	424.20	5.9463	0.01579	296.48	422.78	5.8553
150	0.02140	306.69	435.08	5.9723	0.01619	304.31	433.79	5.8816
160	0.02192	314.43	445.95	5.9977	0.01658	312.12	444.78	5.9073
170	0.02244	322.17	456.80	6.0224	0.01698	319.93	455.75	5.9323
180	0.02296	329.91	467.65	6.0467	0.01737	327.74	466.70	5.9568
190	0.02347	337.64	478.49	6.0703	0.01776	335.54	477.65	5.9807
200	0.02399	345.39	489.32	6.0935	0.01816	343.35	488.58	6.0040

continued next page…

Appendix K.2 - Superheated Nitrogen (N$_2$)

	P = 6000 kPa				P = 8000 kPa			
	v	u	h	s	v	u	h	s
T(°C)	m^3/kg	kJ/kg	kJ/kg	kJ/kg-K	m^3/kg	kJ/kg	kJ/kg	kJ/kg-K
210	0.02451	353.13	500.16	6.1161	0.01855	351.15	499.51	6.0269
220	0.02502	360.88	510.99	6.1383	0.01894	358.96	510.44	6.0492
230	0.02553	368.64	521.83	6.1601	0.01932	366.77	521.36	6.0712
240	0.02604	376.41	532.67	6.1814	0.01971	374.59	532.28	6.0927
250	0.02656	384.18	543.51	6.2023	0.02010	382.42	543.20	6.1137
260	0.02707	391.97	554.36	6.2229	0.02049	390.25	554.13	6.1344
270	0.02758	399.77	565.22	6.2431	0.02087	398.10	565.06	6.1547
280	0.02809	407.58	576.09	6.2629	0.02126	405.95	575.99	6.1747
290	0.02859	415.41	586.97	6.2824	0.02164	413.82	586.94	6.1943
300	0.02910	423.25	597.86	6.3015	0.02202	421.70	597.89	6.2136
310	0.02961	431.10	608.76	6.3204	0.02241	429.59	608.85	6.2325
320	0.03012	438.97	619.67	6.3389	0.02279	437.50	619.82	6.2512
330	0.03062	446.86	630.60	6.3572	0.02317	445.43	630.81	6.2696
340	0.03113	454.77	641.54	6.3752	0.02355	453.37	641.80	6.2876
350	0.03163	462.69	652.50	6.3929	0.02394	461.32	652.81	6.3055
360	0.03214	470.64	663.47	6.4104	0.02432	469.30	663.84	6.3230
370	0.03264	478.60	674.46	6.4276	0.02470	477.29	674.88	6.3403
380	0.03315	486.58	685.47	6.4446	0.02508	485.31	685.93	6.3574
390	0.03365	494.58	696.50	6.4614	0.02546	493.34	697.00	6.3742
400	0.03416	502.61	707.54	6.4779	0.02584	501.39	708.09	6.3908
410	0.03466	510.65	718.61	6.4942	0.02622	509.46	719.20	6.4072
420	0.03516	518.71	729.69	6.5103	0.02660	517.55	730.32	6.4233
430	0.03567	526.80	740.79	6.5262	0.02698	525.66	741.47	6.4393
440	0.03617	534.91	751.92	6.5419	0.02735	533.80	752.63	6.4550
450	0.03667	543.04	763.06	6.5575	0.02773	541.95	763.81	6.4706
460	0.03717	551.19	774.22	6.5728	0.02811	550.12	775.01	6.4860
470	0.03767	559.36	785.41	6.5879	0.02849	558.32	786.23	6.5012
480	0.03818	567.56	796.61	6.6029	0.02887	566.54	797.47	6.5162
490	0.03868	575.78	807.84	6.6177	0.02924	574.78	808.73	6.5311
500	0.03918	584.02	819.08	6.6324	0.02962	583.04	820.00	6.5458

adapted from NIST data   The reference state for all nitrogen properties is h = 0 and s = 0 for the saturated liquid at its normal boiling point which is -195.8°C.

Appendix L.1 - Saturated Oxygen (O_2)

T °C	P kPa	v^L m³/kg	v^V m³/kg	u^L kJ/kg	u^V kJ/kg	h^L kJ/kg	h^V kJ/kg	s^L kJ/kg-K	s^V kJ/kg-K
-220	0.10	0.000763	139.7500	-62.28	167.57	-62.28	181.38	-0.8875	3.6968
-215	0.45	0.000775	33.9020	-53.92	170.79	-53.92	185.89	-0.7371	3.3868
-210	1.55	0.000789	10.5600	-45.55	173.99	-45.54	190.38	-0.5990	3.1368
-205	4.43	0.000803	3.9858	-37.16	177.17	-37.16	194.83	-0.4712	2.9327
-200	10.81	0.000818	1.7477	-28.77	180.33	-28.76	199.22	-0.3525	2.7642
-195	23.29	0.000834	0.8625	-20.38	183.43	-20.36	203.51	-0.2415	2.6232
-190	45.37	0.000851	0.4677	-11.97	186.42	-11.94	207.64	-0.1372	2.5035
-185	81.45	0.000869	0.2735	-3.54	189.27	-3.47	211.55	-0.0387	2.4005
-180	136.66	0.000888	0.1701	4.94	191.92	5.07	215.16	0.0549	2.3103
-175	216.80	0.000909	0.1111	13.50	194.33	13.69	218.42	0.1444	2.2302
-170	328.12	0.000931	0.0756	22.15	196.47	22.45	221.27	0.2304	2.1579
-165	477.26	0.000956	0.0531	30.92	198.30	31.38	223.66	0.3135	2.0915
-160	671.11	0.000983	0.0384	39.85	199.77	40.51	225.54	0.3944	2.0297
-155	916.79	0.001014	0.0283	48.98	200.84	49.91	226.82	0.4735	1.9709
-150	1221.60	0.001050	0.0213	58.36	201.42	59.64	227.41	0.5516	1.9139
-145	1593.10	0.001091	0.0162	68.07	201.43	69.81	227.17	0.6294	1.8573
-140	2039.30	0.001140	0.0124	78.23	200.69	80.55	225.90	0.7078	1.7994
-135	2568.80	0.001202	0.0095	89.02	198.95	92.11	223.27	0.7884	1.7378
-130	3191.50	0.001285	0.0072	100.82	195.71	104.92	218.65	0.8740	1.6684
-125	3919.30	0.001411	0.0053	114.48	189.79	120.01	210.57	0.9708	1.5821
-120	4771.00	0.001692	0.0035	133.83	175.95	141.90	192.56	1.1073	1.4380
-118.57	5042.80	0.002170	0.0022	151.00	151.00	161.94	161.94	1.2341	1.2341

Appendix L.2 - Superheated Oxygen (O$_2$)

	P = 100 kPa (-183.09°C)				P = 200 kPa (-175.91°C)			
	v	u	h	s	v	u	h	s
T(°C)	m^3/kg	kJ/kg	kJ/kg	kJ/kg-K	m^3/kg	kJ/kg	kJ/kg	kJ/kg-K
sat	0.2266	190.31	212.96	2.3647	0.1197	193.9	217.9	2.2442
-180	0.2351	192.41	215.92	2.3969	-	-	-	-
-170	0.2623	199.07	225.30	2.4926	0.1281	197.99	223.61	2.3018
-160	0.2892	205.70	234.62	2.5788	0.1421	204.78	233.20	2.3905
-150	0.3159	212.32	243.90	2.6575	0.1558	211.53	242.69	2.4709
-140	0.3424	218.92	253.16	2.7297	0.1694	218.23	252.11	2.5444
-130	0.3689	225.49	262.38	2.7965	0.1829	224.89	261.46	2.6121
-120	0.3952	232.06	271.58	2.8586	0.1962	231.52	270.77	2.6750
-110	0.4215	238.61	280.76	2.9167	0.2096	238.13	280.04	2.7336
-100	0.4478	245.15	289.93	2.9712	0.2228	244.72	289.28	2.7886
-90	0.4740	251.69	299.09	3.0227	0.2361	251.29	298.50	2.8404
-80	0.5002	258.22	308.24	3.0713	0.2493	257.86	307.71	2.8893
-70	0.5264	264.75	317.39	3.1175	0.2624	264.41	316.90	2.9357
-60	0.5525	271.28	326.53	3.1614	0.2756	270.97	326.08	2.9798
-50	0.5786	277.81	335.68	3.2033	0.2887	277.52	335.26	3.0219
-40	0.6047	284.35	344.82	3.2434	0.3018	284.07	344.43	3.0621
-30	0.6308	290.88	353.96	3.2818	0.3149	290.62	353.60	3.1006
-20	0.6569	297.42	363.11	3.3187	0.3280	297.17	362.78	3.1376
-10	0.6830	303.97	372.27	3.3542	0.3411	303.73	371.96	3.1732
0	0.7091	310.52	381.43	3.3883	0.3542	310.30	381.14	3.2074
10	0.7351	317.09	390.60	3.4213	0.3673	316.88	390.33	3.2405
20	0.7612	323.67	399.78	3.4532	0.3803	323.47	399.53	3.2724
30	0.7872	330.26	408.98	3.4840	0.3934	330.07	408.74	3.3033
40	0.8133	336.87	418.19	3.5139	0.4064	336.68	417.97	3.3332
50	0.8393	343.49	427.42	3.5429	0.4195	343.31	427.21	3.3623
60	0.8653	350.13	436.66	3.5711	0.4325	349.96	436.47	3.3905
70	0.8914	356.79	445.93	3.5985	0.4456	356.63	445.74	3.4179
80	0.9174	363.48	455.22	3.6252	0.4586	363.32	455.04	3.4446
90	0.9434	370.18	464.53	3.6512	0.4716	370.03	464.36	3.4707
100	0.9695	376.91	473.86	3.6765	0.4847	376.77	473.70	3.4960

Appendix L - Properties of Oxygen

	P = 500 kPa (-164.34 °C)				P = 1000 kPa (-153.53 °C)			
	v	u	h	s	v	u	h	s
T(°C)	m³/kg	kJ/kg	kJ/kg	kJ/kg-K	m³/kg	kJ/kg	kJ/kg	kJ/kg-K
sat	0.05086	198.51	223.94	2.0832	0.02600	201.1	227.1	1.9540
-160	0.05360	201.77	228.56	2.1248	-	-	-	-
-150	0.05965	209.00	238.83	2.2118	0.02730	204.09	231.39	1.9896
-140	0.06550	216.07	248.82	2.2898	0.03070	212.08	242.78	2.0786
-130	0.07121	223.01	258.62	2.3608	0.03389	219.63	253.53	2.1564
-120	0.07682	229.86	268.27	2.4260	0.03695	226.94	263.89	2.2264
-110	0.08236	236.65	277.83	2.4864	0.03992	234.07	273.98	2.2903
-100	0.08784	243.38	287.30	2.5428	0.04282	241.07	283.89	2.3492
-90	0.09328	250.07	296.71	2.5956	0.04567	247.98	293.65	2.4040
-80	0.09869	256.74	306.08	2.6454	0.04849	254.83	303.31	2.4554
-70	0.10407	263.38	315.41	2.6925	0.05127	261.62	312.90	2.5037
-60	0.10942	270.00	324.72	2.7372	0.05404	268.38	322.41	2.5495
-50	0.11476	276.62	334.00	2.7798	0.05678	275.10	331.88	2.5929
-40	0.12009	283.23	343.27	2.8204	0.05951	281.81	341.32	2.6342
-30	0.12540	289.83	352.53	2.8593	0.06222	288.49	350.72	2.6737
-20	0.13070	296.43	361.77	2.8966	0.06493	295.17	360.10	2.7115
-10	0.13599	303.03	371.02	2.9324	0.06762	301.84	369.46	2.7478
0	0.14127	309.63	380.27	2.9669	0.07030	308.51	378.81	2.7827
10	0.14655	316.24	389.51	3.0001	0.07298	315.17	388.15	2.8163
20	0.15182	322.86	398.76	3.0322	0.07565	321.84	397.49	2.8487
30	0.15708	329.49	408.02	3.0633	0.07831	328.52	406.83	2.8800
40	0.16234	336.13	417.29	3.0934	0.08097	335.20	416.17	2.9103
50	0.16759	342.78	426.58	3.1226	0.08363	341.89	425.52	2.9397
60	0.17284	349.45	435.87	3.1509	0.08628	348.60	434.88	2.9683
70	0.17808	356.14	445.18	3.1784	0.08893	355.33	444.25	2.9960
80	0.18333	362.85	454.51	3.2052	0.09157	362.07	453.64	3.0229
90	0.18857	369.58	463.86	3.2313	0.09421	368.83	463.04	3.0492
100	0.19380	376.33	473.23	3.2568	0.09685	375.61	472.46	3.0748

	P = 2000 kPa (-140.41 °C)				P = 4000 kPa (-124.49 °C)			
	v	u	h	s	v	u	h	s
T(°C)	m³/kg	kJ/kg	kJ/kg	kJ/kg-K	m³/kg	kJ/kg	kJ/kg	kJ/kg-K
sat	0.01263	200.78	226.05	1.8042	0.00512	188.93	209.42	1.5715
-140	0.01274	201.27	226.75	1.8095	-	-	-	-
-130	0.01496	211.53	241.45	1.9161	-	-	-	-
-120	0.01686	220.28	254.01	2.0009	0.00620	199.84	224.64	1.6726
-110	0.01861	228.38	265.60	2.0743	0.00768	213.70	244.44	1.7980
-100	0.02025	236.10	276.60	2.1398	0.00883	224.23	259.56	1.8881
-90	0.02183	243.56	287.22	2.1994	0.00984	233.47	272.84	1.9627
-80	0.02337	250.83	297.57	2.2544	0.01077	242.01	285.10	2.0278
-70	0.02487	257.98	307.71	2.3056	0.01165	250.10	296.70	2.0864
-60	0.02634	265.02	317.70	2.3536	0.01249	257.90	307.85	2.1400
-50	0.02779	272.00	327.58	2.3989	0.01330	265.48	318.67	2.1896
-40	0.02922	278.92	337.36	2.4418	0.01409	272.91	329.26	2.2360
-30	0.03064	285.79	347.07	2.4825	0.01486	280.21	339.65	2.2797
-20	0.03204	292.63	356.72	2.5214	0.01562	287.42	349.89	2.3210
-10	0.03344	299.44	366.32	2.5586	0.01637	294.56	360.02	2.3602
0	0.03483	306.24	375.89	2.5943	0.01710	301.64	370.05	2.3976
10	0.03620	313.02	385.43	2.6286	0.01783	308.67	380.00	2.4334
20	0.03757	319.80	394.95	2.6617	0.01855	315.68	389.89	2.4677
30	0.03894	326.57	404.45	2.6935	0.01927	322.66	399.73	2.5007
40	0.04030	333.34	413.94	2.7244	0.01998	329.61	409.53	2.5325
50	0.04166	340.12	423.43	2.7542	0.02069	336.56	419.30	2.5633
60	0.04301	346.90	432.92	2.7831	0.02139	343.50	429.05	2.5930
70	0.04436	353.70	442.41	2.8112	0.02208	350.44	438.77	2.6217
80	0.04570	360.50	451.90	2.8384	0.02278	357.38	448.49	2.6496
90	0.04704	367.32	461.41	2.8650	0.02347	364.32	458.20	2.6767
100	0.04838	374.16	470.92	2.8908	0.02416	371.27	467.90	2.7031

adapted from NIST data The reference state for all oxygen properties is h = 0 and s = 0 for the saturated liquid at its normal boiling point which is -182.96°C.

Appendix M.1 - Vapor-Liquid Saturated Carbon Dioxide (CO_2)

Temperature Entry

T °C	P kPa	v^L m³/kg	v^V m³/kg	u^L kJ/kg	u^V kJ/kg	h^L kJ/kg	h^V kJ/kg	s^L kJ/kg-K	s^V kJ/kg-K
-55	553.97	0.000853	0.068151	82.62	393.23	83.09	430.99	0.5352	2.1300
-50	682.34	0.000866	0.055789	92.35	394.61	92.94	432.68	0.5794	2.1018
-45	831.84	0.000880	0.046046	102.14	395.83	102.87	434.13	0.6228	2.0747
-40	1004.50	0.000896	0.038284	112.00	396.87	112.90	435.32	0.6656	2.0485
-35	1202.40	0.000912	0.032035	121.95	397.71	123.05	436.23	0.7079	2.0230
-30	1427.80	0.000930	0.026956	132.01	398.33	133.34	436.82	0.7498	1.9980
-25	1682.70	0.000949	0.022789	142.20	398.71	143.79	437.06	0.7914	1.9732
-20	1969.60	0.000969	0.019343	152.54	398.79	154.45	436.89	0.8328	1.9485
-15	2290.80	0.000992	0.016467	163.07	398.55	165.34	436.27	0.8742	1.9237
-10	2648.70	0.001017	0.014048	173.83	397.93	176.52	435.14	0.9157	1.8985
-5	3045.90	0.001046	0.011996	184.86	396.84	188.05	433.38	0.9576	1.8725
0	3485.10	0.001078	0.010241	196.24	395.20	200.00	430.89	1.0000	1.8453
5	3969.50	0.001116	0.008724	208.07	392.85	212.50	427.48	1.0434	1.8163
10	4502.20	0.001161	0.007399	220.50	389.57	225.73	422.88	1.0884	1.7847
15	5087.10	0.001218	0.006222	233.79	384.99	239.99	416.64	1.1359	1.7489
20	5729.10	0.001293	0.005149	248.46	378.36	255.87	407.87	1.1877	1.7062
25	6434.20	0.001408	0.004120	265.73	367.92	274.78	394.43	1.2485	1.6498
30	7213.70	0.001686	0.002898	292.40	344.23	304.55	365.13	1.3435	1.5433
30.978	7377.30	0.002139	0.002139	316.47	316.47	332.25	332.25	1.4336	1.4336

Pressure Entry

P kPa	T °C	v^L m³/kg	v^V m³/kg	u^L kJ/kg	u^V kJ/kg	h^L kJ/kg	h^V kJ/kg	s^L kJ/kg-K	s^V kJ/kg-K
600	-53.12	0.000858	0.063134	86.28	393.77	86.80	431.65	0.5520	2.1192
700	-49.37	0.000868	0.054430	93.58	394.77	94.19	432.87	0.5849	2.0984
800	-46.01	0.000878	0.047828	100.17	395.59	100.87	433.86	0.6141	2.0801
900	-42.94	0.000887	0.042640	106.19	396.28	106.99	434.65	0.6405	2.0638
1000	-40.12	0.000895	0.038453	111.76	396.84	112.66	435.30	0.6646	2.0491
1200	-35.06	0.000912	0.032099	121.84	397.70	122.93	436.22	0.7075	2.0233
1400	-30.58	0.000927	0.027497	130.83	398.27	132.13	436.77	0.7450	2.0009
1600	-26.56	0.000943	0.024002	139.01	398.62	140.52	437.02	0.7785	1.9809
1800	-22.89	0.000957	0.021254	146.55	398.78	148.27	437.04	0.8089	1.9628
2000	-19.50	0.000971	0.019033	153.58	398.79	155.52	436.85	0.8369	1.9461
2200	-16.36	0.000986	0.017199	160.19	398.65	162.36	436.49	0.8630	1.9305
2400	-13.42	0.001000	0.015656	166.45	398.40	168.85	435.97	0.8873	1.9158
2600	-10.65	0.001014	0.014340	172.41	398.03	175.05	435.32	0.9103	1.9018
2800	-8.03	0.001028	0.013202	178.13	397.56	181.01	434.53	0.9321	1.8884
3000	-5.55	0.001043	0.012207	183.63	396.99	186.75	433.61	0.9529	1.8754
3200	-3.19	0.001057	0.011329	188.94	396.32	192.32	432.57	0.9729	1.8628
3400	-0.93	0.001072	0.010548	194.09	395.56	197.74	431.42	0.9920	1.8505
3600	1.23	0.001087	0.009848	199.11	394.70	203.02	430.15	1.0106	1.8384
3800	3.30	0.001103	0.009215	204.00	393.74	208.19	428.76	1.0286	1.8264
4000	5.30	0.001119	0.008640	208.80	392.69	213.27	427.25	1.0461	1.8145
5000	14.28	0.001209	0.006383	231.82	385.74	237.87	417.66	1.1289	1.7544
6000	21.98	0.001332	0.004742	254.86	374.87	262.85	403.32	1.2102	1.6862
7000	28.68	0.001567	0.003289	282.91	353.89	293.88	376.91	1.3093	1.5844
7377.30	30.978	0.002139	0.002139	316.47	316.47	332.25	332.25	1.4336	1.4336

adapted from NIST data

The reference state for all carbon dioxide data is h = 200 kJ/kg and s = 1 kJ/kg-K for the saturated liquid at 0°C. This choice is common and results in h > 0 for all commonly encountered states.

Appendix M.2 - Superheated Carbon Dioxide (CO_2)

T(°C)	P = 100 kPa				P = 200 kPa			
	v m³/kg	u kJ/kg	h kJ/kg	s kJ/kg-K	v m³/kg	u kJ/kg	h kJ/kg	s kJ/kg-K
40	0.58911	459.81	518.72	2.7814	0.29330	459.22	517.88	2.6486
50	0.60818	466.60	527.42	2.8088	0.30292	466.05	526.64	2.6761
60	0.62724	473.49	536.21	2.8356	0.31253	472.97	535.48	2.7031
70	0.64628	480.47	545.09	2.8618	0.32213	479.98	544.41	2.7295
80	0.66531	487.54	554.07	2.8876	0.33171	487.08	553.42	2.7554
90	0.68432	494.70	563.13	2.9129	0.34128	494.26	562.52	2.7808
100	0.70333	501.95	572.28	2.9378	0.35084	501.53	571.70	2.8057
110	0.72233	509.28	581.52	2.9622	0.36040	508.89	580.97	2.8302
120	0.74132	516.70	590.84	2.9862	0.36994	516.33	590.32	2.8543
130	0.76031	524.21	600.24	3.0098	0.37948	523.85	599.75	2.8780
140	0.77928	531.79	609.72	3.0331	0.38901	531.45	609.25	2.9013
150	0.79826	539.46	619.28	3.0559	0.39854	539.13	618.84	2.9242
160	0.81722	547.20	628.92	3.0784	0.40806	546.88	628.50	2.9468
170	0.83619	555.02	638.64	3.1006	0.41758	554.72	638.23	2.9690
180	0.85515	562.92	648.43	3.1225	0.42709	562.62	648.04	2.9909
190	0.87410	570.88	658.29	3.1440	0.43660	570.60	657.92	3.0124
200	0.89305	578.92	668.23	3.1652	0.44611	578.65	667.87	3.0337
210	0.91200	587.03	678.23	3.1861	0.45561	586.77	677.89	3.0546
220	0.93095	595.21	688.31	3.2068	0.46511	594.96	687.98	3.0753
230	0.94989	603.46	698.45	3.2271	0.47461	603.22	698.14	3.0957
240	0.96883	611.78	708.66	3.2472	0.48410	611.54	708.36	3.1158
250	0.98777	620.16	718.93	3.2671	0.49359	619.92	718.64	3.1357
260	1.00670	628.60	729.27	3.2866	0.50308	628.37	728.99	3.1553
270	1.02560	637.11	739.67	3.3060	0.51257	636.89	739.40	3.1746
280	1.04460	645.67	750.13	3.3250	0.52205	645.46	749.87	3.1937
290	1.06350	654.30	760.65	3.3439	0.53154	654.10	760.40	3.2126
300	1.08240	662.99	771.23	3.3625	0.54102	662.79	770.99	3.2312
310	1.10140	671.74	781.87	3.3809	0.55050	671.54	781.64	3.2496
320	1.12030	680.54	792.57	3.3991	0.55998	680.35	792.34	3.2678
330	1.13920	689.40	803.32	3.4171	0.56946	689.21	803.10	3.2858
340	1.15810	698.31	814.13	3.4349	0.57893	698.13	813.92	3.3036
350	1.17700	707.28	824.99	3.4524	0.58841	707.11	824.79	3.3212

Appendix M.2 - Superheated Carbon Dioxide (CO_2)

	P = 300 kPa				P = 400 kPa			
	v	u	h	s	v	u	h	s
T(°C)	m³/kg	kJ/kg	kJ/kg	kJ/kg-K	m³/kg	kJ/kg	kJ/kg	kJ/kg-K
40	0.19469	458.63	517.04	2.5701	0.14538	458.04	516.19	2.5139
50	0.20117	465.50	525.85	2.5978	0.15029	464.94	525.06	2.5417
60	0.20763	472.45	534.74	2.6249	0.15518	471.93	534.00	2.5690
70	0.21408	479.49	543.71	2.6514	0.16005	479.00	543.02	2.5956
80	0.22051	486.62	552.77	2.6774	0.16491	486.15	552.12	2.6218
90	0.22694	493.83	561.91	2.7030	0.16976	493.39	561.29	2.6474
100	0.23335	501.12	571.12	2.7280	0.17460	500.70	570.54	2.6725
110	0.23975	508.49	580.42	2.7526	0.17943	508.10	579.87	2.6972
120	0.24615	515.95	589.80	2.7767	0.18425	515.57	589.28	2.7214
130	0.25254	523.49	599.25	2.8005	0.18907	523.13	598.76	2.7452
140	0.25892	531.11	608.78	2.8238	0.19388	530.76	608.31	2.7687
150	0.26530	538.80	618.39	2.8468	0.19868	538.47	617.94	2.7917
160	0.27168	546.57	628.07	2.8694	0.20348	546.25	627.64	2.8143
170	0.27804	554.41	637.83	2.8917	0.20828	554.11	637.42	2.8367
180	0.28441	562.33	647.65	2.9136	0.21307	562.04	647.26	2.8586
190	0.29077	570.32	657.55	2.9352	0.21785	570.04	657.18	2.8803
200	0.29713	578.38	667.52	2.9565	0.22264	578.11	667.16	2.9016
210	0.30348	586.51	677.55	2.9775	0.22741	586.24	677.21	2.9226
220	0.30983	594.70	687.65	2.9982	0.23219	594.45	687.33	2.9433
230	0.31618	602.97	697.82	3.0186	0.23696	602.72	697.51	2.9638
240	0.32252	611.30	708.05	3.0387	0.24173	611.06	707.75	2.9839
250	0.32887	619.69	718.35	3.0586	0.24650	619.46	718.06	3.0038
260	0.33521	628.15	728.71	3.0782	0.25127	627.92	728.43	3.0235
270	0.34154	636.67	739.13	3.0976	0.25603	636.45	738.86	3.0428
280	0.34788	645.25	749.61	3.1167	0.26080	645.04	749.35	3.0620
290	0.35422	653.89	760.15	3.1356	0.26556	653.68	759.91	3.0809
300	0.36055	662.59	770.75	3.1543	0.27032	662.39	770.51	3.0996
310	0.36688	671.34	781.41	3.1727	0.27507	671.15	781.18	3.1180
320	0.37321	680.16	792.12	3.1909	0.27983	679.97	791.90	3.1362
330	0.37954	689.03	802.89	3.2089	0.28458	688.84	802.67	3.1543
340	0.38587	697.95	813.71	3.2267	0.28934	697.77	813.50	3.1721
350	0.39219	706.93	824.59	3.2443	0.29409	706.75	824.39	3.1897

Appendix M.2 - Superheated Carbon Dioxide (CO_2)

	P = 500 kPa				P = 517.74 kPa (-56.568 °C)			
	v	u	h	s	v	u	h	s
T(°C)	m³/kg	kJ/kg	kJ/kg	kJ/kg-K	m³/kg	kJ/kg	kJ/kg	kJ/kg-K
sat	-	-	-	-	0.07270	392.77	430.41	2.1391
-50	-	-	-	-	0.07559	397.20	436.33	2.1660
-40	-	-	-	-	0.07988	403.85	445.20	2.2049
-30	-	-	-	-	0.08405	410.44	453.96	2.2417
-20	-	-	-	-	0.08815	417.01	462.65	2.2767
-10	-	-	-	-	0.09218	423.61	471.33	2.3103
0	-	-	-	-	0.09616	430.23	480.02	2.3427
10	-	-	-	-	0.10010	436.91	488.74	2.3741
20	-	-	-	-	0.10401	443.65	497.50	2.4045
30	-	-	-	-	0.10788	450.45	506.31	2.4340
40	0.11579	457.44	515.33	2.4698	0.11174	457.33	515.18	2.4628
50	0.11976	464.38	524.26	2.4978	0.11557	464.28	524.12	2.4909
60	0.12370	471.40	533.25	2.5252	0.11939	471.31	533.12	2.5184
70	0.12763	478.51	542.32	2.5520	0.12319	478.42	542.20	2.5452
80	0.13155	485.69	551.46	2.5783	0.12698	485.60	551.34	2.5715
90	0.13545	492.94	560.67	2.6040	0.13075	492.87	560.56	2.5972
100	0.13935	500.28	569.96	2.6292	0.13452	500.21	569.85	2.6225
110	0.14324	507.70	579.32	2.6540	0.13828	507.63	579.22	2.6472
120	0.14712	515.19	588.75	2.6783	0.14203	515.13	588.66	2.6715
130	0.15099	522.77	598.26	2.7022	0.14577	522.70	598.17	2.6954
140	0.15485	530.41	607.84	2.7257	0.14950	530.35	607.76	2.7189
150	0.15871	538.14	617.49	2.7487	0.15323	538.08	617.41	2.7420
160	0.16257	545.93	627.22	2.7715	0.15696	545.88	627.14	2.7647
170	0.16642	553.80	637.01	2.7938	0.16068	553.75	636.94	2.7871
180	0.17026	561.74	646.87	2.8158	0.16440	561.69	646.80	2.8091
190	0.17410	569.75	656.80	2.8375	0.16811	569.70	656.74	2.8308
200	0.17794	577.83	666.80	2.8589	0.17182	577.78	666.74	2.8522
210	0.18178	585.98	676.87	2.8799	0.17552	585.93	676.81	2.8732
220	0.18561	594.19	687.00	2.9007	0.17922	594.15	686.94	2.8940
230	0.18944	602.47	697.19	2.9211	0.18292	602.43	697.14	2.9144
240	0.19326	610.82	707.45	2.9413	0.18662	610.78	707.40	2.9346
250	0.19709	619.23	717.77	2.9612	0.19031	619.19	717.72	2.9546
260	0.20091	627.70	728.15	2.9809	0.19401	627.66	728.10	2.9742
270	0.20473	636.23	738.60	3.0003	0.19770	636.19	738.55	2.9936
280	0.20855	644.82	749.10	3.0194	0.20138	644.79	749.05	3.0128
290	0.21236	653.48	759.66	3.0384	0.20507	653.44	759.61	3.0317
300	0.21618	662.19	770.27	3.0571	0.20876	662.15	770.23	3.0504
310	0.21999	670.95	780.95	3.0755	0.21244	670.92	780.91	3.0689
320	0.22380	679.78	791.68	3.0938	0.21612	679.74	791.64	3.0871
330	0.22761	688.66	802.46	3.1118	0.21980	688.62	802.42	3.1051
340	0.23142	697.59	813.30	3.1296	0.22348	697.56	813.26	3.1230
350	0.23523	706.57	824.19	3.1472	0.22716	706.54	824.15	3.1406

Appendix M.2 - Superheated Carbon Dioxide (CO_2)

T(°C)	P = 600 kPa (-53.12 °C)				P = 800 kPa (-46.01 °C)			
	v m³/kg	u kJ/kg	h kJ/kg	s kJ/kg-K	v m³/kg	u kJ/kg	h kJ/kg	s kJ/kg-K
sat	0.06313	393.77	431.65	2.1192	0.04783	395.59	433.86	2.0801
-50	0.06435	395.93	434.54	2.1323	-	-	-	-
-40	0.06816	402.75	443.65	2.1722	0.04966	399.96	439.69	2.1054
-30	0.07185	409.48	452.59	2.2098	0.05261	407.06	449.15	2.1452
-20	0.07546	416.16	461.44	2.2454	0.05546	414.03	458.40	2.1824
-10	0.07900	422.84	470.24	2.2795	0.05823	420.94	467.52	2.2178
0	0.08248	429.54	479.03	2.3123	0.06094	427.82	476.57	2.2516
10	0.08592	436.28	487.83	2.3440	0.06360	434.72	485.60	2.2840
20	0.08933	443.07	496.67	2.3746	0.06623	441.63	494.62	2.3153
30	0.09271	449.92	505.54	2.4044	0.06883	448.60	503.66	2.3456
40	0.09607	456.83	514.47	2.4334	0.07140	455.61	512.73	2.3751
50	0.09940	463.82	523.46	2.4616	0.07395	462.68	521.84	2.4037
60	0.10272	470.88	532.51	2.4892	0.07648	469.81	530.99	2.4316
70	0.10602	478.01	541.62	2.5161	0.07900	477.01	540.21	2.4589
80	0.10931	485.22	550.80	2.5425	0.08150	484.27	549.47	2.4855
90	0.11258	492.50	560.05	2.5684	0.08399	491.61	558.80	2.5115
100	0.11585	499.86	569.37	2.5937	0.08647	499.02	568.20	2.5371
110	0.11911	507.30	578.77	2.6185	0.08895	506.50	577.65	2.5621
120	0.12236	514.81	588.23	2.6429	0.09141	514.05	587.18	2.5866
130	0.12560	522.40	597.76	2.6668	0.09386	521.67	596.77	2.6107
140	0.12883	530.07	607.37	2.6904	0.09631	529.37	606.42	2.6343
150	0.13206	537.80	617.04	2.7135	0.09876	537.14	616.14	2.6576
160	0.13529	545.61	626.79	2.7363	0.10119	544.97	625.93	2.6804
170	0.13851	553.49	636.60	2.7587	0.10363	552.88	635.78	2.7029
180	0.14173	561.45	646.48	2.7807	0.10606	560.86	645.70	2.7251
190	0.14494	569.47	656.43	2.8024	0.10848	568.90	655.68	2.7469
200	0.14815	577.56	666.45	2.8238	0.11090	577.01	665.73	2.7683
210	0.15135	585.72	676.53	2.8449	0.11332	585.19	675.84	2.7895
220	0.15455	593.94	686.67	2.8657	0.11573	593.43	686.01	2.8103
230	0.15775	602.23	696.88	2.8862	0.11815	601.73	696.25	2.8309
240	0.16095	610.58	707.15	2.9064	0.12056	610.10	706.55	2.8511
250	0.16414	619.00	717.48	2.9263	0.12296	618.53	716.90	2.8711
260	0.16733	627.47	727.87	2.9460	0.12537	627.02	727.32	2.8908
270	0.17052	636.01	738.33	2.9654	0.12777	635.58	737.79	2.9103
280	0.17371	644.61	748.84	2.9846	0.13017	644.19	748.32	2.9295
290	0.17690	653.27	759.41	3.0036	0.13257	652.86	758.91	2.9485
300	0.18008	661.98	770.03	3.0223	0.13497	661.58	769.56	2.9672
310	0.18327	670.76	780.72	3.0407	0.13736	670.37	780.26	2.9857
320	0.18645	679.59	791.45	3.0590	0.13976	679.20	791.01	3.0040
330	0.18963	688.47	802.25	3.0770	0.14215	688.10	801.82	3.0221
340	0.19281	697.41	813.09	3.0949	0.14454	697.04	812.68	3.0399
350	0.19598	706.40	823.99	3.1125	0.14693	706.04	823.59	3.0576

Appendix M.2 - Superheated Carbon Dioxide (CO_2)

	P = 1000 kPa (-40.12 °C)				P = 2000 kPa (-19.50 °C)			
	v	u	h	s	v	u	h	s
T(°C)	m³/kg	kJ/kg	kJ/kg	kJ/kg-K	m³/kg	kJ/kg	kJ/kg	kJ/kg-K
sat	0.03845	396.84	435.30	2.0491	0.01903	398.79	436.85	1.9461
-40	0.03849	396.94	435.42	2.0497	-	-	-	-
-30	0.04101	404.49	445.50	2.0920	-	-	-	-
-20	0.04342	411.80	455.21	2.1312	-	-	-	-
-10	0.04574	418.96	464.70	2.1679	0.02051	407.57	448.58	1.9915
0	0.04799	426.05	474.04	2.2028	0.02193	416.14	460.00	2.0341
10	0.05020	433.11	483.31	2.2361	0.02326	424.33	470.84	2.0731
20	0.05236	440.17	492.53	2.2681	0.02453	432.27	481.32	2.1095
30	0.05449	447.25	501.74	2.2990	0.02575	440.07	491.57	2.1438
40	0.05660	454.36	510.96	2.3289	0.02693	447.78	501.65	2.1766
50	0.05868	461.52	520.19	2.3579	0.02809	455.45	511.63	2.2079
60	0.06074	468.73	529.47	2.3862	0.02922	463.09	521.54	2.2381
70	0.06279	475.99	538.78	2.4137	0.03034	470.74	531.41	2.2673
80	0.06482	483.32	548.14	2.4406	0.03143	478.40	541.26	2.2956
90	0.06684	490.71	557.55	2.4669	0.03252	486.08	551.11	2.3231
100	0.06885	498.17	567.01	2.4926	0.03359	493.80	560.97	2.3499
110	0.07085	505.69	576.54	2.5178	0.03465	501.56	570.85	2.3760
120	0.07284	513.28	586.12	2.5425	0.03570	509.36	580.76	2.4015
130	0.07482	520.94	595.76	2.5667	0.03674	517.22	590.69	2.4265
140	0.07680	528.67	605.47	2.5905	0.03777	525.12	600.66	2.4509
150	0.07877	536.47	615.24	2.6138	0.03880	533.07	610.68	2.4749
160	0.08074	544.33	625.07	2.6368	0.03983	541.08	620.73	2.4984
170	0.08270	552.26	634.96	2.6594	0.04084	549.15	630.84	2.5214
180	0.08465	560.26	644.92	2.6816	0.04186	557.27	640.98	2.5441
190	0.08661	568.33	654.94	2.7035	0.04286	565.45	651.18	2.5663
200	0.08856	576.46	665.02	2.7250	0.04387	573.69	661.43	2.5882
210	0.09050	584.66	675.16	2.7462	0.04487	581.99	671.73	2.6098
220	0.09244	592.91	685.36	2.7671	0.04587	590.34	682.07	2.6310
230	0.09438	601.24	695.62	2.7877	0.04686	598.75	692.47	2.6518
240	0.09632	609.62	705.94	2.8080	0.04786	607.21	702.92	2.6724
250	0.09826	618.07	716.32	2.8281	0.04885	615.73	713.43	2.6927
260	0.10019	626.57	726.76	2.8478	0.04984	624.31	723.98	2.7127
270	0.10212	635.14	737.26	2.8673	0.05082	632.94	734.58	2.7324
280	0.10405	643.76	747.81	2.8866	0.05180	641.63	745.24	2.7518
290	0.10597	652.44	758.42	2.9056	0.05279	650.37	755.94	2.7710
300	0.10790	661.18	769.08	2.9244	0.05377	659.17	766.70	2.7899
310	0.10982	669.97	779.80	2.9429	0.05475	668.01	777.50	2.8086
320	0.11174	678.82	790.57	2.9612	0.05572	676.91	788.36	2.8271
330	0.11366	687.73	801.39	2.9793	0.05670	685.86	799.26	2.8453
340	0.11558	696.68	812.26	2.9972	0.05767	694.87	810.21	2.8633
350	0.11750	705.69	823.19	3.0149	0.05865	703.92	821.21	2.8811

Appendix M.2 - Superheated Carbon Dioxide (CO_2)

T(°C)	P = 5000 kPa (14.284 °C)				P = 10000 kPa			
	v m³/kg	u kJ/kg	h kJ/kg	s kJ/kg-K	v m³/kg	u kJ/kg	h kJ/kg	s kJ/kg-K
sat	0.00638	385.74	417.66	1.7544	-	-	-	-
20	0.00711	396.83	432.38	1.8051	0.00117	231.02	242.70	1.1251
30	0.00806	411.13	451.44	1.8691	0.00130	258.65	271.62	1.2220
40	0.00885	422.90	467.13	1.9200	0.00159	297.13	313.04	1.3563
50	0.00954	433.46	481.15	1.9641	0.00260	358.05	384.07	1.5795
60	0.01017	443.33	494.19	2.0039	0.00345	390.53	425.02	1.7045
70	0.01077	452.75	506.59	2.0405	0.00404	410.29	450.65	1.7804
80	0.01133	461.87	518.54	2.0749	0.00451	425.72	470.85	1.8384
90	0.01188	470.78	530.16	2.1073	0.00493	439.03	488.31	1.8872
100	0.01240	479.56	541.55	2.1383	0.00530	451.11	504.14	1.9302
110	0.01291	488.23	552.76	2.1679	0.00565	462.38	518.88	1.9692
120	0.01340	496.83	563.84	2.1965	0.00598	473.09	532.86	2.0052
130	0.01389	505.39	574.82	2.2240	0.00629	483.41	546.28	2.0389
140	0.01436	513.92	585.73	2.2508	0.00658	493.44	559.28	2.0708
150	0.01483	522.44	596.59	2.2767	0.00687	503.25	571.95	2.1011
160	0.01529	530.96	607.41	2.3020	0.00715	512.90	584.37	2.1301
170	0.01575	539.49	618.22	2.3267	0.00742	522.42	596.58	2.1580
180	0.01619	548.04	629.01	2.3508	0.00768	531.85	608.64	2.1849
190	0.01664	556.60	639.79	2.3743	0.00794	541.20	620.56	2.2109
200	0.01708	565.20	650.58	2.3974	0.00819	550.51	632.39	2.2362
210	0.01751	573.82	661.38	2.4199	0.00844	559.78	644.13	2.2607
220	0.01795	582.47	672.20	2.4421	0.00868	569.02	655.82	2.2847
230	0.01838	591.16	683.04	2.4639	0.00892	578.26	667.45	2.3080
240	0.01880	599.89	693.90	2.4852	0.00916	587.48	679.04	2.3308
250	0.01922	608.66	704.78	2.5062	0.00939	596.71	690.61	2.3531
260	0.01965	617.47	715.69	2.5269	0.00962	605.94	702.16	2.3750
270	0.02006	626.31	726.63	2.5472	0.00985	615.18	713.69	2.3964
280	0.02048	635.20	737.61	2.5672	0.01008	624.44	725.22	2.4175
290	0.02090	644.13	748.61	2.5870	0.01030	633.71	736.75	2.4381
300	0.02131	653.11	759.65	2.6064	0.01053	643.01	748.28	2.4584
310	0.02172	662.12	770.73	2.6256	0.01075	652.33	759.81	2.4784
320	0.02213	671.19	781.84	2.6444	0.01097	661.67	771.36	2.4980
330	0.02254	680.29	792.99	2.6631	0.01119	671.05	782.92	2.5173
340	0.02295	689.44	804.17	2.6815	0.01140	680.45	794.49	2.5364
350	0.02335	698.62	815.39	2.6996	0.01162	689.88	806.08	2.5551

adapted from NIST data The reference state for all carbon dioxide data is h = 200 kJ/kg and s = 1 kJ/kg-K for the saturated liquid at 0°C. This choice is common and results in h > 0 for all commonly encountered states.

Appendix N.1 - Saturated Methane (CH_4)

T °C	P kPa	v^L m³/kg	v^V m³/kg	u^L kJ/kg	u^V kJ/kg	h^L kJ/kg	h^V kJ/kg	s^L kJ/kg-K	s^V kJ/kg-K
-180	15.90	0.002231	3.007100	-63.57	429.41	-63.53	477.22	-0.6199	5.1853
-175	28.24	0.002265	1.775500	-46.64	436.63	-46.58	486.76	-0.4429	4.9910
-170	47.23	0.002301	1.108100	-29.59	443.65	-29.49	495.99	-0.2735	4.8208
-165	75.09	0.002340	0.724660	-12.41	450.44	-12.24	504.85	-0.1108	4.6704
-160	114.29	0.002380	0.492910	4.92	456.94	5.19	513.28	0.0459	4.5363
-155	167.57	0.002423	0.346650	22.42	463.13	22.82	521.22	0.1973	4.4156
-150	237.84	0.002469	0.250770	40.11	468.95	40.69	528.60	0.3440	4.3058
-145	328.17	0.002519	0.185830	58.01	474.36	58.84	535.34	0.4866	4.2050
-140	441.77	0.002572	0.140540	76.16	479.30	77.30	541.38	0.6257	4.1111
-135	581.92	0.002631	0.108130	94.60	483.70	96.13	546.63	0.7618	4.0228
-130	752.01	0.002695	0.084400	113.37	487.50	115.39	550.97	0.8956	3.9384
-125	955.50	0.002766	0.066665	132.53	490.59	135.17	554.29	1.0276	3.8566
-120	1195.90	0.002846	0.053154	152.17	492.86	155.58	556.43	1.1585	3.7759
-115	1477.00	0.002936	0.042678	172.40	494.15	176.74	557.19	1.2893	3.6949
-110	1802.60	0.003042	0.034416	193.37	494.25	198.85	556.28	1.4209	3.6117
-105	2176.70	0.003167	0.027786	215.31	492.81	222.21	553.29	1.5548	3.5238
-100	2604.00	0.003323	0.022363	238.60	489.32	247.25	547.56	1.6935	3.4278
-95	3089.50	0.003526	0.017814	263.91	482.82	274.80	537.86	1.8408	3.3174
-90	3639.90	0.003822	0.013837	292.80	471.16	306.71	521.52	2.0062	3.1791
-85	4264.80	0.004396	0.00993	331.00	446.44	349.75	488.81	2.2242	2.9632
-82.586	4599.20	0.006148	0.00615	387.32	387.32	415.59	415.59	2.5624	2.5624

Appendix N.2 - Superheated Methane (CH$_4$)

	P = 100 kPa (-161.64 °C)				P = 200 kPa (-152.53 °C)			
	v	u	h	s	v	u	h	s
T(°C)	m³/kg	kJ/kg	kJ/kg	kJ/kg-K	m³/kg	kJ/kg	kJ/kg	kJ/kg-K
sat	0.5572	454.84	510.56	4.5787	0.2944	466.06	524.94	4.3601
-160	0.5664	457.55	514.19	4.6111	-	-	-	-
-150	0.6215	473.87	536.02	4.7959	0.3018	470.37	530.72	4.4075
-140	0.6757	489.95	557.52	4.9638	0.3303	487.08	553.14	4.5826
-130	0.7294	505.90	578.84	5.1182	0.3583	503.46	575.11	4.7417
-120	0.7828	521.76	600.03	5.2613	0.3858	519.64	596.79	4.8881
-110	0.8358	537.57	621.15	5.3949	0.4130	535.69	618.29	5.0241
-100	0.8887	553.35	642.22	5.5202	0.4400	551.67	639.66	5.1513
-90	0.9414	569.12	663.26	5.6384	0.4668	567.60	660.96	5.2708
-80	0.9940	584.90	684.30	5.7502	0.4935	583.51	682.21	5.3838
-70	1.0465	600.70	705.36	5.8565	0.5201	599.43	703.44	5.4910
-60	1.0989	616.55	726.44	5.9578	0.5466	615.37	724.69	5.5931
-50	1.1513	632.46	747.59	6.0548	0.5730	631.36	745.97	5.6906
-40	1.2036	648.45	768.80	6.1478	0.5994	647.42	767.30	5.7842
-30	1.2558	664.54	790.12	6.2373	0.6257	663.58	788.72	5.8741
-20	1.3080	680.74	811.55	6.3237	0.6520	679.84	810.24	5.9608
-10	1.3602	697.09	833.11	6.4072	0.6783	696.24	831.89	6.0447
0	1.4123	713.60	854.83	6.4882	0.7045	712.79	853.69	6.1260
10	1.4644	730.29	876.74	6.5670	0.7307	729.53	875.66	6.2050
20	1.5165	747.18	898.84	6.6437	0.7569	746.46	897.83	6.2820
30	1.5686	764.29	921.15	6.7185	0.7830	763.60	920.20	6.3570
40	1.6206	781.64	943.70	6.7917	0.8092	780.98	942.81	6.4304
50	1.6727	799.24	966.51	6.8634	0.8353	798.61	965.66	6.5022
60	1.7247	817.11	989.58	6.9337	0.8614	816.50	988.78	6.5726
70	1.7767	835.26	1012.93	7.0028	0.8875	834.68	1012.17	6.6418
80	1.8287	853.71	1036.58	7.0707	0.9135	853.15	1035.86	6.7099
90	1.8806	872.46	1060.53	7.1376	0.9396	871.93	1059.85	6.7768
100	1.9326	891.54	1084.80	7.2035	0.9657	891.02	1084.15	6.8429
110	1.9846	910.94	1109.40	7.2685	0.9917	910.44	1108.78	6.9080
120	2.0365	930.68	1134.33	7.3328	1.0177	930.19	1133.74	6.9723
130	2.0885	950.76	1159.61	7.3963	1.0438	950.29	1159.05	7.0359
140	2.1404	971.19	1185.23	7.4591	1.0698	970.74	1184.70	7.0987
150	2.1924	991.98	1211.21	7.5212	1.0958	991.54	1210.71	7.1609
160	2.2443	1013.13	1237.55	7.5827	1.1218	1012.70	1237.07	7.2225
170	2.2962	1034.64	1264.26	7.6437	1.1478	1034.23	1263.80	7.2835
180	2.3481	1056.52	1291.33	7.7041	1.1738	1056.12	1290.89	7.3439
190	2.4000	1078.77	1318.77	7.7640	1.1998	1078.38	1318.35	7.4039
200	2.4519	1101.39	1346.58	7.8234	1.2258	1101.01	1346.18	7.4633
210	2.5038	1124.38	1374.77	7.8823	1.2518	1124.02	1374.38	7.5223
220	2.5557	1147.75	1403.33	7.9408	1.2778	1147.40	1402.96	7.5809
230	2.6076	1171.50	1432.26	7.9989	1.3038	1171.15	1431.91	7.6390
240	2.6595	1195.62	1461.58	8.0566	1.3298	1195.28	1461.24	7.6967
250	2.7114	1220.12	1491.26	8.1139	1.3558	1219.79	1490.94	7.7540
260	2.7633	1244.99	1521.33	8.1708	1.3817	1244.67	1521.02	7.8110
270	2.8152	1270.24	1551.76	8.2274	1.4077	1269.93	1551.47	7.8676
280	2.8671	1295.87	1582.57	8.2836	1.4337	1295.56	1582.30	7.9238
290	2.9190	1321.86	1613.76	8.3395	1.4596	1321.56	1613.49	7.9797
300	2.9708	1348.23	1645.31	8.3950	1.4856	1347.94	1645.06	8.0353

Appendix N - Properties of Methane

	P = 500 kPa (-137.80 °C)				P = 1000 kPa (-124.01 °C)			
	v	u	h	s	v	u	h	s
T(°C)	m³/kg	kJ/kg	kJ/kg	kJ/kg-K	m³/kg	kJ/kg	kJ/kg	kJ/kg-K
sat	0.12497	481.31	543.79	4.0716	0.06370	491.11	554.81	3.8406
-130	0.13493	495.41	562.88	4.2088	-	-	-	-
-120	0.14715	512.81	586.39	4.3675	0.06673	499.32	566.04	3.9149
-110	0.15897	529.76	609.24	4.5121	0.07376	518.54	592.30	4.0811
-100	0.17052	546.42	631.68	4.6456	0.08035	536.77	617.12	4.2287
-90	0.18186	562.89	653.82	4.7699	0.08665	554.41	641.06	4.3632
-80	0.19306	579.24	675.76	4.8866	0.09275	571.66	664.42	4.4873
-70	0.20413	595.52	697.58	4.9967	0.09871	588.67	687.37	4.6032
-60	0.21511	611.77	719.32	5.1012	0.10455	605.51	710.06	4.7123
-50	0.22602	628.02	741.03	5.2007	0.11030	622.26	732.57	4.8154
-40	0.23686	644.31	762.74	5.2959	0.11599	638.97	754.96	4.9136
-30	0.24766	660.66	784.49	5.3872	0.12161	655.69	777.30	5.0074
-20	0.25840	677.10	806.31	5.4751	0.12719	672.45	799.64	5.0975
-10	0.26912	693.66	828.22	5.5600	0.13273	689.29	822.02	5.1842
0	0.27980	710.36	850.25	5.6422	0.13823	706.24	844.46	5.2679
10	0.29045	727.22	872.44	5.7220	0.14370	723.32	867.02	5.3490
20	0.30107	744.26	894.80	5.7996	0.14915	740.56	889.71	5.4278
30	0.31168	761.51	917.35	5.8752	0.15458	757.99	912.57	5.5044
40	0.32227	778.98	940.12	5.9491	0.15998	775.63	935.62	5.5792
50	0.33284	796.70	963.12	6.0214	0.16537	793.50	958.87	5.6523
60	0.34339	814.68	986.37	6.0923	0.17074	811.62	982.36	5.7239
70	0.35393	832.93	1009.90	6.1618	0.17610	830.00	1006.10	5.7941
80	0.36446	851.47	1033.70	6.2302	0.18145	848.66	1030.11	5.8631
90	0.37498	870.31	1057.80	6.2975	0.18679	867.61	1054.39	5.9309
100	0.38549	889.46	1082.20	6.3638	0.19211	886.86	1078.97	5.9976
110	0.39599	908.94	1106.90	6.4292	0.19743	906.43	1103.86	6.0635
120	0.40648	928.75	1132.00	6.4938	0.20274	926.33	1129.07	6.1284
130	0.41696	948.89	1157.40	6.5575	0.20804	946.56	1154.60	6.1925
140	0.42744	969.39	1183.10	6.6206	0.21333	967.13	1180.46	6.2559
150	0.43791	990.23	1209.20	6.6829	0.21862	988.05	1206.67	6.3186
160	0.44838	1011.40	1235.60	6.7447	0.22390	1009.32	1233.22	6.3806
170	0.45884	1033.00	1262.40	6.8058	0.22918	1030.94	1260.13	6.4420
180	0.46929	1054.90	1289.60	6.8664	0.23445	1052.93	1287.39	6.5028
190	0.47975	1077.20	1317.10	6.9265	0.23972	1075.29	1315.01	6.5631
200	0.49019	1099.90	1345.00	6.9861	0.24498	1098.01	1342.99	6.6229
210	0.50063	1122.90	1373.20	7.0452	0.25024	1121.09	1371.34	6.6822
220	0.51107	1146.30	1401.90	7.1038	0.25550	1144.55	1400.05	6.7410
230	0.52151	1170.10	1430.90	7.1620	0.26075	1168.38	1429.14	6.7994
240	0.53194	1194.30	1460.20	7.2198	0.26600	1192.59	1458.59	6.8573
250	0.54237	1218.80	1490.00	7.2773	0.27125	1217.16	1488.41	6.9149
260	0.55280	1243.70	1520.10	7.3343	0.27649	1242.11	1518.60	6.9721
270	0.56322	1269.00	1550.60	7.3909	0.28173	1267.42	1549.16	7.0288
280	0.57364	1294.60	1581.50	7.4473	0.28697	1293.11	1580.09	7.0853
290	0.58406	1320.70	1612.70	7.5032	0.29221	1319.17	1611.38	7.1413
300	0.59448	1347.10	1644.30	7.5588	0.29744	1345.60	1643.04	7.1971

Appendix N - Properties of Methane

	P = 2000 kPa (-107.28 °C)				P = 4000 kPa (-87.04 °C)			
T(°C)	v m³/kg	u kJ/kg	h kJ/kg	s kJ/kg-K	v m³/kg	u kJ/kg	h kJ/kg	s kJ/kg-K
sat	0.03063	493.68	554.95	3.5646	0.01158	459.41	505.74	3.0689
-100	0.03423	512.21	580.67	3.7165	-	-	-	-
-90	0.03844	534.21	611.09	3.8874	-	-	-	-
-80	0.04222	554.36	638.80	4.0347	0.01541	501.17	562.80	3.3708
-70	0.04575	573.46	664.95	4.1668	0.01850	533.11	607.13	3.5949
-60	0.04911	591.91	690.12	4.2877	0.02096	558.48	642.30	3.7640
-50	0.05234	609.93	714.62	4.4000	0.02311	581.08	673.51	3.9072
-40	0.05548	627.69	738.65	4.5054	0.02509	602.15	702.49	4.0342
-30	0.05855	645.28	762.38	4.6051	0.02694	622.29	730.06	4.1500
-20	0.06156	662.78	785.91	4.6999	0.02871	641.82	756.68	4.2573
-10	0.06453	680.26	809.31	4.7906	0.03042	660.97	782.65	4.3579
0	0.06745	697.77	832.66	4.8777	0.03208	679.88	808.18	4.4531
10	0.07034	715.34	856.02	4.9616	0.03369	698.65	833.41	4.5439
20	0.07320	733.03	879.43	5.0429	0.03527	717.37	858.45	4.6308
30	0.07604	750.85	902.93	5.1217	0.03682	736.11	883.39	4.7144
40	0.07886	768.84	926.57	5.1984	0.03835	754.91	908.31	4.7953
50	0.08166	787.03	950.35	5.2732	0.03986	773.82	933.25	4.8737
60	0.08444	805.44	974.33	5.3463	0.04135	792.88	958.26	4.9499
70	0.08721	824.09	998.52	5.4178	0.04282	812.11	983.39	5.0242
80	0.08997	843.00	1022.93	5.4879	0.04428	831.55	1008.67	5.0969
90	0.09271	862.17	1047.60	5.5568	0.04573	851.21	1034.13	5.1680
100	0.09545	881.64	1072.54	5.6245	0.04717	871.12	1059.80	5.2377
110	0.09817	901.41	1097.75	5.6912	0.04860	891.30	1085.69	5.3062
120	0.10089	921.48	1123.26	5.7570	0.05002	911.76	1111.83	5.3735
130	0.10360	941.88	1149.08	5.8218	0.05143	932.52	1138.23	5.4398
140	0.10630	962.61	1175.22	5.8858	0.05284	953.57	1164.91	5.5052
150	0.10900	983.68	1201.68	5.9491	0.05423	974.95	1191.88	5.5697
160	0.11169	1005.09	1228.47	6.0117	0.05563	996.65	1219.15	5.6334
170	0.11437	1026.85	1255.60	6.0736	0.05701	1018.68	1246.73	5.6963
180	0.11705	1048.96	1283.07	6.1349	0.05840	1041.04	1274.63	5.7586
190	0.11973	1071.43	1310.89	6.1956	0.05977	1063.75	1302.85	5.8202
200	0.12240	1094.26	1339.06	6.2558	0.06115	1086.81	1331.40	5.8812
210	0.12507	1117.45	1367.59	6.3155	0.06252	1110.21	1360.29	5.9416
220	0.12773	1141.01	1396.48	6.3747	0.06389	1133.97	1389.51	6.0014
230	0.13039	1164.94	1425.72	6.4334	0.06525	1158.09	1419.08	6.0608
240	0.13305	1189.23	1455.33	6.4916	0.06661	1182.56	1448.99	6.1197
250	0.13570	1213.89	1485.29	6.5495	0.06797	1207.39	1479.25	6.1781
260	0.13836	1238.91	1515.63	6.6069	0.06932	1232.58	1509.86	6.2360
270	0.14100	1264.31	1546.32	6.6639	0.07067	1258.13	1540.82	6.2935
280	0.14365	1290.07	1577.37	6.7206	0.07202	1284.04	1572.13	6.3507
290	0.14630	1316.20	1608.79	6.7769	0.07337	1310.31	1603.79	6.4074
300	0.14894	1342.70	1640.57	6.8328	0.07471	1336.95	1635.80	6.4637

adapted from NIST data — The reference state for all methane properties is h = 0 and s = 0 for the saturated liquid at its normal boiling point which is -161.48°C.

Appendix O.1 - Saturated Hydrogen (H_2)

T °C	P kPa	v^L m³/kg	v^V m³/kg	u^L kJ/kg	u^V kJ/kg	h^L kJ/kg	h^V kJ/kg	s^L kJ/kg-K	s^V kJ/kg-K
-259	8.21	0.013013	6.988400	-52.66	344.23	-52.56	401.61	-2.9757	29.1210
-258	13.90	0.013156	4.384300	-45.46	349.65	-45.27	410.60	-2.4835	27.6070
-257	22.19	0.013310	2.899900	-37.89	354.82	-37.60	419.17	-2.0001	26.2830
-256	33.72	0.013477	2.002400	-29.94	359.69	-29.49	427.22	-1.5222	25.1080
-255	49.20	0.013657	1.432300	-21.58	364.20	-20.91	434.67	-1.0478	24.0530
-254	69.36	0.013852	1.054900	-12.78	368.29	-11.82	441.46	-0.5755	23.0940
-253	94.93	0.014065	0.796100	-3.52	371.92	-2.19	447.50	-0.1035	22.2130
-252	126.69	0.014299	0.613130	6.23	375.03	8.04	452.71	0.3699	21.3940
-251	165.41	0.014555	0.480320	16.51	377.56	18.92	457.01	0.8465	20.6250
-250	211.88	0.014839	0.381660	27.37	379.44	30.51	460.31	1.3283	19.8940
-249	266.90	0.015156	0.306840	38.87	380.60	42.91	462.49	1.8176	19.1920
-248	331.28	0.015512	0.249030	51.08	380.92	56.22	463.42	2.3174	18.5080
-247	405.85	0.015918	0.203590	64.11	380.29	70.57	462.92	2.8312	17.8350
-246	491.47	0.016387	0.167290	78.10	378.52	86.15	460.74	3.3638	17.1610
-245	589.02	0.016942	0.137810	93.24	375.35	103.22	456.52	3.9223	16.4730
-244	699.48	0.017618	0.113440	109.86	370.36	122.19	449.70	4.5175	15.7530
-243	823.93	0.018481	0.092852	128.51	362.85	143.74	439.35	5.1684	14.9730
-242	963.64	0.019673	0.074887	150.30	351.40	169.26	423.57	5.9139	14.0780
-241	1120.30	0.021622	0.058079	178.37	332.13	202.59	397.20	6.8644	12.9170
-240.01	1296.40	0.031988	0.031988 critical point						

Appendix O.2 - Superheated Hydrogen (H_2)

T(°C)	P = 100 kPa (-252.83 °C)				P = 200 kPa (-250.24 °C)			
	v m³/kg	u kJ/kg	h kJ/kg	s kJ/kg-K	v m³/kg	u kJ/kg	h kJ/kg	s kJ/kg-K
sat	0.7596	372.50	448.47	22.0660	0.4028	379.06	459.62	20.0660
-250	0.8919	392.14	481.33	23.5810	0.4092	380.97	462.81	20.2050
-240	1.3325	457.52	590.77	27.5170	0.6481	451.57	581.19	24.4750
-230	1.7579	521.04	696.83	30.3140	0.8679	516.92	690.50	27.3600
-220	2.1778	584.00	801.78	32.5020	1.0817	580.88	797.22	29.5840
-210	2.5951	647.10	906.60	34.3090	1.2928	644.60	903.15	31.4110
-200	3.0108	711.05	1012.10	35.8600	1.5023	708.98	1009.40	32.9730
-190	3.4257	776.60	1119.20	37.2320	1.7110	774.83	1117.00	34.3510
-180	3.8401	844.35	1228.40	38.4710	1.9191	842.82	1226.60	35.5960
-170	4.2540	914.74	1340.10	39.6110	2.1268	913.39	1338.70	36.7390
-160	4.6676	987.99	1454.70	40.6710	2.3342	986.79	1453.60	37.8020
-150	5.0810	1064.20	1572.30	41.6660	2.5414	1063.09	1571.40	38.7990
-140	5.4942	1143.20	1692.60	42.6060	2.7484	1142.23	1691.90	39.7400
-130	5.9074	1225.00	1815.70	43.4970	2.9553	1224.07	1815.10	40.6320
-120	6.3204	1309.20	1941.30	44.3450	3.1621	1308.44	1940.90	41.4810
-110	6.7333	1395.90	2069.20	45.1540	3.3688	1395.12	2068.90	42.2900
-100	7.1462	1484.60	2199.20	45.9270	3.5754	1483.91	2199.00	43.0640
-90	7.5590	1575.20	2331.10	46.6680	3.7820	1574.59	2331.00	43.8050
-80	7.9718	1667.50	2464.70	47.3780	3.9886	1666.96	2464.70	44.5160
-70	8.3845	1761.40	2599.80	48.0600	4.1951	1760.84	2599.90	45.1980
-60	8.7972	1856.60	2736.30	48.7160	4.4016	1856.07	2736.40	45.8540
-50	9.2099	1952.90	2873.90	49.3470	4.6080	1952.47	2874.10	46.4860
-40	9.6225	2050.40	3012.60	49.9550	4.8144	2049.92	3012.80	47.0940
-30	10.0350	2148.70	3152.20	50.5410	5.0208	2148.29	3152.50	47.6800
-20	10.4480	2247.80	3292.60	51.1070	5.2272	2247.45	3292.90	48.2460
-10	10.8600	2347.70	3433.70	51.6530	5.4336	2347.31	3434.00	48.7930
0	11.2730	2448.10	3575.40	52.1820	5.6399	2447.78	3575.80	49.3220
10	11.6860	2549.10	3717.60	52.6930	5.8463	2548.76	3718.00	49.8330
20	12.0980	2650.50	3860.30	53.1880	6.0526	2650.20	3860.70	50.3290
30	12.5110	2752.30	4003.40	53.6680	6.2589	2752.02	4003.80	50.8080
40	12.9230	2854.40	4146.80	54.1340	6.4653	2854.18	4147.20	51.2740
50	13.3360	2956.90	4290.40	54.5850	6.6716	2956.61	4290.90	51.7260
60	13.7480	3059.50	4434.30	55.0240	6.8779	3059.28	4434.90	52.1640
70	14.1610	3162.40	4578.40	55.4500	7.0842	3162.15	4579.00	52.5910
80	14.5730	3265.40	4722.70	55.8640	7.2904	3265.19	4723.30	53.0050
90	14.9860	3368.60	4867.10	56.2680	7.4967	3368.37	4867.70	53.4080
100	15.3980	3471.90	5011.70	56.6600	7.7030	3471.67	5012.30	53.8010

Appendix O - Properties of Hydrogen

	P = 500 kPa (-245.91 °C)				P = 1000 kPa (-241.76 °C)			
	v	u	h	s	v	u	h	s
T(°C)	m³/kg	kJ/kg	kJ/kg	kJ/kg-K	m³/kg	kJ/kg	kJ/kg	kJ/kg-K
sat	0.16430	378.29	460.44	17.0980	0.07078	347.71	418.49	13.8310
-240	0.23542	431.54	549.24	20.0650	0.09122	383.27	474.50	15.5730
-230	0.33349	504.03	670.77	23.2780	0.15476	480.43	635.19	19.8590
-220	0.42396	571.31	783.30	25.6250	0.20471	554.71	759.42	22.4530
-210	0.51141	637.01	892.72	27.5120	0.25106	624.12	875.18	24.4500
-200	0.59728	702.72	1001.36	29.1090	0.29572	692.18	987.90	26.1070
-190	0.68223	769.52	1110.63	30.5090	0.33941	760.63	1100.05	27.5440
-180	0.76658	838.22	1221.51	31.7680	0.38250	830.56	1213.06	28.8270
-170	0.85052	909.35	1334.61	32.9210	0.42519	902.63	1327.82	29.9970
-160	0.93418	983.20	1450.29	33.9910	0.46758	977.23	1444.82	31.0800
-150	1.01760	1059.87	1568.68	34.9930	0.50977	1054.52	1564.29	32.0920
-140	1.10090	1139.32	1689.77	35.9390	0.55179	1134.49	1686.28	33.0440
-130	1.18410	1221.43	1813.45	36.8340	0.59369	1217.03	1810.73	33.9450
-120	1.26710	1306.02	1939.57	37.6860	0.63550	1302.01	1937.50	34.8010
-110	1.35010	1392.90	2067.94	38.4980	0.67722	1389.22	2066.43	35.6160
-100	1.43300	1481.86	2198.35	39.2730	0.71888	1478.46	2197.34	36.3950
-90	1.51580	1572.69	2330.62	40.0160	0.76048	1569.55	2330.03	37.1400
-80	1.59870	1665.20	2464.53	40.7280	0.80205	1662.29	2464.34	37.8540
-70	1.68140	1759.21	2599.93	41.4110	0.84357	1756.50	2600.07	38.5390
-60	1.76420	1854.54	2736.63	42.0680	0.88506	1852.02	2737.09	39.1970
-50	1.84690	1951.05	2874.50	42.7000	0.92653	1948.70	2875.23	39.8310
-40	1.92960	2048.59	3013.38	43.3090	0.96797	2046.40	3014.37	40.4410
-30	2.01220	2147.04	3153.17	43.8960	1.00940	2144.99	3154.38	41.0290
-20	2.09490	2246.29	3293.73	44.4630	1.05080	2244.36	3295.15	41.5960
-10	2.17750	2346.22	3434.98	45.0100	1.09220	2344.41	3436.59	42.1440
0	2.26010	2446.75	3576.83	45.5390	1.13360	2445.05	3578.61	42.6740
10	2.34280	2547.80	3719.18	46.0510	1.17490	2546.21	3721.13	43.1860
20	2.42540	2649.30	3861.97	46.5460	1.21630	2647.80	3864.07	43.6820
30	2.50790	2751.17	4005.15	47.0260	1.25760	2749.77	4007.38	44.1630
40	2.59050	2853.38	4148.64	47.4920	1.29890	2852.05	4151.00	44.6290
50	2.67310	2955.86	4292.40	47.9440	1.34030	2954.61	4294.88	45.0810
60	2.75570	3058.57	4436.40	48.3830	1.38160	3057.40	4438.99	45.5210
70	2.83820	3161.49	4580.59	48.8090	1.42290	3160.39	4583.28	45.9470
80	2.92080	3264.57	4724.95	49.2240	1.46420	3263.53	4727.73	46.3620
90	3.00330	3367.78	4869.44	49.6270	1.50550	3366.81	4872.31	46.7660
100	3.08580	3471.11	5014.04	50.0200	1.54680	3470.20	5017.00	47.1590

Appendix O - Properties of Hydrogen

	P = 2000 kPa				P = 4000 kPa (-87.04 °C)			
	v	u	h	s	v	u	h	s
T(°C)	m³/kg	kJ/kg	kJ/kg	kJ/kg-K	m³/kg	kJ/kg	kJ/kg	kJ/kg-K
-250	0.01424	16.80	45.29	0.8448	0.01424	16.80	45.28	0.8448
-240	0.01911	156.13	194.34	6.0695	0.01911	156.13	194.34	6.0695
-230	0.06455	422.10	551.19	15.6060	0.06455	422.10	551.19	15.6060
-220	0.09538	519.03	709.78	18.9310	0.09538	519.03	709.78	18.9310
-210	0.12126	597.53	840.05	21.1810	0.12126	597.53	840.05	21.1810
-200	0.14527	670.84	961.38	22.9650	0.14527	670.84	961.38	22.9650
-190	0.16829	742.80	1079.40	24.4770	0.16829	742.80	1079.37	24.4770
-180	0.19070	815.26	1196.70	25.8090	0.19070	815.26	1196.66	25.8090
-170	0.21272	889.27	1314.70	27.0130	0.21272	889.27	1314.70	27.0130
-160	0.23445	965.40	1434.30	28.1190	0.23445	965.40	1434.30	28.1190
-150	0.25598	1043.90	1555.90	29.1490	0.25598	1043.93	1555.89	29.1490
-140	0.27736	1124.90	1679.60	30.1150	0.27736	1124.93	1679.64	30.1150
-130	0.29861	1208.30	1805.60	31.0270	0.29861	1208.35	1805.57	31.0270
-120	0.31977	1294.10	1933.60	31.8910	0.31977	1294.07	1933.62	31.8910
-110	0.34086	1381.90	2063.70	32.7140	0.34086	1381.93	2063.65	32.7140
-100	0.36188	1471.80	2195.50	33.4980	0.36188	1471.75	2195.52	33.4980
-90	0.38286	1563.30	2329.10	34.2480	0.38286	1563.35	2329.06	34.2480
-80	0.40379	1656.50	2464.10	34.9660	0.40379	1656.54	2464.11	34.9660
-70	0.42468	1751.20	2600.50	35.6540	0.42468	1751.15	2600.52	35.6540
-60	0.44554	1847.00	2738.10	36.3160	0.44554	1847.04	2738.13	36.3160
-50	0.46638	1944.10	2876.80	36.9510	0.46638	1944.05	2876.81	36.9510
-40	0.48720	2042.10	3016.40	37.5630	0.48720	2042.05	3016.44	37.5630
-30	0.50799	2140.90	3156.90	38.1530	0.50799	2140.92	3156.90	38.1530
-20	0.52877	2240.50	3298.10	38.7220	0.52877	2240.55	3298.08	38.7220
-10	0.54953	2340.80	3439.90	39.2720	0.54953	2340.84	3439.90	39.2720
0	0.57028	2441.70	3582.30	39.8030	0.57028	2441.70	3582.26	39.8030
10	0.59101	2543.10	3725.10	40.3160	0.59101	2543.06	3725.09	40.3160
20	0.61174	2644.80	3868.30	40.8130	0.61174	2644.84	3868.32	40.8130
30	0.63246	2747.00	4011.90	41.2950	0.63246	2746.98	4011.90	41.2950
40	0.65317	2849.40	4155.80	41.7620	0.65317	2849.44	4155.77	41.7620
50	0.67387	2952.20	4299.90	42.2150	0.67387	2952.15	4299.89	42.2150
60	0.69456	3055.10	4444.20	42.6550	0.69456	3055.09	4444.21	42.6550
70	0.71525	3158.20	4588.70	43.0820	0.71525	3158.21	4588.71	43.0820
80	0.73593	3261.50	4733.30	43.4980	0.73593	3261.48	4733.34	43.4980
90	0.75661	3364.90	4878.10	43.9020	0.75661	3364.88	4878.10	43.9020
100	0.77728	3468.40	5023.00	44.2950	0.77728	3468.39	5022.95	44.2950

adapted from NIST data   The reference state for all hydrogen properties is h = 0 and s = 0 for the saturated liquid at its normal boiling point which is -252.78°C for hydrogen.

Thermodynamics for Chemical Engineering Thermodynamics **Appendix P - Properties of R-11**

Appendix P.1 - Saturated R-11 Temperature Increment - SI Units

T	P	v^L	v^V	h^L	h^V	s^L	s^V
°C	kPa	m³/kg	m³/kg	kJ/kg	kJ/kg	kJ/kg-K	kJ/kg-K
-110	0.01	0.000566	1419.2	109.81	337.07	0.5795	1.9724
-100	0.03	0.000572	402.55	117.55	341.35	0.6255	1.9181
-90	0.08	0.000579	133.79	125.42	345.77	0.6697	1.8728
-80	0.23	0.000586	50.774	133.41	350.31	0.7122	1.8351
-70	0.57	0.000593	21.547	141.50	354.96	0.7530	1.8037
-60	1.28	0.000601	10.054	149.66	359.71	0.7922	1.7777
-50	2.65	0.000609	5.0875	157.89	364.55	0.8299	1.7561
-40	5.09	0.000617	2.7595	166.18	369.48	0.8663	1.7382
-30	9.19	0.000625	1.5892	174.54	374.48	0.9014	1.7236
-20	15.73	0.000634	0.96383	182.96	379.53	0.9353	1.7118
-10	25.67	0.000642	0.61140	191.44	384.64	0.9681	1.7023
0	40.20	0.000652	0.40328	200.00	389.77	1.0000	1.6947
10	60.68	0.000662	0.27518	208.64	394.92	1.0310	1.6889
20	88.68	0.000672	0.19342	217.36	400.07	1.0612	1.6845
30	125.96	0.000683	0.13950	226.18	405.20	1.0907	1.6813
40	174.43	0.000694	0.10290	235.11	410.29	1.1196	1.6790
50	236.14	0.000707	0.077410	244.15	415.32	1.1479	1.6776
60	313.29	0.000720	0.059238	253.32	420.25	1.1756	1.6767
70	408.18	0.000734	0.046010	262.62	425.07	1.2030	1.6764
80	523.23	0.000749	0.036198	272.08	429.76	1.2299	1.6764
90	660.94	0.000766	0.028794	281.70	434.27	1.2564	1.6766
100	823.89	0.000784	0.023118	291.50	438.57	1.2827	1.6769
110	1014.80	0.000804	0.018703	301.51	442.63	1.3088	1.6771
120	1236.40	0.000826	0.015223	311.75	446.40	1.3347	1.6772
130	1491.60	0.000852	0.012442	322.27	449.82	1.3606	1.6770
140	1783.40	0.000881	0.010192	333.10	452.81	1.3865	1.6763
150	2115.10	0.000915	0.008346	344.33	455.26	1.4127	1.6748
160	2490.30	0.000957	0.006807	356.08	456.98	1.4393	1.6722
170	2913.10	0.001011	0.005497	368.57	457.68	1.4668	1.6679
180	3388.80	0.001088	0.004345	382.27	456.71	1.4963	1.6605
190	3924.90	0.001219	0.003253	398.49	452.24	1.5303	1.6464
197.96	4407.60	0.001805	0.001805	428.64	428.64	1.5933	1.5933

adapted from NIST data   The reference state for R-11 properties in SI units is h = 200kJ/kg and s = 1kJ/kg-K for the saturated liquid at 0°C

Appendix P.1 - Saturated R-11 Temperature Increment - English Units

T	P	v^L	v^V	h^L	h^V	s^L	s^V
°F	psia	ft³/lbm	ft³/lbm	BTU/lbm	BTU/lbm	BTU/lbm-°F	BTU/lbm-°F
-160	0.00160	0.009097	14649	-23.150	74.13	-0.06483	0.25977
-140	0.00643	0.009215	3882.0	-19.425	76.20	-0.05280	0.24633
-120	0.02152	0.009338	1232.5	-15.632	78.34	-0.04129	0.23536
-100	0.06181	0.009466	454.14	-11.784	80.54	-0.03029	0.22640
-80	0.15633	0.009598	189.37	-7.8915	82.80	-0.01976	0.21911
-60	0.35535	0.009735	87.583	-3.9625	85.11	-0.00967	0.21319
-40	0.73793	0.009877	44.203	0.0000	87.46	0.00000	0.20840
-20	1.4188	0.010025	24.022	3.9951	89.85	0.00930	0.20457
0	2.5537	0.010179	13.903	8.0241	92.27	0.01825	0.20153
20	4.3427	0.010341	8.4906	12.090	94.72	0.02690	0.19916
40	7.0306	0.010511	5.4294	16.195	97.17	0.03528	0.19734
60	10.906	0.010690	3.6116	20.344	99.64	0.04340	0.19599
80	16.301	0.010880	2.4851	24.543	102.09	0.05131	0.19501
100	23.581	0.011082	1.7605	28.795	104.53	0.05902	0.19435
120	33.150	0.011299	1.2787	33.108	106.94	0.06656	0.19393
140	45.439	0.011532	0.94890	37.486	109.30	0.07394	0.19370
160	60.905	0.011784	0.71720	41.937	111.60	0.08119	0.19361
180	80.029	0.012059	0.55060	46.470	113.83	0.08832	0.19362
200	103.31	0.012361	0.42827	51.092	115.96	0.09535	0.19368
220	131.28	0.012696	0.33674	55.815	117.97	0.10231	0.19376
240	164.46	0.013073	0.26704	60.654	119.85	0.10921	0.19381
260	203.43	0.013502	0.21309	65.627	121.55	0.11608	0.19379
280	248.77	0.014000	0.17065	70.760	123.04	0.12296	0.19364
300	301.12	0.014596	0.13670	76.094	124.26	0.12989	0.19329
320	361.18	0.015335	0.10903	81.695	125.10	0.13695	0.19262
340	429.78	0.016315	0.08590	87.691	125.40	0.14428	0.19144
360	507.98	0.017782	0.06571	94.383	124.74	0.15224	0.18928
380	597.49	0.020726	0.04578	102.820	121.59	0.16203	0.18438
388.32	639.23	0.028358	0.029473	183.940	184.86	0.38025	0.38133

The reference state for R-11 properties in English Units is h = 0 and s = 0 for the saturated liquid at -40°F

Appendix P.1 - Saturated R-11 Pressure Increment - SI Units

P	T	v^L	v^V	h^L	h^V	s^L	s^V
kPa	°C	m³/kg	m³/kg	kJ/kg	kJ/kg	kJ/kg·K	kJ/kg·K
2	-53.98	0.000605	6.61390	154.61	362.61	0.8151	1.7642
3	-48.16	0.000610	4.52280	159.42	365.46	0.8367	1.7525
4	-43.80	0.000614	3.45530	163.03	367.60	0.8526	1.7446
5	-40.28	0.000616	2.80480	165.95	369.34	0.8653	1.7387
10	-28.49	0.000626	1.46920	175.80	375.23	0.9065	1.7217
15	-20.92	0.000633	1.00740	182.18	379.06	0.9322	1.7128
20	-15.21	0.000638	0.77101	187.01	381.97	0.9511	1.7070
25	-10.57	0.000642	0.62668	190.96	384.35	0.9663	1.7028
30	-6.62	0.000646	0.52910	194.32	386.37	0.9790	1.6995
40	-0.11	0.000652	0.40511	199.90	389.71	0.9996	1.6948
50	5.20	0.000657	0.32934	204.48	392.45	1.0162	1.6915
60	9.72	0.000661	0.27806	208.39	394.78	1.0301	1.6891
70	13.67	0.000665	0.24097	211.83	396.82	1.0422	1.6871
80	17.21	0.000669	0.21286	214.92	398.64	1.0529	1.6856
90	20.41	0.000672	0.19078	217.72	400.28	1.0624	1.6843
100	23.34	0.000676	0.17296	220.30	401.79	1.0711	1.6833
110	26.05	0.000679	0.15828	222.69	403.18	1.0792	1.6824
120	28.57	0.000681	0.14595	224.92	404.47	1.0866	1.6817
130	30.94	0.000684	0.13545	227.02	405.68	1.0935	1.6810
140	33.17	0.000687	0.12640	229.00	406.82	1.0999	1.6805
150	35.27	0.000689	0.11851	230.88	407.90	1.1060	1.6800
160	37.28	0.000691	0.11156	232.67	408.91	1.1118	1.6795
180	41.01	0.000696	0.09991	236.01	410.80	1.1225	1.6788
200	44.43	0.000700	0.09051	239.10	412.53	1.1322	1.6783
220	47.60	0.000704	0.08275	241.97	414.12	1.1411	1.6778
240	50.56	0.000707	0.07624	244.66	415.59	1.1494	1.6775
260	53.33	0.000711	0.07070	247.19	416.97	1.1572	1.6772
280	55.95	0.000714	0.06590	249.65	418.26	1.1644	1.6770
300	58.42	0.000718	0.06173	251.86	419.48	1.1713	1.6768
400	69.21	0.000733	0.04691	261.88	424.70	1.2008	1.6764
500	78.12	0.000746	0.03783	270.29	428.89	1.2248	1.6763
600	85.79	0.000759	0.03167	277.62	432.39	1.2453	1.6765
700	92.55	0.000770	0.02720	284.18	435.38	1.2632	1.6766
800	98.63	0.000781	0.02381	290.15	437.99	1.2791	1.6768
900	104.17	0.000792	0.02114	295.65	440.30	1.2936	1.6770
1000	109.28	0.000802	0.01899	300.78	442.35	1.3069	1.6771
1500	130.31	0.000853	0.01237	322.60	449.92	1.3614	1.6770
2000	146.67	0.000903	0.00892	340.54	454.51	1.4039	1.6754
2500	160.24	0.000959	0.00677	356.37	457.01	1.4399	1.6721
3000	171.92	0.001024	0.00527	371.08	457.65	1.4723	1.6668
4000	191.30	0.001245	0.00310	400.98	451.11	1.5355	1.6434
4407.60	197.96	0.001805	0.00181	428.64	428.64	1.5933	1.5933

Appendix P.1 - Saturated R-11 Pressure Increment - English Units

P	T	v^L	v^V	h^L	h^V	s^L	s^V
psia	°F	ft³/lbm	ft³/lbm	BTU/lbm	BTU/lbm	BTU/lbm-°F	BTU/lbm-°F
1	-30.972	0.009943	33.284	1.7993	88.534	0.0042408	0.20656
2	-8.5706	0.010112	17.45	6.2933	91.230	0.014454	0.20274
3	5.8590	0.010226	11.97	9.2112	92.985	0.020817	0.20077
4	16.771	0.010314	9.1639	11.431	94.319	0.025526	0.1995
5	25.657	0.010388	7.4494	13.246	95.409	0.029298	0.19859
10	55.902	0.010652	3.9147	19.490	99.133	0.041758	0.19623
20	90.874	0.010988	2.0535	26.848	103.42	0.055526	0.19462
30	113.97	0.011232	1.4047	31.801	106.26	0.064303	0.19403
40	131.73	0.011433	1.0709	35.668	108.33	0.070904	0.19377
50	146.37	0.011610	0.8663	38.896	110.04	0.076261	0.19366
60	158.94	0.011770	0.72759	41.700	111.48	0.080805	0.19361
70	170.02	0.011918	0.62710	44.198	112.73	0.084772	0.19361
80	179.97	0.012058	0.55080	46.463	113.83	0.088306	0.19362
90	189.03	0.012191	0.49080	48.545	114.80	0.091504	0.19364
100	197.37	0.012319	0.44235	50.479	115.69	0.094431	0.19367
110	205.12	0.012443	0.40235	52.290	116.49	0.097136	0.19370
120	212.35	0.012563	0.36875	53.997	117.22	0.099655	0.19373
130	219.16	0.012681	0.34010	55.614	117.89	0.10202	0.19376
140	225.59	0.012797	0.31537	57.155	118.51	0.10424	0.19378
150	231.69	0.012910	0.29380	58.628	119.09	0.10634	0.19380
160	237.50	0.013023	0.27481	60.041	119.62	0.10835	0.19381
170	243.04	0.013134	0.25795	61.401	120.12	0.11025	0.19381
180	248.35	0.013245	0.24287	62.713	120.58	0.11208	0.19381
190	253.45	0.013355	0.22931	63.983	121.02	0.11383	0.19381
200	258.36	0.013464	0.21703	65.213	121.42	0.11552	0.19380
250	280.50	0.014014	0.16970	70.892	123.08	0.12313	0.19363
300	299.60	0.014583	0.13731	75.985	124.24	0.12975	0.19330
350	316.47	0.015191	0.11354	80.684	124.99	0.13569	0.19277
400	331.64	0.015866	0.095121	85.123	125.36	0.14117	0.19202
450	345.43	0.016647	0.080204	89.417	125.34	0.14636	0.19098
500	358.08	0.017605	0.067578	93.692	124.87	0.15143	0.18955
550	369.74	0.018882	0.056307	98.127	123.75	0.15661	0.18751
600	380.52	0.020858	0.045184	103.10	121.42	0.16235	0.18415
630.00	386.54	0.023429	0.036621	107.39	118.02	0.16729	0.17985

adapted from NIST data

Appendix P - Properties of R-11

Appendix P.2 - Superheated R-11 - SI Units

P = 50 kPa (5.20 °C)

T(°C)	v (m³/kg)	u (kJ/kg)	h (kJ/kg)	s (kJ/kg-K)
sat	0.32934	375.98	392.45	1.6915
20	0.34839	383.67	401.09	1.7218
40	0.37367	394.22	412.90	1.7608
60	0.39864	404.98	424.91	1.7979
80	0.42343	415.98	437.15	1.8336
100	0.44810	427.20	449.61	1.8679
120	0.47269	438.65	462.28	1.9010
140	0.49723	450.31	475.17	1.9330
160	0.52172	462.16	488.24	1.9639
180	0.54618	474.19	501.50	1.9938
200	0.57061	486.39	514.92	2.0228
220	0.59502	498.74	528.49	2.0509
240	0.61941	511.24	542.21	2.0781
260	0.64379	523.87	556.06	2.1046
280	0.66815	536.62	570.02	2.1303
300	0.69250	549.48	584.11	2.1553
320	0.71683	562.45	598.29	2.1796
340	0.74116	575.52	612.58	2.2033

P = 75 kPa (15.49 °C)

T(°C)	v (m³/kg)	u (kJ/kg)	h (kJ/kg)	s (kJ/kg-K)
sat	0.28427	300.48	321.80	1.7859
20	0.22998	383.19	400.44	1.6956
40	0.24724	393.87	412.41	1.7351
60	0.26417	404.71	424.52	1.7726
80	0.28090	415.75	436.82	1.8084
100	0.29751	427.01	449.32	1.8428
120	0.31404	438.48	462.03	1.8760
140	0.33051	450.15	474.94	1.9080
160	0.34693	462.01	488.03	1.9390
180	0.36332	474.06	501.31	1.9689
200	0.37969	486.26	514.74	1.9980
220	0.39603	498.62	528.33	2.0261
240	0.41235	511.13	542.05	2.0534
260	0.42865	523.76	555.91	2.0799
280	0.44494	536.52	569.89	2.1056
300	0.46122	549.39	583.98	2.1306
320	0.47749	562.36	598.17	2.1550
340	0.49374	575.44	612.47	2.1787

P = 100 kPa (23.34 °C)

T(°C)	v (m³/kg)	u (kJ/kg)	h (kJ/kg)	s (kJ/kg-K)
sat	0.17296	384.49	401.79	1.6833
40	0.18400	393.51	411.91	1.7165
60	0.19692	404.43	424.12	1.7543
80	0.20963	415.52	436.48	1.7903
100	0.22221	426.81	449.03	1.8249
120	0.23471	438.30	461.77	1.8582
140	0.24714	449.99	474.71	1.8902
160	0.25954	461.87	487.82	1.9212
180	0.27189	473.92	501.11	1.9512
200	0.28422	486.14	514.56	1.9803
220	0.29653	498.51	528.16	2.0084
240	0.30882	511.02	541.90	2.0357
260	0.32109	523.66	555.77	2.0622
280	0.33334	536.42	569.75	2.0880
300	0.34558	549.29	583.85	2.1130
320	0.35781	562.27	598.06	2.1374
340	0.37003	575.35	612.35	2.1611

P = 200 kPa (44.43 °C)

T(°C)	v (m³/kg)	u (kJ/kg)	h (kJ/kg)	s (kJ/kg-K)
sat	0.09051	394.43	412.53	1.6783
60	0.09592	403.23	422.41	1.7087
80	0.10265	414.56	435.09	1.7456
100	0.10921	426.00	447.84	1.7807
120	0.11568	437.59	460.73	1.8144
140	0.12208	449.35	473.77	1.8467
160	0.12843	461.28	486.97	1.8779
180	0.13474	473.38	500.33	1.9081
200	0.14102	485.63	513.84	1.9373
220	0.14728	498.03	527.49	1.9655
240	0.15351	510.57	541.28	1.9929
260	0.15973	523.24	555.18	2.0195
280	0.16593	536.02	569.21	2.0453
300	0.17212	548.92	583.34	2.0704
320	0.17830	561.92	597.58	2.0948
340	0.18447	575.01	611.91	2.1186

Appendix P.2 - Superheated R-11 - English Units

P = 1 psia (-30.97 °F)

T(°F)	v (ft³/lbm)	u (BTU/lbm)	h (BTU/lbm)	s (BTU/lbm-°F)
sat	33.284	82.37	88.534	0.20656
0	35.754	85.87	92.487	0.21546
40	38.918	90.53	97.738	0.22641
80	42.066	95.37	103.16	0.23685
120	45.206	100.37	108.74	0.24683
160	48.342	105.52	114.48	0.25639
200	51.476	110.82	120.35	0.26558
240	54.607	116.24	126.36	0.27442
280	57.738	121.79	132.48	0.28292
320	60.868	127.44	138.71	0.29112
360	63.997	133.18	145.03	0.29904
400	67.126	139.02	151.45	0.30668
440	70.254	144.94	157.95	0.31407
480	73.382	150.93	164.52	0.32122
520	76.510	156.99	171.16	0.32814
560	79.638	163.12	177.87	0.33484
600	82.765	169.30	184.63	0.34135
640	85.892	175.54	191.44	0.34766

P = 5 psia (25.66 °F)

T(°F)	v (ft³/lbm)	u (BTU/lbm)	h (BTU/lbm)	s (BTU/lbm-°F)
sat	7.4494	88.512	95.409	0.19859
0	-	-	-	0.20257
40	7.6860	90.254	97.37	0.21321
80	8.3357	95.180	102.90	0.22330
120	8.9768	100.23	108.54	0.23293
160	9.6133	105.41	114.31	0.24215
200	10.247	110.73	120.21	0.25101
240	10.879	116.16	126.23	0.25954
280	11.510	121.71	132.37	0.26776
320	12.140	127.37	138.61	0.27568
360	12.769	133.12	144.94	0.28333
400	13.398	138.96	151.37	0.29072
440	14.027	144.89	157.87	0.29788
480	14.655	150.88	164.45	0.30480
520	15.282	156.95	171.10	0.31152
560	15.910	163.07	177.80	0.31802
600	16.537	169.26	184.57	0.32434
640	17.164	175.50	191.39	-

P = 10 psia (55.90 °F)

T(°F)	v (ft³/lbm)	u (BTU/lbm)	h (BTU/lbm)	s (BTU/lbm-°F)
sat	3.9147	91.884	99.133	0.19623
80	4.1176	94.930	102.56	0.20272
120	4.4471	100.05	108.28	0.21296
160	4.7716	105.27	114.10	0.22267
200	5.0931	110.61	120.04	0.23194
240	5.4128	116.06	126.08	0.24084
280	5.7313	121.62	132.23	0.24939
320	6.0489	127.28	138.48	0.25762
360	6.3658	133.04	144.83	0.26556
400	6.6822	138.89	151.26	0.27322
440	6.9981	144.82	157.78	0.28062
480	7.3137	150.82	164.36	0.28778
520	7.6290	156.89	171.01	0.29472
560	7.9440	163.02	177.73	0.30143
600	8.2588	169.21	184.50	0.30795
640	8.5733	175.45	191.32	0.31427

P = 15 psia (75.71 °F)

T(°F)	v (ft³/lbm)	u (BTU/lbm)	h (BTU/lbm)	s (BTU/lbm-°F)
sat	2.6852	94.111	101.570	0.19519
80	2.7100	94.665	102.190	0.19635
120	2.9364	99.858	108.010	0.20676
160	3.1572	105.12	113.890	0.21656
200	3.3749	110.48	119.860	0.22589
240	3.5905	115.95	125.920	0.23482
280	3.8049	121.53	132.090	0.24339
320	4.0184	127.20	138.360	0.25164
360	4.2312	132.96	144.720	0.25959
400	4.4435	138.82	151.160	0.26727
440	4.6553	144.75	157.680	0.27468
480	4.8667	150.75	164.270	0.28185
520	5.0778	156.83	170.930	0.28879
560	5.2887	162.96	177.650	0.29551
600	5.4993	169.15	184.430	0.30203
640	5.7097	175.40	191.250	0.30835

Appendix P.2 - Superheated R-11 - SI Units

	P = 300 kPa (58.42 °C)				P = 400 kPa (69.21 °C)			
	v	u	h	s	v	u	h	s
T(°C)	m³/kg	kJ/kg	kJ/kg	kJ/kg-K	m³/kg	kJ/kg	kJ/kg	kJ/kg-K
sat	0.06173	400.96	419.48	1.6768	0.04691	405.93	424.70	1.6764
60	0.06212	401.89	420.52	1.6799	-	-	-	-
80	0.06691	413.51	433.58	1.7180	0.04896	412.37	431.96	1.6973
100	0.07150	425.13	446.58	1.7538	0.05260	424.21	445.25	1.7339
120	0.07597	436.84	459.64	1.7879	0.05609	436.07	458.50	1.7685
140	0.08037	448.69	472.80	1.8206	0.05949	448.01	471.81	1.8015
160	0.08471	460.68	486.10	1.8520	0.06284	460.07	485.21	1.8331
180	0.08901	472.83	499.53	1.8823	0.06613	472.27	498.72	1.8636
200	0.09328	485.12	513.11	1.9116	0.06940	484.61	512.36	1.8931
220	0.09752	497.56	526.81	1.9400	0.07263	497.07	526.13	1.9216
240	0.10174	510.12	540.65	1.9675	0.07585	509.67	540.01	1.9492
260	0.10594	522.82	554.60	1.9942	0.07904	522.39	554.01	1.9760
280	0.11013	535.63	568.66	2.0201	0.08222	535.22	568.11	2.0019
300	0.11430	548.54	582.83	2.0452	0.08539	548.16	582.32	2.0271
320	0.11846	561.56	597.10	2.0697	0.08854	561.20	596.62	2.0517
340	0.12262	574.67	611.46	2.0935	0.09169	574.33	611.01	2.0755

	P = 600 kPa (85.79 °C)				P = 800 kPa (98.63 °C)			
	v	u	h	s	v	u	h	s
T(°C)	m³/kg	kJ/kg	kJ/kg	kJ/kg-K	m³/kg	kJ/kg	kJ/kg	kJ/kg-K
sat	0.03167	413.39	432.39	1.6765	0.02381	418.94	437.99	1.6768
100	0.03360	422.19	442.35	1.7037	0.02397	419.83	439.01	1.6795
120	0.03615	434.40	456.09	1.7395	0.02610	432.55	453.43	1.7172
140	0.03858	446.57	469.72	1.7734	0.02808	445.02	467.48	1.7521
160	0.04094	458.80	483.36	1.8056	0.02996	457.45	481.41	1.7850
180	0.04324	471.11	497.06	1.8365	0.03177	469.91	495.33	1.8164
200	0.04550	483.54	510.85	1.8663	0.03354	482.45	509.29	1.8465
220	0.04774	496.09	524.74	1.8950	0.03528	495.09	523.31	1.8756
240	0.04995	508.76	538.73	1.9229	0.03699	507.82	537.41	1.9036
260	0.05214	521.53	552.82	1.9498	0.03868	520.66	551.60	1.9307
280	0.05431	534.42	567.00	1.9759	0.04035	533.60	565.88	1.9570
300	0.05647	547.40	581.28	2.0013	0.04201	546.63	580.24	1.9825
320	0.05862	560.48	595.65	2.0259	0.04365	559.75	594.67	2.0073
340	0.06075	573.65	610.10	2.0499	0.04529	572.96	609.19	2.0313

Appendix P.2 - Superheated R-11 - English Units

	P = 20 psia (90.87 °F)				P = 30 psia (113.97 °F)			
	v	u	h	s	v	u	h	s
T(°F)	ft³/lbm	BTU/lbm	BTU/lbm	BTU/lbm-°F	ft³/lbm	BTU/lbm	BTU/lbm	BTU/lbm-°F
sat	2.0535	95.819	103.42	0.19462	1.4047	98.416	106.22	0.19403
120	2.1804	99.659	107.73	0.20225	1.4230	99.234	107.14	0.19563
160	2.3497	104.97	113.67	0.21215	1.5414	104.65	113.21	0.20576
200	2.5155	110.36	119.67	0.22154	1.6557	110.10	119.30	0.21527
240	2.6793	115.84	125.77	0.23050	1.7677	115.62	125.44	0.22432
280	2.8417	121.43	131.95	0.23910	1.8782	121.24	131.67	0.23297
320	3.0031	127.11	138.23	0.24737	1.9877	126.94	137.98	0.24128
360	3.1639	132.89	144.60	0.25534	2.0964	132.73	144.37	0.24927
400	3.3241	138.74	151.05	0.26302	2.2046	138.60	150.84	0.25698
440	3.4838	144.68	157.58	0.27044	2.3123	144.54	157.39	0.26442
480	3.6432	150.69	164.18	0.27762	2.4196	150.56	164.00	0.27162
520	3.8022	156.77	170.85	0.28457	2.5266	156.65	170.68	0.27858
560	3.9610	162.90	177.57	0.29129	2.6333	162.79	177.42	0.28532
600	4.1195	169.10	184.36	0.29782	2.7397	168.99	184.21	0.29185
640	4.2779	175.34	191.19	0.30415	2.8460	175.24	191.05	0.29819

	P = 40 psia (131.73 °F)				P = 60 psia (158.94 °F)			
	v	u	h	s	v	u	h	s
T(°F)	ft³/lbm	BTU/lbm	BTU/lbm	BTU/lbm-°F	ft³/lbm	BTU/lbm	BTU/lbm	BTU/lbm-°F
sat	1.0709	100.40	108.33	0.19377	0.7276	103.40	111.48	0.19361
160	1.1364	104.31	112.73	0.20103	0.7294	103.55	111.66	0.19389
200	1.2253	109.83	118.91	0.21069	0.7938	109.25	118.07	0.20392
240	1.3116	115.40	125.11	0.21983	0.8549	114.92	124.42	0.21327
280	1.3963	121.04	131.38	0.22854	0.9140	120.63	130.78	0.22211
320	1.4799	126.76	137.72	0.23689	0.9717	126.40	137.19	0.23055
360	1.5626	132.56	144.14	0.24491	1.0286	132.23	143.66	0.23864
400	1.6448	138.45	150.63	0.25265	1.0849	138.14	150.20	0.24643
440	1.7265	144.40	157.19	0.26011	1.1406	144.13	156.80	0.25393
480	1.8078	150.43	163.82	0.26732	1.1959	150.17	163.46	0.26117
520	1.8887	156.52	170.51	0.27429	1.2508	156.28	170.18	0.26818
560	1.9694	162.68	177.26	0.28104	1.3055	162.45	176.95	0.27495
600	2.0498	168.88	184.07	0.28759	1.3599	168.67	183.78	0.28152
640	2.1301	175.14	190.92	0.29394	1.4141	174.94	190.65	0.28789

Appendix P.2 - Superheated R-11 - SI Units

	P = 1000 kPa (109.28 °C)				P = 1500 kPa (130.31 °C)			
	v	u	h	s	v	u	h	s
T(°C)	m³/kg	kJ/kg	kJ/kg	kJ/kg-K	m³/kg	kJ/kg	kJ/kg	kJ/kg-K
sat	0.01899	423.36	442.35	1.6771	0.01237	431.37	449.92	1.6770
120	0.02000	430.46	450.46	1.6980	-	-	-	-
140	0.02173	443.34	465.07	1.7343	0.01309	438.32	457.95	1.6966
160	0.02334	456.02	479.36	1.7680	0.01442	451.99	473.61	1.7337
180	0.02487	468.65	493.53	1.8000	0.01561	465.23	488.64	1.7676
200	0.02635	481.32	507.67	1.8306	0.01672	478.30	503.39	1.7994
220	0.02780	494.05	521.85	1.8599	0.01779	491.34	518.02	1.8297
240	0.02921	506.87	536.08	1.8882	0.01881	504.39	532.61	1.8587
260	0.03060	519.77	550.37	1.9155	0.01981	517.48	547.20	1.8866
280	0.03197	532.77	564.74	1.9420	0.02079	530.64	561.82	1.9135
300	0.03333	545.85	579.18	1.9676	0.02175	543.86	576.48	1.9396
320	0.03467	559.02	593.69	1.9925	0.02269	557.15	591.18	1.9648
340	0.03600	572.26	608.27	2.0167	0.02362	570.51	605.93	1.9892

	P = 2000 kPa (146.67 °C)				P = 3000 kPa (171.92 °C)			
	v	u	h	s	v	u	h	s
T(°C)	m³/kg	kJ/kg	kJ/kg	kJ/kg-K	m³/kg	kJ/kg	kJ/kg	kJ/kg-K
sat	0.00892	436.67	454.51	1.6754	0.00527	441.85	457.65	1.6668
160	0.00979	446.95	466.53	1.7036	-	-	-	-
180	0.01089	461.25	483.02	1.7408	0.00582	449.95	467.40	1.6885
200	0.01185	474.95	498.65	1.7746	0.00682	466.63	487.08	1.7310
220	0.01275	488.40	513.89	1.8061	0.00761	481.60	504.42	1.7669
240	0.01359	501.76	528.94	1.8360	0.00831	495.90	520.82	1.7995
260	0.01440	515.09	543.88	1.8646	0.00895	509.89	536.73	1.8300
280	0.01518	528.43	558.80	1.8921	0.00955	523.74	552.38	1.8588
300	0.01594	541.81	573.70	1.9185	0.01012	537.52	567.88	1.8863
320	0.01669	555.24	588.62	1.9441	0.01068	551.27	583.29	1.9127
340	0.01742	568.72	603.56	1.9689	0.01121	565.02	598.66	1.9382

adapted from NIST data The reference state for R-11 properties in SI units is $h = 200$ kJ/kg and $s = 1$ kJ/kg-K for the saturated

Appendix P.2 - Superheated R-11 - English Units

	P = 80 psia (179.97 °F)				P = 100 psia (197.37 °F)			
	v	u	h	s	v	u	h	s
T(°F)	ft³/lbm	BTU/lbm	BTU/lbm	BTU/lbm-°F	ft³/lbm	BTU/lbm	BTU/lbm	BTU/lbm-°F
sat	0.5508	105.67	113.83	0.19362	0.4424	107.49	115.69	0.19367
200	0.5768	108.61	117.15	0.19874	0.4453	107.89	116.14	0.19436
240	0.6258	114.41	123.68	0.20835	0.4876	113.86	122.89	0.20430
280	0.6723	120.20	130.16	0.21735	0.5269	119.75	129.50	0.21349
320	0.7174	126.02	136.65	0.22590	0.5645	125.63	136.08	0.22215
360	0.7614	131.90	143.18	0.23406	0.6009	131.55	142.68	0.23040
400	0.8048	137.84	149.76	0.24190	0.6366	137.52	149.31	0.23830
440	0.8475	143.84	156.40	0.24945	0.6716	143.55	155.99	0.24590
480	0.8899	149.91	163.09	0.25673	0.7062	149.64	162.72	0.25321
520	0.9318	156.03	169.84	0.26376	0.7404	155.79	169.50	0.26028
560	0.9735	162.22	176.64	0.27056	0.7742	161.98	176.32	0.26711
600	1.0149	168.45	183.49	0.27715	0.8079	168.23	183.19	0.27372
640	1.0561	174.74	190.38	0.28354	0.8412	174.53	190.11	0.28012

	P = 200 psia (258.36 °F)				P = 300 psia (299.60 °F)			
	v	u	h	s	v	u	h	s
T(°F)	ft³/lbm	BTU/lbm	BTU/lbm	BTU/lbm-°F	ft³/lbm	BTU/lbm	BTU/lbm	BTU/lbm-°F
sat	0.2170	113.38	121.42	0.19380	0.1373	116.610	124.240	0.19330
280	0.2319	117.00	125.59	0.19952	-	-	-	-
320	0.2563	123.41	132.90	0.20914	0.1492	120.440	128.730	0.19914
360	0.2784	129.65	139.96	0.21797	0.1685	127.350	136.710	0.20913
400	0.2992	135.84	146.92	0.22627	0.1853	133.920	144.210	0.21806
440	0.3190	142.03	153.85	0.23415	0.2006	140.360	151.510	0.22636
480	0.3383	148.25	160.78	0.24168	0.2150	146.760	158.710	0.23419
520	0.3571	154.50	167.73	0.24892	0.2289	153.150	165.870	0.24165
560	0.3755	160.79	174.70	0.25590	0.2422	159.550	173.010	0.24879
600	0.3936	167.12	181.70	0.26263	0.2552	165.970	180.150	0.25567
640	0.4114	173.49	188.73	0.26914	0.2679	172.420	187.300	0.26229

The reference state for R-11 properties in English Units is $h = 0$ and $s = 0$ for the saturated liquid at -40°F

Subject Index

Acentric factor
    definition , 36
    table of values for, 298 - 300
    use in generalized correlations, 39 - 39

Activity and activity coefficients
    analytical representation of, 226 - 37
    coefficient, definition, 182
    definition, 182
    from experimental data, 237 - 242
    reaction equilibrium constant, 259

Adiabatic flame temperature, 66 - 67
Adiabatic process, ideal gas, 44 - 45
Antoine equation, 108 - 109
Antoine constants, table, 305 - 306
Azeotropes, 241 - 242
Benedict-Webb-Rubin equation of state, 38
Bernoulli's equation, 127 - 128
Bubble point, 184 - 186
Carlson and Colburn, method of, 237 - 238
Carnot cycle, 84 - 86

Chemical potential
    definition, 177
    ideal gas, 180
    relationship to fugacity, 180 - 181
    relationship to free energy of reaction, 258 - 259

Chemical reactions
    equilibrium constant for, 259 - 260
    extent of reaction, 256
    heats of formation, reaction and combustion, 62 - 65
    reaction coordinate for, 256
    standard Gibbs Free Energy of, 258

Chemical reaction equilibria, 256 - 275
    calculation of equilibrium constant, 158 - 159
    effect of pressure on, 261
    effect of temperature on, 258, 261
    in heterogeneous systems, 262 - 263
    for multiple reactions, 263 - 265

Clapeyron equation, 106 - 107
Clausius-Clapeyron equation, 107
Combustion, standard heat of, 65 - 66

Compressibility factor
    critical compressibility factor, 35
    definition, 32 - 33
    generalized correlation for, 38 - 40

Compressors, 133 - 137
    multistage, 135 - 137

Consistency tests in VLE, 242 - 245
    differential test, 242 - 243
    integral test, 243 - 245

Continuity equation, definition, 120 - 121
Conversion factors, 295

Corresponding states
    compressibility factor using, 39 - 40
    residual properties using, 105 - 106
    theorem of, 38 - 40

Critical point, 25 - 26
    table of critical properties, 298

Density, definition, 6

Dew point, definition, 184
    calculation of, 185 - 186

Diesel engine, 153 - 154

Efficiency
    Brayton Cycle, 156
    Carnot cycle, 87
    Diesel engine, 154
    of heat engines, 87
    Otto engine, 149 - 153
    Steam Rankine cycle, 147
    thermal, of compressors, 137
    thermal, of turbines, 137

Energy
    conservation of, (First Law of Thermodynamics) 8 - 9,121
    internal, 9
    kinetic, definition, 5
    potential, definition, 5

Engines
    Carnot, 87
    heat, 80 - 84
    internal combustion, 149 - 154

Enthalpy
    definition, 12
    effect of T and P on, 100
    excess, 231, 245
    ideal gas, 42
    of combustion, 65 - 66
    of formation, 62
    of reaction, 63
    residual
        definition, 101
        from corresponding states, 105 - 106
        generalized correlation for, 105 - 106

Entropy
    definition, 87
    effect of T and P on, 100
    ideal gas, 89
    and irreversibility, 88, 126
    residual, 101 - 102,
    generalized correlation for, 105 - 106
    statistical interpretation of, 77

Equation of state, definition, 31
Equilibrium
    criteria for, 176
    Equilibrium constant for chemical reaction, 194
    phase, 181
Euler's Theorem, 223
Exact differentials, 98 - 99
Excess properties, 226 - 228
Extensive property, 5
Extent of reaction, definition, 256
First Law of Thermodynamics
    for closed systems, 10
    for flow processes, 11
Flame temperature, adiabatic, 66 - 67
Flash calculation for VLE, 188 - 189
Flow processes
    Bernoulli' equation, 127 - 128
    continuity equation, 120
    energy balance for, 121
    maximum velocity in a pipe, 128 - 129
Friction factor for flow in pipes, 128
Fugacity
    calculation of for pure components, 189 - 192
    calculation of using generalized correlations, 190
    calculation of using EOS for mixtures, 204 - 205
    definition, 180
    in chemical reaction equilibrium, 260
    of a component in a gas mixture, 205
    in ideal gas mixtures, 183
Fugacity coefficient
    calculation of, 205
    definition, 182
    generalized correlation for, 190
Gas constant, 30, 295
Generalized Stoichiometry
    for multiple reactions, 255
    for single reactions, 61 - 62, 255
Gibbs-Duhem equation, 224
Gibbs free energy
    definition, 98
    and equilibrium, 176
    of reaction, standard, 257 - 258
    for change of phase, 106
Heat
    of combustion, 65
    of formation, 62
    of reaction, 63
    sign convention for, 10
Heat capacity
    at constant P or V, 17
    definition, 16
    table, as function of T, 296 - 297
    of liquids, table, 297
Heat engine, 80 - 81
Helmholtz Free Energy, definition, 98
Henry's Law, 245 - 246
    and Raoult's law, 246
Ideal gas
    definition, 29 - 30
    enthalpy changes for, 42
    entropy changes for, 88 - 89
    fugacity of, 180
    heat capacities of, table, 296 - 297
    internal energy changes for, 42
    property changes of mixing for, 192 - 193
    temperature scale, 30
Ideal gas law, 31
Ideal solutions, 227
    and Lewis fugacity rule, 206
    and Raoult's law, 183
Independent set of chemical reactions, 255
Intensive property, 5 - 6
Internal energy
    definition, 9
    of ideal gases, 42
Irreversibility, 12, 124
Isentropic process, 134
Isobaric process, 43
Isothermal process, 42
K values for VLE, 188
Kinetic energy, definition, 5
Lewis fugacity rule, 206
Liquid-liquid equilibrium, 276 - 285
Liquids
    fugacities of, 191
    volume correlation for, 40
Lost work, 90 - 92
Margules 2-suffix equation, 234
Margules 3-suffix equation, 234
Maximum velocity in a pipe, 128 - 129
Maxwell relations
    for pure components, 98 - 99
    for mixtures, 179
Mollier diagram, for steam, 327
Partial molar properties, 178
    excess, 228 - 229
    method of tangent intercepts for, 223
Partial pressure, 181, 184
Path functions, 87
PH diagrams
    ammonia, 359
    R-134a, 334
    R-23, 344
    steam, 326
Phase diagrams, 184, 185, 196, 232, 242, 276

Phase equilibria, criteria for, 106
Phase rule, 28
Pitzer correlations, 38, 39
Potential energy, 5
Pressure
    critical, 26
    partial, 181
    pseudo critical, 211
    reduced, 26
Pressure composition diagram, 184
Pressure-enthalpy diagram
    for ammonia, 359
    for R-134a, 334
    for R-23, 344
    for steam, 326
Pressure volume diagram, 27
Properties
    critical, 26
    extensive and intensive, 5
    reduced, 26
    residual, 101 - 103
Pseudo critical pressure and temperature, 211
PVT relationships, 27
    equations of state, 31
    for gas mixtures, 206
    generalized, 210 - 212
Quality, 28
Rackett equation, 42
Raoult's Law, 183 - 187
    modified, 231 - 232
Reaction coordinate, 255 - 257
Redlich-Kister expansion, 233
Redlich-Kwong equation of state, 37
Regular solution, 230
Relative volatility, 187
Residual properties, 101 - 105
Reversibility, 12 - 13
Saturated liquid and vapor, 27
Second law of thermodynamics
    and the definition of entropy, 87
    in analysis of processes, 124
    statement of, 80
Second virial coefficient correlation, 39
Shaft work, 10
Sonic velocity in flow, 129
Standard Gibbs free energy change
    of reactions, 257 - 258
    effect of temperature on, 257
Standard heat
    effect of temperature on, 64
    of formation, 62
    of reaction, 63
Standard state, 180
    for chemical reaction, 259

    for components in solution, 181, 182
State function, 16
Statistical thermodynamics, 59
Steady state flow processes, 127 - 138
Steam
    Mollier diagram for, 327
    PH diagram for, 326
    TS diagram for, 328
    tables, 309
Steam turbine, 137 - 138
Surroundings, 8
System, 8
Temperature
    ideal gas scale, 29 - 31
    critical, 26
    pseudo critical, 211
    reduced, 26
    thermal energy (thermodynamic) scale, 80
Temperature composition diagram, 185
Temperature entropy diagram, 328
Temperature scales
    ideal gas, 29 - 31
    thermodynamic, 80
Thermodynamic consistency, 242 - 245
Thermodynamic properties
    of Ammonia, 345
    of Carbon Dioxide, 380
    of Hydrogen, 391
    of Methane, 387
    of Nitrogen, 360
    of Oxygen, 376
    of R-11, 395
    of R-134a, 329
    of R-23, 335
    of Steam, 309
Third law of thermodynamics, 88
Tie line, 186
Triple point, 3, 25
Turbine, 137 - 138
Two phase systems,
    Clapeyron equation, 106 - 107
    quality in, 28
Units and conversion factors, 295
Universal gas constant
    definition, 30
    values for, 295
Unsteady state flow processes, 120, 121, 122
Van der Waals equation of state, 35 - 36
van Laar equation, 236
    determination of constants for, 239
    linear regression form, 240
Van't Hoff equation, 197
Vapor liquid equilibrium,
    and infinite dilution activity coefficients,

        234
        by Raoult's law, 183
        for ideal solutions, 183
        flash calculation, 188 - 189
        fundamental problem of, 182 - 183
        K values for, 187 - 188
        phase diagrams for binary systems, 184, 185, 232
        phi-gamma formulation for, 183
        and thermodynamic consistency tests, 242 - 245

Vapor pressure, 26
        Antoine's equation, 107 - 108
        from equation of state, 287 - 288
        table of constants for calculation of, 305

Vaporization, heat of, 9
        from Clapeyron equation, 106 - 108

Velocity
        maximum value in pipes, 128 - 129
        sonic, 129

Virial coefficients, 33
        generalized correlation of second, 39
        for mixtures, 207, 210

Virial equation of state, 33

Volume
        critical, 26
        reduced, 26

Wilson's equation, 236

Work
        adiabatic compression, 44, 133
        definition, 4
        isothermal compression, 42, 133
        lost, 90 - 92, 126
        requirements for compressors, 133
        shaft, 10
        sign convention for, 10

www.ingramcontent.com/pod-product-compliance
Lightning Source LLC
Chambersburg PA
CBHW080753300426
44114CB00020B/2726